Chromatographic Techniques in the Forensic Analysis of Designer Drugs

T0253466

CHROMATOGRAPHIC SCIENCE SERIES

A Series of Textbooks and Reference Books

Editor:
Nelu Grinberg

Founding Editor:
Jack Cazes

Dynamics of Chromatography: Principles and Theory, J. Calvin Giddings

Gas Chromatographic Analysis of Drugs and Pesticides, Benjamin J. Gudzinowicz

Principles of Adsorption Chromatography: The Separation of Nonionic Organic Compounds, Lloyd R. Snyder

Multicomponent Chromatography: Theory of Interference, Friedrich Helfferich and Gerhard Klein

Quantitative Analysis by Gas Chromatography, Josef Novák

High-Speed Liquid Chromatography, Peter M. Rajcsanyi and Elisabeth Rajcsanyi

Fundamentals of Integrated GC-MS (in three parts), Benjamin J. Gudzinowicz, Michael J. Gudzinowicz, and Horace F. Martin

Liquid Chromatography of Polymers and Related Materials, Jack Cazes

GLC and HPLC Determination of Therapeutic Agents (in three parts),
Part 1 edited by Kiyoshi Tsuji and Walter Morozowich, Parts 2 and 3 edited by Kiyoshi Tsuji

Biological/Biomedical Applications of Liquid Chromatography, edited by Gerald L. Hawk

Chromatography in Petroleum Analysis, edited by Klaus H. Altgelt and T. H. Gouw

Biological/Biomedical Applications of Liquid Chromatography II, edited by Gerald L. Hawk

Liquid Chromatography of Polymers and Related Materials II, edited by Jack Cazes and Xavier Delamare

Introduction to Analytical Gas Chromatography: History, Principles, and Practice, John A. Perry

Applications of Glass Capillary Gas Chromatography, edited by Walter G. Jennings

Steroid Analysis by HPLC: Recent Applications, edited by Marie P. Kautsky

Thin-Layer Chromatography: Techniques and Applications, Bernard Fried and Joseph Sherma

Biological/Biomedical Applications of Liquid Chromatography III, edited by Gerald L. Hawk

Liquid Chromatography of Polymers and Related Materials III, edited by Jack Cazes

Biological/Biomedical Applications of Liquid Chromatography, edited by Gerald L. Hawk

Chromatographic Separation and Extraction with Foamed Plastics and Rubbers, G. J. Moody and J. D. R. Thomas

Analytical Pyrolysis: A Comprehensive Guide, William J. Irwin

Liquid Chromatography Detectors, edited by Thomas M. Vickrey

High-Performance Liquid Chromatography in Forensic Chemistry, edited by Ira S. Lurie and John D. Wittwer, Jr.

Chromatographic Techniques in the Forensic Analysis of Designer Drugs

Edited by
Teresa Kowalska
Mieczysław Sajewicz
Joseph Sherma

CRC Press
Taylor & Francis Group
Boca Raton London New York

CRC Press is an imprint of the
Taylor & Francis Group, an **informa** business

CRC Press
Taylor & Francis Group
6000 Broken Sound Parkway NW, Suite 300
Boca Raton, FL 33487-2742

First issued in paperback 2020

ISBN-13: 978-0-367-57228-0 (pbk)
ISBN-13: 978-1-138-03328-3 (hbk)

Library of Congress Cataloging-in-Publication Data

Names: Kowalska, Teresa, editor. | Sajewicz, Mieczyslaw, editor. | Sherma, Joseph, editor.
Title: Chromatographic techniques in the forensic analysis of designer drugs / editors, Teresa Kowalska, Mieczyslaw Sajewicz, Joseph Sherma.
Other titles: Chromatographic science. 0069-3936
Description: Boca Raton : Taylor & Francis/CRC Press, 2018. | Series: Chromatographic science series | Includes bibliographical references.
Identifiers: LCCN 2017041428 | ISBN 9781138033283 (hardback : alk. paper)
Subjects: | MESH: Designer Drugs--analysis | Forensic Toxicology | Chromatography--methods | Mass Spectrometry--methods
Classification: LCC RA1228 | NLM QV 55 | DDC 614/.13--dc23
LC record available at https://lccn.loc.gov/2017041428

Visit the Taylor & Francis Web site at
http://www.taylorandfrancis.com

and the CRC Press Web site at
http://www.crcpress.com

Contents

SECTION I

SECTION II

Preface

According to Wikipedia, a designer drug (or new psychoactive substance [NPS] as designated by the European Union) is a structural or functional analog of a controlled substance that has been designed to mimic the pharmacological effects of the original drug, while avoiding classification as illegal and/or detection in standard drug tests. This book is the first and only one devoted to comprehensive coverage of the main analytical methods that can be employed for qualitative identification and quantification of designer drug samples of forensic interest (drug form and biological samples): high-performance column liquid chromatography (HPLC) and gas chromatography (GC) equipped with a variety of detectors, especially those based on mass spectrometry (MS); off-line application of one-dimensional and two-dimensional nuclear magnetic resonance (NMR); other spectrometric methods; and thin-layer chromatography (TLC). The presentation is systematic according to the analytical method type and drug chemical class. In addition, some basic information on the chemistry of designer drugs and legislative issues related to their regulation are covered.

The book includes 22 chapters in two sections. Section I contains Chapters 1 through 12, which provide general information on legal concepts and legal issues regarding regulation of designer drugs, the way they have been developed in the United States of America (USA) and European Union countries; biological precursors of certain classes of the designer drugs (as not all of them have precursors in the kingdom of plants); fundamentals of the best-performing instrumental techniques for forensic analysis (column liquid chromatography–mass spectrometry [LC-MS], GC-MS, ambient plasma ionization MS, and Raman spectrometry); methods for the preparation of biological material (e.g., human blood, urine, and tissues) for toxicological and forensic analysis; electrochemical generation of drug metabolites for analysis by LC-MS; and discrimination of designer drug isomers by GC-MS, LC-MS, capillary electrophoresis, Raman spectrometry, and Fourier transform infrared spectrometry.

In Section II, Chapters 13 through 21 present an overview of the analytical methods for different classes of designer drugs, i.e., cannabinoids, cathinones, phenylethylamine derivatives, piperazines, fentanyl and analogs, tryptamines, new opioids, and benzodiazepines in commercial product samples and in material collected as forensic evidence. Finally, Chapter 22 provides a review of the techniques of TLC, including sample preparation, sample and standard solution application, plate development with the mobile phase, zone detection, and densitometry, as well as a number of important applications for separation, identification, and quantification of synthetic and natural designer drugs.

The book was prepared with the aim of becoming the critical reference on the analysis of designer drugs by chromatography and selected complementary methods for analytical chemists, toxicologists, forensic scientists, and graduate and undergraduate students in a variety of related college and university courses, as well as lawyers and government employees involved in drug regulation. The editors thank the expert authors who contributed chapters and Barbara (Glunn) Knott, Senior Editor, Analytical, Industrial, and Physical Chemistry, at Taylor & Francis and Danielle Zarfati, Editorial Assistant, Chemical and Life Sciences, for their unfailing support in all aspects of the proposal, editorial, and production processes. The editors also thank Lukasz Komsta for designing the book's cover.

About the Editors

Teresa Kowalska earned an MSc in chemistry from Pedagogical High School in Katowice, Poland, in 1968, a PhD in physical chemistry from the University of Silesia, Katowice, Poland, in 1972, and a DSc in physical chemistry from the Maria Curie-Skłodowska University in Lublin, Poland, in 1988. In the 1974–1975 school year, Professor Kowalska stayed for 12 months as a British Council postdoctoral fellow at the Chemistry Faculty of Salford University, Salford, Lancashire, United Kingdom, under the supervision of the late Professor Hans Suschitzky, head of the Organic Chemistry Department there. Professor Kowalska has authored and coauthored over 300 original research and review papers in approximately 30 different peer-reviewed chemistry journals, 16 invited book chapters, and over 500 conference papers (both lectures and posters) included in the programs of scientific conferences at home and abroad (also on numerous personal invitations). Moreover, Professor Kowalska coedited with Professor Joseph Sherma *Preparative Layer Chromatography* and *Thin Layer Chromatography in Chiral Separations and Analysis*, with Professor Monika Waksmundzka-Hajnos and Professor Joseph Sherma *Thin Layer Chromatography in Phytochemistry*, and with Professor Joseph Sherma and Professor Mieczysław Sajewicz *Planar Chromatography–Mass Spectrometry*, all published in the *Chromatographic Science Series* by CRC Press/Taylor & Francis Group. Last, Professor Kowalska (together with Professor Mieczysław Sajewicz) acts as coeditor-in-chief of the international chromatography journal *Acta Chromatographica*, published by Akademiai Kiado, Budapest, Hungary. She also acts as an editorial board member for several chromatography journals, including the *Journal of Liquid Chromatography & Related Technologies* (published by Taylor & Francis), *Journal of Chromatographic Science* (published by Oxford University Press), and the *Journal of Planar Chromatography—Modern TLC* (published by Akademiai Kiado). On an invitation of approximately 20 internationally recognized analytical chemistry and separation science journals, Professor Kowalska has prepared several hundred peer reviews of manuscript submissions.

Professor Kowalska's main research interests focus on the applications of the thin-layer, high-performance liquid and gas chromatography to physicochemical problems (such as extraction of thermodynamic information from gas chromatographic data, enantioseparations mechanisms, and tracing the nonlinear chemical reactions with the aid of liquid chromatographic techniques). She is also interested in analytical applications of liquid chromatographic techniques to food chemistry (mainly to the chemistry of medicinal plants and meat products). In the field of nonlinear reaction mechanisms, Professor Kowalska has established fruitful collaboration with the world's leading research group in this area, headed by Professor Irving R. Epstein from the Chemistry Department of Brandeis University, Waltham, Massachusetts.

In parallel with her research, Professor Kowalska has been an academic teacher for over four decades and in this capacity she has taught undergraduate courses in general chemistry and undergraduate and postgraduate courses in fundamentals and applications of the chromatographic techniques. Professor Kowalska has supervised approximately 80 MSc and 14 PhD theses (in the field of analytical chemistry and chromatographic science), and has also served as a jury member for over 40 doctorates at home and abroad. She is actively involved in the EU undergraduate students international exchange program Socrates, encouraging her own students to pursue a one-semester training experience in a foreign institution, and also in hosting foreign undergraduate students and postdoctoral fellows in her laboratory. For her continuous and dedicated engagement in teaching chemistry, Professor Kowalska has received many awards at the local and national levels.

Mieczysław Sajewicz earned an MSc and a PhD in chemistry from the University of Silesia, Katowice, Poland, in 1978 and 1989, respectively. The main area of his MSc and PhD studies was analytical chemistry and, more precisely, application of gas chromatography to studying factors governing separation quality (e.g., isomerism of analytes, polarity of stationary phases, and working parameters of the gas chromatographic system as a whole). In 2013, Professor Sajewicz earned a DSc in pharmacy from Collegium Medicum, Jagiellonian University, Kraków, Poland. His DSc dissertation focused on spontaneous nonlinear processes (oscillatory chiral conversion and oscillatory condensation) running in aqueous and nonaqueous solutions of profen drugs and the other low-molecular-weight chiral carboxylic acids (derived from acetic, propionic, and butyric acid).

Professor Sajewicz has authored and coauthored over 150 original research papers and over 350 conference papers, presented at conferences and congresses at home and abroad. Moreover, Professor Sajewicz coedited with Professor Teresa Kowalska and Professor Joseph Sherma *Planar Chromatography–Mass Spectrometry*, published in the *Chromatographic Science Series* by CRC Press/Taylor & Francis Group. He has coauthored one encyclopedia entry in *Encyclopedia of Chromatography* (2nd edition, revised and expanded; 2005) and two book chapters (Chapter 2 in *Preparative Layer Chromatography*, edited by Professor Kowalska and Professor Sherma with CRC Press/Taylor & Francis Group in 2006, and Chapter 9 in *Thin Layer Chromatography in Chiral Separations and Analysis*, also edited by Professor Kowalska and Professor Sherma with CRC Press/Taylor & Francis Group in 2007). Since 1992, Professor Sajewicz has been an editorial board member for *Acta Chromatographica* and, in 2002, he became coeditor-in-chief of the same journal. He also acts as an editorial board member for the analytical chemistry section of *The Scientific World Journal*.

Since 1980, Professor Sajewicz has been a member of the Organizing Committee of the annual all-Polish Symposium on Chromatographic Methods of Investigating the Organic Compounds, organized by the Institute of Chemistry, University of Silesia, and has been cochairman of the Scientific and Organizing Committee of the same scientific event since 2002.

Professor Sajewicz has wide teaching experience at the undergraduate and graduate university levels. He has run courses in chemical calculus, several laboratory courses in planar, high-performance liquid and gas chromatography, and MSc lecture courses on selected chromatographic techniques and on applications thereof to environmental analysis. He has supervised and cosupervised approximately 70 MSc theses, and has supervised one PhD thesis and cosupervised four PhD theses. For his continuous and dedicated engagement in teaching chemistry, Professor Sajewicz has received numerous awards at the local and national levels.

Joseph Sherma received a BS in Chemistry from Upsala College, East Orange, New Jersey, in 1955 and a PhD degree in analytical chemistry from Rutgers, the State University, New Brunswick, New Jersey, in 1958 under the supervision of the renowned ion exchange chromatography expert William Rieman III. Professor Sherma is currently John D. and Francis H. Larkin Professor Emeritus of Chemistry at Lafayette College, Easton, Pennsylvania; he taught courses in analytical chemistry for more than 40 years, was head of the Chemistry Department for 12 years, and continues to supervise research students at Lafayette. During sabbatical leaves and summers, Professor Sherma did research in the laboratories of the eminent chromatographers Harold Strain, Gunter Zweig, James Fritz, and Joseph Touchstone.

Professor Sherma has authored, coauthored, edited, or coedited 870 publications, including research papers and review articles in approximately 55 different peer-reviewed analytical chemistry, chromatography, and biological journals; approximately 40 invited book chapters; and more than 70 books and U.S. government agency manuals in the areas of analytical chemistry and chromatography.

In addition to his research in the techniques and applications of thin-layer chromatography (TLC), Professor Sherma conducted a very productive interdisciplinary research program in the use of analytical chemistry to study biological systems with Bernard Fried, Kreider Professor Emeritus of Biology at Lafayette College for more than 30 years. He wrote the book *Thin Layer Chromatography* (first to fourth editions) and edited the *Handbook of Thin Layer Chromatography* (first to third editions), both published by Marcel Dekker, Inc., as well as editing *Practical Thin Layer Chromatography* for CRC Press. Professor Sherma and Dr. Zweig wrote a book titled *Paper Chromatography* for Academic Press and the first two volumes of the *Handbook of Chromatography* series for CRC Press, and coedited with him 22 more volumes of the chromatography series and 10 volumes of the series *Analytical Methods for Pesticides and Plant Growth Regulators* for Academic Press. After Dr. Zweig's death, Professor Sherma edited five additional volumes of the chromatography handbook series and two volumes in the pesticide series. The pesticide series was completed under the title *Modern Methods of Pesticide Analysis* for CRC Press, with two volumes coedited by Thomas Cairns. Three books on quantitative TLC and advances in TLC were edited jointly with Professor Touchstone for Wiley-Interscience. For the CRC/Taylor & Francis Group *Chromatographic Science Series*, Professor Sherma coedited with Professor Kowalska *Preparative Layer Chromatography* and *Thin Layer Chromatography in Chiral Separations and Analysis*, with Professor Kowalska and Professor Monika Waksmundzka-Hajnos *Thin-Layer Chromatography in Phytochemistry*, with Professor Waksmundzka-Hajnos *High Performance Liquid Chromatography in Phytochemical Analysis*, with Professor Lukasz Komsta and Professor Waksmundzka-Hajnos *Thin Layer Chromatography in Drug Analysis*, with Professor Tomasz Tuzimski *High Performance Liquid Chromatography in Pesticide Residue Analysis*, and with Professor Kowalska and Professor Mieczyslaw Sajewicz *Planar Chromatography–Mass Spectrometry*.

A book titled *Chemometrics in Chromatography*, coedited with Professor Komsta and Professor Yvan Vander Heyden, is in press, and books titled *High Performance Liquid Chromatography in Phytochemical Analysis* (2nd Ed.) coedited with Dr. Lukasz Ciesla and Professor Waksmundzka-Hajnos and *Determination of Target Xenobiotics and Unknown Compounds in Food, Environmental, and Biological Samples* coedited with Professor Tuzimski are in preparation for the *Chromatographic Science Series*.

Professor Sherma served for 23 years as editor for residues and trace elements of the *Journal of AOAC International* and is currently that journal's Acquisitions Editor. He has guest edited with Professor Fried 19 annual special issues on TLC of the *Journal of Liquid Chromatography & Related Technologies* and special sections of issues of the *Journal of AOAC International* on specific subjects in all areas of analytical chemistry. For 12 years, he wrote an article on modern analytical instrumentation for each issue of the *Journal of AOAC International*. Professor Sherma has written biennial reviews of planar chromatography published in the American Chemical Society journal *Analytical Chemistry*, *Journal of AOAC International*, and *Central European Journal of Chemistry* since 1970 and biennial reviews of pesticide analysis by TLC since 1982 in the *Journal of Liquid Chromatography & Related Technologies*, the *Journal of Environmental Science and Health, Part B*, and the *Journal of AOAC International*. He is on the editorial boards of the *Journal of Planar Chromatography–Modern TLC*, *Acta Chromatographica*, *Journal of Environmental Science and Health, Part B*, and *Journal of Liquid Chromatography & Related Technologies*.

Professor Sherma was the recipient of the 1995 ACS Award for Research at an Undergraduate Institution, sponsored by Research Corporation for Science Advancement. The first 2009 issue, Volume 12, of the journal *Acta Universitatis Cibiensis, Seria F, Chemia* was dedicated in honor of Professor Sherma's teaching, research, and publication accomplishments in analytical chemistry and chromatography.

Contributors

Piotr Adamowicz
Department of Forensic Toxicology
Institute of Forensic Research
Kraków, Poland

Beril Anilanmert
Institute of Forensic Sciences
İstanbul University
Cerrahpaşa
İstanbul, Turkey

Bogumiła Byrska
Alcohol and Drugs Analysis Section
Department of Forensic Toxicology
Institute of Forensic Research
Kraków, Poland

Thomas A. Brettell
Department of Chemical and Physical
   Sciences
Cedar Crest College
Allentown, Pennsylvania

Sena Çağlar Andaç
Department of Analytical Chemistry
Faculty of Pharmacy
İstanbul University
İstanbul, Turkey

Fatma Çavuş Yonar
Institute of Forensic Sciences
İstanbul University
Cerrahpaşa
İstanbul, Turkey

Rafał Celiński
Toxicology Laboratory ToxLab
Katowice, Poland

Michael D. Cole
Department of Biomedical and Forensic
   Sciences
Faculty of Science and Technology
Anglia Ruskin University
Cambridge, United Kingdom

Iris Djordjević
Faculty of Veterinary Medicine
University of Belgrade
Belgrade, Serbia

Gregory W. Endres
Consultant New Designer Drug
   Legislation
PinPoint Testing, LLC
AR Children's Hospital Research
   Institute
Little Rock, Arkansas

Yuji Fujita
Division of Emergency Medicine
Department of Emergency, Disaster
   and General Medicine
Iwate Medical University School of
   Medicine
Morioka, Iwate, Japan

Bogna Geppert
Department of Forensic Medicine
Poznań University of Medical Sciences
Poznań, Poland

Łukasz Komsta
Department of Medicinal Chemistry
Faculty of Pharmacy
Medical University of Lublin
Lublin, Poland

Teresa Kowalska
Institute of Chemistry
University of Silesia
Katowice, Poland

Chipo Kuleya
Department of Biomedical and Forensic
    Sciences
Faculty of Science and Technology
Anglia Ruskin University
Cambridge, United Kingdom

Milena Majchrzak
Institute of Chemistry
University of Silesia
and
Toxicology Laboratory ToxLab
Katowice, Poland

Przemysław Mielczarek
Department of Biochemistry and
    Neurobiology
Faculty of Materials Science and
    Ceramics
AGH University of Science and
    Technology
Kraków, Poland

Slobodan M. Milosavljević
Faculty of Chemistry
University of Belgrade
Belgrade, Serbia

Akira Namera
Department of Forensic Medicine
Graduate School of Biomedical and
    Health Sciences
Hiroshima University
Hiroshima, Japan

Ali Acar Özdemir
Alvi Medica
Trakya Serbest Bölgesi
Çatalca-İstanbul, Turkey

Takeshi Saito
Department of Emergency and Critical
    Care Medicine
Tokai University School of Medicine
Kanagawa, Japan

Mieczysław Sajewicz
Institute of Chemistry
University of Silesia
Katowice, Poland

Karolina Sekuła
Department of Forensic Toxicology
Institute of Forensic Research
Kraków, Poland

Joseph Sherma
Department of Chemistry
Lafayette College
Easton, Pennsylvania

Marek Smoluch
Department of Biochemistry and
    Neurobiology
Faculty of Materials Science and
    Technology
AGH University of Science and
    Technology
Kraków, Poland

Karina Sommerfeld-Klatta
Department of Toxicology
Poznań University of Medical Sciences
Poznań, Poland

Jon E. Sprague
Director and BCI Eminent Scholar
Ohio Attorney General's Center for the
    Future of Forensic Science
Bowling Green State University
Bowling Green, Ohio

Roman Stanaszek
Alcohol and Drugs Analysis Section
Department of Forensic Toxicology
Institute of Forensic Research
Kraków, Poland

Artur Teżyk
Department of Forensic Medicine
Poznań University of Medical Sciences
Poznań, Poland

Vlatka Vajs
Center for Chemistry
Institute for Chemistry, Technology and
    Metallurgy
University of Belgrade
Belgrade, Serbia

Ljubodrag Vujisić
Faculty of Chemistry
University of Belgrade
Belgrade, Serbia

Katarzyna Wicha-Komsta
Private Pharmacy
Pulawy, Poland

Matthew R. Wood
Laboratory Director
Ocean County Sheriff's Office
Forensic Science Laboratory
Toms River, New Jersey

Travis J. Worst
Forensic Science Instructor
Department of Chemistry
Bowling Green State University
Bowling Green, Ohio

Dariusz Zuba
Research Director
Institute of Forensic Research
Kraków, Poland

Section I

1 Introduction

Teresa Kowalska, Mieczysław Sajewicz, and Joseph Sherma

CONTENTS

1.1 DESIGNER DRUGS: HISTORICAL BACKGROUND AND PRESENT STATUS

It is not an exaggeration that, since time immemorial, Nature, especially the plant kingdom, has been the primary source of medicines for all living organisms on Earth, most probably even long before the appearance of humans. The earliest documents in support of this presumption are believed to be between 4,000 and 5,000 years old and originate from such ancient civilizations as those of Mesopotamia, Egypt, India, and China. An ancient medicinal legacy of the Mesopotamian kingdoms (Sumer, Babylon, and Assyria) was apparently transferred to ancient Egypt and then to Europe according to two written milestone relics, the Papyrus Ebbers from *ca.* 1550 BC [1] and the treatise known as *Corpus Hippocraticum*, ascribed to the European father of medicine, Hippocrates of Kos (*ca.* 460–*ca.* 370 BC) [2,3]. Both of these documents are in fact herbal pharmacopeias that list several hundred medicinal plants and give strict recommendations regarding their applicability and modes of usage. Ancient medicinal traditions of India and China, specifically Ayurvedic medicine [4] and Traditional Chinese Medicine (TCM) [5], are very much alive today in the respective two cultures, with a vigorous aim to move beyond their regional barriers in order to enrich the global arsenal of curative plants and to make them available worldwide, and also in search of new templates for the design of new synthetic drugs. A famous approach to medicine through use of curative plants that originates from the Islamic Golden Age was described in the writings of the great Persian polymath Avicenna of Bukhara (Ibn Sina, *ca.* 980—June, 1037) and entered medieval Europe owing to the translation and popularization of his opus by the Benedictine Order [6]. Apart from the aforementioned, there are long-established and widely recognized great medicinal traditions, with practically each local community on all continents developing its own curative system by taking advantage of the indigenous flora. Owing to the considerable progress of analytical chemistry witnessed over the last four decades at least, these local medicinal systems have started being systematically investigated, and research has been reported covering practically all interesting regions of the world, from Mexico and the other Mesoamerican and South American countries [7],

3

via Africa [8], Madagascar [9], and Australia [10], to the Pacific Islands [11,12] and beyond.

For most of history, however, there were not sufficient analytical tools available to permit determination of the contents of medicinal plants and define individual active ingredients able to cure an organism or to alleviate the most disturbing disease symptoms, such as pain or elevated temperature. A significant breakthrough in modern pharmacology began only by the end of the eighteenth century with an extract (instead of a whole plant infusion) derived from foxglove (*Digitalis* L.) and applied in the treatment of heart conditions [13]. Later, two pharmacologically active glycosides, digoxin and digitoxin, were identified in and isolated from the foxglove extract, thus considerably contributing to the progress in cardiology. The next big leap in extracting active ingredients from plant material took place around 1804 with the isolation of morphine, the most active alkaloid out of many contained in the straw of the opium poppy [14]. Morphine is considered by many as the first medicine in pure form to be separated from a botanical matrix, and it proved very effective in the treatment of acute and chronic pain, thus allowing fast progress of surgery. Later, chemical modifications of morphine resulted in a number of semisynthetic analgesic compounds from the family of opioids (e.g., hydromorphone, oxycodone, and heroin). Soon after its isolation, morphine and later its derivatives were found to have a high potential for addiction and abuse, thereby becoming a convincing symbol of certain ambiguities inherent in practically all drugs.

It seems highly possible that psychoactive substances of natural origin have been with mankind since prehistoric times. Paleolithic rock art and shamanic imagery suggest that humans have been using mind-altering substances for thousands of years. Some anthropologists even suggest that these substances may have played an evolutionary role in the mental development of humans [15]. On the other hand, there is firm evidence that the same natural psychoactive substances have been included in spiritual practices and ancient rituals of many different cultures worldwide. Some of these practices have survived until our time (e.g., among the tribes of the South American and Mesoamerican Indians), thus enabling diverse anthropologic and medical surveys of this phenomenon [16,17]. Two plants that are particularly popular for their hallucinogenic properties in the aforementioned regions are the psilocybin mushroom (also known as the psychedelic or "magic" mushroom, *Psilocibe semilanceata*) [18] and sage of the diviners (*Salvia divinorum*) [19]. Use of psychoactive substances for ritual and religious purposes has been regulated in each culture by complicated rules prescribed by custom and tradition, and, hence, their intake has remained under at least some control in these cases. However, in many different parts of the world, an uncontrolled approach toward consumption of psychoactive substances also took place, as documented over the centuries since the Neolithic era [20]. There have been multiple reasons for such consumption, e.g., as an antidepressant or painkiller, to tame hunger in periods of famine, to alter a person's consciousness, etc. The most impressive record of massive consumption of a mind-altering substance for a nonritual reason prior to the twentieth century probably is that of opium, which has been abundantly produced in the Middle East, but even more so in the so-called "north belt" ranging from Afghanistan and Pakistan, via North India, to Burma. Throughout the nineteenth century, intake of opium became so commonplace with its inhabitants that it brought the Chinese Empire to the verge of demographic ruin. In order to illustrate the consumption level of opium in China in the nineteenth century, it is enough to compare its annual import of opium by 1858 (equal to *ca.* 4,550 tons) with the very similar amount that represents annual global opium production within the decade surrounding the year 2000 [21].

The end of the nineteenth century and the whole of the twentieth century were hallmarked by the dynamic development of sciences and, at the same time, by a long list of military conflicts on a local and global scale. Developing medical science found itself in growing need of anesthetics and analgesics not only because of surgery, but also for the treatment of pain for cancer, psychiatric, and other patients. At the same time, psychiatrists and psychologists focused on the response of the human nervous system to mind-altering compounds, which eventually led to establishment of anesthesiology as a new branch of medicine in its own right, dedicated to relieving pain and inducing anesthesia during surgery and some other medical procedures. For neurologists and surgeons, it soon became clear that the assortment of available psychoactive compounds was insufficient and more sophisticated fine-tuning of a human nervous system with a new generation of effective drugs was needed. This demand became a challenge for synthetic chemists, who soon delivered a successful response [22]. It was probably around this time that the drug design concept was first conceived, gaining a novel dimension with the development of molecular biology and bioinformatics [23,24]. As a by-product of modern military conflicts, war medicine started developing dynamically, for obvious reasons being largely dependent on anesthetics and analgesics. According to available records, the American Civil War was the first "modern" large-scale military conflict that consumed considerable amounts of opium as a first-choice means for pain management [25], and it is claimed that, in that war, the Union Army used 2.8 million ounces of opium tincture and powder and about 500,000 opium pills [26]. In World War II, opium was regarded by many countries as a key strategic resource, and threat of a blockade of the usual opium transportation routes resulted in the hasty shipment of increased loads of this particular commodity either from India (by the Allies) or from Turkey (by the Nazis) [27].

In the Western world, introduction of anesthetics and analgesics to wider medicinal practice started early in the nineteenth century, in a small and overdue manner through dental surgery [28]. Initially, these were compounds administered by inhalation only (such as ethyl ether or, later, chloroform). Also, laudanum (i.e., an alcohol tincture containing approximately 10% powdered opium, equivalent to 1% morphine) was available in pharmacies without any prescription, and its use was advocated for a wide range of medical conditions [29]. For example, it was recommended to women with psychosomatic symptoms and menstrual cramps at that time diagnosed as neurasthenia, but it is also reported to have been spoon-fed to babies. On the other hand, in the late nineteenth and early twentieth centuries, relatively large groups of the male population were treated with opium in military hospitals on the frontlines of increasingly more frequent armed conflicts. As a consequence, societies were becoming accustomed to slowly spreading addiction to mind-altering preparations and, in certain instances, with an abuse thereof. A separate group was that of artists and intellectuals, attracted by the relatively easy availability of psychoactive substances and always eager to experiment with them, in a search for a new means of inner stimulation. One example is the well-documented dedication of the artistic community, particularly in French circles, before and soon after World War I to absinthe and its assumed influence on the creativity of the Impressionists, Post-Impressionists, Surrealists, Modernists, and Cubists, as reflected in their paintings, poetry, and prose. After 1988, the European Union (EU) countries absolved absinthe of an earlier "infamy" and ban due to its supposedly exaggerated performance as a psychoactive liquor [30], and the alcoholic beverage quickly returned to shelves throughout the EU. The truth is that it contains an extract from grand wormwood (*Artemisia absinthium*) and can be responsible for mind-altering effects in the first instance caused by α-thujone (provided the content of this monoterpene in the drink surpasses the neurotoxic level of 35 mg L^{-1} [31]). Also, many famous scientists—among them, for example, Sigmund Freud and William Stewart Halsted—experimented in certain periods of their lives with psychoactive substances, as documented by Markel [32]. The steadily expanding spectrum of relatively easily available psychoactive substances, combined with an insufficient pharmaceutical prescription system, resulted in the first restrictive legal regulations in many different countries of the world. These legislative moves certainly contributed to an improvement of pharmaceutical law, yet the Pandora's Box, handsomely stuffed with psychoactive substances, had already been opened.

The moral shock of World War II and then the slowly improving postwar economic conditions of Western societies, on the one hand, and an oppressively technicized and materialistic modern world, on the other, triggered the anxieties of youth in the United States during the early 1960s, which then spread around the world to become a part of the world's intellectual and artistic stir, later referred to as the hippie movement. Around that time, a massive public longing for recreational drugs first came to pass, largely inspired by the well-documented experiments of hippie communities with mind-altering substances. As recreational drugs, one understands both legal and illegal psychoactive substances that alter one's mental state through changes to the central nervous system in order to create positive emotions and feelings. In parallel with the division of psychoactive substances into legal and illegal, another classification system was coined that divides them into hard and soft drugs, which is, however, less clear. Although this second classification tends to address the drugs' harmfulness, with hard and soft drugs meaning the harmful and the less-harmful (or even harmless) ones, respectively, it is controversial and contested by many; one can hardly imagine harmless drugs. Moreover, how can the harmfulness of a drug be defined? Is a drug harmful if it develops strong addiction without causing any considerable physical damage, or another one that does not develop strong addiction but instead damages inner organs? This evident ambiguity and confusion in classification is reflected in legal approaches to the drug problem, and, as a result, different countries have adopted different legislative systems. The legislative issues referring to recreational drugs will be discussed in some later chapters of this book, and now we will focus on so-called designer drugs, which are the main subject matter of this book.

As mentioned before, in the second half of the twentieth century people in more advanced societies worldwide developed a need for easily available recreational drugs in an effort to temporarily alleviate the challenges and tensions of everyday life. Intensifying intercontinental tourism enhanced this tendency by documented examples of whole societies in developing countries making recreational use of local herbs known for containing psychoactive components (e.g., the marijuana-containing cannabis plant in the Indian subcontinent and most of the South Asian region, the cocaine-containing coca leaves in Latin America, the amphetamine-like cathinone-containing khat leaves in the Arabian Peninsula and the Horn of Africa, etc.). An important demand placed on recreational drugs was that their usage should not conflict with legal regulations of a given country placed on psychoactive substances. In short, this is how the designer drugs concept was created. According to the generally accepted view, designer drugs are structural or functional analogs of controlled (either illegal or legal) drugs that have been devised to mimic the pharmacological effects of known and classified drugs, yet escape both classification as illegal and detection in the standard drug tests [33–35]. In many cases, designer drugs involve the innovative structural modification of long-known psychoactive compounds, and in some of them these modifications are carried out on templates derived from plant material. In other cases, the chemical structures of novel designer drugs are by no means modeled by Nature, yet tend to imitate the activity of already-known drugs. The EU has designated all of these compounds as new psychoactive substances (NPSs), and, basically, they are prepared as brand-new chemical structures in clandestine laboratories. However, those who synthesize designer drugs are usually acquainted with some earlier scientific investigations that have been carried out in official research laboratories in their effort to discover more potent derivatives of the already-known psychoactive substances that have fewer harmful side effects and do not end up on the official lists of illegal products. Thus, on certain occasions these earlier syntheses are repeated in the clandestine laboratories to produce designer drugs that are not brand-new chemical structures and do not end up on any regulated list. As the efficacy and safety of these substances have not been sufficiently evaluated in any animal or human tests, the use of designer drugs may result in unexpected health-endangering side effects and, in extreme cases, in casualties [36].

For the aforementioned reasons, designer drugs pose very serious problems for criminal investigators, toxicologists, and analytical chemists. Upon finding a suspicious-looking chemical of unknown origin or in a case involving intoxication or death that is apparently caused by poisoning with a designer drug, a criminal investigator immediately initializes an inquiry procedure. As a result, a toxicological laboratory begins working on a sample collected from the crime scene, which is either an unknown chemical or biological evidence (mostly, a body fluid or a tissue sample). The task of identifying the chemical structure of an unknown substance is within the competence domain of analytical chemists, and in a case of biological evidence a fairly complicated preliminary sample-preparation step is needed. General strategies of analytical procedures applicable to forensic evidence of the kind discussed are provided in a number of toxicology handbooks and monographs (e.g., 37–39). An additional challenge faced by a forensic analyst working on identification of a designer drug is that the substance in question can be a brand-new compound originating from a clandestine laboratory, and it has never before been reported in the chemical or toxicological literature. Hence, it does not appear on any official list of designer drugs, and in such cases a toxicological laboratory is expected to perform the pioneering tasks of properly documenting identification of this substance, reporting on its existence, and proving its basic physicochemical characteristics to an appropriate legal organization. It must be kept in mind that the present scene of designer drugs is undergoing very dynamic changes, and on pace with the gradual incrimination of older drugs recognized for their health-endangering effects and put on the illicit drug list, brand-new and not-properly-tested compounds continuously emerge on the designer drug market [40,41].

Due to their excellent performance, high-performance column liquid chromatography (HPLC) and gas chromatography (GC) implemented with efficient instrumental detection systems, especially mass spectrometry (MS), are the major analytical techniques of choice in toxicological analysis today, and for the identification and quantification of designer drugs as well. High-performance thin-layer chromatography (HPTLC) with densitometric detection and/or hyphenated with MS detection using recently developed interfaces can also be successfully used for the same purpose. The main goal of this book is to discuss the possibilities of individual chromatographic techniques and of the auxiliary spectrometric techniques in detection, identification, and quantification of designer drugs, both in their drug form and contained in biological samples, in the context of their molecular structure determined by belonging to a given chemical class. In our view, the advantage of this systematic presentation is not undermined by the steadily expanding pool of the designer drugs within each individual class. Moreover, the book provides basic information regarding legislative issues with respect to designer drugs and an insight into the biological precursors of some of them.

There has been no previous complete discussion of the chromatographic analysis of designer drugs as is found in this book. It updates and expands on the limited information given in the few review articles published earlier in the field [42–46].

1.2   CONTENTS OF THE BOOK

Our book comprises 22 chapters that are divided into Sections I and II. Section I contains Chapters 1 through 12, which provide an overview of legislation on the federal and state level in the United States of America (U.S.) referring to designer drugs, and legal issues in some other countries regarded as the trendsetters in solving the designer drug problem (basically, in the EU countries). Moreover, in Section I insight is offered from a botanical perspective on the natural precursors of certain designer drugs, fundamentals of instrumental techniques that perform the best in the field of forensic analysis are discussed, and principles of the preparation of biological material for the demands of toxicological analysis are summarized. Section II contains Chapters 13 through 22, all of which (except Chapter 22) focus on an overview of the sample-preparation techniques and the hyphenated chromatographic approaches most successfully applied to the analysis of different classes of designer drugs (e.g., cannabinoids, cathinones, phenylethylamine derivatives, opioids, piperazines, etc.) in traded samples and in materials collected as judicial evidence. Chapter 22 provides a review of TLC applied to the analysis of designer drugs.

After this introductory chapter that begins Section I of the book, Chapter 2, titled "Designer Drugs: Legislating for Tomorrow Through the Use of Class Definitions and Pharmacophore Principles," introduces conceptual problems with establishing efficient legislative measures that might successfully prohibit the uncontrolled spreading of the designer drugs that are not included in any chemical or pharmaceutical register, or even bar future spreading of those that have not yet been synthesized. The authors focus on the approaches that are currently considered in the U.S. based on analytical recognition of a pharmacophore motif in the chemical structure of an unknown sample that is identical with the structural motif of a recognized class of drugs as a sufficient requirement to undertake preventive and/or penalizing measures in an attempt to protect society from the harmful effects of designer drugs.

In Chapter 3, titled "Alternative Legislation with Designer Drugs," a difference is presented between the designer drug problem as viewed by the American legislative system, which accommodates it within the framework of common law, and the legislative systems of European countries, most of which, with certain exceptions like the United Kingdom (UK), operate within the framework of civil law, also called the Romano-Germanic legal system. This system is based on fixed codes and statutes, where—contrary to common law—the decisions of the judges do not affect the laws of a country. The European approach (assumed by all member countries of the EU) is that classification of a given psychoactive substance as a designer drug (or otherwise) strictly depends on the chemical structure of an involved active compound. This approach is known as a generic approach, and the author illustrates his discussion with interesting practical examples (e.g., emphasizing that lists of illicit psychoactive compounds can differ from one European country to another).

In Chapter 4, titled "Biological Background of Designer Drugs," various biological aspects related to the designer drugs are discussed, ranging from human physiology (explaining the role of neurotransmitters, physiological action of drugs, and physiological meaning of addiction) to those classes of designer drugs that are synthetic variations of the prototypes originating from the kingdom of plants. Moreover, the physiological impact of designer drugs from these different classes on the human nervous system is discussed.

Chapter 5, titled "Liquid Chromatography–Mass Spectrometry in the Analysis of Designer Drugs," provides a comprehensive review of different variants of the LC-MS technique applied to identification and quantification of designer drugs in seized samples and biological materials collected from living and deceased abusers. The author comprehensively presents individual LC techniques and their advantages within the broad background of the approaches commonly applied in the analysis of the designer drugs for toxicological and judicial purposes. She convincingly points to their role as the techniques of first choice due to outperforming the tandem GC techniques. Chapter 5 ends with a list of analytical issues that are still weak points in the efforts to combat the manufacture and trafficking of designer drugs, and the author also presents her views on different actions that need to be undertaken to overcome present-day shortcomings in this respect.

Chapter 6, titled "Ambient Plasma Ionization Techniques in Forensic Analysis," presents the potential of several MS techniques as emerging tools for the fast screening of chemical substances suspected of being psychoactive and, therefore, hazardous, and also of legal evidence of various kinds. So far, the most popular in this respect is DART (direct analysis in real time) MS, but the authors also point out some other ambient plasma ionization techniques that are currently gaining an increased interest in forensic analysis (e.g., FAPA [flowing atmospheric pressure afterglow] MS and DBDI [dielectric barrier discharge ionization] MS). A selection of interesting application examples is also provided. The main advantage of all these MS variants is that they are extremely fast and, hence, they can provide results literally within seconds. In other words, they can become an attractive alternative for the LC-MS and GC-MS approaches in the future because of their unrivaled speed of delivering a discriminative result capable of saving lives in critical intoxication cases. For the time being, DART, FAPA, and DBDI MS still have problematic quantification capabilities, and, until this handicap is overcome, they can be applied as the screening forerunners of LC-MS and GC-MS as they are presented in this chapter.

Chapter 7, titled "Application of LC-QTOFMS for the Identification of the Structure of New Psychoactive Substances," demonstrates the excellent performance of LC coupled with quadrupole time-of-flight tandem mass spectrometry (QTOFMS) using a hybrid spectrometer composed of a Q mass filter and TOF analyzer. The authors extensively discuss the advantages of such an instrumental setup over the traditional LC–electrospray ionization (EI) MS system, which is nowadays available in most toxicological laboratories worldwide as standard equipment but in critical experimental cases can provide an incorrect identification of structurally related compounds. A common trick in devising novel designer drugs is modifying the original structures by introducing to them chlorine or bromine atoms; the authors explain how analysts can benefit from tracing the characteristic isotopic distribution of these two halogens, available with use of the LC-QTOFMS system but often impossible to obtain with use of the traditional LC-EI-MS setup. The contents of Chapter 7 are well illustrated, with a good number of practical examples taken from the literature that demonstrate how on different occasions LC-QTOFMS outperforms LC-EI-MS.

Chapter 8, titled "Rapid Generation of Metabolites by Electrochemical Method," presents an innovative concept to rapidly obtain the metabolites of different chemical compounds in the course of oxidation induced by an electrochemical method, as an alternative to a tedious and time-consuming *in vitro* procedure that employs liver microsomes of experimental animals. This latter approach is a biochemical process that in chemical terms means an oxidation of the precursor with the P450 contained in the microsomes. Electrochemical oxidation reliably mimics an outcome of the biochemical process, its main advantages being that it avoids usage of an animal tissue and is incomparably faster. Further LC-MS identification of the electrochemically generated metabolites can provide a rapid answer as to the chemical nature of a toxic agent. With such preliminary knowledge, detoxification of a vitally endangered and unconscious subject can be beneficially accelerated, and it can also prove helpful in the postmortem identification of a toxic agent. The authors comprehensively discuss this innovative concept by providing a convincing comparison of the results obtained for XLR-11 ((1-[5′-fluoropentyl]indol-3-yl)-(2,2,3,3-tetramethyl-cyclopropyl)methanone) and, more specifically, the metabolites of this designer drug generated by its incubation with rat liver microsomes and, alternatively, using the electrochemical method.

Chapter 9, titled "Gas Chromatography–Mass Spectrometry in the Analysis of Designer Drugs," presents general thinking on the advantages and drawbacks of this hyphenated analytical technique, which undoubtedly is an important approach in the analysis of this group of chemical compounds. The main drawback of using the GC-MS technique is possible thermal degradation of the designer drugs, which leads to falsification of analytical response. This threat is explained and convincingly illustrated by a number of well-selected practical examples. Moreover, two different strategies for circumventing certain interpretational problems in the GC-MS analysis of the designer drugs are extensively discussed. These strategies are derivatization of an analyzed material, which enhances both volatility of the low-volatile designer drugs and specificity of the respective mass spectra, and retention time locking (RTL) as a helpful analytical tool in the identification of isomers.

Chapter 10, titled "NMR Spectroscopy in the Analysis of Illegal Drugs," presents information on invaluable performance of ^{1}H and ^{13}C nuclear magnetic resonance (NMR) spectroscopy in identification of illegal psychoactive compounds both in street samples and those originating from pathomorphologic examinations. This technique proves itself practically unsurpassable in discriminating among different isomeric structures. Although the NMR spectroscopic detector integrated with LC as a hyphenated LC-NMR analytical system is technically possible and commercially available (although quite expensive and burdened with certain technical limitations), so far no reports have been published on its usage for solving designer drug-related problems. However, its complementary role in providing efficient off-line assistance in solving difficult forensic analyses is well documented in the chapter, and, moreover, the authors provide a selection of interesting practical examples originating from their own long-term experience.

Chapter 11, titled "Preparation of Biological Material for Toxicological Analysis," provides an overview of sample-preparation approaches in the field of toxicological and forensic analysis related mainly to identification and quantification of designer drugs in human bodies. As biological materials of analytical importance, the authors include samples collected from living and deceased subjects, i.e., body fluids (blood, urine, and the vitreous humor from the eyeball) and also the tissues of inner organs originating from the autopsy of cadavers at different decomposition stages (liver, kidney, stomach, brain, guts, etc.). The chapter is structured in such a way as to first introduce fast screening methods (i.e., those that allow an analyst to get a preliminary hint as to a possible kind of intoxication), and then the authors discuss the advantages of liquid–liquid extraction (LLE) and solid-phase extraction (SPE) applied to body fluids. Then the approaches to preparation of samples for the analysis of internal organs and hair (a significant source of forensic evidence) are presented, and the chapter ends with innovative implementation of nonmodified and modified QuEChERS (**Qu**ick, **E**asy, **Ch**eap, **E**ffective, **R**ugged, **S**afe) techniques used to prepare biological material for hyphenated chromatographic analysis.

Chapter 12, titled "Discrimination among Designer Drug Isomers by Chromatographic and Spectrometric Methods," starts with an exhaustive reminder of the concepts of isomerism and isomers in chemistry, and of possible harmful physiological effects related to the antipodes of chiral drugs in general, and designer drugs in particular. Then the authors discuss possibilities of discriminating the designer drug isomers by GC-MS and LC-MS, which depend not only on a chromatographic technique alone but also on the class and chemical structure of individual designer drugs. Depending on such classification, different MS techniques are recommended as those that allow the best establishment of the chirality of a considered designer drug. Although MS techniques are known as the tools of first choice, the authors present advantages of some subsidiary spectroscopic techniques employed in the on- and off-line modes as well; NMR spectroscopy, Fourier transform infrared (FTIR) spectroscopy, and capillary electrophoresis (CE). Section I ends with Chapter 12, and this section provides a general understanding of the legislation within different legal systems referring to designer drugs and offers a brief overview of analytical techniques most frequently applied in forensic and toxicological analysis targeting the designer drugs.

Section II opens with Chapter 13, titled "Mass Spectrometric Analysis Applied for the Determination of Synthetic Cannabinoids." As a matter of fact, the boom with synthetic cannabinoids, which started some ten years ago, turned this particular class of designer drug into one of the most frequently abused NPSs. The authors discuss the background of a rather sudden and, at the same time, very abundant appearance of synthetic cannabinoids on the designer drug market worldwide, and then they provide an up-to-date overview of the most frequently used analytical techniques for the identification of this class of compounds in seized samples and biological matrices. As gold standards, gas chromatography–mass spectrometry (GC-MS) and liquid chromatography–mass spectrometry (LC-MS) are introduced, and the authors not only highlight their analytical potential but also present possible pitfalls when using each method. At the end, perspectives for the future detection, identification, and quantification of cannabinoids are provided.

Chapter 14, titled "Cathinone Derivatives and their Analysis," contains considerable comprehensive information about this specific class of designer drugs with a clear botanical provenance, which in the course of almost two decades have become among the world's most wanted illicit psychoactive substances. The precursor of the whole class, i.e., cathinone, shows structural similarity to amphetamine, and the physiological effects of the cathinone derivatives in many ways resemble that of their natural analog. An important property of the basic cathinone template is that it can be imaginatively modified in many ways to yield an unexpectedly vast number of different psychoactive derivatives and to escape legal control, leading to a high degree of required analytical resources. The authors of this chapter present the possibilities of the most efficient instrumental tools in the analysis of cathinones and illustrate their presentation with a selection of practical examples.

Chapters 15 and 16, entitled, respectively, "2C Derivatives of Phenylethylamines and Their Analysis" and "NBOMe Derivatives of Phenylethylamines and Their Analysis," provide a thorough and comprehensive overview of chromatographic approaches to the analysis of phenylethylamine derivatives, which are designer drugs known for their strong hallucinogenic properties. The physiological action of phenylethylamine derivatives is recognized as that of the partial serotonin 5-HT$_{2A}$ receptor agonists, and although it resembles the physiological action of

LSD (lysergic acid diethylamide), the hazardous properties of phenylethylamines are incomparably more serious. Casualties reported as caused by the designer phenylethylamines are numerous, whereas lethal action of LSD is hardly documented. Detailed description of the most important analytical procedures is provided in both chapters.

Chapter 17, titled "Forensic Analysis of Piperazines," covers a class of designer drugs that mimics the physiological effects of amphetamine, which so far neither is abundantly represented nor endangers public health on a global scale. For these reasons, piperazines currently are not under international scrutiny, although in certain parts of the world they are monitored by specialized centers for drugs and drug addiction (e.g., in the EU, Australia, Canada, Japan, New Zealand, and U.S.). However, the chemical structures of the benzylpiperazine and phenylpiperazine backbones leave the doors wide open for possible future modifications, and, therefore, a number of efficient analytical (mostly chromatographic) techniques have been adapted to trace and quantify piperazines in various matrices of toxicological and forensic importance. These techniques are comprehensively discussed in this chapter.

Chapter 18, titled "Chromatographic Analysis of Fentanyl and Its Analogs," provides a comprehensive overview of the origin of a group of synthetic piperidine analogs, initially designed as relatively lipophilic compounds to enable effective passage of the blood–brain barrier, and introduced to medicinal practice as a fast-acting and potent group of analgesic drugs. Primarily, this characteristic was intended to lower the amounts of analgesia required in legal and controlled medical treatments. High potency combined with an ease of synthesis soon made this class of compounds very attractive to illicit manufacturers and was also the reason for huge numbers of fatal poisonings with fentanyls worldwide. The authors provide a review of analytical approaches to identification and quantification of fentanyls in seized and postmortem samples, providing useful details regarding particularly successful analytical procedures. They take into account GC with MS and other detection systems, and LC in combination with a vast number of MS types and configurations. Finally, the authors mention successful analytical application of TLC for the separation of 18 fentanyl analogs.

Chapter 19, titled "Modern Techniques for the Identification of Tryptamines," provides an informative introduction to the subject matter of designer tryptamines, starting from their natural precursors present in certain mushrooms growing in South America, Mexico, and the U.S., and recognized for their psychedelic properties in the Indian cultures of the Pre-Columbian era. Then the author reminds readers of the quite-specific circumstances that triggered wider interest in designer tryptamines, i.e., the publication of the book in 1997 by Shulgin and Shulgin titled *TIHKAL: The Continuation*, with TIHKAL being an acronym for "Tryptamines I Have Known and Loved." Although by the end of 2016 a relatively low number of 33 tryptamines had been encountered on the designer drugs market and consequently reported to the European Monitoring Centre for Drugs and Drug Addiction (EMCDDA), a considerable number of casualties attributed to tryptamines points to a necessity to watchfully monitor this class of drugs. The author comprehensively presents the GC and LC methods of first choice to identify and quantify tryptamines, pointing out the important nuances (e.g., discriminating between the isomers) and the complementary role of these two main chromatographic techniques. Finally, the author provides an overview of the auxiliary analytical techniques helpful in solving identification and quantification problems with tryptamines.

Chapter 20, titled "Toxicological Analysis of New Opioids," provides even more information than promised by its title alone. Firstly, an introduction is given on physiological impact of opioids on the human organism, which corresponds well with the general overview of the mechanisms of physiology of addictions provided in Chapter 4 of Section I ("Biological Background of Designer Drugs"). Then the authors discuss applicability of different hyphenated chromatography systems to the detection and quantification of opioids contained in legal evidence of various kinds. A real highlight of this chapter is the tabulated, systematically arranged, and abundant information on chromatographic details of the approaches employed by various authors for the analysis of new opioids derived from the literature.

Chapter 21, titled "Toxicological Analysis of Designer Benzodiazepines," is authored by the same team of specialists as the previous chapter, and it provides a discussion of major analytical problems related to the detection and identification of designer benzodiazepines, due to their close structural resemblance to a vast group of legal benzodiazepine medicines, and also due to the similarity of certain known physiological effects evoked by legal and illicit representatives of this group of drugs. The authors make the point that due to low volatility of benzodiazepines, for this particular class of compounds LC-MS is the analytical method of first choice. MS techniques possess a higher identification power than the chromatographic techniques due to structural similarity between the legal and illicit drugs and, hence, the similar chromatographic retention parameters dictate use of the MS detector.

Chapter 22, titled "Thin-Layer Chromatography in the Analysis of Designer Drugs," is the last one in the book, and its aim is to supplement the analytical, chromatographic overview provided in this monograph by describing the techniques of TLC, including sample preparation, sample and standard solution application, plate development with the mobile phase, zone detection, and densitometry, as well as a number of important applications for separation, identification, and quantification. The topics covered are analysis of human samples, determination of designer drugs that are naturally present or added to herbal samples, chiral determination, quantification of MDMA (3,4-methylenedioxymethamphetamine; ecstasy/molly) in seized tablets, TLC coupled with FTIR spectroscopy and MS, and isolation by preparative layer chromatography (PLC).

In conclusion, it can be stated that this book in a comprehensive manner covers a wide variety of issues related to designer drugs, which encompass a historical survey of human attitudes toward the mind-altering substances up to modern trends in using recreational drugs and the contemporary legal efforts to successfully bar an uncontrolled spreading thereof in an attempt to protect society from casualties or serious health damages. However, the main accent is laid on a survey of the hyphenated analytical techniques with a predominant role played by LC and GC in identification of known designer drugs and characterization of unknown chemical structures belonging to the main chemical classes in all types of legal evidence samples. As editors, we hope that this book will attract a wide readership, in the first instance among analytical chemists working in toxicological laboratories, but also among lawyers engaged in designer drug-related cases.

REFERENCES

1. *The Papyrus Ebers: The Greatest Egyptian Medical Document*; translated by Ebbell, B., 1937. Levin & Munksgaard, Copenhagen.
2. Adams, F., 1891. *The Genuine Works of Hippocrates*, William Wood & Co., New York.
3. Garrison, F.H., 1966. *History of Medicine*, W.B. Saunders & Co., Philadelphia.
4. Meulenbeld, G.J., 1999. *A History of Indian Medical Literature*, Egbert Forsten, Groningen.
5. Unschuld, P.U., 1985. *Medicine in China: A History of Ideas*, University of California Press, Berkeley, California.
6. McGinnis, J., 2010. *Avicenna*, Oxford University Press, Oxford.
7. Cetto, A.A. and Heinrich, M., 2016. Introduction to the special issue: The centre of the Americas—An ethnopharmacology perspective, *J. Ethnopharm.*, 187: 239–240.
8. Mahomoodally, M.F., 2013. Traditional medicines in Africa: An appraisal of ten potent African medicinal plants, *J. Evid. Based Complementary Altern. Med.*, http://dx.doi .org/10.1155/2013/617459.
9. Beaujard, P., 1988. Plantes et medecine traditionnelle dans le Sud-Est de Madagascar, *J. Ethnopharm.*, 23: 165–266.
10. Byard, R., 1988. Traditional medicine of aboriginal Australia, *Can. Med. Assoc. J.*, 139: 792–794.
11. Dasilva, E.J., Murukesan, V.K., Nandwani, D., Taylor, M., and Josekutty, P.C., 2004. The Pacific Islands: A biotechnology resource bank of medicinal plants and traditional intellectual property, *World J. Microbiol. Biotechnol.*, 20: 903–924.
12. Morrison, J., Geraghty, P., and Crowl, L, 1994. *Science of Pacific Island Peoples: Fauna, Flora, Food and Medicine*, Vol. 3, Institute of Pacific Studies, The University of the South Pacific, Suva, Fiji.
13. Goldthorp, W.O., 2009. Medical classics: An account of the foxglove and some of its medicinal uses by William Withering, published 1785, *Brit. Med. J.*, 338: b2189.
14. Courtwright, D.T., 2009, *Forces of Habit Drugs and the Making of the Modern World*, 1st Edition, Harvard University Press, Cambridge, Massachusetts, pp. 36–37.
15. Murray, J.D., 2003. Shamanism and rock art, Chapter 12.5, in: *Mathematical Biology. II. Spatial Models and Biomedical Applications*, 3rd Edition, Springer, New York, pp. 657–659.
16. Dobkin de Rios, M., and Grob, C.S., 2015. Ritual uses of psychoactive drugs, in: *Encyclopedia of Psychopharmacology*, Eds Stolerman, I.P. and Price, L.H., Springer, Berlin, pp. 1474–1479.
17. Dobkin de Rios, M., 2009. *Psychedelic Journey of Marlene Dobkin de Rios: 45 Years with Shamans, Ayahuasqueros, and Ethnobotanists*, Park Street Press, Rochester, Vermont.
18. Metzner, R., Ed., 2005. *Sacred Mushroom of Visions: Teonanácatl: A Sourcebook on the Psilocybin Mushroom*, 2nd Edition, Park Street Press, Rochester, Vermont.
19. Prisinzano, T.E., 2005. Psychopharmacology of the hallucinogenic sage Salvia divinorum, *Life Sci.*, 78: 527–531.
20. Merlin, M.D., 2003. Archeological evidence for the tradition of psychoactive plant use in the Old World, *Econ. Bot.*, 57: 295–323.
21. Global opium production, *The Economist*, June 24, 2010 (Source: United Nations Office on Drugs and Crime).
22. Jones, A.W., 2011. Early drug discovery and the rise of pharmaceutical chemistry, *Drug Test Anal.*, 3: 337–344.
23. Zupan, J. and Gasteiger, J., 1999. *Neural Networks in Chemistry and Drug Design: An Introduction*, 2nd Edition, Wiley–VCH, Weinheim.

24. Madsen, U., Krogsgaard-Larsen, P., Liljefors, T., 2002. *Textbook of Drug Design and Discovery*, Taylor & Francis, Washington, DC.

25. Chisholm, J.J., 1861. *A Manual of Military Surgery*, West & Johnston, Richmond.

26. Schiff, P.L., Jr., 2002. Opium and its alkaloids, *Am. J. Pharm. Edu.*, 66: 186–194.

27. Stolberg, V.B., 2016. *Painkillers: History, Science and Issues*, Greenwood, California; p. 105.

28. Desai, S.P., Desai, M.S., and Pandav, C.S., 2007. The discovery of modern anaesthesia—Contributions of Davy, Clarke, Long, Wells and Morton, *Indian J. Anaesth.*, 51: 472–478.

29. Davenport-Hines, R., 2004. *The Pursuit of Oblivion: A Global History of Narcotics*, W.W. Norton & Co., New York (Chapter One, *Early History*).

30. Council Directive (EEC) No 88/388 on the approximation of the laws of the Member States relating to flavourings for use in foodstuffs and to source materials for their production. 1988, *Off. J. Europ. Comm.*, L184, pp. 61–66.

31. Padosch, S.A., Lachenmeier, D.W., and Kröner, A.U., 2006. Absinthism: A fictitious 19th century syndrome with present impact. *Subst. Abuse Treat. Prev. Policy*, 1: 14; doi: 10.1186/1747-597x-1-14.

32. Markel, H., 2012. *An Anatomy of Addiction: Sigmund Freud, William Halsted, and the Miracle Drug, Cocaine*, Vintage Books, A Division of Random House, Inc., New York.

33. Valter., K. and Arrizabalaga, P., 1998. *Designer Drugs Directory*, Elsevier, Amsterdam.

34. Jenkins, P., 1999. *Synthetic Panics: The Symbolic Politics of Designer Drugs*, New York University Press, New York.

35. Wohlfarth, A. and Weinmann, W., 2010. Bioanalysis of new designer drugs, *Bioanal.*, 2: 965–979.

36. Reneman, L., 2003. Designer drugs: How dangerous are they? *J. Neural Transm. Suppl.*, 66: 61–83.

37. Siegel, J.A., Ed., 2016. *Forensic Chemistry: Fundamentals and Applications*, Wiley-Blackwell, Chichester, UK.

38. Kobilinsky, L.F., Ed., 2012. *Forensic Chemistry Handbook*, Wiley, Hoboken, New Jersey.

39. Bell, S., 2014. *Forensic Chemistry*, 2nd Edition., Pearson, Harlow, UK.

40. Weaver, M.F., Hopper, J.A., and Gunderson, E.W., 2015. Designer drugs 2015: Assessment and management, *Addict. Sci. Clin. Pract.*, 10: 8; doi: 10.1186/s13722-015-0024-7.

41. Rácz, J. and Csák, R., 2014. Emergence of novel psychoactive substances among clients of a needle exchange program in Budapest, Hungary, *Orv. Hetil*, 155: 1383–1394; in Hungarian, abstract in English.

42. Namera, A., Naksmoto, A., Akihiro, T., and Nagao, M., 2011. Colorimetric detection and chromatographic analyses of designer drugs in biological matrices: A comprehensive review, *Forensic Toxicol.*, 29: 1–14.

43. Schwaninger, A.E., Meyer, M.R., and Maurer, H.H., 2012. Chiral drug analysis using mass spectrometric detection relevant to research and practice in clinical and forensic toxicology, *J. Chromatogr. A*, 1268: 122–135.

44. Plotka, J., Malgorzata, B., Biziuk, M., and Morrison, C., 2011. Common methods for the chiral determination of amphetamines and related compounds I. Gas, liquid and thin layer chromatography, *TrAC, Trends Anal. Chem.*, 30: 1139–1158.

45. Shima, N., Katagi, M., and Tsuchihashi, H., 2009. Direct analysis of conjugate metabolites of methamphetamine, 3.4-methylenedioxymethamphetamine, and their designer drugs in biological fluids, *J. Health Sci.*, 55: 495–502.

46. Thevis, M. and Schanzer, W., 2007. Mass spectrometry in sports drug testing: Structure characterization and analytical assays, *Mass Spectrom. Rev.*, 26: 79–107.

2 Designer Drugs
Legislating for Tomorrow Through the Use of Class Definitions and Pharmacophore Principles

Gregory W. Endres, Travis J. Worst, and Jon E. Sprague

CONTENTS

2.1 INTRODUCTION

In order to circumvent the law, clandestine laboratories modify the chemical structure of drugs with known psychoactive effects to generate new psychoactive substances (NPSs). The NPS agents are commonly referred to as synthetic agents, such as the synthetic cathinones ("bath salts"), synthetic cannabinoids ("spice") and synthetic opioids. Because of the packaging and distribution techniques used by illicit manufacturers to circumvent regulations, the designer substances are sold in head shops and over the Internet as noncontrolled substances. In most cases, the products are labelled as "not for human consumption." Because of the novelty of their chemical structures, these agents are not always recognized as controlled substances and efforts to regulate these agents prove to be difficult due to the ever-changing nature of the chemical modifications. Here, we outline how we can legislate for tomorrow through the use of class definitions and pharmacophore principles.

2.2 FEDERAL LEGISLATION

The Controlled Substances Act was adopted in 1970 to stem the illegal importation, manufacture, distribution, possession, and improper use of controlled substances that have a substantial and detrimental effect on the health and general welfare of the American people (Pub. L. 91-513, 84 Stat. 1242) [1]. Controlled substances may be designated as such by listing by name through legislative processes or by meeting certain criteria of the Controlled Substances Act to the satisfaction of the court. Specifically named controlled substances are grouped within five designated "schedules" based on abuse potential, safety, and accepted medical use (Table 2.1). Several amendments have been made to this law, including the Federal Analog Act, adopted in 1986, which states that a chemical that is "substantially similar" to a controlled substance listed in schedule I or II is to be treated as if it were also specified as a scheduled substance when other criteria pertaining to its pharmacological activity are met or if it was represented to possess such activity. These similar substances are often called "designer drugs" as they were presumably designed to circumvent regulations to produce a "legal" form of the scheduled item. More recent updates to the Controlled Substances Act have been made, including legislation in June 2012, when

TABLE 2.1
Drug Enforcement Administration Drug Schedules

Schedule	Definition	Example Drugs
I	Drugs with no currently accepted medical use and a high potential for abuse	Heroin, lysergic acid diethylamide (LSD), marijuana (cannabis), 3,4-methylenedioxymethamphetamine (ecstasy), methaqualone, and peyote
II	Drugs with a high potential for abuse, with use potentially leading to severe psychological or physical dependence. These drugs are also considered dangerous	Combination products with less than 15 milligrams of hydrocodone per dosage unit, cocaine, methamphetamine, methadone, hydromorphone, meperidine, oxycodone, and fentanyl
III	Drugs with a moderate to low potential for physical and psychological dependence. Schedule III drug-abuse potential is less than schedule I and schedule II drugs but more than schedule IV	Products containing less than 90 milligrams of codeine per dosage unit, ketamine, anabolic steroids, testosterone
IV	Drugs with a low potential for abuse and low risk of dependence	Alprazolam, diazepam, tramadol
V	Drugs with lower potential for abuse than schedule IV and consist of preparations containing limited quantities of certain narcotics. Schedule V drugs are generally used for antidiarrheal, antitussive, and analgesic purposes	Cough preparations with less than 200 milligrams of codeine per 100 milliliters

Source: https://www.dea.gov/druginfo/ds.shtml.

26 synthetic cannabinoids were named as schedule I compounds. Although it may be less ambiguous to regulate these NPSs individually (specifying each by name), new substances with no legitimate use nor medical purpose emerge at rates that far exceed the rate at which new laws that designate these substances as illegal can be adopted. Therefore, an alternate mechanism is required to address these examples and to protect the public by providing a deterrence.

On the federal level, a controlled substance may be designated as such in two ways: (1) by specifically naming individual substances through legislative processes, or (2) by determination that a substance is an "analog" of another substance that is specifically named as a controlled substance. Analog determination is defined by the following rules adopted in 21 U.S.C. § 802(32):

- A. Except as provided in subparagraph (C), the term *controlled substance analogue* means a substance—
    - i. The chemical structure of which is substantially similar to the chemical structure of a controlled substance in schedule I or II;
    - ii. Which has a stimulant, depressant, or hallucinogenic effect on the central nervous system that is substantially similar to or greater than the stimulant, depressant, or hallucinogenic effect on the central nervous system of a controlled substance in schedule I or II; or
    - iii. With respect to a particular person, which such person represents or intends to have a stimulant, depressant, or hallucinogenic effect on the central nervous system that is substantially similar to or greater than the stimulant, depressant, or hallucinogenic effect on the central nervous system of a controlled substance in schedule I or II.

In short, the Federal Analog Act states that a substance is an analog if it is "substantially similar" to an illegal substance and either has a substantially similar effect *or* was intended to have such an effect. Therefore, pharmacological data (prong ii) may be helpful, but not required, in analog determination if it can be shown that it was represented or intended to have such an effect. Therefore, the primary requirement is "substantial similarity" between a substance listed in schedule I or schedule II and the substance of concern. "Similar" is a relative term used in this case for the comparison of two chemical structures. The word "substantial" simply qualifies the degree to which the two substances are similar and requires that a more compelling case be made that they are similar rather than dissimilar. Since the decision of whether two substances are substantially similar rests with a judge or jury with no scientific training, the information pertaining to the substances must be presented in a fashion that is both accurate and in a manner that is understood by the layperson. A variety of methodologies may be used in considering structural similarity and it is the role of the court to ensure that the methods and opinions presented are derived from sound, established scientific principles that are well-recognized by those trained in the art.

Given the limitless structural complexity of chemical substances, there is no single rule or formula that will provide an unambiguous yes or no answer that will apply to all examples. Rather, arguments are presented to the court that support or dispel the argument that the two substances are substantially similar. Forensic scientists and expert witness testimony are often relied upon to assist both the prosecution and defense. One may argue that an easier and more reliable approach to the determination is to use a defined formula and value in the assessment of structural similarity such as that described by Rogers and Tanimoto [2], developed for computational medicinal chemistry purposes. The Tanimoto coefficient (Tc) may be used to quantify molecular similarity and is generally defined as Tc = $c/(a + b - c)$, where a and b represent the total attributes for molecule A and molecule B, respectively, and c represents the shared attributes between the substances. The similarity in this function of shared and unique features between two molecules (A,B) is numerically expressed within the extremes of a unit interval $0 <$ Tc(A,B) < 1. The use of the Tanimoto coefficient and other such calculations have their place for comparative purposes and may neutralize the subjectivity in the assessment of analog similarity, however their use as a definitive measure of the extent ("substantial" or otherwise) of the similarity to a regulated substance is best avoided. Such use of a strict definition based on a calculated value, if used as a threshold for the determination substantial similarity, may inadvertently provide a roadmap to the illicit chemist who would have free reign to manufacture and sell NPSs that are designed to fall just outside of this value.

2.3 STATE LAWS

The Constitution of the United States is the guiding document of this country, from statement of rights to formation of the government and dictating how laws are made. The hierarchy is established in Article VI, which reads, "This Constitution, and the laws of the United States which shall be made in pursuance thereof; and all treaties made, or which shall be made, under the authority of the United States, shall be the supreme law of the land; and the judges in every state shall be bound thereby, anything in the Constitution or laws of any State to the contrary notwithstanding." This section of the Constitution essentially describes federal laws, including drug laws, as the supreme rule of the land, taking precedent over state laws if the two contradict. This is likely the reason that state laws have a long history of following the lead of federal laws. So much so that most states have a rule in place that if the federal drug law changes, state law will comply immediately. For the state of Ohio, this rule can be found in Ohio Revised Code 3719.43 (2015), stating:

> When pursuant to federal drug abuse control laws the attorney general of the United States adds a compound, mixture, preparation, or substance to a schedule of the laws, transfers any of the same between one schedule of the laws to another, or removes a compound, mixture, preparation, or substance from the schedules of the laws then such addition, transfer, or removal is automatically effected in the corresponding schedule or schedules in section 3719.41 of the Revised Code, subject to amendment pursuant to section 3719.44 of the Revised Code [3]. Ohio's rule is specific to the attorney general of the United States, but is later clarified to include any agent of the attorney general, such as the Drug Enforcement Agency (DEA).

The advent of the designer drug craze has required laws at the state level to move faster than at the federal level to control substances. To address this situation, states have developed a variety of methods to make compounds illegal. In Colorado and

Ohio, the Board of Pharmacy has the emergency power to schedule compounds. In other states, such as Florida, this duty falls under the State Attorney General. Specific committees have also been created, such as the Controlled Substances Board in Wisconsin, to advise the appropriate legislating body on what actions to take. And there are some states that leave this power up to the actual legislating body to pass laws and make those dangerous compounds illegal.

The recent appearance of the designer drug craze, including synthetic cannabinoids, substituted cathinones, and more recently the fentanyl-like opioids, has made the job of emergency-scheduling compounds by the DEA and individual states far more difficult. The past approach of singling out a drug and going through the process of making it illegal gives the illicit drug designers enough time to modify the structure to a level where the new drug is no longer illegal. This has forced a new era of drug laws that includes an entire class of drugs, rather than a single drug.

2.4 CLASS DEFINITIONS AND PHARMACOPHORES

Federal and state laws have varied greatly using individual naming (precise) and/or class definition (specific) models (Figure 2.1). In 1970, the Controlled Substances Act named individual agents and then was modified in 1986 to include "analogs." The Synthetic Drug Abuse Prevention Act of 2012 permanently added "cannabimimetic agents" to the Controlled Substances Act. Cannabimimetic agents "mimic" the effects of cannabinoids. Under the Synthetic Drug Abuse Prevention Act of 2012, cannabimimetic agents are defined as one of five structural classes (i.e., a class definition model). Many states modified this class definition approach to capture other newly identified synthetic drugs. However, the clandestine laboratories quickly modified the synthetic drugs to avoid scheduling under the class definition. This led the State of Ohio to the more generalizable approach of scheduling based on the use of pharmacophores in drug design.

Drugs induce their pharmacologic and toxicologic response through interactions with protein targets. Receptors, enzymes, and reuptake proteins are common drug targets and are composed of amino acids. The amino acids have a charge at

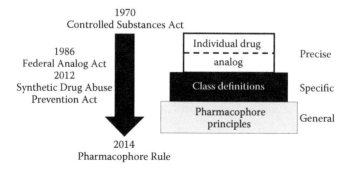

FIGURE 2.1 Timeline of the progression of the laws affiliated with designer drugs. Through the years the laws have gone from being precise based on the specific name of an agent to being more generalized to encompass the principles of drug design.

physiological pH and thus can participate in chemical interactions with drug molecules. Drugs interact with these biological targets through any of the following potential methods: hydrogen bonds, ionic bonds, π-stacking, and hydrophobic effects (for a review, see [4]). During the drug discovery process, chemical structures are modified in an attempt to maximize these types of interactions with the biological target. In cases where an endogenous ligand has been identified, the structural requirements for binding to the protein target are known. The structural portion or portions of a drug molecule responsible for producing a pharmacological response are referred to as a pharmacophore (for a review, see [5]). For example, the neurotransmitter dopamine has a known pharmacophore; the phenethylamine portion of the molecule (Figure 2.2) is the "core" scaffold to which functional groups can be added. In this case, the 3,4-dihydroxy substitutions give dopamine its classical recognition as a catechol. Amphetamine that binds to the dopamine transporter (drug target) to facilitate the release of dopamine has the same phenethylamine backbone. In the case of amphetamine, the functional group addition is a methyl group off the α-carbon. In general, the addition of a methyl group increases lipophilicity and the likelihood of a drug crossing membranes. Figure 2.3 provides some of the common functional groups added to the core scaffold of a pharmacophore.

Functional groups can participate in hydrogen bond donating (HBD) or hydrogen bond accepting (HBA). From a purely chemical perspective, functional groups are very different from each other. However, from a pharmacological perspective they contribute to HBD or HBA, allowing drugs to interact with receptors. The "substantially similar" issue that arises from the Federal Analog Act results from a lack of consideration of drug synthesis principles.

HO / HO — Dopamine   NH$_2$ NH$_2$ — Amphetamine

FIGURE 2.2 Endogenously produced dopamine has the same phenethylamine pharmacophore (blue) as exogenously administered amphetamine.

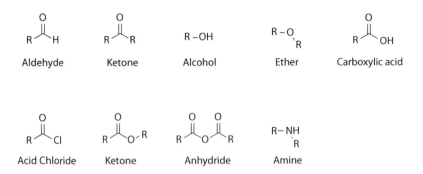

FIGURE 2.3 Common drug functional groups added to core structures.

One method that works particularly well in the determination of structural similarity or lack thereof is the use of pharmacophores to compare the commonality between two substances. For example, a chemical substance may be divided into multiple subgroups. In the comparison of two chemical structures, the similarity or identicality of subgroups (pharmacophores) provides support for an opinion that they are substantially similar. Likewise, if the two substances are not composed in such a way that they may be divided into similar subgroups, or if the subgroups themselves are not similar, this would provide support that the two substances are not substantially similar. Multicomponent pharmacophore models are routinely used in the drug discovery process and such models have been applied to NPSs [6–8]. This is illustrated in Figure 2.4 for the comparison of CP-47,497, Δ9-tetrahydrocannabinol (THC), and JWH-018. A three-component system is used to subdivide each substance in this example, however using two different models. CP-47,497 and THC both use a system based on the separate (**A**, **B**, and **C**) rings of the tetrahydrocannabinol structure of Δ9-THC. It may be argued that CP-47,497 is substantially similar to Δ9-THC based on the ability to describe and depict both substances within the same pharmacophore model and the presence of two components (**A** and **C**) that bear structural similarity to the analogous components to the schedule I substance, Δ9-THC. In contrast, the structure of JWH-018 does not fit the same tricyclic model and may instead be divided into an independent three-component model based on the "head," "core," and "tail" subunits. It may be argued that JWH-018 is *not* substantially similar to Δ9-THC based on the differing pharmacophore models and lack of structural similarity between the individual subcomponents.

Another example is provided in Figure 2.5 between the synthetic cannabinoids UR-144 and JWH-018, and the essential amino acid tryptophan. Each molecule in this example fits a three-component (head/core/tail) model and each example possesses an identical core indole subunit (highlighted in blue). When simple and common substructures are utilized for comparative purposes in this strategy, the potential for overextension of the model exists. Simple pharmacophores like the indole subcomponent are ubiquitously present in nature; however, this does not imply that all indole-containing substances may be deemed illegal. Rather, fitting the context of the act requires that the structural similarity is complimented with similar pharmacology (prong ii), or represented or intended to possess such activity

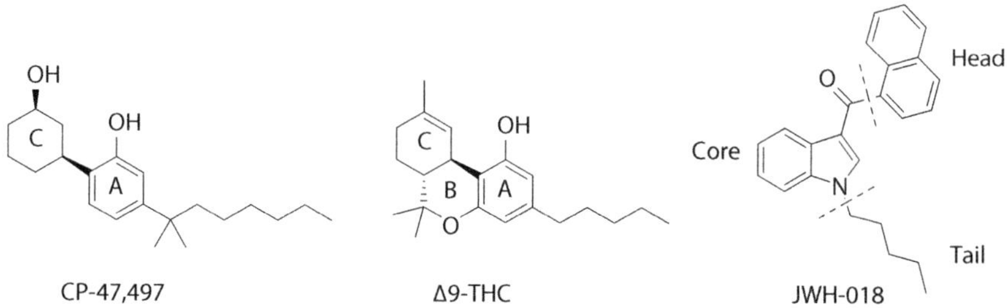

FIGURE 2.4 Three-component system (head, core, tail) used to subdivide the synthetic cannabinoids.

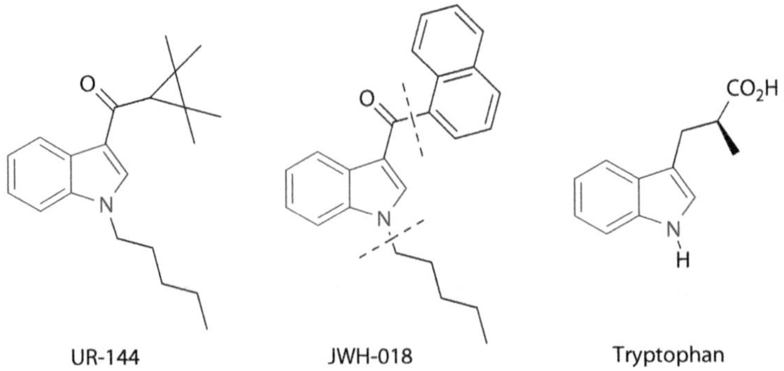

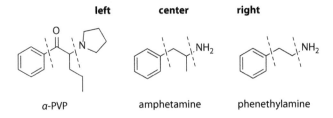

UR-144   JWH-018   Tryptophan

FIGURE 2.5 Expansion of the three-component system (head, core, tail) to the essential amino acid tryptophan.

as described in prong iii. Tryptophan does not meet these requirements and therefore any argument that tryptophan is an analog of a controlled substance would be out of context and excluded from consideration.

The pharmacophore-based approach may be employed for any or all examples in addition to cannabinoids. Figure 2.6 depicts an analogous example between three phenethylamine-based substances: alpha-PVP (pyrrolidinopentiophenone), amphetamine, and phenethylamine. In this example, it may be argued that alpha-PVP, the cathinone-based NPS also known as "flakka," is substantially similar to amphetamine by virtue of a common three-component pharmacophore model with similar (as well as identical) subcomponents. Phenethylamine, a substance that is found in living systems as well as food products, also fits this pharmacophore model. However, schedule I analog determination based on substantial similarity between phenethylamine and amphetamine in the same "left/center/right" model will also require that the conditions of prongs (ii) or (iii) of the analog act are met.

In 2014, the State of Ohio enacted Ohio Administrative Code 4729-11-02 [3], "The Pharmacophore Rule." The pharmacophore rule applied the general principles of drug design outlined above to the synthetic cathinones, cannabinoids and opioids. The synthetic cathinones (bath salts) are often described as having one pharmacophore with functional groups being added to skirt the Federal Analog Act. New chemical entities with variation to known pharmacophores are designed and synthesized as part of the legal drug discovery process. These modifications are tested for differences in both

left   center   right

α-PVP   amphetamine   phenethylamine

FIGURE 2.6 Application of the pharmacophore principles to three phenethylamine-based substances: alpha-PVP, amphetamine, and phenethylamine.

pharmacological and toxicological effects under strict guidelines to ensure safety before approval. New drugs of abuse are also created through the modification of known drugs that provide a desired effect but with the intent to circumvent regulations and without regard for potential toxicity. As described above, pharmacophore-based legislation provides context to the issue and guidance to the question and determination of "structural similarity." This is accomplished by a simple comparison of two substances and determining structural similarity based on the commonality of individual subgroups between scheduled (listed) and unlisted drugs of abuse.

For purposes of illustration, let us compare two synthetic cathinones (bath salts), methylone and 3′,4′-methylenedioxy-α-pyrrolidinopropiophenone (MDPPP; Figure 2.7).

Based on the Federal Analog Act, these two compounds may or may not be considered "substantially similar" and this is subject to opinion. However, if the molecules were classified based on the pharmacophore requirements for binding to a receptor (in this case the dopamine transporter), they would both be classed as schedule I compounds. Furthermore, based on the presence of the common cathinone "core" of methylone, known structural pharmacophores of a specifically named schedule I substance, MDPPP would also be classified as a schedule I compound based on the structural portions of a drug molecule meeting the pharmacophore requirements for drug action. In regards to the synthetic cathinones, the Ohio Revised Code simply states "Except as otherwise provided in section 3719.41 of the Revised Code, any compound that contains the structural requirements of the cathinone pharmacophore, as identified by a report from an established forensic laboratory, is a schedule I controlled substance." The exception in this case includes the antidepressant bupropion, which also contains a cathinone core.

With regard to the synthetic cannabinoids, the Ohio Revised Codes states:

> Any compound that meets at least three of the following pharmacophore requirements to bind at the CB1 and CB2 receptors, as identified by a report from an established forensic laboratory, is a schedule I controlled substance hallucinogen:
>
> (1) A chemical scaffold consisting of substituted or nonsubstituted ring structures that facilitate binding of required elements (such as: indole compounds, indazoles, benzimidazoles or other ring types);
> (2) Alkyl or aryl side chain off the chemical scaffold providing hydrophobic interaction with the CB1 and CB2 receptors;
> (3) Carbonyl or ester or equivalent for hydrogen bonding;
> (4) Cyclohexane, naphthalene ring, substituted butanamide or equivalent for steric requirements for CB1 and CB2 receptor binding.

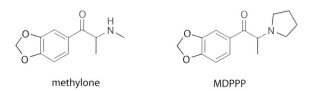

methylone   MDPPP

FIGURE 2.7   Methylone and MDPPP. The cathinone pharmacophore is highlighted in blue.

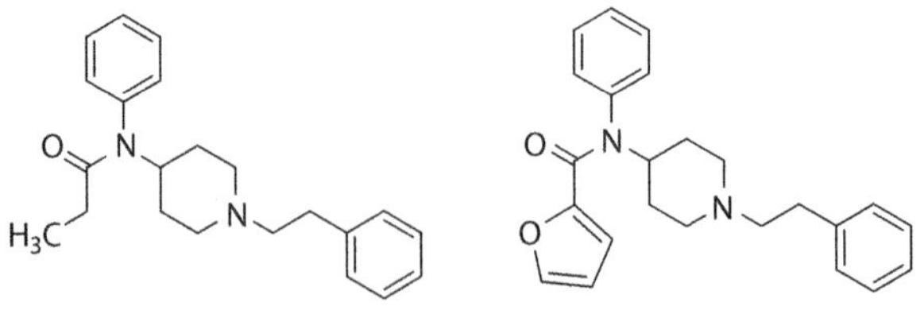

Fentanyl   Furanylfentanyl

FIGURE 2.8   Fentanyl and furanylfentanyl meet the requirements of the pharmacophore rule.

The Ohio Administrative Code 4729-11-02 also addresses the synthetic fentanyl agents. The analgesic and drug-seeking effects of fentanyl are mediated through its interaction with the μ-opioid receptors. The pharmacophore for fentanyl-related agents is a five-, six-, or seven-membered ring containing a nitrogen (Figure 2.8). A nitrogen-containing functional group with a polar group attached is required for hydrogen binding to the μ-opioid receptor. Finally, an aryl or aryl substitution attached to the ring for steric requirement for binding. Figure 2.8 demonstrates how furanylfentanyl meets the pharmacophore requirement for binding to the μ-opioid receptors.

2.5   CONCLUSION

Drug policies that focus on drug prohibition, such as those outlined in this chapter, have recently come under much scrutiny [9]. The Johns Hopkins–*Lancet* Commission on Drug Policy and Health raised concerns over the lack of scientific evidence on drug use and dependence in drug policies [9]. The policies outlined here take into consideration the general principles of drug design but do not consider the neurobiology of drug use and dependence. In 2011, the Substance Abuse Mental Health Services Administration (SAMHSA) reported that synthetic cathinones accounted for 22,904 hospital emergency department (ED) visits [10]. In 2010, synthetic cannabinoids accounted for 11,406 ED visits [11]. Synthetic opioids have additionally been identified in postmortem overdose cases [12]. When clandestine laboratories modify the drug molecules to circumvent the law, they also modify the toxicological profile of the drug. Because these new chemical agents have not been pharmacologically characterized, the user is unwillingly and unknowingly playing a chemical game of "Russian Roulette." The policies enacted concerning synthetic designer drugs were introduced in an attempt to protect society from the toxicological effects of designer drugs.

REFERENCES

1. The Controlled Substance Act of 1970. Pub. L 91-513, 84 Stat. 1242, enacted 27 October 1970; 21 Code of Federal Regulation, Section 801 et. seq., Part 1300 to end. *United States Government Publishing Office.* https://www.gpo.gov/fdsys/pkg/STATUTE-84/pdf/STATUTE-84-Pg1236.pdf (accessed 3 January 2017).
2. Rogers, D.J., and Tanimoto, T.T. 1960. A computer program for classifying plants, *Science.* 132(3434): 1115–1118.

3. Ohio Revised Code Section 3719.43. *LexisNexis Academic.* http://www.lexisnexis.com/hottopics/lnacademic/ (accessed 3 January 2017).

4. Jordon, A.M., and Roughley, S.D. 2009. Drug discovery chemistry: A primer for the non-specialist, *Drug Discovery Today.* 14: 731–744.

5. Worst, T.J., and Sprague, J.E. 2015. The "pharmacophore rule" and the "spices," *Forensic Toxicol.* 33(1): 170–173.

6. *Cayman Currents* Issue 26, October 2015. Caymanchem.com/syncanflipbook (accessed 3 January 2017).

7. EMCDDA Perspectives on Drugs: Synthetic Cannabinoids in Europe, 2015. emcdda .europa.eu/topics/pods/synthetic-cannabinoids (accessed 3 January 2017).

8. Shevyrin, V., Melkozerov, V., Endres, G.W., Shafran, Y., and Morzherin, Y. 2016. On a new cannabinoid classification system: A sight on the illegal market of novel psychoactive substances, *Cannabis Cannabinoid Res.* 1(1): 186–194.

9. Csete, J., Kamarulzaman, A., Kazatchkine, M. et al. 2016. Public health and international drug policy, *Lancet.* 387: 1427–1482.

10. Drug Abuse Warning Network (DAWN). 'Bath Salts' were involved in over 20,000 drug-related emergency department visits in 2011. *The DAWN Report,* 17 September 2013.

11. Drug Abuse Warning Network (DAWN). Drug-related emergency department visits involving synthetic cannabinoids. *The DAWN Report,* 4 December 2012.

12. Fort, C., Curtis, B., Nichols, C., and Niblo, C. 2016. Acetyl fentanyl toxicity: Two case reports, *J Anal Toxicol.* 40(9): 754–757.

3 Alternative Legislation with Designer Drugs

Dariusz Zuba

CONTENTS

3.1 INTRODUCTION

It is well known that different countries have different legal systems in place. Common law, which is the legal system used in the United States, originated during the Middle Ages in England, and from there propagated to the colonies of the British Empire. Today, one-third of the world's population lives in common-law jurisdictions or in systems mixed with civil law. Still in force in the United Kingdom, common law is mainly based on the idea of precedent: when a court makes a decision about a case, that decision becomes a part of the law of the country. However, it is possible for governments to add to or modify common law: for example, the Misuse of Drugs Act 1971 is an Act of the Parliament of the United Kingdom (UK), which represents action in line with treaty commitments under the conventions of the United Nations (UN). A list of prohibited drugs is given in Schedule 2 to this Act. It also defines offenses, including, among others, unlawful possession of a controlled drug, possession of a controlled drug with intent to supply it, supplying or offering to supply a controlled drug (even where no charge is made for the drug) and allowing premises you occupy or manage to be used unlawfully for the purpose of producing or supplying controlled drugs. In 2016, new legislation concerning designer drugs came into effect in the UK. The Psychoactive Substances Act 2016 is a law intended to restrict the production, sale, and supply of a new class of psychoactive substances often referred to as "legal highs." The bill was given Royal Assent on January 28, 2016, and came into force on May 26, 2016 across the entire United Kingdom.

An alternative to common law is civil law, often called a continental or Romano-Germanic legal system. A civil law system is based on concepts, categories, and rules derived from Roman law, sometimes largely supplemented or modified by local custom or culture. Civil law is based on fixed codes and statutes. In this legal system, the decisions of judges do not affect the laws of a country. Originally, civil law was a common legal system in much of Europe, but during the seventeenth century it became fractured into separate national systems.

Today, the law of Europe is diverse and changing fast. Each European country has its own legal system; however, the 28 Member States of the European Union (EU) are influenced by EU law. The EU has developed an internal single market through a standardized system of laws that apply in all Member States. EU policies aim to ensure the free movement of people, goods, services, and capital within the internal market, and enact legislation in justice and home affairs. It maintains common policies on trade, agriculture, fisheries, and regional development, but not on drug control.

From the early 1990s, many so-called "designer drugs" were regularly discovered in Europe. They were usually psychotropic substances related to amphetamine, and their appearance raised questions about possible health risks and other potential problems. The free movement of people across Europe could lead to unintentional violation of the law, as drugs subject to control vary from jurisdiction to jurisdiction and from country to country. For example, cocaine is regulated almost everywhere, while khat is regulated in some countries, but not others. Most jurisdictions also regulate prescription drugs—medicinal drugs not considered dangerous but that can only be supplied to holders of a medical prescription—and sometimes drugs available without prescription but only from an approved supplier such as a pharmacy. It was agreed in the 1990s that progress in knowledge about the drug market in Europe could be made by sharing information and by establishing a risk-assessment procedure and a mechanism for their eventual EU-wide control. In 1993 the EU established an agency called the European Monitoring Centre for Drugs and Drug Addiction (EMCDDA). This agency exists to provide the EU and its Member States with a factual overview of European drug problems and a solid evidence base to support the drugs debate. Today it offers policymakers the data they need for drawing up informed drug laws and strategies. It also helps professionals and practitioners working in the field to pinpoint best practice and new areas of research [1]. In 1997, more than ten years before an enormous increase in the number of new psychoactive substances (NPSs) marketed around the world, the Council of the EU adopted joint action, leading to the introduction of a three-step approach: exchange of information by the early-warning system (EWS), risk assessment, and a procedure for bringing specific new synthetic drugs under control [2]. The EWS is a very efficient tool for sharing information about newly marketed designer drugs among forensic and clinical laboratories, as well as other institutions involved in drug analysis. In 2005, The Council of the EU defined a new psychoactive substance as "a new narcotic or psychotropic drug, in pure form or in preparation, that is not controlled by the United Nations (UN) drug conventions, but which may pose a public health threat comparable to that posed by substances listed in these conventions" [3]. The aforementioned UN conventions are discussed in detail in the next paragraph.

3.2 THE INDIVIDUAL DRUG LISTING SYSTEM

Most countries in the world use an individual listing system to control narcotic drugs and psychoactive substances, including designer drugs. International drug control is subject to consensus-based multilateral agreements and existing law in this field is regulated by UN conventions. Member States that follow the conventions are required to establish criminal penalties to combat and punish unauthorized trade in controlled substances. As it is a general principle that criminal law must be certain, the substances subject to such penalties must be clearly specified, which means listing them individually, or in some cases in tightly defined groups according to their chemical structure [4].

International cooperation in drug control and counteracting drug addiction started over 100 years ago, in 1909, when the international community met in Shanghai to discuss the Chinese opium epidemic. The first international drug convention, the International Opium Convention of The Hague, was signed in 1912 and entered into force in 1915. International drug control became one of the tasks of the League of Nations in 1920. Under its auspices, three main conventions in 1925, 1931, and 1936 were developed [5]. These provided the groundwork for the practical operation of the international drug control system.

After World War II, international drug control became the responsibility of the UN and a number of protocols were established in order to improve the system, the most important being the 1953 Opium Protocol [5]. The next milestones were three conventions still in force at present: the 1961 Single Convention on Narcotic Drugs (which was subsequently amended by a Protocol in 1972), the 1971 Convention on Psychotropic Substances, and the 1988 United Nations Convention against Illicit Traffic in Narcotic Drugs and Psychotropic Substances. The principal objectives of the Conventions are to limit the possession, use, trade, distribution, import, export, manufacture, and production of drugs exclusively to medical and scientific purposes, and to address drug trafficking through international cooperation to deter and discourage drug traffickers. These Conventions were adopted by most Member States of the UN and therefore their national legislation on drugs was subsequently prepared on the basis of regulations included in these Conventions.

According to the First Convention on Narcotic Drugs of 1961 [6], 116 drugs were classified as narcotics. It included mainly plant-based products, such as cannabis, coca (and cocaine, which is obtained from it), as well as opium and poppy ingredients (morphine, codeine). The list of narcotic drugs may be amended after notification by a Member State or the World Health Organization (WHO). If notified, the substance is assessed by the WHO, and if it finds that the substance is liable to similar abuse and productive of similar ill effects as the controlled narcotic drug(s) on the list or is convertible into such a drug, it may be added to the list. The current list of narcotic drugs under international control is known as the International Narcotics Control Board (INCB) Yellow List [7].

The Convention has four Schedules of controlled substances, ranging from Schedule I (most restrictive) to Schedule IV (least restrictive). Schedule I includes drugs claimed to create a serious risk to public health, whose therapeutic value is not currently acknowledged by the Commission on Narcotic Drugs. It includes, among others, common drugs of abuse such as heroin (diacetylmorphine) and cocaine, morphine and many of its derivatives, as well as many other agonists of opioid receptors, e.g., fentanyl. Schedule II includes mainly codeine and its several derivatives, while Schedule III includes medicinal preparations of codeine and other derivatives of morphine, as well as medicines with low doses of cocaine, opium, or morphine. Schedule IV contains mainly derivatives of fentanyl. The list of narcotic drugs is amended relatively rarely, and the Schedules cannot ban many newly discovered substances, since the scope of the Convention was limited to drugs with cannabis-, coca-, and opium-like effects. Of the designer drugs marketed in the last ten years, only one substance, AH-7921, has been added to Schedule I (in 2016). This substance is a potential analgesic drug with addictive properties and has been synthesized to mimic the effects of morphine or fentanyl. AH-7921 was first synthesized in the mid-1970s, but it only recently (in 2012) appeared on the drug market and has led to many intoxications and fatalities [8].

However, UN Member States are free to broaden the list of substances in their national legislation. Any amendment to the list of individual substances controlled at the national level requires, in most cases, a process that encompasses a health risk assessment based on scientific data and human experience data that, in the case of NPSs, is often scarce. Unfortunately, such a process is time-consuming and usually takes several months [4]. It is a common rule used by national legislators for classification of new designer drugs that, as these substances are usually not used in medicine, they are classified to Schedule I. The question is whether a substance may cause dependence and to what extent, as many have not been subjected to toxicological study and, commonly, designer drugs are first synthesized just before market introduction. The only verifiable information is available from addiction treatment centers, but at least several weeks are needed to collect reliable data. The largest group of designer drugs to have been added to the list of narcotic drugs as a part of UN Conventions are synthetic cannabinoids. Other groups are compounds acting on opioid receptors, including designer fentanyls and designer benzodiazepines, which could be considered as narcotics from the pharmacological point of view (although many derivatives of benzodiazepine are classified by the UN as psychotropic substances). New derivatives of fentanyl or benzodiazepine are also subject to control according to the American analog law, because their chemical structures are substantially similar to the chemical structures of parent compounds and have a depressant effect on the central nervous system that is substantially similar to or greater than the depressant effect of these substances. A more complex issue is with synthetic cannabinoids, whose chemical structures are often different to those of controlled substances.

The Convention on Psychotropic Substances of 1971 includes four Schedules [9], which are amended more often and to which substances are more regularly added. Article 2 of the Convention sets out a process for adding additional drugs to the Schedules. The WHO communicates an assessment of the substance, including the extent or likelihood of abuse, the degree of seriousness to public health and social problems, and the degree of usefulness of the substance in medical therapy, together with recommendations on control measures, if any, that would be appropriate in the light of its assessment, if it finds:

a. That the substance has the capacity to produce
  i.   1. A state of dependence, and
       2. Central nervous system stimulation or depression, resulting in hallucinations or disturbances in motor function or thinking or behavior or perception or mood, or
  ii. Similar abuse and similar ill effects as a substance in Schedule I, II, III or IV, and
b. That there is sufficient evidence that the substance is being or is likely to be abused so as to constitute a public health and social problem warranting the placing of the substance under international control.

The Commission on Narcotic Drugs makes the final decision on whether to add the drug to a Schedule, "taking into account the communication from the World Health Organization, whose assessments shall be determinative as to medical and scientific matters, and bearing in mind the economic, social, legal, administrative and other factors it may consider relevant."

The list of psychotropic substances under international control is presented in the so-called INCB Green List [10]. Schedule I includes cathinone, a parent substance of the largest chemical group of designer drugs marketed in recent years. Many countries have also added simple derivatives of cathinone to this Schedule, indicating its high risk of causing dependence and lack of medicinal use. Several designer drugs marketed in the 1990s and subjected to risk assessment by the European Monitoring Centre for Drugs and Drug Addiction (EMCDDA) were also placed on Schedule I. The more complex situation is with the derivatives of pyrovalerone, which is a psychoactive drug with stimulant effects, and is used for the clinical treatment of chronic fatigue or lethargy and as an anorectic or appetite suppressant for weight loss purposes. This substance is listed in Schedule IV. In many countries, designer drugs from this family have also been listed in Schedule IV, although they are not used for medicinal purposes.

It is worth mentioning that many countries have individually decided to extend control measures to a substance's isomers, esters and ethers, and salts. But a number of countries have stated that an individual listing system is not effective, as a legislative procedure may take several months. In the meantime, new designer drugs are marketed, while the others are withdrawn. The "shelf life" of many designer drugs on the market is no longer than several months, and it is not realistic to perform a complex toxicological study, which usually takes two to three years, to find the reliable proof that the substance should be taken into control within this time. To account for the rapid emergence of designer drugs, many countries have introduced alternative ways to speed up their ordinary legislative procedure, through temporary or rapid procedures. A temporary (emergency) ban is an accelerated procedure to quickly introduce temporary restrictions on a substance for a limited period of time (usually for one or two years) while the legislative process is completed or a rigorous assessment of the risks is conducted and a final decision to control the substance is made. If there is no decision to control the substance, the temporary ban expires [4]. But many countries have decided to change their approach to drug control legislation beyond the individual list of substances, e.g., by introducing analog and generic legislation or through the use of medicinal prod-uct law to counteract the designer drugs problem. The analog approach has been discussed in Chapter 2, while other approaches are discussed below.

3.3   THE GENERIC APPROACH

The generic approach to designer drugs is based on the similarity of the chemical struc-ture of a substance to known illicit drug(s), but, in contrast to analog law existing in the U.S., the generic legislation describes clusters of compounds strictly. When analyzing the structures of designer drugs, it is clear that many are very close to the structures of controlled drugs. A great variety of chemical substituents can be inserted on virtually any atom (carbon or nitrogen) of the basic skeleton of the parent compound [11]. The modification of designer drugs usually concerns the substitution of a hydrogen atom by another atom, e.g., chlorine, fluorine, bromine, or iodine, by the hydroxy, alkoxy, or nitro groups, prolongation of an alkyl chain, cyclization, or a change in the order or atoms (formation of isomers). Sometimes one phenyl ring is substituted by another, e.g., naphthyl or thienyl. According to the generic approach concept, all possible substitu-ents have to be included in the definition of the cluster. In addition, the putative salts and esters should also be included. Hence, all possible chemical derivatives, that is, all variants of the parent compound, are described in the cluster to be jointly banned [11].

Many generic definitions on NPSs focus on derivatives of synthetic drugs, mainly amphetamine and amphetamine-type stimulants (methamphetamine, MDMA—3,4-methylenedioxymethamphetamine). Definitions relate to common chemical structures of parent compounds, such as phenethylamine, cathinone, tryptamine, piperazine, fentanyl, etc. The generic legislation describes clusters of compounds that have simi-larity of chemical structure to parent compounds. The situation is more complex for synthetic cannabinoids, because their chemical structures are significantly different from the structure of delta-9-tetrahydrocannabinol (delta-9-THC), the main psycho-active component of the cannabis plant. Therefore, their definitions refer to the com-mon chemical skeleton of different classes of synthetic cannabinoids.

The generic approach has been introduced in many countries around the world, but there are significant differences between countries. The differences include, among others, the range of chemical groups covered by the generic definition as well as the number and kind of substituents.

One of the most recent and interesting generic definitions of designer drugs was introduced in Germany in 2016 [12]. For three groups of substances, that is, phenethylamines, cathinones, and cannabinoids, the controlled structures were defined very strictly. The structures of the first two classes were described by divid-ing a model molecule into two structural fragments: the ring system with potential substituent(s) and the side chain. The maximum molecular mass was set at 500 Da. Twenty-two ring systems were defined as being subject to control. The ring substitu-ents included halogen atoms, functional groups with a defined maximum length of carbon chain (e.g., alkoxy group of up to six carbon atoms). Location of the atom or group was not defined, which allows control of positional isomers. In turn, the 2-aminoethyl side chain can be substituted by different groups or ring systems. Substances in which the nitrogen atom is a part of a cyclic system (for example, pyrrolidinyl or piperidinyl) are also included. The difference between cathinones and phenethylamines is the presence of a carbonyl group located on the C2 atom, that is, in beta position to the nitrogen atom. Therefore, cathinones are often called beta-keto derivatives of phenethylamines (or simply bk derivatives).

Controlled synthetic cannabinoids were defined by dividing their structures into four parts. The core structure is linked at a defined position via the bridge to another ring system or group of atoms; specific side chains can be substituted to the core ring.

The German generic law is full of chemical expressions. There has been lively discussion about its compatibility with the German Constitution, especially concerning whether ordinary people are able to understand it, but the Act was finally approved in November 2016.

In turn, the British Misuse of Drugs Act 1971 contains substances and products by their chemical names, followed by the generic definition of several chemical groups, including phenethylamines, tryptamines, and fentanyls [13]. Several years ago, the Advisory Council on the Misuse of Drugs (ACMD), which is responsible for recommendation of classification of new or existing drugs that may be misused, considered a number of options for the control of cathinone derivatives, including the listing of named substances, several generic definitions, and combinations of these approaches. The ACMD recommended generic definition of cathinones as: those obtained from 2-amino-1-phenyl-1-propanone by substitution in the phenyl ring to any extent with alkyl, alkoxy, alkylenedioxy, haloalkyl, or halide substituents; by substitution at the 3-position with an alkyl substituent; by substitution at the nitrogen atom with alkyl or dialkyl groups; or by inclusion of the nitrogen atom in a cyclic structure [14]. The primary definition of cathinone derivatives came into action, but it turned out that people working in clandestine laboratories produced new compounds that were not covered by such a definition. An example was naphyrone, a drug derived from cathinone, in which a phenyl ring was substituted by a naphthyl ring.

Such problems prompted a change to the approach to control designer drugs. Last year, the Psychoactive Substances Act 2016 came into force in the UK [15]. In this Act, "psychoactive substance" means any substance that is capable of producing a psychoactive effect in a person who consumes it, and is not an exempted substance. The exempted substances include controlled drugs (within the meaning of the Misuse of Drugs Act 1971), medicinal products, alcohol and alcoholic products that do not contain any psychoactive substance, nicotine and tobacco products and caffeine or caffeine products, as well as food.

3.4   THE MEDICINAL PRODUCT APPROACH

Drug design is not only used in clandestine laboratories. It is also a branch of pharmacology aimed at finding new medications based on the knowledge of a biological target. Ligand-based drug design relies on knowledge of other compounds that bind to the biological target of interest. These other molecules may be used to derive a so-called pharmacophore model that defines the minimum necessary structural characteristics a molecule must possess in order to bind to the target. Recommendations to the interpretation of "substantial similarity" existing in the U.S. analog definition of designer drugs refer to pharmacophore models.

The modelings are performed at universities, but also by people working for criminal groups. The illegal production and introduction into the market of simple derivatives of approved medicines is a well-known phenomenon. Such illegally manufactured substances are also called designer drugs, or designer medicines, and products are often offered in online shops. Market introduction of substances that have not been tested may lead to serious consequences, including nonlethal and lethal intoxications. Similar problems have been identified due to the marketing of derivatives of illegal drugs, and a rapid increase in their popularity has caused harm to public health. One response to regulating the supply of designer drugs has been their classification as medicines. At least seven European countries have used the harmonized EU definition of medicinal products to fight against this threat. In Directive 2001/83/EC of the European Parliament and of the Council of 6 November 2001 on the Community code relating to medicinal products for human use, medicinal products are defined as

a. Any substance or combination of substances presented as having properties for treating or preventing disease in human beings; or
b. Any substance or combination of substances which may be used in or administered to human beings either with a view to restoring, correcting or modifying physiological functions by exerting a pharmacological, immunological or metabolic action, or to making a medical diagnosis. [16]

This definition of a medicinal product appeared not to require such a product to have beneficial effects on human health, therefore there has been room for countries to use this legislation to respond to designer drugs. When a national medicines agency classifies a new psychoactive substance as a medicinal product, it can then demand a license for any importation, marketing, or distribution of that product. However, in July 2014 the Court of Justice of the European Union (CJEU) ruled that such an interpretation of the harmonized EU definition of medicinal products was not correct. The main reasoning of the CJEU decision was that although the wording "modifying physiological functions" does not expressly clarify whether the alteration must be positive or negative, the intent of the legislator clearly was to include only substances that have a beneficial effect on human beings. This intention is clear because "modifying" is employed in connection with the wording "restoring [and] correcting physiological functions," which implies beneficial effects. The term "modifying" must therefore be interpreted as meaning an alteration accompanied by beneficial effects to human health [4].

EU Member States that have used medicinal law to combat designer drugs have declared that legislative initiatives have been adopted or are being planned in response to the judgment. Germany has decided to use a generic approach to control synthetic cannabinoids and compounds derived from 2-phenylethylamine, as presented in the previous paragraph. The Netherlands is planning legislative action with a view to reconciling legal certainty and clarity with the need to avoid being one step behind when the chemical composition of substances changes slightly. In Finland, since 2014 new psychoactive substances have been covered under the Finnish Narcotics Act and listed in the government regulation on the consumer market of psychoactive substances. In France, designer drugs are submitted for regular assessment and classified as narcotics. In 2015, seven families of synthetic cannabinoids were defined as such as per the decree of the Minister for Health [4].

3.5 CONCLUSIONS

Control of psychoactive substances is the subject of international treaties, which are signed to protect public health. The main documents regulating drug control are the United Nations Conventions, which are aimed at limitation of the possession, use, trade, distribution, import, export, manufacture, and production of drugs exclusively to medical and scientific purposes. As it is a general principle that criminal law must be certain, the substances subject to such penalties must be clearly specified, which means listing them individually. Therefore, the Conventions include lists of controlled narcotic drugs and psychotropic substances. These lists may be amended, but this process must be preceded by a risk assessment performed by the WHO. A similar process is performed in Europe by the EMCDDA, in cooperation with Europol, for the Council of the EU. Sometimes this process is also carried out by Member States, but, unfortunately, this approach is currently not very effective. The main reason for the low efficiency of such an approach is that the number of designer drugs introduced to the drug market has significantly increased in recent years. More than ten years ago, when the system was formed, less than five new designer drugs per year were identified and it was possible to assess the risk associated with their marketing. However, in recent years, the number of new psychoactive substances is around 100 per year, meaning that a new designer drug enters the market every three to four days.

The low efficiency of an individual drug control system has led to its modification. In order to accelerate legislative processes, some countries have introduced temporary control regimes, allowing time for investigation of the need for permanent control. Another response to the threat of new designer drugs has been to manage them under medicinal product laws, which are based on a harmonized EU definition of a medicinal product. But the use of these laws was challenged in 2014 by the Court of Justice of the European Union (CJEU), which ruled that designer cannabinoids are not medicinal products if they do not have beneficial effects on human health. Consequently, the prosecution of designer drugs cases based on medicinal laws has become more difficult. Therefore, most countries have changed their approach in order to control these substances, and have typically adopted a generic approach, based on group definitions of substances under control. This approach also has its opponents, as the generic definitions are couched in the language of technical chemistry and are often difficult for both legal professionals and common people to understand. However, a generic approach is one of the most efficient, allowing for limitation of the rapid increase in new designer drugs and being in agreement with the requirements of criminal law.

Other existing laws are focused on consumer or health protection, or have introduced innovative new laws to address these substances—in a few cases even defining a psychoactive substance by its effect, rather than by its chemical structure. However, different countries use different definitions, and it is out of the scope of this book to present these in detail.

REFERENCES

1. European Monitoring Centre for Drugs and Drug Addiction. www.emcdda.europa.eu /about/mission (accessed 3 March 2017).
2. European Monitoring Centre for Drugs and Drug Addiction. 2007. Early-warning system on new psychoactive substances—Operating guidelines, Office for Official Publications of the European Communities, Luxembourg.
3. Council Decision 2005/387/JHA of 10 May 2005 on the information exchange, risk-assessment and control of new psychoactive substances, OJ L 127, 20.5.2005, 32–37.
4. European Monitoring Centre for Drugs and Drug Addiction and Eurojust. 2016. *New psychoactive substances in Europe: Legislation and prosecution—Current challenges and solutions*, EMCDDA–Eurojust joint publication, Publications Office of the European Union, Luxembourg.
5. United Nations Office on Drugs and Crime. 2008. A century of international drug control. www.unodc.org/documents/data-and-analysis/Studies/100_Years_of_Drug_Control.pdf (accessed 3 March 2017).
6. International Narcotics Control Board. 1961. Single Convention on Narcotic Drugs. www.incb.org/incb/en/narcotic-drugs/1961_Convention.html (accessed 3 March 2017).
7. International Narcotics Control Board. 2016. Yellow List: List of Narcotic Drugs under International Control. www.incb.org/incb/en/narcotic-drugs/Yellowlist_Forms/yellow -list.html (accessed 3 March 2017).
8. Katselou, M., Papoutsis, I., Nikolaou, P., Spiliopoulou, C., and Athanaselis, S. 2015. AH-7921: The list of new psychoactive opioids is expanded. *Forensic Toxicol.* 33(2): 195–201.
9. United Nations. 1971. Convention on Psychotropic Substances. www.unodc.org/pdf /convention_1971_en.pdf (accessed 3 March 2017).
10. International Narcotics Control Board. 2016. Green List: List of Psychotropic Substances under International Control. www.incb.org/incb/en/psychotropic-substances/green-lists .html (accessed 3 March 2017).
11. van Amsterdam, J., Nutt, D., and van den Brink, W. 2013. Generic legislation of new psychoactive drugs. *J. Psychopharmacol.* 27(3): 317–24.
12. Gesetz zur Bekämpfung der Verbreitung neuer psychoaktiver Stoffe. 2016. Bundesgesetzblatt Tail I Nr. 55 (25 November 2016).
13. Misuse of Drugs Act. 1971. www.legislation.gov.uk/ukpga/1971/38/contents (accessed 3 March 2017).
14. Advisory Council on the Misuse of Drugs. 2010. Consideration of the cathinones. www .gov.uk/government/publications/acmd-report-on-the-consideration-of-the-cathinones (accessed 3 March 2017).
15. Psychoactive Substances Act. 2016. www.legislation.gov.uk/ukpga/2016/2/contents /enacted (accessed 3 March 2017).
16. Directive 2001/83/EC of the European Parliament and of the Council of 6 November 2001 on the Community code relating to medicinal products for human use, OJ L 311, 28.11.2001, p. 67.

4 Biological Background of Designer Drugs

Łukasz Komsta and Katarzyna Wicha-Komsta

CONTENTS

4.1   INTRODUCTION

The human *central nervous system* (CNS), consisting of the brain and spinal cord, is a very complex structure. Its proper function depends on dozens of intercorrelated factors. Besides memory, knowledge, and skills, the state of the CNS corresponds with emotional life. Emotions and psychological conditions are regulated by substances called *neurotransmitters.*

Neurotransmitters are *endogenous* substances, i.e., they are synthesized inside the body. Their action is performed mainly by chemical interaction with *receptors*—complex protein structures inside the cell membranes. Receptors act as a switch, which can be switched only by an appropriate substance, called a *ligand* [1,2]. When a ligand creates a chemical complex with the receptor, the mode of further action depends on the receptor type. In the case of *ionotropic* receptors, the conformation of the receptor's proteins is changed, which results in the opening of *ion channels* located inside the receptor. Ion channels are made of special tunnel-like proteins and their opening allows the ions to flow through the membrane *freely*, i.e., in the passive mode—according to the concentration gradient, without any active energetic support. In the case of *metabotropic* receptors, which do not contain any ion channel, binding of a ligand initiates a cascade of biochemical processes, changing the metabolism inside the cell (they can also indirectly open other ion channels in the cell membrane).

A ligand that switches (activates) the receptor is called its *agonist*. For most receptors, there are also other known classes of ligands, representing the opposite behavior. An *antagonist* binds to the receptor and does not activate it, blocking the receptor to be activated by a natural endogenous agonist. A *reverse agonist* can bind to a receptor and perform the opposite action to the agonist. Receptors also have active sites for the substances that can change their sensitivity. These substances are called *modulators* and they do not bind to the main active site of the receptor [1,2].

Several neurotransmitters undergo regulation by their *reuptake* [3,4]. The substance released to the synaptic cleft is transported back to the neuron by special transport proteins. This allows the reuse of neurotransmitters without their needing to resynthesize. There are substances that increase the activity of neurotransmitters by inhibiting their reuptake.

The receptors differ also in their structure between body organs, so the ligand of only one receptor subtype can induce specific effects in the organs having this subtype of the receptors. For example, histamine is responsible mainly for inflammation and allergic reactions [5]. It also activates excretion of hydrochloric acid in the stomach by a subtype of histamine receptors called H_2. Antagonists of these receptors are antiacids, not showing any other action related to histamine.

In normal conditions, the neurotransmitter concentration varies according to emotional life. The body has the ability to keep their concentration within the safe limits [6]. Almost all psychological diseases, such as depression, schizophrenia, insomnia, abnormal anxiety, and neurosis are caused by impaired equilibrium between neurotransmitter activity. The imbalance is also responsible for diseases not strictly perceived as psychological, but related to the CNS, such as Parkinson's disease. The imbalance can have various reasons. For instance, a hypoactivity of a particular neurotransmitter can, besides improper concentration of the neurotransmitter itself, be caused by the abnormally low amount (concentration) of receptors or a significant change in receptor sensitivity. The latter can also be caused by a long exposure to addictive substances.

4.2   NEUROTRANSMITTERS

The division of neurotransmitters connected with the central nervous system (and therefore important for designer drugs action) is presented in Figure 4.1. This is a

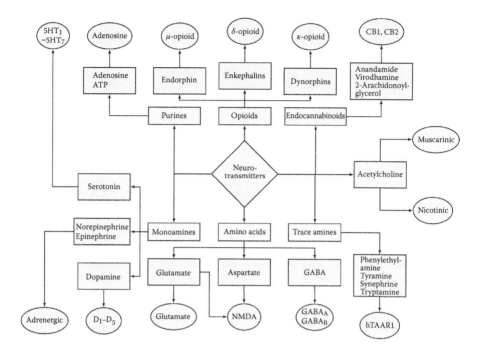

FIGURE 4.1 Classification of selected neurotransmitters. The receptors they are acting on are shown inside the ellipses.

basis for the classification of designer drugs in this chapter. They are described in the context of chemical similarity to a particular neurotransmitter, or sole affinity to particular receptors. However, the biological processes are complex and some drugs cannot be easily classified to one of these groups. The summary of the division is given below; detailed description is given later in the chapter.

1. *Acetylcholine* is a ligand for nicotinic receptors in the brain and muscarinic receptors in various organs. The psychotic effects are very limited. Nicotine is not a designer drug candidate, nor are muscarinic ligands (muscarine, atropine, hyoscyamine, etc.).
2. *Adenosine* has its own receptors, which are mainly connected with relaxation and inducing of sleep. Purines (caffeine, theobromine, theophylline) are their antagonists, but these are not currently considered a base of designer drugs.
3. *Dopamine* has its own receptors, D_1–D_5, and plays a very important role in the feeling of desire and pleasure, and in the development of addiction. Ephedrine, cathinones, and amphetamines exhibit strong chemical similarity to dopamine and to another endogenous substance, phenylethylamine. In this chapter, they are described as dopamine analogs, but their pharmacological action is much wider. Cocaine, its derivatives, and phenidates also exhibit a similar action. There is also a strong similarity to benzylpiperazines.

4. *Serotonin* (5-hydroxytryptamine, 5-HT), acting on 5-HT_1–5-HT_7 receptors, is responsible for pleasure and relaxation; the lack of this neurotransmitter is considered to be the main factor in depression. The main substances with a serotoninergic mode of action are tryptamines and ergot alkaloids.
5. *Opioids* mediate the feeling of pain, vomiting reflex, cough reflex, and many other physiological phenomena. Morphine and codeine are natural opioid receptor agonists, whereas most designer drugs are based on a synthetic morphine analog, fentanyl.
6. *Cannabinoids*, acting on CB1 and CB2 receptors, are connected with appetite, vomiting reflex, the feeling of pain, and the mood and memory processes. *Cannabis* alkaloids are the natural agonists and the base for designer ligands.
7. *Amino acids.* Gamma-butyric acid (GABA) is an agonist of GABA receptors. Benzodiazepines are synthetic drugs interacting with the GABA_A receptor, responsible for many processes, including mood changes and aggression. Ketamine and its designer derivatives interact with NMDA receptors (activated endogenously by glutamate and aspartate), causing general anesthesia.

4.2.1 Drug Chemistry and Action

The affinity to a particular receptor is not the only condition for a chemical compound to be a good drug candidate. During the development of new drugs, many active substances are finally rejected, due to various reasons. One of the main reasons is toxicity, when the side (bad) effects are much greater than the therapeutic action. The optimal substance can be found by making a series of similar compounds (derivatives of the same basic structure) and the evaluation of them. There are many *in silico* (computer) methods that allow screening of the possible substances using various computational algorithms. However, the final investigation must always be done with model organisms or cell cultures (*in vivo*). A small change in the drug molecule can heavily alter pharmacological and physiological parameters of the drug, including the mode of action. This rule applies both to fully synthetic drugs and those derived from natural compounds.

The fate of a drug inside the human body is abbreviated to LADME (liberation, absorption/administration, distribution, metabolism, excretion; see Figure 4.2) and all these processes must be taken into account during drug design. The pharmacokinetic profile (the concentration of the drug in blood as the function of time) must exceed the therapeutic concentration, not exceeding the toxic one (Figure 4.3) [7,8]. This curve is modeled by a system of differential kinetic equations similar to that of chemical kinetics processes. Various kinetic models are used in practice, depending on all LADME steps and penetration of the drug to different *compartments* (parts of the body with constant drug concentration). The simplest case is the first-order exponential decay: when the drug is administered immediately (intravenously), the transfer between various compartments can be neglected and the elimination is solely dependent on the concentration in the full therapeutic range. In the case of oral

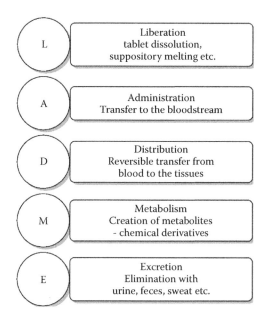

FIGURE 4.2 Disposition of a pharmaceutical compound within the human body is often memorized as the "LADME" abbreviation.

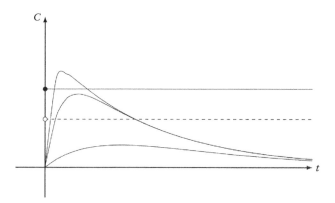

FIGURE 4.3 Impact of pharmacokinetics on drug action. The concentration of a drug (C) as the function of time (t) must exceed the therapeutic concentration (horizontal dashed line), not exceeding the toxic concentration (horizontal solid line). Only the middle profile satisfies this requirement. The chemical structure of a drug is the main factor determining the concentration profile, and small substitutions can alter the curve shape substantially.

administration, the curve is analogous to the concentration of intermediate product reaction in chemical kinetics.

The basic requirements for the substance to be a good drug candidate are gathered into Lipiński's famous rule of five: no more than five hydrogen bond donors, no more than ten hydrogen bond acceptors, the molecular mass less than 500 Da and the octanol–water partition coefficient log P (lipophilicity) not greater than 5. *Lipophilicity* is one of the most important parameters determining drug action, as it influences penetration through cell membranes [8,9]. When log P is too low, the drug does not penetrate into the cells; however, too-high lipophilicity impairs the absorption and distribution of the substance. Strongly lipophilic substances are insoluble in physiological fluids. The substituents involved in new designer drugs often change the lipophilicity significantly compared to the parent compound.

The structure of a pharmaceutical formulation influences the *liberation* step. It is the only way to change the pharmacokinetics without modifying the chemical structure. The *sustained release* formulations are used when the prolongation of therapeutic effect (with subsequent lowering of concentration) is needed. It is hard to believe that the black market takes into account this phase when making tablets or other formulations of designer drugs.

Absorption of a drug inside the intestinal tract (or in another part of the body, depending on the route of administration) depends on various parameters: solubility, lipophilicity, and acidic/basic properties. The drugs can be weak acids or bases and they can exist in ionized or free form, depending on the pH. Absorption generally occurs when the drug is in free form, i.e., when dissociation of the drug does not occur. The drug must be designed to be soluble in small amounts in its free form. The absorption of acidic drugs occurs in acidic conditions (in the stomach), whereas basic drugs are absorbed in basic conditions (in the duodenum). In general, *free* transport occurs according to the concentration gradient. Exceptionally, active transport (involving energy stored in ATP) can occur.

The drug is then *distributed* to all parts of the body with the blood and it can optionally be caught by blood proteins. The affinity of the drug to these proteins depends significantly on chemical structure and alters its pharmacokinetic profile. Only unbound drugs can penetrate to the tissues and the action depends on the concentration of an unbound drug. When the drug has a high protein affinity, its *elimination rate* (the kinetic parameter of blood decay) can be very low and it can last inside the body for a very long time.

Another important parameter is the ability to bypass the *blood–brain barrier*. These two parts of the body are separated by a highly selective membrane, which provides natural protection for the brain against toxins. Drugs that act on the central nervous system have to be able to penetrate this barrier. Passing this barrier is much more difficult than penetration of the drug into all other tissues. When therapy is used for pregnant women, another important barrier is the blood–placenta barrier, which protects the fetus from toxins.

Drugs are *xenobiotics*—substances that should not be present inside the body in normal physiological conditions. Therefore, the body treats a drug as a toxin. Drugs which cannot be easily excreted in unchanged form undergo *metabolism* inside the liver [10]. Metabolism is a defense mechanism created in the evolution process to eliminate toxins from the body. Substances administered orally pass through the liver before they reach the majority of tissues. This phenomenon is called *first-pass metabolism*. To avoid drug decomposition and a need to take higher doses, other routes of administration have to be used.

The first phase of metabolism is based on small modifications to a drug molecule, catalyzed by various liver enzymes. A drug can undergo oxidation, methylation, hydrolysis, or other decomposition to form a *metabolite*. The metabolism process is very important in the analysis of designer drugs in biological material as, besides the original drug, metabolites can (and should) be determined. The metabolic pathway of a particular drug must be investigated and known to be sure which compounds are metabolites. There are some *in vitro* or even *in silico* methods to investigate possible metabolism of a newly synthetized compound. If the first phase metabolite cannot be easily excreted, the second phase occurs. A drug can undergo acetylation or conjugation with various substances, such as glucuronic acid. It is important that metabolites can be active, inactive, or even toxic. Some drugs are not active themselves as they are *prodrugs*; they must undergo metabolism, and all action is caused by the metabolites.

The *excretion* of a drug can be done in various ways, most often with urine, but some drugs are excreted with feces or bile. Rarely can a drug be excreted with saliva, sweat, or breathing air. Therefore, besides the blood, the main biological material to detect designer drugs (and/or their metabolites) is urine.

The development of new designer drugs is not oriented to improve pharmacological properties, but to create another substance by any means, in order to bypass legal regulations against forbidden substances [11]. Taking this fact into account, designer drugs can vary substantially in their toxicity, metabolism, pharmacokinetics, and bioavailability [12]. It is not recommended to assume any parameter of a new designer drug based on similar known substances [13,14].

4.2.2 ADDICTION

Many drugs affect the *reward system*, a brain structure responsible for the feeling of liking and fulfillment of craving (desire) [15,16]. The reward system plays a positive role in human life when it works properly: many everyday activities, such as sexual activity, eating food when being hungry, drinking water when feeling thirst, and interacting with children (parental instinct), are "rewarding" [17]. The main aim of this system is to satisfy needs induced by the lack of *homeostasis*—the equilibrium of body processes. There is also a *punishment* system, which is responsible for negative emotional reactions. It is believed that the dopamine system is the main part of all factors involved in reward [18] and *psychological* addiction [19]. Most designer drugs only cause psychological dependence. However, several of them (mostly opioid derivatives) can lead to *physiological* dependence, when a drug changes the biochemical pathways; this process cannot be easily reversed.

Under special conditions, the reward and punishment systems can be involved in the *conditioning* process [20–22]. In this case a neutral stimulus (situation or behavior), occurring together with a rewarding stimulus, becomes also rewarding or is treated by the brain as an intermediate step to be rewarded. This phenomenon is not connected with consciousness and causes unconscious *habit*, which can be *reinforced* (more and more "fixated") when the conditioning is repeated many times [23–25]. Conditioning is believed to be one of the main factors of the psychological *addiction* process [20,21,26]. The reinforcement can be positive (when the aim is to get a reward) or negative (when the aim is to avoid a punishment, for example to reduce fear, anxiety, or *withdrawal* symptoms) [27]. In many cases, the development of addiction is a consequence of both processes (a reward and a punishment) repeated many times. Many authors treat reward and reinforcement as connected processes, but occurring separately at various stages of seeking a reward and then "consuming" it. The role of avoiding stress (understood as a punishment) is also very important [28].

The stimulus does not need to be very strong to induce addiction—the best examples are behavioral addictions (shopping, watching TV, washing hands, etc.). The key factor is *sensitization*. This phenomenon, opposite to *tolerance*, increases the response to a stimulus when exposition is not continuous (tolerance decreases the response to a continuous stimulus). Sensitization is very important in the case of designer drugs, especially acting on the dopamine system. Once developed, sensitization can last for a very long time, being responsible for *compulsive* behaviors during searching for a reward. Sensitization and compulsive reactions are often *triggered* by some special factors or circumstances. When the surroundings and conditions are different, they can simply be hidden. Sensitization is responsible for *craving* behavior and occurs mainly in the parts of the reward system that are responsible not for "liking" the reward, but for "wanting" to be rewarded. It is believed that the dopaminergic system is the main substrate of such changes. The case of "wanting to be rewarded, just before being rewarded" as giving much more pleasure than the reward itself is very accurately described by Winnie the Pooh: "although eating honey was a very good thing to do, there was a moment just before you began to eat it which was better than when you were." The activity leading to a positive *reinforcement* induces defense reactions from the nervous system (tolerance) and in many cases the response for the stimulus becomes lower when the drug is taken often. Very prolonged exposure to the stimulus and very strong stimuli, called *superstimuli*, can cause permanent morphological changes in neurons. Stopping of the exposure leads to *withdrawal* symptoms.

4.3 ALKALOIDS AS PLANT METABOLITES

Plants did not develop any excretory system and they can excrete only limited number of substances (mainly water, oxygen, and carbon dioxide). The other metabolites have to be stored inside their cells in vacuoles, often in special storage organs (roots, corms, tubers, rhizomes, fruits). Storage organs are evolutionally developed for storing primary metabolites: the substances that can be further reused (carbohydrates, proteins, fats, nucleic acids, etc.). They are essential to the plant's survival during its life cycle. For example, biennial plants develop a foliage during the first year of growth and "focus" on storing the primary metabolites of photosynthesis in the root. These metabolites are then reused in the second year as a "backup food" to develop the quickly flowering stem and then end the life cycle.

The compounds being drug candidates are the *secondary metabolites*. They are not necessary in life processes; however, they were developed during evolution to protect plants from natural dangers or to lure insects by their smell [29]. Almost all compounds investigated by pharmacognosy are secondary metabolites: flavonoids, steroids, saponins, terpenes, coumarins, lignans, tannins, etc. From designer drugs' perspective, the most important metabolites are *alkaloids*.

The definition of alkaloid is still discussed and the main doubts are the differentiation between alkaloids and some complex amines [30]. In general, they are secondary metabolites containing one or more *basic* nitrogen atoms. Almost all alkaloids have strong pharmacological action and they exhibit significant toxicity. There are thousands of alkaloids and only several groups are used as precursors of designer drugs as they act on the central nervous system [31–33].

In most cases, the biosynthesis of alkaloids starts from an amino acid as the nitrogen source [34,35]:

1. Ornithine and arginine is the starting point in atropine synthesis—it can be converted to putrescine, which is converted to N-methyl-Δ1-pyrrolium cation. It can be directly condensed to finally form nicotine, but it also can undergo conversion to hygrine, then recyclized to tropinone. Further processes lead to tropine, littorine, and finally hyoscyamine/atropine. Cocaine is synthesized with a different path: the N-methyl-Δ1-pyrrolium cation is condensed with acetyl-SCoA, then cyclized and condensed with benzoyl-CoA.
2. Phenylalanine is the source of nitrogen for opium alkaloids (morphine, codeine), ephedrine, and cathinone. During the opium alkaloids pathway, phenylalanine is converted to DOPA, coclaurine, reticuline, and salutaridinol, forming thebaine. Thebaine is next converted to neopinone, codeinone, codeine, and finally morphine. An additional route can occur from thebaine to morphine, with oripavine and morphinone as the intermediate products. The main ephedrine and cathinones biosynthesis pathway starts from phenylalanine, transforming to benzaldehyde. It undergoes condensation with pyruvate, forming 1-hydroxy-1-phenylpropan-2-one or 1-phenylpropane-1,2-dione. This compound is then converted to norephedrine and finally methylated to ephedrine and it is suspected that cathinone is an intermediate product of this reaction [30]. It is also known that the transition steps from phenylalanine to 1-phenylpropane-1,2-dione can occur independently with another β-oxidative pathway, with the participation of acetyl-CoA.
3. Tryptophan is the substrate in ergot alkaloid biosynthesis. It is substituted with 4-dimethylallyl substituent, then undergoes cyclization forming chanoclavine. Chanoclavine is oxidized to aldehyde derivative, then reduced to agroclavine. Agroclavine then becomes elymoclavine, paspalic acid, and finally lysergic acid, a substrate for all ergot alkaloids mentioned in this chapter.
4. Xanthosine (a nucleotide synthesized together with all purines) is the substrate for the synthesis of caffeine, theobromine, and theophylline.

4.4   ADENOSINE

Adenosine is a purine derivative being also a nucleoside. It decreases the heart rate, widens coronary arteries, decreases dopaminergic activity in the CNS, and inhibits the action of various neurons (the role connected with sleep induction). It also constricts the muscles in the walls of bronchioles and relaxes cardiac muscles. So far, no ligands are available as designer drugs.

Three important natural substances are antagonists of adenosine receptors (so they induce reverse phenomena) [36–38]:

1. Caffeine [36,39,40], a natural alkaloid occurring in coffee plants: *Coffea arabica* L. and *Coffea canephora* Pierre ex A. Froehner (synonym *Coffea robusta* L. Linden) [41], tea plant *Camellia sinensis* (L.) Kuntze (synonym: *Thea sinensis* L.) [42], cola nuts (*Cola acuminata* (P.Beauv.) Schott & Endl. and other *Cola* species) [43], various *Ilex* species, especially mate (*Ilex paraguariensis* A.St.-Hil.) [44] and guarana (*Paullinia cupana* Kunth) [45,46].
2. Theobromine [36,37] and theophylline [47], natural alkaloids occurring in large amounts in cacao seeds (*Theobroma cacao* L.) [48]. A small amount of theobromine is also found in the above caffeine-rich plants.

4.5   OPIOIDS

Endogenous opioids (endorphins, encephalin, dynorphin) are neurotransmitters responsible for pain suppression, antidepressant action, depression of the respiratory system (including cough reflex reduction), diarrhea suppression, and reduction of the vomiting reflex [49]. Each action is associated with one or several opioid receptor types. Besides three main types of receptors (μ, δ, and κ), several subtypes are classified. Chemical modification of the opioid molecule can change the affinity to a particular type of receptor. Many synthetic opioid drugs are designed to possess only one mode of action (for example, loperamide is antidiarrheal, dextromethorphan is antitussive).

Morphine, codeine, and thebaine are natural agonists of opioid receptors occurring in opium poppy (*Papaver somniferum* L.) [50]. It is an annual plant, cultivated in many cultivars all over the world. Its natural origin is not fully known; however, it is believed that the plant comes from the Eastern Mediterranean region. The word "opium" concerns the dried juice from immature capsules (fruits). Besides opium alkaloids, poppies contain isochinoline alkaloids (papaverine with antispasmodic action, noscapine with antitussive action). Due to the risk of abuse, the cultivation of poppies requires special licenses in many countries. Wild poppies (*Papaver rhoeas* L., *Papaver argemone* L., and *Papaver dubium* L.) do not contain any significant amount of opioids and their traditional medicinal use is restricted to isoquinoline substances (especially rhoeadine with mild sedative action).

Heroin (diacetylmorphine) is the main product of opium acetylation. This substance is widely abused due to its ease of production from opium in a home environment. However, no significant interest in developing heroin or morphine analogs as designer drugs is observed. The main effort made by the black-market producers is put on fentanyl—a morphine synthetic analog [51–54] (Table 4.1). This drug is commonly used in anesthesia during surgery and exhibits strong action (many times stronger than morphine), together with reduced amnesia after sleep. It undergoes an extensive first-pass metabolism, mainly detachment of both substituents attached to nitrogens and hydroxylation on the end of the alkyl chain. Although many pharmacological properties of fentanyl are improved comparing to morphine, this drug (and

TABLE 4.1
Structural Similarities Between Morphine (left), Methadone (middle), and Fentanyl Derivatives (right)

	R1	R2	R3
3-MF	C_2H_5	CH_3	H
3-MBF	C_3H_7	CH_3	
Butyrfentanyl		H	
4-FBF			F
4-MeO-BF			OCH_3
Acetylfentanyl	CH_3		H
Acrylfentanyl	$CHCH_2$		
Fentanyl	C_2H_5		
Furanylfentanyl	furan		

its designer analogs) still have side effects. One of the most important is respiratory depression, which can be life-threatening in the case of overdose.

4.6 CANNABINOIDS

There are two types of cannabinoid receptors: CB1 and CB2. They are activated by endogenous ligands: anandamide (both types) and 2-AG (2-arachidonyl glycerol, CB1 only) [55]. The CB1 receptor is located in the brain (mainly in the hippocampus, cerebellum, and striatum), while the CB2 receptor can be found in the spleen and in hemopoietic cells [56].

The activation of CB receptors causes a series of physiological effects: suppression of the vomiting reflex, increase of appetite, reduction of pain sensitivity, good mood, psychological relaxation (resembling the effect of ethanol), reduction of muscle tension, increased sensitivity to stimuli, but with simultaneous suppression of memory and cognitive abilities. Also, heart rate increases with a simultaneous (!) decrease of blood pressure.

Natural agonists of cannabinoid receptors occur in plants of the genus *Cannabis* [57]. The taxonomy of this genus changed over time and now it is considered to be monotypic, i.e., containing only one species: *Cannabis sativa* L. All taxa considered previously as distinct species, such as *Cannabis indica* Lam., *Cannabis chinensis*

Delile, *Cannabis americana* Pharm. ex Wehmer, or *Cannabis ruderalis* Janisch., are now treated as subspecies of *C. sativa* (some authors aggregate them to three species: *C. sativa*, *C. indica*, and *C. ruderalis*). The *Cannabis* genus is believed to have been used since ancient times, both due to its pharmacologic properties and for its production of hemp (a soft fiber from the plant stem).

Cannabis contains more than 60 known cannabinoids [58]; however, its action is caused mainly by tetrahydrocannabinol (THC), cannabinol, and cannabidiol (the last one has increased affinity to CB2 receptors) [50]. The other important compounds are spermidine alkaloids, several amines and amides, several quaternary bases (including muscarine), amino acids, proteins, glycoproteins, enzymes and sugars, steroids (mainly campesterol, ergosterol, and sitosterol), and terpenes.

Psychoactive cannabinoids produce all the effects associated with the activation of CB receptors. It should be underlined that the resultant inhibition of cognitive deficiencies is prolonged and seems to persist for a long time after use. The receptor activation increases the activity of dopaminergic neurons in the ventral tegmental area–mesolimbic pathway and this mechanism is responsible for reinforcing and addictive properties. The metabolism of cannabinoids is based on hydroxylation of the CH_3 substituent in the phenyl ring to CH_2OH, followed by its oxidization to the carboxylic acid.

Chronic use of cannabinoids causes tolerance to antinociception, anticonvulsant activity, catalepsy, depression of locomotor activity, and hypotension. This phenomenon concerns a small ratio of people, mainly those using huge amounts of marijuana. There are serious doubts about the existence of real addiction. The withdrawal symptoms are weak and similar to that of ethanol and opioids: nausea, hyperactivity, irritation, tachycardia, and sweating. No compulsive actions are observed.

The use of cannabinoids in therapy of various diseases is still very controversial; however, in many countries they are used as drugs and the greatest hope is connected with their role to treat cancer. There are efforts to produce synthetic cannabinoids that could act as drugs but without the typical undesired psychotic and mental effects [59].

The structure of designer synthetic cannabinoids is highly differentiated and it cannot be easily summarized as the base skeleton with some substituents. Most of them are significantly more potent agonists than natural cannabinoids: they have greater binding affinity to the cannabinoid CB1 receptor than THC and greater affinity to the CB1 than the CB2; some selective strong CB2 ligands are also reported. They can be divided into [60] indoles (adamantoylindoles, aminoalkylindoles, benzoylindoles, naphtoylindoles, naphtylmethylindoles, and phenylacethylindoles), cyclohexylphenols, dibenzopyranes, indazoles, naphtylmethylindenes, and naphtoylpyrroles.

4.7   SEROTONIN

Serotonin (5-hydroxytryptamine, 5-HT) is a neurotransmitter and a peripheral hormone. The serotoninergic system is very complex and its receptors play various, often opposite, roles. The imbalance of the serotoninergic system is responsible for anxiety, mood disorders, depression, and migraines. There are seven main types of serotonin receptors $5\text{-}HT_1$–$5\text{-}HT_7$, further divided to the letter-labelled subtypes.

The hyperactivity of the serotonin system (the overdose of the serotonin receptor agonists or the reuptake inhibitors) leads to serotonin syndrome: headache, hallucinations, fever, insomnia, sweating, hypertension, vomiting, diarrhea, and convulsions. The designer and natural ligands of serotonin receptors are tryptamines and ergot alkaloids.

4.7.1 TRYPTAMINES

Psilocybin and psilocin are naturally occurring tryptamine alkaloids in various mushrooms from the *Psilocybe* genus (mainly *Psilocybe cubensis* (Earle) Singer and *Psilocybe semilanceata* (Fr.) P.Kumm.). Psilocybin is quickly converted by the body to psilocin, which has various effects similar to LSD and mescaline. Mescaline is a natural phenylethylamine derivative, but it acts on 5-HT$_2$ receptors. It occurs in several plants, including *Lophophora williamsii* (Lem. ex Salm-Dyck) J.M. Coult., *Echinopsis pachanoi* (Britton & Rose) Friedrich & G.D. Rowley and *Trichocereus bridgesii* (Salm-Dyck) Britton & Rose.

The modification of the tryptamine skeleton can alter the pharmacological properties; for example, sumatriptan is an antimigraine drug with suppressed psychoactive properties. Probably the most interesting natural derivative is ibogaine, an alkaloid from *Tabernanthe iboga* Baill. (synonym: *Iboga vateriana* Braun-Blanq. & K. Schum). It is a ligand of many receptor types in the CNS, used in treatment of a wide spectrum of addictions.

Designer tryptamines (Table 4.2) are the agonists of many subtypes of serotonin receptors [61]. They have lower affinity to 5-HT$_{2A}$ receptors compared to

---

TABLE 4.2
Structure of Tryptamine Derivatives

	R1	R2	R3	Ring
NMT	CH$_3$	H	H	(none)
Tryptamine	H			
Serotonin				5-hydroxy
DMT	CH$_3$	CH$_3$		(none)
Melatonin	H	COCH$_3$		5-methoxy
Psilocin	CH$_3$	CH$_3$		4-hydroxy
Tryptophan	H	H	-COOH	(none)
Benzofurans		(various)		O instead of N

phenethylamines. However, the affinity to 5-HT$_{1A}$ receptors is still high and these receptors are mainly responsible for the hallucinogenic effects. Some alpha-methylated tryptamines also exhibit stimulant activity (by strong inhibition of dopamine reuptake), occurring mainly in lower doses. They undergo extensive first-pass metabolism and are almost inactive when administered orally. Dimethyltryptamine (DMT) is very quickly eliminated from the blood (in 1h), mainly to 3-indoleacetic acid after oxidative deamination. Incorporation of a substituent in fifth position increases the clinical effects and lowers the metabolism—the compound must undergo O-demethylation, 6-hydroxylation, and N-dealkylation. In general, metabolism of tryptamine derivatives is very complex and there are several alternative pathways [62].

4.7.2 ERGOT ALKALOID DERIVATIVES

The natural source of ergot alkaloids (Table 4.3) is the sclerotium of ergot fungus, *Claviceps purpurea* (Fr.) Tul. [63,64]. It is a parasite of grass plants, mainly rye (*Secale cereale* L.). The life cycle of ergot is perfectly matched with the cycle of the host plant: the ascospore acts as a pollen grain and the infection develops inside a flower; the shape of the sclerotium is similar to the fruit (grain). Before the development of effective fungicides, ergots caused many intoxications, which have now

TABLE 4.3
Structure of Ergot Alkaloids and Derivatives

	R1	R2	R3
Lysergic acid	H	OH	CH$_3$
Ergine (LSA)		NH2	
Ergotamine		(complex substituent)	
Ergometrine		-NHCH(CH$_3$)CH$_2$OH	
Methysergide	CH$_3$	NHCH(C$_2$H$_5$)CH$_2$OH	
1P-LSD	COC$_2$H$_5$	N(C$_2$H$_5$)2	
LSD	H		
AL-LAD			CH$_3$CHCH$_2$
ETH-LAD			C$_2$H$_5$
PRO-LAD			C$_3$H$_7$

almost disappeared. However, ergot is still cultivated under controlled conditions as the source of alkaloids for medical purposes. The sclerotium contains ergometrine, ergometrinine, ergotamine, ergotaminine, ergocornine, ergocorninine, ergocriptine, ergocriptinine, ergocristine, ergocristinine, ergosine, and ergosinine in various proportions according to development conditions. Ergot alkaloids, in pure form and as derivatives, are used widely in migraine treatment. Their abuse started in the 1930s after synthesis of lysergic acid diethylamide (LSD, LSD-25) by Albert Hoffman in Sandoz. LSD is one of the strongest known drugs as it can give the pharmacological effect in amazingly low doses.

The action of ergot derivatives is connected with the inhibition of serotonin receptors, but the mode of action and pharmacological effects vary between the derivatives. The location of receptors also matters, which makes the whole problem difficult to describe. For example, ergotamine is an antagonist of 5-HT_{1A} and 5-HT_{1D}, being simultaneously an agonist of 5-HT_{1B}. LSD-25 activates 5-HT_{2B}, 5-HT_{2C}, 5-HT_{7}, and 5-HT_{2A} in CNS, but inhibits peripheral 5-HT_{2A}. This is the reason why migraine drugs do not exhibit a hallucinogenic effect, whereas LSD is primarily hallucinogenic. LSD changes perception of all senses, changing heard sounds and inducing fake visions, sounds, and smells. Many cases of synesthesia (hearing of colors, seeing sounds, etc.) have been reported. The overall perception of hallucinations is mainly positive (a good trip), whereas bad trips also occur, mainly in the case of people with anxiety. The hallucinations can be induced spontaneously a long time after use without another LSD administration; this phenomenon is called a *flashback*. Metabolism of LSD is based on the creation of 2-oxo derivatives, N-desmethyl derivatives, and conjugation of them with glucuronic acid [62].

Designer lysergamides are available on the black market. The most widely known compounds are 1P-LSD (1-propionyl-LSD), AL-LAD (6-allyl-6-nor-LSD), and LSZ (2,4-dimethylazetidide) [65,66].

4.8 AMINO ACIDS

There are three neurotransmitters belonging to this group: glutamate, gamma-aminobutyric acid (GABA), and aspartate. GABA and glutamate have their own dedicated receptor families. Additionally, aspartate and glutamate activate the NMDA receptors.

GABA exists in almost half of synapses present in the central nervous system. Its action is very complex and GABA activity alters many physiological processes: motoric activity, appetite, aggression, sexual activity, mood changes, thermoregulation, and pain sensitivity. GABA receptors divide to two classes: $GABA_A$ and $GABA_B$. The second class is located mainly outside the nervous system and its selective agonist, baclofen, does not have any psychotic effect (besides slight sedation) and is used as a muscle-relaxing agent. The psychotic effects are connected only with $GABA_A$ receptors. Besides active sites for binding GABA itself, the $GABA_A$ receptor has active sites for binding benzodiazepines and barbiturates.

NMDA receptors are believed to be critical in *synaptic plasticity*, a cellular mechanism for learning and memory. Their inhibition induces general anesthesia.

4.8.1 BENZODIAZEPINES

Benzodiazepines consist of a large group of drugs used as sedatives, anxiolytics, and antiepileptic agents. By modifying the chemical structure, it is possible to create drugs that act mainly as sedatives (estazolam, flunitrazepam), anxiolytics (alprazolam, clorazepate), and antiepileptics (clonazepam). Their action is also affected by complex metabolism, when some derivatives are metabolites of the other ones, exhibiting the pharmacological activity. It is also possible to regulate the time of their action, from very short to extremely prolonged.

Historically, the first designer benzodiazepines (not approved for medicinal use, but sold on the black market) were pyrazolam, diclazepam, and flubromazepam, then clonazolam, deschloroetizolam, flubromazolam, nifoxipam, and meclonazepam [67].

4.8.2 KETAMINE AND PHENCYCLIDINE ANALOGS

Ketamine is an old drug used mainly in anesthesia. It has no natural precursors and its action is connected mainly with blocking NMDA receptors [68]. The recreational use of ketamine causes altered perception followed by a general lack of responsive awareness. When administered in an improper environment it puts the recreational user at a significant risk of personal harm. The pharmacological action depends strictly on dosage: at low doses ketamine exhibits opposite (stimulant) effect; the psychedelic and dissociative effects predominate at higher doses. Metabolism is based mainly on demethylation and hydroxylation. Several ketamine designer derivatives have appeared on the black market [69]: 2-fluorodeschloroketamine, 2-fluoroketamine, methoxetamine, ethoxmethamine, methoxyketamine.

Phencyclidine was used as a drug with a similar dissociative and anesthetic action, now withdrawn from medicinal use. More than ten designer derivatives were sold for nonmedical and illicit use until the 1990s [69], including methoxyeticyclidine, hydroxyphencyclidine, hydroxyeticyclidine, and methoxydine. Several similar diarylethylamines, such as ephenidine, diphenidine, or methoxphenidine, are also (despite the fact that they were synthesized many years ago and then abandoned) sold on the black market.

4.9   ACETYLCHOLINE

Acetylcholine acts on two types of cholinergic receptors: muscarine and nicotine. Nicotine receptors are divided into two main types, neuronal and muscle receptors, and the role of neuronal receptors is still very poorly known [70,71]. It is believed that they participate in the memory process, as the antagonists of these receptors impair memory in animals; this action is reversed during nicotine administration. The action of muscarinic receptors is better known [72]. They are responsible for many physiological reactions: increasing the action of exocrine glands, lowering heart rate, inducing spasms of smooth muscles and spasms of the uterus, and diastole of blood vessels. The cholinergic system, associated with these receptors, acts as counterbalance to the adrenergic (noradrenaline) system.

Nicotine occurs naturally in *Nicotiana* plants (especially *Nicotiana tabacum* L. and *Nicotiana rustica* L.) [73], *Duboisia hopwoodii* (F. Muell.) F. Muell.

[74], and *Asclepias syriaca* L. [75]. Muscarine can be found in various mushrooms, where it is responsible for their toxicity, for example, *Amanita muscaria* (L.) Lam., *Mycena pura* (Pers.) P. Kumm, and different species from the genera *Inocybe* and *Clitocybe* [76].

Cholinergic receptors cannot create any visible emotional and psychotic effects when they are activated or blocked; moreover, their activation can lead to a toxic effect (the example of muscarine in mushrooms). This is the main reason why they are not considered a target for new designer drugs.

Some tropane alkaloids, such as atropine, hyoscyamine, hyoscine (scopolamine), and belladonnine (Table 4.4), exhibit strong anticholinergic activity. They occur mainly in plants of the *Solanaceae* family [73]: *Atropa belladonna* L., *Datura*

TABLE 4.4
Structure of Phenylethylamine Derivatives

	R1	R2	R3	R4	Ring
Methylphenidate	COOCH$_3$	Closed to piperidine ring		H	none
Amphetamine	H	CH$_3$	H		
Cathine (norephedrine, norpseudoephedrine)	OH				
Cathinone	=O				
Ephedrine (pseudoephedrine)	OH		CH$_3$		
Methylephedrine (methylpseudoephedrine)			CH$_3$		
Methcathinone	=O			H	
Metamphetamine	H				
MDMA					3,4-methylenedioxy
Mephedrone	=O				4-methyl
Phenethylamine	H	H	H		
2C-x derivatives					2,5-dimethoxy
Epinephrine	OH		CH$_3$		3,4-dihydroxy
Norepinephrine			H		
Dopamine	H				
Tyramine					3-hydroxy
Mescaline					3,4,5-trimethoxy
Benzofurans		(various)			Furan attached to benzene
Methiopropamines					Tiophene instead of benzene

genus (mainly *Datura innoxia* Mill. and *Datura stramonium* L.), *Mandragora* genus (mainly *Mandragora officinarum* L.), *Hyoscyamus* genus (mainly *Hyoscyamus niger* L.), and *Brugmansia* genus, in various proportions. Cocaine is the only tropane alkaloid that is used as the base of designer drugs; however, its mechanism of action is different (as shown in the next section).

4.10 DOPAMINE

Dopamine is one of the most complex and important neurotransmitters and its action cannot easily be summarized [77]. The lack of balance of the dopaminergic system is one of the main factors in schizophrenia (in the case of its hyperactivity) and Parkinson's disease (hypoactivity). As all monoamine systems are interconnected and interact with each other, the dopaminergic system is also believed to play an important role (together with serotonin) in depression. The role of dopamine is also crucial for addiction and memory [18,23,78].

All phenylethylamine derivatives (Table 4.4) and benzylpiperazines exhibit strong structural similarity to dopamine and this is the reason why they are described in this section. Similar mechanisms are also observed in the case of cocaine. However, their mechanism of action is complex and much wider than if they were simple dopamine analogs.

4.10.1 Ephedrine

The genus *Ephedra* (family *Ephedraceae*) contains more than 40 species, the most important of them being *Ephedra sinica* Stapf, used in Traditional Chinese Medicine (TCM) for thousands of years, and *Ephedra gerardiana* Wall. ex Stapf, used in Indian folk medicine [79]. They are herbaceous perennials growing in South Europe, Asia, North Africa, and both Americas, with a strong pine odor and astringent taste. The plants contain ephedrine, methylephedrine, and norephedrine, being isomers with an R configuration of carbon attached to the OH group and S configuration of the other asymmetric carbon. These compounds have their (S,S) analogs, also occurring naturally in *Ephedra* and called pseudoephedrine, methylpseudoephedrine, and norpseudoephedrine, respectively. Several species also contain ephedroxane, with the side chain closed to the oxazolidone ring, exhibiting anti-inflammatory activity, as the chemical structure is similar to nonsteroid anti-inflammatory drugs. The other classes of alkaloids (macrocyclic spermine family) are also present. The other constituents include flavones, flavanols, tannins, carboxylic acids, volatile terpenes, and N-methylbenzamine; similar to ephedrine, but with a shorter chain.

Ephedrine, the main active substance, acts as an agonist of the adrenergic receptors, both the α and β family. Moreover, it causes the release of norepinephrine in synapses. Receptor binding stimulates the CNS (as it penetrates the blood–brain barrier easily), also increasing heart rate, the constriction of blood vessels, and bronchodilation [79]. This is why ephedrine is used in many countries as an ingredient in nasal drops or cough syrups. There is a tendency in many countries to withdraw these pharmaceuticals from the market as a potential source of abuse. It can be abused directly or converted by amateur chemists to amphetamine, methcathinone, or other drugs. Ephedrine also reduces appetite, which is the reason for abuse through self-medication aimed at weight loss. Its action on the excretion system causes retention of urine. The consequences of overdose are related to pharmacological action: hypertension, palpitations, arrhythmias, strokes, cardiac arrests, psychosis, urine retention, insomnia, anxiety, and dry exfoliating skin. Ephedrine is metabolized to norephedrine (the opposite reaction to biosynthesis in plants).

4.10.2 CATHINONES

Khat is the name of *Catha edulis* (Vahl) Endl. (family *Celastraceae*), which is an evergreen shrub native to the Horn of Africa and Arabia [80] that has been cultivated for a long time, mainly in Ethiopia and Yemen. It is cultivated in well-drained soil, and can be harvested two to three years after planting by breaking off young branches [30]. Many authors believe that khat was used in ancient Egypt; however, the first verified written source comes from the fourteenth century [81]. The chewing of khat leaves is popular among local people during gatherings, both on happy and sad occasions, and the number of daily users of khat is believed to be about 20 million [82]. Rarely, it is also used as a material for tea-type infusions or smoked [82,83]. Historically, it was suspected in the nineteenth century that khat contained caffeine. It took until the 1930s to isolate the main active ingredient of the plant, cathine (norpseudoephedrine) [30,84]. Young leaves also contain cathinone, which is converted to cathine during growth. The other complex alkaloid groups are cathedulins and phenylpentenylamines (merucathine, merucathinone, pseudomerucathinone). Khat also contains flavonoids (dihydromyricetin, kaempferol, myricetin, quercetin, and their glycosides), sterols and triterpenes, volatiles, amino acids, and vitamins [30]. Cathinone can convert *in vivo* to isocathinones or cyclic pyrazines [84].

Cathinones are less lipophilic than amphetamines due to the presence of keto moiety, and require higher doses to achieve the same effect [84]. They can be administered orally, intravenously, or by insufflation, and are often taken in conjunction with other drugs to enhance the effect; which often results in dangerous interactions [85–87]. They are absorbed both in the buccal cavity and the small intestine. Cathine and cathinone reach their blood concentration peak about two hours after administration [84]; they are also present in breast milk [82]. Cathinone derivatives undergo extensive (stereoselective) phase I metabolism: demethylation, reduction of ketone to alcohol, hydroxylation of methyl groups to alcohol, followed by their oxidization to carboxylic acid. Similar chemical processes can occur during long-term storage [82,84,86]. Methylenedioxy derivatives are demethylated to two hydroxyl groups, which can further undergo methylation. Fluorinated derivatives act longer due to the difficulty of breaking the fluorine bond. The elimination of phase I metabolites occurs with urine in a free form, or after phase II, mainly glucuronidation.

Its action is similar to ephedrine: positive inotropic and chronotropic cardiac effects, arterial constriction, mydriasis, euphoria, hyperactivity, logorrhea, hyperthermia, anorexia, mouth dryness. It inhibits neural uptake of noradrenaline [31]. Although tolerance to amphetamine can develop easily, khat chewing does not result in easy dependence, probably due to low amounts taken during chewing [80].

All cathinone derivatives cause enhanced release of dopamine and noradrenaline, inhibiting their reuptake [86,88,89], whereas their effect on serotonin is significantly weaker [84]. It is suggested [82] that psychostimulation is mediated primarily by the meso-striato-cortico limbic dopaminergic pathway. The psychotic symptoms are similar to those of amphetamine and cocaine (including reduction of hunger); however, there is no evidence that chewing khat in Yemen has increased psychotic disorders among chewers, compared to nonchewers. Some sources (see [82]) suggest that a depressive mood is induced during the beginning of use. Although there is no clear agreement about the addiction risk of these substances, it is clearly shown that some individuals involved in regular khat chewing or cathinone derivatives taking have a compulsive need for repeating self-administration of these substances. Surveys among mephedrone users indicate strong craving in 80% of them [84]. There are conflicting reports about the existence and mechanism of the withdrawal syndrome [82]. Cathinone and its derivatives, when investigated on animals, share some effects with amphetamine (hypermotility, stereotyped movements, responding reinforced by the suppression of food), with cocaine (self-administration, conditioned taste aversion), and with both (analgesia, anorexia, conditioned place preference) [82].

4.10.3 AMPHETAMINES

Amphetamines were first synthesized in the late nineteenth century in Germany. A short time later, methamphetamine was synthesized from ephedrine. They exhibit the similarity to cathine, ephedrine, phenylethylamine, and partially to dopamine.

Amphetamines are strong central nervous system stimulants [90]. They release and inhibit reuptake of dopamine and noradrenaline as the substrates of the monoamine transporter in the dopaminergic, noradrenergic, and adrenergic neurons. They have no direct affinity to the adrenergic and dopaminergic receptors. Amphetamines improve motoric activity, induce euphoria and excitement, and reduce the feeling of hunger and the need for sleep. The action lasts several hours after administration and changes to the opposite symptoms: depression and anxiety. Taking another dose to avoid the bad symptoms is one of the main factors in the development of addiction. Additionally, the repeated administration of amphetamines induces tolerance and the euphoric symptoms decrease. This is another addiction factor, encouraging increase of the dose. A very prolonged use with an increased dosage leads to amphetamine psychosis with symptoms very similar to schizophrenia.

Amphetamines are metabolized in two main pathways: the O-demethylenation to dihydroxy derivatives (catechols), which are then methylated to methoxy derivatives or the degradation of the side chain to N-dealkyl and deaminooxo metabolites. There is evidence that the third toxic route exists and occurs only after high dosages: the production of neurotoxic trihydroxyamphetamines [91].

Many amphetamine derivatives are therapeutically used as sympathomimetics, anorectics, nonopioid analgesics, antiparkinsonians, or vasodilators, for example, amphetaminil, fenethylline, fencamine, fenfluramine, sibutramine, selegiline, and prenylamine. Methylphenidate, a derivative with the second closed ring, is used in the treatment of attention-deficit/hyperactivity disorder (ADHD).

4.10.4 DESIGNER MODIFICATIONS OF PHENYLETHYLAMINE DERIVATIVES

All phenylethylamine derivatives are modified in a similar way in designer drug synthesis. The simplest designer modifications are based on the ring substitution at the fourth position, forming bromo, chloro, methyl, fluoro, iodo, and methoxy derivatives. 2,5-Dimethoxy derivatives (named 2C-x) are synthesized as well [92]. The benzene ring can be changed to naphthalene, benzodihydrodifuran, benzofuran, or other similar heterocyclic structures. Various substituents can be attached to the nitrogen atom, optionally closing both substituents to form the pyrrolidine ring. Many designer derivatives have the 3,4-methylenedioxy moiety; the amphetamine molecule modified in this way is called MDMA ("ecstasy") [93,94], whereas the corresponding ketone is named methylone [95]. These derivatives act also on the serotonin system, inducing release and reducing reuptake. Designer cathinone derivatives are often offered as "bath salts" [96–100] and their number constantly increases [86].

4.10.5 PIPERAZINES

Piperazines are often labelled as "herbal" or "natural" preparations, but they are a solely artificial group of drugs (there are no direct analogs in nature, but they also exhibit chemical similarity to phenylethylamines) [101]. They can be divided into 1-benzyl- and 1-phenyl-piperazine derivatives [61]. They were first developed to act against parasites and further research was focused on their antidepressant action. Nowadays they are not used in official medicine, with several exceptions (for example, cyclizine used in nausea treatment). Until the 1990s they were not known on the black market and their abuse only started seriously around 2004.

The action of benzylpiperazines is very similar to amphetamine derivatives; people often cannot recognize the difference. They increase extracellular dopamine levels in the central nervous system, increasing also serotonin and noradrenaline. All these neurotransmitters increase their action by an enhanced release and inhibition of their reuptake. On the contrary, phenylpiperazines act mainly directly (postsynaptically) at serotonin directions and presynaptically by reversal of the transporter of serotonin reuptake [61]. Therefore, the symptoms of acute intoxication of both groups are different: benzylpiperazines give sympathomimetic symptoms (hypertension, risk of stroke, anxiety, delusions, mydriasis, paranoia), whereas phenylpiperazines give symptoms similar to serotonin syndrome (nausea, anxiety, and headache).

4.10.6 COCAINE

Cocaine is a natural alkaloid occurring in almost 20 different *Erythroxylum* species, mainly *Erythroxylum coca* Lam. (*Erythroxylaceae*) and *Erythroxylum novogranatense* (D.Morris) Hieron., growing in South America [102]. Coca leaves have been chewed by Indians to prevent hunger and to increase endurance since ancient times. They contain

TABLE 4.5
Structure of Tropane Alkaloids

	R1	R2
Cocaine	COOCH$_3$	OCOPh
Benzoylecgonine	COOH	OCOPh
Atropine	H	OCO(CH$_3$OH)Ph
Egconine	COOH	OH

cocaine as the main alkaloid (Table 4.5), together with several similar compounds: cinnamoylcocaine, benzoylegzonine, methylegconine, methylegconidine, pseudotropine, tropanocaine, dihydroxytropane. Several pyrrolidine alkaloids (hygrine, hygroline, cuscohygroline) and trace amounts of nicotine are also present.

Cocaine acts in a similar manner to amphetamines, with a lower tendency to induce paranoia and hallucination. It is often sniffed, or smoked as a free base (called "crack"). Overdose results in convulsion. Peripheral action causes tachycardia, vasospasms, and hypertension. A strong psychological addiction is observed. It undergoes metabolism based on hydrolysis. Physiological addiction is generally considered to be impossible; however, there is still open discussion about it. Beside pure cocaine, a 4-fluorinated derivative is often sold on the black market.

Cocaine exhibits a local anesthetic effect and its analogs without the tropane moiety are used in local anesthesia (benzocaine, procaine, tetracaine, lidocaine, etc.). Some of them are sold as illegal drugs (meprylcaine, nitracaine, dimethocaine).

4.11   STEROIDS

Designer steroids are used mainly as anabolics, with the main use in sport as performance-enhancing substances. They are all analogs of naturally occurring hormones: testosterone and dihydrotestosterone [103]. The examples are dimethazine, methylclostebol, mentabolan, methoxygonadiene, methylepithiostanol, and methylstenbolone [104]. The pharmacology of these drugs is almost identical to natural hormones: they induce "anabolism"—building the tissues from substrates (the growth of bone and increase of muscle mass).

Recently, a new group of drugs called selective androgen receptor modulators (SARMs) has entered the market [105,106]. They are designed to separate anabolic and androgenic effects, maintaining only the first. They are used to maintain desirable muscle weight in many diseases, while reducing androgenic side effects. As their use in muscle building is safer than steroids themselves, the interest of the black market in them is significant. However, no significant trend in the synthesis of modified molecules is observed.

REFERENCES

1. Kenakin, T. 2008. Overview of receptor interactions of agonists and antagonists, *Current Protocols in Pharmacology*, 42 (Suppl.): 4.1.1–4.1.24.
2. Hoyer, D., and H. W. G. M. Boddeke. 1993. Partial agonists, full agonists, antagonists: dilemmas of definition, *Trends in Pharmacological Sciences*, 14 (7): 270–275.
3. Rudnick, G., and J. Clark. 1993. From synapse to vesicle: The reuptake and storage of biogenic amine neurotransmitters, *BBA—Bioenergetics*, 1144 (3): 249–263.
4. Fuller, R. W., and D. T. Wong. 1977. Inhibition of serotonin reuptake, *Federation Proceedings*, 36 (8): 2154–2158.
5. Shahid, M., T. Tripathi, F. Sobia, S. Moin, M. Siddiqui, and R. A. Khan. 2009. Histamine, histamine receptors, and their role in immunomodulation: An updated systematic review, *Open Immunology Journal*, 2 (1): 9–41.
6. LeDoux, J. E. 1995. Emotion: Clues from the brain, *Annual Review of Psychology*, 46: 209–235.
7. Ruiz-Garcia, A., M. Bermejo, A. Moss, and V. G. Casabo. 2008. Pharmacokinetics in drug discovery, *Journal of Pharmaceutical Sciences*, 97 (2): 654–690.
8. Testa, B., P. Crivori, M. Reist, and P.-A. Carrupt. 2000. The influence of lipophilicity on the pharmacokinetic behavior of drugs: Concepts and examples, *Perspectives in Drug Discovery and Design*, 19: 179–211.
9. Arnott, J. A., and S. L. Planey. 2012. The influence of lipophilicity in drug discovery and design, *Expert Opinion on Drug Discovery*, 7 (10): 863–875.
10. Xu, C., C. Y.-T. Li, and A.-N. T. Kong. 2005. Induction of phase I, II and III drug metabolism/transport by xenobiotics, *Archives of Pharmacal Research*, 28 (3): 249–268.
11. Carroll, F. I., A. H. Lewin, S. W. Mascarella, H. H. Seltzman, and P. A. Reddy. 2012. Designer drugs: A medicinal chemistry perspective, *Annals of the New York Academy of Sciences*, 1248 (2): 18–38.
12. Bialer, P. A. 2002. Designer drugs in the general hospital, *Psychiatric Clinics of North America*, 25 (1): 231–243.
13. Staack, R. F., and H. H. Maurer. 2005. Metabolism of designer drugs of abuse, *Current Drug Metabolism*, 6 (3): 259–274.
14. Meyer, M. R., and H. H. Maurer. 2010. Metabolism of designer drugs of abuse: An updated review, *Current Drug Metabolism*, 11 (5): 468–482.
15. Giannantonio, M. Di, G. Davide, R. Giovannangelo, D. Leonetti, and M. Nacci. 1998. Reward system: A biological and psychological perspective, *Italian Journal of Psychiatry and Behavioural Sciences*, 8 (1): 36–40.
16. Wise, R. A. 1981. Action of drugs of abuse on brain reward systems, *Pharmacology Biochemistry and Behavior*, 13 (Suppl. 1): 213–223.
17. Olsen, C. M. 2011. Natural rewards, neuroplasticity, and non-drug addictions, *Neuropharmacology*, 61 (7): 1109–1122.
18. Berridge, K. C., and T. E. Robinson. 1998. What is the role of dopamine in reward: Hedonic impact, reward learning, or incentive salience?, *Brain Research Reviews*, 28 (3): 309–369.
19. Nutt, D. J., A. Lingford-Hughes, D. Erritzoe, and P. R. A. Stokes. 2015. The dopamine theory of addiction: 40 years of highs and lows, *Nature Reviews Neuroscience*, 16 (5): 305–312.
20. West, R. 2001. Theories of addiction, *Addiction*, 96 (1): 3–13.

21. Lazić, N. 1997. The general theory of systems and the addiction diseases, *Alcoholism*, 33 (1-2): 45–53.

22. Thauberger, P., L. Vaselenak, and L. Pagliaro. 1989. The need for a shift in paradigm in the theory of addiction, *Proceedings of the Western Pharmacology Society*, 32: 27–31.

23. Dayan, P. 2009. Dopamine, reinforcement learning, and addiction, *Pharmacopsychiatry*, 42 Suppl 1: 56–65.

24. Everitt, B. J., and T. W. Robbins. 2005. Neural systems of reinforcement for drug addiction: From actions to habits to compulsion, *Nature Neuroscience*, 8 (11): 1481–1489.

25. Self, D. W., and E. J. Nestler. 1995. Molecular mechanisms of drug reinforcement and addiction, *Annual Review of Neuroscience*, 18: 463–495.

26. Volkow, N. D., and R. D. Baler. 2014. Addiction science: Uncovering neurobiological complexity, *Neuropharmacology*, 76 (Part B): 235–249.

27. Koob, G. F. 2013. Negative reinforcement in drug addiction: The darkness within, *Current Opinion in Neurobiology*, 23 (4): 559–563.

28. Koob, G. F., C. L. Buck, A. Cohen, S. Edwards, P. E. Park, J. E. Schlosburg, B. Schmeichel et al. 2014. Addiction as a stress surfeit disorder, *Neuropharmacology*, 76 (Part B): 370–382.

29. Waterman, P. G. 1992. Roles for secondary metabolites in plants, *Ciba Foundation symposium*, 171: 255–269; discussion 269.

30. Getasetegn, M. 2016. Chemical composition of Catha edulis (khat): A review, *Phytochemistry Reviews*, 15 (5): 907–920.

31. Carlini, E. A. 2003. Plants and the central nervous system, *Pharmacology Biochemistry and Behavior*, 75 (3): 501–512.

32. Richardson III, W. H., C. M. Slone, and J. E. Michels. 2007. Herbal drugs of abuse: An emerging problem, *Emergency Medicine Clinics of North America*, 25 (2): 435–457.

33. Ujváry, I. 2014. Psychoactive natural products: Overview of recent developments, *Annali dell'Istituto Superiore di Sanita*, 50 (1): 12–27.

34. Ziegler, J., and P. J. Facchini. 2008. Alkaloid biosynthesis: Metabolism and trafficking, *Annual Review of Plant Biology*, 59: 735–769.

35. Kutchan, T. M. 1995. Alkaloid biosynthesis—The basis of metabolic engineering of medicinal plants, *Plant Cell*, 7 (7): 1059–1070.

36. Ashihara, H., H. Sano, and A. Crozier. 2008. Caffeine and related purine alkaloids: Biosynthesis, catabolism, function and genetic engineering, *Phytochemistry*, 69 (4): 841–856.

37. Anaya, A. L., R. Cruz-Ortega, and G. R. Waller. 2006. Metabolism and ecology of purine alkaloids, *Frontiers in Bioscience*, 11 (Suppl. 1): 2354–2370.

38. Ralevic, V., and G. Burnstock. 1998. Receptors for purines and pyrimidines, *Pharmacological Reviews*, 50 (3): 413–492.

39. Ferré, S. 2016. Mechanisms of the psychostimulant effects of caffeine: Implications for substance use disorders, *Psychopharmacology*, 233 (10): 1963–1979.

40. Stephenson, P. E. 1977. Physiologic and psychotropic effects of caffeine on man: A review, *Journal of the American Dietetic Association*, 71 (3): 240–247.

41. Patay, É. B., T. Bencsik, and N. Papp. 2016. Phytochemical overview and medicinal importance of *Coffea* species from the past until now, *Asian Pacific Journal of Tropical Medicine*, 9 (12): 1127–1135.

42. Mohanpuria, P., V. Kumar, and S. K. Yadav. 2010. Tea caffeine: Metabolism, functions, and reduction strategies, *Food Science and Biotechnology*, 19 (2): 275–287.

43. Burdock, G. A., I. G. Carabin, and C. M. Crincoli. 2009. Safety assessment of kola nut extract as a food ingredient, *Food and Chemical Toxicology*, 47 (8): 1725–1732.

44. Heck, C. I., and E. G. De Mejia. 2007. Yerba mate tea (*Ilex paraguariensis*): A comprehensive review on chemistry, health implications, and technological considerations, *Journal of Food Science*, 72 (9): R138–R151.

45. Schimpl, F. C., J. F. Da Silva, J. F. D. C. Gonçalves, and P. Mazzafera. 2013. Guarana: Revisiting a highly caffeinated plant from the Amazon, *Journal of Ethnopharmacology*, 150 (1): 14–31.

46. Ravi Subbiah, M. T. 2005. Guarana consumption: A review of health benefits and risks, *Alternative and Complementary Therapies*, 11 (4): 212–213.

47. Hendeles, L., and M. Weinberger. 1983. Theophylline: A 'state of the art' review, *Pharmacotherapy*, 3 (1): 2–44.

48. Rusconi, M., and A. Conti. 2010. Theobroma cacao L., the Food of the Gods: A scientific approach beyond myths and claims, *Pharmacological Research*, 61 (1): 5–13.

49. Foley, K. M. 1993. Opioids, *Neurologic Clinics*, 11 (3): 503–522.

50. Calixto, J. B., A. Beirith, J. Ferreira, A. R. S. Santos, V. C. Filho, and R. A. Yunes. 2000. Naturally occurring antinociceptive substances from plants, *Phytotherapy Research*, 14 (6): 401–418.

51. Suzuki, J., and S. El-Haddad. 2017. A review: Fentanyl and non-pharmaceutical fentanyls, *Drug and Alcohol Dependence*, 171: 107–116.

52. Davis, M. P. 2011. Fentanyl for breakthrough pain: A systematic review, *Expert Review of Neurotherapeutics*, 11 (8): 1197–1216.

53. Vučković, S., M. Prostran, M. Ivanović, Lj. Došen-Mićović, Z. Todorović, Z. Nešić, R. Stojanović, N. Divac, and Ž. Miković. 2009. Fentanyl analogs: Structure-activity-relationship study, *Current Medicinal Chemistry*, 16 (19): 2468–2474.

54. Clotz, M. A., and M. C. Nahata. 1991. Clinical uses of fentanyl, sufentanil, and alfentanil, *Clinical Pharmacy*, 10 (8): 581–593.

55. Pamplona, F. A., and R. N. Takahashi. 2012. Psychopharmacology of the endocannabinoids: Far beyond anandamide, *Journal of Psychopharmacology*, 26 (1): 7–22.

56. Ameri, A. 1999. The effects of cannabinoids on the brain, *Progress in Neurobiology*, 58 (4): 315–348.

57. Szulakowska, A., and H. Milnerowicz. 2007. Cannabis sativa in the light of scientific research, *Advances in Clinical and Experimental Medicine*, 16 (6): 807–815.

58. Turner, C. E., M. A. Elsohly, and E. G. Boeren. 1980. Constituents of cannabis sativa L. XVII. A review of the natural constituents, *Journal of Natural Products (Lloydia)*, 43 (2): 169–234.

59. Seely, K. A., P. L. Prather, L. P. James, and J. H. Moran. 2011. Marijuana-based drugs: Innovative therapeutics or designer drugs of abuse?, *Molecular Interventions*, 11 (1): 36–51.

60. Castaneto, M. S., D. A. Gorelick, N. A. Desrosiers, R. L. Hartman, S. Pirard, and M. A. Huestis. 2014. Synthetic cannabinoids: Epidemiology, pharmacodynamics, and clinical implications, *Drug and Alcohol Dependence*, 144: 12–41.

61. Hill, S. L., and S. H. L. Thomas. 2011. Clinical toxicology of newer recreational drugs, *Clinical Toxicology*, 49 (8): 705–719.

62. Araújo, A. M., F. Carvalho, M. L. Bastos, P. Guedes de Pinho, and M. Carvalho. 2015. The hallucinogenic world of tryptamines: An updated review, *Archives of Toxicology*, 89 (8): 1151–1173.

63. Wallwey, C., and S.-M. Li. 2011. Ergot alkaloids: Structure diversity, biosynthetic gene clusters and functional proof of biosynthetic genes, *Natural Product Reports*, 28 (3): 496–510.

64. Flieger, M., M. Wurst, and R. Shelby. 1997. Ergot alkaloids: Sources, structures and analytical methods, *Folia Microbiologica*, 42 (1): 3–30.

65. Brandt, S. D., P. V. Kavanagh, F. Westphal, S. P. Elliott, J. Wallach, T. Colestock, T. E. Burrow et al. 2017. Return of the lysergamides. Part II: Analytical and behavioural characterization of N6-allyl-6-norlysergic acid diethylamide (AL-LAD) and (2'S,4'S)-lysergic acid 2,4-dimethylazetidide (LSZ), *Drug Testing and Analysis*, 9 (1): 38–50.

66. Brandt, S. D., P. V. Kavanagh, F. Westphal, A. Stratford, S. P. Elliott, K. Hoang, J. Wallach, and A. L. Halberstadt. 2016. Return of the lysergamides. Part I: Analytical and behavioural characterization of 1-propionyl-d-lysergic acid diethylamide (1P-LSD), *Drug Testing and Analysis*, 8 (9): 891–902.
67. Moosmann, B., L. A. King, and V. Auwärter. 2015. Designer benzodiazepines: A new challenge, *World Psychiatry*, 14 (2): 248.
68. Wolff, K., and A. R. Winstock. 2006. Ketamine: From medicine to misuse, *CNS Drugs*, 20 (3): 199–218.
69. Morris, H., and J. Wallach. 2014. From PCP to MXE: A comprehensive review of the non-medical use of dissociative drugs, *Drug Testing and Analysis*, 6 (7-8): 614–632.
70. Gotti, C., M. Zoli, and F. Clementi. 2006. Brain nicotinic acetylcholine receptors: Native subtypes and their relevance, *Trends in Pharmacological Sciences*, 27 (9): 482–491.
71. Gotti, C., and F. Clementi. 2004. Neuronal nicotinic receptors: From structure to pathology, *Progress in Neurobiology*, 74 (6): 363–396.
72. Karczmar, A. G. 2009. Story of muscarinic receptors, alkaloids with muscarinic significance and of muscarinic functions and behaviors, *Annual Review of Biomedical Sciences*, 11: 1–50.
73. Shah, V. V., N. D. Shah, and P. V. Patrekar. 2013. Medicinal plants from solanaceae family, *Research Journal of Pharmacy and Technology*, 6 (2): 143–151.
74. Moyano, E., S. Fornalé, J. Palazón, R. M. Cusidó, N. Bagni, and M. T. Piñol. 2002. Alkaloid production in Duboisia hybrid hairy root cultures overexpressing the pmt gene, *Phytochemistry*, 59 (7): 697–702.
75. Marion, L. 1939. The occurrence of l-nicotine in Asclepias syriaca L., *Canadian Journal of Research*, 17 (1): 21–22.
76. Persson, H. 2007. Mushrooms, *Medicine*, 35 (12): 635–637.
77. Beaulieu, J.-M., and R. R. Gainetdinov. 2011. The physiology, signaling, and pharmacology of dopamine receptors, *Pharmacological Reviews*, 63 (1): 182–217.
78. Nieoullon, A. 2002. Dopamine and the regulation of cognition and attention, *Progress in Neurobiology*, 67 (1): 53–83.
79. Abourashed, E. A., A. T. El-Alfy, I. A. Khan, and L. Walker. 2003. Ephedra in perspective: A current review, *Phytotherapy Research*, 17 (7): 703–712.
80. Al-Hebshi, N. N., and N. Skaug. 2005. Khat (Catha edulis): An updated review, *Addiction Biology*, 10 (4): 299–307.
81. De Felice, L. J., R. A. Glennon, and S. S. Negus. 2014. Synthetic cathinones: Chemical phylogeny, physiology, and neuropharmacology, *Life Sciences*, 97 (1): 20–26.
82. Feyissa, A. M., and J. P. Kelly. 2008. A review of the neuropharmacological properties of khat, *Progress in Neuro-Psychopharmacology and Biological Psychiatry*, 32 (5): 1147–1166.
83. Graziani, M., M. S. Milella, and P. Nencini. 2008. Khat chewing from the pharmacological point of view: An update, *Substance Use and Misuse*, 43 (6): 762–783, 832, 834, 836.
84. Kelly, J. P. 2011. Cathinone derivatives: A review of their chemistry, pharmacology and toxicology, *Drug Testing and Analysis*, 3 (7-8): 439–453.
85. Schifano, F., A. Albanese, S. Fergus, J. L. Stair, P. Deluca, O. Corazza, Z. Davey et al. 2011. Mephedrone (4-methylmethcathinone; 'Meow meow'): Chemical, pharmacological and clinical issues, *Psychopharmacology*, 214 (3): 593–602.
86. Dargan, P. I., R. Sedefov, A. Gallegos, and D. M. Wood. 2011. The pharmacology and toxicology of the synthetic cathinone mephedrone (4-methylmethcathinone), *Drug Testing and Analysis*, 3 (7-8): 454–463.
87. Dybdal-Hargreaves, N. F., N. D. Holder, P. E. Ottoson, M. D. Sweeney, and T. Williams. 2013. Mephedrone: Public health risk, mechanisms of action, and behavioral effects, *European Journal of Pharmacology*, 714 (1-3): 32–40.

88. Coppola, M., and R. Mondola. 2012. Synthetic cathinones: Chemistry, pharmacology and toxicology of a new class of designer drugs of abuse marketed as "bath salts" or "plant food", *Toxicology Letters*, 211 (2): 144–149.

89. Iversen, L., M. White, and R. Treble. 2014. Designer psychostimulants: Pharmacology and differences, *Neuropharmacology*, 87: 59–65.

90. Sulzer, D., M. S. Sonders, N. W. Poulsen, and A. Galli. 2005. Mechanisms of neurotransmitter release by amphetamines: A review, *Progress in Neurobiology*, 75 (6): 406–433.

91. Kraemer, T., and H. H. Maurer. 2002. Toxicokinetics of amphetamines: Metabolism and toxicokinetic data of designer drugs, amphetamine, methamphetamine, and their N-alkyl derivatives, *Therapeutic Drug Monitoring*, 24 (2): 277–289.

92. Boer, D. De, and I. Bosman. 2004. A new trend in drugs-of-abuse: The 2C-series of phenethylamine designer drugs, *Pharmacy World and Science*, 26 (2): 110–113.

93. Climko, R. P., H. Roehrich, D. R. Sweeney, and J. Al-Razi. 1986. Ecstasy: A review of MDMA and MDA, *International Journal of Psychiatry in Medicine*, 16 (4): 359–372.

94. Rattray, M. 1991. Ecstasy: Towards an understanding of the biochemical basis of the actions of MDMA, *Essays in Biochemistry*, 26: 77–87.

95. Bossong, M. G., J. P. Van Dijk, and R. J. M. Niesink. 2005. Methylone and mCPP, two new drugs of abuse?, *Addiction Biology*, 10 (4): 321–323.

96. Baumann, M. H., J. S. Partilla, and K. R. Lehner. 2013. Psychoactive "bath salts": Not so soothing, *European Journal of Pharmacology*, 698 (1-3): 1–5.

97. German, C. L., A. E. Fleckenstein, and G. R. Hanson. 2014. Bath salts and synthetic cathinones: An emerging designer drug phenomenon, *Life Sciences*, 97 (1): 2–8.

98. Lewin, A. H., H. H. Seltzman, F. I. Carroll, S. W. Mascarella, and P. A. Reddy. 2014. Emergence and properties of spice and bath salts: A medicinal chemistry perspective, *Life Sciences*, 97 (1): 9–19.

99. Murphy, C. M., A. R. Dulaney, M. C. Beuhler, and S. Kacinko. 2013. "Bath salts" and "plant food" products: The experience of one regional US poison center, *Journal of Medical Toxicology*, 9 (1): 42–48.

100. Gunderson, E. W., M. G. Kirkpatrick, L. M. Willing, and C. P. Holstege. 2013. Substituted cathinone products: A new trend in "bath salts" and other designer stimulant drug use, *Journal of Addiction Medicine*, 7 (3): 153–162.

101. Arbo, M. D., M. L. Bastos, and H. F. Carmo. 2012. Piperazine compounds as drugs of abuse, *Drug and Alcohol Dependence*, 122 (3): 174–185.

102. Novák, M., C. A. Salemink, and I. Khan. 1984. Biological activity of the alkaloids of *Erythroxylum coca* and *Erythroxylum novogranatense*, *Journal of Ethnopharmacology*, 10 (3): 261–274.

103. Kazlauskas, R. 2010. Designer steroids, *Handbook of Experimental Pharmacology*, 195: 155–185.

104. Rahnema, C. D., L. E. Crosnoe, and E. D. Kim. 2015. Designer steroids—Over-the-counter supplements and their androgenic component: Review of an increasing problem, *Andrology*, 3 (2): 150–155.

105. Zhang, X., J. C. Lanter, and Z. Sui. 2009. Recent advances in the development of selective androgen receptor modulators, *Expert Opinion on Therapeutic Patents*, 19 (9): 1239–1258.

106. Zhang, X., and Z. Sui. 2013. Deciphering the selective androgen receptor modulators paradigm, *Expert Opinion on Drug Discovery*, 8 (2): 191–218.

5 Liquid Chromatography– Mass Spectrometry in the Analysis of Designer Drugs

Beril Anilanmert

CONTENTS

5.1 INTRODUCTION

Due to the persistent manufacture of new, unknown substances, the detection of new designer drugs is a challenge for clinical and forensic toxicologists [1]. Standard screening procedures might fail because of cross-reactions in immunoassays or the absence of recently discovered unknown substances in the library used. Nevertheless, many method-development and validation studies on metabolism, screening, and structure-profiling of these drugs and case reports concentrating on a single or a few of these compounds have been published. Since these drugs are not detected by the standard screening tests, there is an underestimation of the real picture about the use of these new drugs [2]. The legal status of these substances changes regularly as the new psychoactive drugs are introduced to the market by the clandestine laboratories, with only slight changes in the general structure of the group formula. Because of this, in some cases, world is directed to application of generic classifications and scientists are looking for the detection of the group structure from which some designer drugs are created with small modifications, when possible.

New psychoactive substances (NPSs) have been designed to mimic widespread illicit drugs, such as cannabis, cocaine, ecstasy, and LSD [3]. Black science laboratories continuously produce new chemicals to replace those that are taken under schedules via legislation, which means that the chemical structures of the drugs are constantly changing to try to stay ahead of the law. NPSs are being introduced to the illicit drug market at a surprising rate (Figure 5.1) [4]. One of the biggest misconceptions about NPSs is caused by the advertisement of these as "legal highs," which does not mean they are safe. These are chosen, utilizing the

FIGURE 5.1 Some samples of NPSs. (Reprinted from Dunn, T. N. 2016. Prison Drugs Scandal: One in 10 prisoners are high on dangerous designer drugs as deadly new epidemic sweeps jail, https://www.thesun.co.uk/news/1828789/one-in-10-prisoners-are-high-on-dangerous-designer-drugs-as-deadly-new-epidemic-sweeps-jail/, accessed 20 April 2017, Copyright: City of Edinburgh Council.)

results of some pharmaceutical research (for example publications and scientific patents), to provide desired psychoactive and/or physiological effects [5]. These are sold without the approval of the FDA or Ministry of Health in many countries and, actually, this alone is enough to make them illegal to be used in humans. But the black science laboratories are intelligent enough to deceptively label the drug preparations they produce as "not for human consumption," to stay out of the chamber of legislation. Besides the rate at which new drugs emerge, it is difficult to know the effects of these drugs and the type of effects according to the dose ranges. They are unregulated and untested. The dosage inside the package is not typically consistent with the recommended dosage printed on the label. Development of new production processes, products, and marketing opportunities by the illicit manufacturers or sellers makes the situation more difficult to cope with [6]. The suppliers adapt rapidly to control measures and rapidly introduce legal alternatives to drugs taken under control and, unfortunately, limited information can be provided about their short- and long-term effects, metabolism, and pharmacokinetics over such short periods. The molecular structures of some designer drugs are given in Figure 5.2.

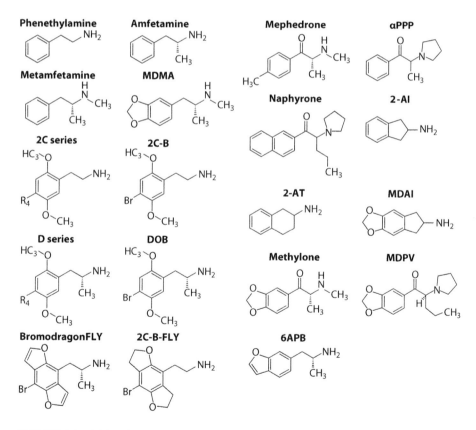

FIGURE 5.2 Molecular structures of some designer drugs. (Reprinted from Hill, S. L. and Thomas, S. H., 2011, *Clin Toxicol*, 49:705–719.)

5.1.1 COLOR TESTS

In the screening tests, 2,4-dinitrophenylhydrazine reagent, which reacts with a keto moiety, reacts with synthetic cannabimimetics, such as the naphthoylindole, phenylacetylindole, benzoylindole, and cyclopropylindole classes, either in powder form or adsorbed onto plant material, and, if the sample is positive, the color turns from yellow to orange [7]. Marquis reagent, which reacts with all nitrogen-containing drugs, shows if the material is positive for cyclohexylphenols and the JWH series. Dragendorff reagent also works with the JWH analogs; however, its LOD concentration is higher. Although it is possible to detect synthetic cannabinoids with these color reagents, more sensitive tests are necessary to detect low concentrations. ELISAs developed in-house can be calibrated at 5 ng/mL with the 5-OH and 4-OH metabolites of JWH-018 and JWH-250, respectively, and evaluated for synthetic cannabinoid screening in urine. Some commercial immunoassay kits, such as Drug-Check K2/Spice Test, DrugSmart Cassette, and RapiCard InstaTest, have been developed for easy and fast screening of these drugs in urine, without the requirement of color reagents. However, new designer drugs such as QUPIC and AB-CHMINACA cannot be detected.

5.1.2 IMMUNOASSAYS

Immunoassays are the classical screening techniques for traditional designer drugs, where positive results are confirmed using GC-MS or LC-MS/MS. There are some studies on immunoassays to also detect NPSs in recent years, but many new designer drugs are not detected using the existing standard immunoassays for routine drug screening [8]. Although immunoassays are available for more common analogs of synthetic cannabinoids, e.g., JWH-018 [5], they frequently display high cross-reactivity for many new synthetic cannabinoids. Furthermore, even if the new immunoassays are produced, the list of detectable synthetic cannabinoids always remains inadequate and outdated as the new ones are continuously introduced to the market. According to one of the studies, where the immunoassay cross-reactivities were investigated [9], 34–46% of amphetamine-type stimulants were found to show response on standard immunoassay screening tests, while the detection highly varied depending on the specific substance, concentration, and the manufacturer. Standard immunoassay tests utilized in emergency medicine generally do not allow the detection of bath salts. However, there are case reports of synthetic cathinones causing false positive results for some drugs, such as amphetamines and PCP (phencyclidine), because of structural similarity. At 5,000 ng/mL, MDPV did not cross-react in CEDIA, AxSYM, and EMIT, while mephedrone cross-reacted in CEDIA [10]. Higher concentrations are more prone to cause false positives.

Recently there have been new approaches, such as 2D molecular similarity calculations predicting which drugs and metabolites should be tested in cross-reactivity studies, designing experiments and predicting antigens from the point of cross-reactivity for a wider range of designer drugs [8]. Virtual chemical libraries used for molecular similarity analysis were reported to be built using parent drugs and metabolites identified in the package inserts and literature references and from additional published literature on pharmacokinetics of amphetamine-type stimulants and synthetic cannabinoids.

Although there is an attempt to keep the immunoassay technique alive with recent improvements and approaches, regarding all the disadvantages, it is obvious that immunoassays seem to be no longer adequate in comprehensive toxicological screening.

5.1.3 Prominent MS Techniques in Screening, Identification, and Confirmation

In addition to immunoassays and colorimetric tests, gas chromatography/mass spectrometry (GC/MS) or LC/MS/MS can be used for screening and identification of NPSs and their metabolites. They are generally used in confirmation of positive immunoassay screening results or in the detection of drugs known to be undetectable or not readily detected by immunoassays [8].

Bromo-DragonFLY is a new hallucinogenic NPS, a phenethylamine analog, with effects continuing up to three days [11]. Its psychoactive effects are similar to those of lysergic acid diethylamide (LSD). The users will often combine Bromo-DragonFLY with LSD, cannabis, 2C-B, ketamine, methylone, amphetamines, alprazolam, cocaine, and alcohol. Bromo-DragonFLY is not detectable with commercial immunoassay techniques, although GC-MS or LC-MS/MS may confirm exposure.

Mass spectrometric methods are more prone to keeping up with the emerging designer drug market [5], because new MS methods are easily developed in laboratory to target new analytes, the compounds are identified via analysis of biological matrices, sold products, human liver microsome and hepatocytes (especially in metabolite identification) using instruments such as high-resolution mass spectrometers (HRMSs). In routine forensic analysis, samples have been extracted and separation performed through gas chromatography (GC) and then analyzed using MS, for years [12]. GC/MS has been the gold standard because most psychoactive drug molecules are volatile and can be fragmented so that a very nice fingerprint is obtained. To get even better resolution, forensic scientists can add a second MS, making it GC/MS/MS. In recent years, GC-MS and/or LC-MS/MS have been the most used techniques, as is evident both from the literature and from the routine laboratory reports in the NPS analysis. Many types of synthetic cannabinoids, sometimes even their positional isomers and metabolites, can be detected by GC-MS [13,14] or LC-MS/MS [11]. It is also possible to identify specific synthetic cathinones present in blood, urine, and the other bodily fluids using techniques such as GC-MS [9]. The value of testing for cathinones is greater in monitoring for bath salt use in a drug rehabilitation program and in epidemiological research than for the acute management of patients.

The GC-MS methods can be combined with an automated SPME technique where the extraction is performed with a pen-like instrument with a fiber coating at the end, and the adsorbed analytes during the extraction are desorbed at the injection port of the instrument [15]. The SPME technique can be routinely used in combination with gas chromatography, high-performance liquid chromatography, and capillary electrophoresis, and it speeds up the method and places no restriction on MS.

Capillary electrophoresis (CE) combined with MS can also be successfully applied to the analysis of some new designer drugs. For example, a CE-MS method is described in the literature for the identification, screening, and quantification in plasma samples of four 2,5-methylenedioxy derivatives of 4-thioamphetamine (ALEPH series), structurally similar to the phenethylamines of the 2C-T series (2,5-methylenedioxy-derivatives of 4-thiophenethylamine) [16]. CE-MS is one of the strong hyphenated systems [17]. However, poor concentration sensitivity, method robustness, migration time variability, and lack of standardized operating conditions has restricted its use. But the recent advances in CE-MS interface designs and method development now provide a new perspective for this technique. For example, in recent years, the electrokinetically pumped sheath-flow ESI interface with nL/min flow rate and the sheathless interface with porous tip, manufactured for CE-MS instruments, provide a better robustness [18]. However, its use is recommended as a complementary technique to LC-MS/MS.

High-resolution mass spectrometry (HRMS) methods have started to be used more often in the determination of NPSs, but they also have limitations, such as the inability to distinguish position-isomers or to clarify the structures of new, unknown compounds [19]. However, these are overcome in many studies, using a good separation technique (like chiral columns, or a suitable GC method) and other instruments, such as NMR spectroscopy and FTIR spectroscopy, as identification and/or cutting-edge bioinformatics tools.

MALDI HRMS is also used, as reported in the literature, in the fast detection of unknown substances in biological matrices, including blood and urine. Ostermann et al. performed MALDI-HRMS and MS/MS using the MALDI LTQ Orbitrap XL mass spectrometer. The scan range was m/z 100–600. The obtained drug samples were dissolved in methanol before being combined with a 1.5-μL solid or liquid MALDI matrix. Seventy-four street drug samples were analyzed with the MALDI full-scan HRMS. Forty-nine molecules were identified and confirmed with MS/MS. Considerable adulteration was observed when the spectra for some of the samples were investigated. For example, a sample contained ecstasy (MDMA), cathinones (methylone and butylone), MDMA precursor (safrole), an alkaloid drug (scopolamine), and a contaminant (strychnine). The spectra for a sample sold as cocaine revealed an additional cocaine metabolite (benzoylecgonine), in addition to some other molecules (caffeine, phenacetine, cis-/trans-cinnamoylcocaine, levamisole, lidocaine, and procaine).

In recent years, new approaches, such as the DART ionization source coupled with the TOF-MS, have been developed for the rapid detection and identification of these emerging psychoactive substances with minimum or no sample preparation or chromatographic separation [20]. In the DART source, metastable species are generated from a heated gas after a glow discharge and these heated species ionize the analyte via desorption from the surface and detect using the TOF-MS module. Ion mobility spectrometry (IMS), which identifies analytes in ten seconds according to the size and shape of ions, may also be used in the DART systems. In this technique, ions in a group produced by the ionization source as radioactive ^{63}Ni are separated according to their collisional cross section by traveling through the drift region under the influence of an electrical field. In 2015, Gwak and Almirall used a ^{63}Ni-IMS and DART ionization source combined with a QTOF-MS technique for rapid screening and identification of 35 new designer drugs. The sample was introduced to the system via a linear rail system with the 1D transmission mode with a mesh strip. LODs were 40 to 80 pg. However, LC-MS/MS is still a common choice because of its availability and separation quantification capability.

5.1.4 MULTITECHNIQUES IN DESIGNER DRUG ANALYSIS

Multitechniques are used in some studies for the characterization for new designer drugs [21]. For example, the DALT template is one of the many potential but less-explored substitution patterns. Brandt et al. have described the analytical characterization of 17 DALTs, including DALT, 2-phenyl-, 4-acetoxy-, 4-hydroxy-, 4,5-ethylenedioxy-, 5-methyl-, 5-methoxy-, 5-methoxy-2-methyl-, 5-ethoxy-, 5-fluoro-, 5-fluoro-2-methyl-, 5-chloro-, 5-bromo-, 5,6-methylenedioxy-, 6-fluoro-, 7-methyl, and 7-ethyl-DALT, respectively. The DALTs were characterized by NMR, GC, EI/CI, MS, low- and high-mass accuracy MS/MS, diode array detection, and GC solid-state IR analysis, respectively. As the MS/MS system, a hybrid system consisting of an LTQ coupled to an Orbitrap Fourier transform mass spectrometer was used for accurate mass measurements. There are other previous analytical characterization studies in the literature but only a few on DALT derivatives, with GC-MS, IR, LC-PDA, LC-MS, GC-EI-IT-MS, LC-QqQ-MS, ^{1}H NMR and ^{13}C NMR, elemental analysis, etc.

These types of multitechnique characterization studies are important in providing a collection of spectral and chromatographic data for the research communities who face the challenge of encountering newly emerging substances where analytical data are not available. Application samples are given in Section 5.2.9.

5.2 LC-MS/MS ANALYSIS OF NPSs

New designer drugs brought significant challenges for drug monitoring applications. Emerging drug analogs generally contain minor structural modifications that exclude them from spectral libraries, and therefore these compounds essentially become moving targets [22]. Besides GC-MS, other information obtained by TOF-MS or tandem MS is required for identifying the molecular structure [7]. A drug standard/ real drug or library database is required for identification. GC-MS was accepted as a gold standard in both screening and confirmation; however, high-resolution LC-MS/ MS techniques started to take its place because of higher sensitivity, less false negatives, and the chance of reliable identification of the accurate molecular formula of more designer drugs and the targeted and more accurate MRM screening with LC-QTOF-MS [7,23]. Specific identification of the conjugates and functional groups can be predicted in the subsequent MS fragmentation modes where several metabolites at fragmentation lose a distinct group [23].

Hyphenated techniques such as LC-MS/MS and GC-MS require a clean-up step before analysis, especially with complex matrices like blood [24]. Sample pretreatment is necessary for an effective purification from undesired matrix components and therefore in order to enhance sensitivity during analysis. Many of the sample preparation methods, like LLE, SPE solid–liquid extraction, dilution, etc., are compatible with LC-MS/MS. If the nonpolar extraction solvents immiscible with water are used, the extract should be evaporated under the stream of nitrogen and the residue should be dissolved in mobile phase or an LC-MS/MS-compatible solvent like methanol, acetonitrile, etc. While working with matrices like plasma and serum, protein precipitation is required [25]. The simplest way for protein precipitation is to use acetonitrile and methanol. There are newly emerging sample preparation techniques, like DLLME and SLE, which require very small sample amounts, and are green and fast. DLLME is a very rapid, cheap, and more green extraction technique that is performed with only microliters of organic solvents [24]. DLLME can be used as an alternative method to classical LLE or SPE techniques due to its speed, very low volumes of organic solvents, cheapness, and ability to simultaneously extract a wide scale of analytes from different chemical classes, in case a good repeatability is maintained.

Analysts may bring together both sample preparation (extraction) and sample injection in one device, using SPME or other online extraction techniques with the GC-MS or LC-MS/MS instruments for the detection of some designer drugs, so that all steps, such as extraction, sample enrichment, and/or derivatization and injection to the chromatography instrument are gathered in one step and one device, which simplifies the whole procedure [15]. SPME technique uses a fused-silica fiber that is coated on the outside with an appropriate stationary phase. When it is immersed in the sample, the analytes are directly extracted to the fiber coating. There are automated SPME systems coupled also with LC-MS/MS for multiple-sample analysis [26].

The "dilute and shoot" approach is another way to reduce the time and work in LC-MS/MS. The sample to be analyzed by means of LC-MS/MS is diluted (so that the matrix effect is decreased with the codilution of matrix components) and injected directly to LC-MS/MS [27]. It is better to use a guard column to protect the analytical column. However, in this technique ion suppression is most likely to occur, especially if a sample contains high concentrations of basic components (in positive ion mode; acidic in negative ion mode) that elute in the same time window as the analyte, and have a similar mass. Optimization of chromatography might help, through shifting analyte peaks out of the maximum ion suppression time window. It may also prevent interfering peaks, thus further increasing sensitivity.

Supported liquid extraction (SLE) is a fast, simple load-wait-elute procedure, which provides cleaner extracts than other simple techniques such as "dilute and shoot" or protein precipitation. However, in the literature, generally similar molecules with close polarities and pKa values are preferred in analysis with simple extraction methods.

The simultaneous extraction with high recoveries and analysis of both polar and nonpolar analytes with varying pH's in a wide range is a challenge in sample extraction techniques. Our 2016 method, developed using LC-MS/MS, 4-step liquid–liquid extraction, had to be applied, to be able to extract eight date-rape drugs from a wide range of polarity and pK_a, and it increased the window of detection in 1 mL urine [28]. Ketamine, phenytoin, and group representatives from z-drugs, barbiturates, and GHB (γ-hydroxybutyric acid) were determined. Guard column was applied to protect the analytical column. LODs were low and the recoveries were high (only the recovery of the very polar analyte GHB was 71%). Since some drugs included ester and amide functional groups that easily hydrolyze, hydrolysis of the conjugates of these drugs was performed using a weak acid, HCOOH, at 40°C for 45 minutes; 40.0 mg NaF was used as a preservative for 1 mL urine and vortexed. The first step of extraction was performed using 6 mL ethyl acetate:dichlorometane (3:1) mixture for ten minutes, and the second step with 2 mL hexane:ethyl acetate:dichlorometane:dieth ylether (1:1:1:1) for five minutes. After adjusting pH to 11.0 with NH_3, the third and fourth steps of extraction were performed with 3 mL hexane:ethyl acetate:diethylether (1:1:1), then 3 mL hexane:ethyl acetate:dichlorometane:diethylether (1:1:1:1) for five minutes, respectively. The collected organic phase was evaporated to dryness and the residue was dissolved in 400.0 µL methanol and analyzed in LC-MS/MS. A C_{18} poro-shell column (2.7 µm C_{18}, 100 × 3.0 mm at 60°C) was used. The ESI(−) and ESI(+) modes were used simultaneously for the analyses, regarding the molecular structure of each analyte. MS parameters were −5500.0 and +5500.0 V for ESI (corona) needle voltage, −600.0 and +600.0 V for ESI shield voltage. The gas pressures were 55.0 psi for nebulizer gas pressure, 30.0 psi for drying gas, with its temperature set to 350°C. The isocratic 50:50 (A:B, v/v) mobile phase system at 0.3 mL/min flow rate was satisfactory for discriminating each analyte in 5.5 minutes and obtaining the maximum S/N ratio for most of the analytes.

5.2.1 SHORT GLANCE AT LC-MS/MS

LC-MS/MS provides the chance of identification of the analytes in a complex matrix and/or at the trace levels, structural elucidation of unknown molecules through the MS/MS fragmentation to produce product ions from the precursor ion (an analyte of interest), and qualitative and quantitative analysis of similar and nonsimilar analytes with different polarities, pK_a values, etc. [29]. Its limitation is that the analytes should be able to become charged in the ion chamber, which normally requires the molecule to be polar.

In the LC-MS/MS technique, the molecules that are separated according to their physicochemical properties in the HPLC module are passed to a tandem mass detector and analyzed in this module. The mass detector response and the retention time of the compound of interest may then be compared to a reference material or the library data. Combination of the separation and purification power of the chromatography technique and the identification capability of MS/MS module gives advantage for the identification, confirmation, and quantification of analytes with different mass spectra, even when the retention times are the same. This highly specific technique can be successful in the determination of some analytes that cannot be achieved with the other chromatographic techniques. In classical mass spectrometry, while the different ions progress in the magnetic fields, they deviate in different ratios from each other. The deviation ratio depends on the mass and charge of the ion. The light ions deviate more than the heavy ions and the ions with higher charge deviate more than the ions with the lower charge. These two factors were combined in the m/z (mass to charge ratio) variable. In quadrupole mass spectrometry, the combination of a radio frequency (RF) and direct current (DC) field is applied to 4 cylindrical or hyperbolic bar electrodes. The scheme of a quadrupole filter is given in Figure 5.3. Depending on the applied RF and DC voltages, certain ions follow a fixed path along the electrode system (resonant ions), while other ions follow the nonfixed orbits (nonresonant ions) and disappear in the system [30].

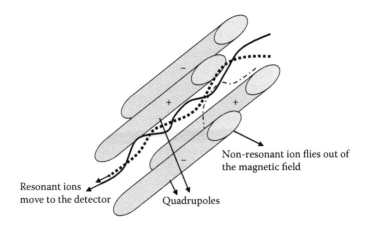

Non-resonant ion flies out of the magnetic field

Resonant ions move to the detector

Quadrupoles

FIGURE 5.3   Quadrupole filter.

Since the four electrodes facing each other are desired for the ions to travel to the nearest point to the center, the closest voltages to provide this condition are selected.

Some of the ionization methods used in LC-MS/MS are EI, CI, FAB, TSP, ESI, and APCI [31]. In most cases, as the molecular weight of the analyte is greater than the solvent impurities, solvent effects are minimum. Techniques such as ESI and APCI are used more often in LC-MS/MS. These are termed "soft ionization," since they produce primarily molecular species with little fragmentation [29]. These techniques are suitable for both large and small molecules with a wide polarity and volatility/nonvolatility range [32]. As a result of ionization, molecular ions (M^+ or M^-), protonated molecules $[M + H]^+$, simple adduct ions $[M + Na]^+$, ions formed by simple losses, e.g., $[M + H - H_2O]^+$, arise. Chromatographic modifiers may change the composition of the chemical ionization plasma in a way that other ions may be formed. For example, introducing ammonium acetate may lead to $(M + NH_4)^+$ in positive-ion mode and $(M+CH_3COO)^-$ in negative-ion mode [31]. In this technique, the signal/noise ratio increases due to the removal of unwanted ions in each MS quadrupole filter (in fact, noise decreases). This makes the technique very sensitive and specific, and allows detections and quantifications at very low concentrations.

ESI, which works with the compounds in a wide range of polarity, is the most used technique in LC-MS/MS (Figure 5.4). The mobile phase that comes out from the column is sprayed with a gas flow at a certain pressure from a metal capillary with a high-voltage charge [31]. Positive ionization involves higher voltage (3–5 kV) with more positive ions in the mobile phase; negative ionization involves a slightly lower voltage and mobile phase that includes negative ions. The analytes in the aerosols sprayed from the needle are released as the high-temperature drying gas dries the mobile phase droplets. The analyte ions freed from the mobile phase pass through the capillary to the MS.

In the APCI technique (Figure 5.5), the HPLC mobile phase is passed through a pneumatic nebulizer, typically heated to 250–400°C [32,33]. After that, this spray passes through a region where the vapor is dried [33]. The neutral species produced pass through the corona discharge unit. In this way, the analyte ionizes through the

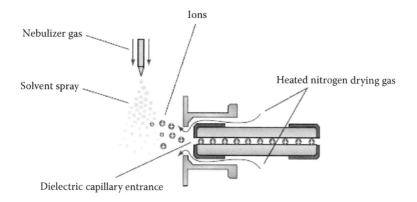

FIGURE 5.4 Electrospray ionization method. (Reprinted from *Basics of LC-MS*. 2001. Agilent Technologies, http://ccc.chem.pitt.edu/wipf/Agilent%20LC-MS%20primer.pdf, accessed April 2017.)

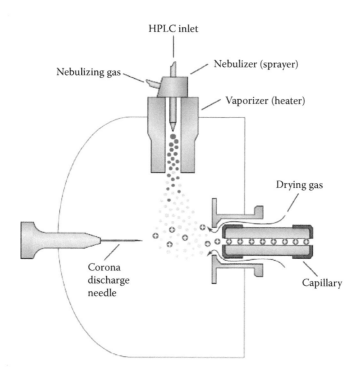

FIGURE 5.5 APCI ionization method. (Reprinted from *Basics of LC-MS*. 2001. Agilent Technologies, http://ccc.chem.pitt.edu/wipf/Agilent%20LC-MS%20primer.pdf, accessed April 2017.)

charge transfer by chemical reactions from the evaporated and ionized solvent ions acting as the reactive gas. The analyte ions are accelerated toward the less charged plate and the capillary through the hole in the middle of the plate.

ESI is more suitable for polarized molecules, which may provide their own charge, and APCI is more suitable for nonpolar species, as the collisions with the reactive gas may provide the molecule with a charge [29]. Buffers in LC-MS/MS should be volatile. The most used acids and buffer systems especially with ESI(+) mode are formic acid (usually 0.1%), acetic acid, formate, and acetate buffers. Ammonium salts are mostly preferred. Salts such as ammonium nitrate, ammonium chloride, and additives such as the 0.1% solutions in chloroform are prone to use in the APCI technique for certain analytes. Ions such as phosphate and sulfate are avoided to protect the MS.

5.2.2 TRIPLE QUADRUPOLE LC-MS/MS

In multiple-stage MSs (Figure 5.6), the sample entering the capillary tube (Q_0) is subjected to primary fragmentation; the first quadrupole filter (Q_1) acts to separate by the ratio m/z (mass/charge). The first quadrupole filter (Q_1) is used to select the precursor ion (e.g., [M+H]$^+$). The desired ion (precursor ion) passes through the filter; the others are removed. The precursor ion undergoes second fragmentation with the high-purity argon or helium gas in the collision cell (Q_2) (CID). Q_3, namely the second quadrupole filter; transmits only one secondary ion detector; the rest is filtered out. Only this secondary ion detector can achieve qualitative determination by observing the secondary ion formed of the primary ion and quantitation using the chromatogram of the secondary ion. This process is called "Multiple Reaction Monitoring" (MRM). The intensity or S/N ratio of the secondary ion formed determines the sensitivity of the MRM process. Although there are many molecules with the same mass/ion ratio (m/z), the molecules can be discriminated from each other with a specific MRM fragmentation in LC-MS/MS, lowering the noise with the two-stage filtering of the unwanted ions, except the target ions (parent and

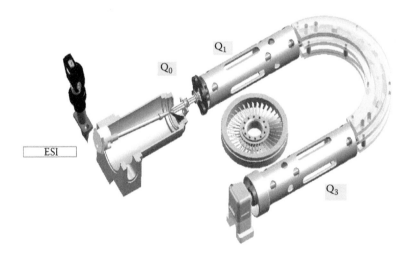

FIGURE 5.6   MS/MS section of a triple-quadrupole LC-MS/MS by Zivak (Turkiye).

daughter ion). The LC-MS/MS technique therefore makes it possible to quantify the substance at very low concentrations, as well as being a test specific to the substance being analyzed. The high specificity of the technique comes from the evaluation of the analyte according to its primary and secondary ions (molecular ion and product ion) in addition to the retention time (t_R). Scanning (scan) mode and SIM mode may be used in LC-MS/MS. In the scan mode, full fragmentation is observed. In the SIM mode, the mass analyzer monitors only a few fragments arising from the product ion. The scan mode is used merely for qualitative analyses [32]. The SIM mode is used for quantitation and monitoring of target compounds and for choosing characteristic ions during the development of the method. Almost all methods use MRM or SRM mode for sensitive determination by LC-MS-MS [7]. The fragmentation pathways of the new compounds can be found in the MS libraries or in new literature. Generally, the $[M+H]^+$ ion is selected as a precursor ion, and one should hold in mind that as the number of product ions that reflect the chemical structures of the analyzed drugs increases, the percentage shot for identifying and confirming the true compound increases. The most stable MRM ion that has the highest S/N that is used in quantification is called "quantifier." The other supportive MRM ions characteristic to the same analyte are called "qualifier 1," "qualifier 2," etc. LC-ESI-MS provides the molecular weight information due to its soft ionization [34].

5.2.3 Most Used LC-HR-MS Techniques in the Analysis of NPSs

LC-HR-MS techniques started to be frequently used in the identification, screening, determination, and quantitation of NPSs in recent years. According to the classification made by Holcapek et al. [35], most Q and IT mass analyzers belong to the low-RP category, TOF-based analyzers to the high-RP, and the ultra-high RP contains two Fourier transform (FT) mass analyzers: Orbitrap and ICR. HRMS techniques like Orbitrap, QTOF, etc. are especially preferred in the identification of new designer drugs, their metabolites, and enantiomeric analysis.

RP is an ability of a mass analyzer to resolve the peaks in mass spectra. Two peaks are regarded as resolved if the valley between them is equal to 10% of the weaker peak intensity when using magnetic or ion cyclotron resonance (ICR) instruments and 50% when using quadrupoles, ion trap, TOF, etc. [36]. The resolving power is calculated via the following equation, where Δm is the smallest mass difference for which two peaks with masses m and $m + \Delta m$ are resolved: $RP = m/\Delta m$. Mass accuracy (MA) is the accuracy of the m/z that the mass analyzer calculated. It is the difference that is observed between the theoretical (theor) and the measured/experimental (exp) m/z: MA (ppm) = $10^6 \times ((m/z)$ exp $- (m/z)$theor$)/(m/z)$theor [35,36]. It is largely linked to the stability and the resolution of the analyzer. A low-resolution instrument cannot provide high accuracy. High MA instruments can be used in the determination of elemental composition.

The resolution of spherical ITs and especially LITs is slightly higher when compared to the Q analyzer, but as the resolution increases the sensitivity decreases [35]. Low-resolution mass analyzers do not have sufficient MA for the determination of elemental formula. The Q rods with ideal hyperbolic profiles provide a higher resolution than the regular round Q rods with the MA, but a lower acquisition speed. ICR, Orbitrap, and TOF analyzers have the best RP and MA, in order. If FT analyzers are used, the image currents should be recorded in sufficient number to receive the best RP and MA values. The TOF mass analyzers have the highest scanning speed and have high selectivity even with overlapping peaks and a high matrix burden. TOF has the advantage of predicting empirical molecular formula from the accurate molecular mass and isotope peak pattern [37]. It is also possible to obtain structural information through the comparison of the accurate masses of the fragment ions and neutral losses with the postulated fragment formulas.

From the literature, it is understood that ion traps, triple quadrupoles, and TOF analyzers or hybrid analyzers are among the most preferred mass analyzer types in the analysis of the new designer drugs.

5.2.3.1 Ion Trap MS

This type of mass analyzer (IT) is composed of a circular-ring electrode with two end caps that together form a chamber [32]. The ions proceed from EI, ESI, or MALDI and are led into IT via an electrostatic lensing system [38]. They are trapped in a chamber with an electromagnetic field (IT) and further collisions occur where daughter ions or fragments emerge. There is an electrostatic ion gate which holds and injects ions into the ion trap using – and + voltage pulses. This system differs from the "beam" instruments (e.g., quadrupoles) where ions continually enter the mass analyzer. The time period that ions are led into the trap, called the "ionization period," is set such that the signal is maximized and space-charge effects, which cause a reduction in performance, are minimized (the space-charge effect is a consequence of the distortion of electrical fields caused by the high quantity of ions in the trap). The kinetic energy of the ions is reduced through colliding with 1 mTorr helium to easily direct the ion populations to the center of the trap. Here the ions are trapped and the oscillating potential (RF) is applied to the ring electrode to focus them right in the center. The ion is trapped according to the mass, charge, size, oscillating frequency, and voltage on the ring electrode. Finally, a field is applied to eject ions selectively from IT [32]. Multistage MS (MS^n) analysis can be made with IT instruments, where the trapped product ion is fragmented again and again to obtain a new fragmentation spectrum with the new product ions in each stage. A fingerprint is obtained with the multistage MS, which provides important structural information on the analyte. The enhanced product ion scan and the MS^n scan can be used in the structural elucidation of unknown drugs and identification of the unknown metabolites of drugs.

Different types of the ion traps modules exist in the market; for example, the spherical ion trap, linear ion trap, Orbitrap, etc. The working mechanism of the quadrupole IT resembles the quadrupole mass filter and the 2D- and 3D-type devices may be employed [36]. In the 3D ion traps, the electrodes are arranged in a sandwich geometry: two end-cap electrodes enclose a ring electrode (Figure 5.7). Different than the 2D instruments, the electric fields are applied in three dimensions to trap the ions. Unlike quadrupole mass filters, they detect unstable ions. For the scanning the RF voltage is applied to obtain a mass-selective instability; as the voltage increases, the ions with the higher *m/z* values are ejected through an end-cap opening. LQT instruments collect and inject pulses of ions coming from continuous sources [39]. This type of instrument collects ions using RF, which is produced as four rods arranged as in Figure 5.8. The ion beam is reflected repeatedly between the two plate lenses at

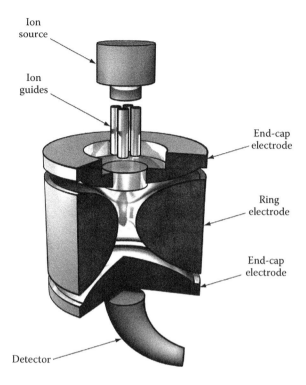

Ion source

Ion guides

End-cap electrode

Ring electrode

End-cap electrode

Detector

FIGURE 5.7 Illustration of the basic components of a 3D ion trap system. (Reprinted from Hart-Smith, G. and Blanksby, S. J. 2012. Mass analysis. In Barner-Kowollik, C., Gruendling, T., Falkenhagen, J. and Weidner, S. Eds., *Mass Spectrometry in Polymer Chemistry*. Wiley-VCH Verlag & Co., Weinheim, pp. 5–32, 20–21, 86–87.)

the ends of the quadrupole, which produce a DC trapping field [39]. Ions are trapped radially in a central section by the RF voltage, and axially by DC potentials applied to end electrodes [36]. The ions are ejected radially through slots in central rods, as an alternating current is applied [36]. As well as such radial ejection devices, axial ejection linear ion traps are also used for increasing the performance of the triple quadrupole MS instruments. Linear quadrupole ion traps have a higher storage volume and sensitivity [39]. However, their performance is closely related to the errors in mechanical arrangement; if the rods of the trap are not exactly parallel, then ions at different positions will experience different field strength, which in turn will cause different ejection times.

Orbitrap is one of the preferred recent technologies in the analysis of new designer drugs. The Orbitrap mass analyzer is composed of a spindle-like central electrode, where the ions are radially trapped, and a barrel-like outer electrode, which is two parts, one for ion excitation and the other for detection (Figure 5.9) [39,40]. A DC voltage is applied between the two axially symmetric electrodes and an electrostatic potential distribution is obtained. This field is the sum of a quadrupole field of the ion trap and a logarithmic field of the cylindrical capacitor. As a result, the pulse variation in the electrostatic field between the rods provides coherent oscillations of the

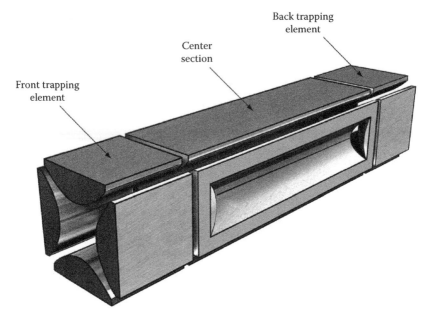

FIGURE 5.8 Illustration of the basic components of a radial ejection linear ion trap. (Reprinted from Hart-Smith, G. and Blanksby, S. J. 2012. Mass analysis. In Barner-Kowollik, C., Gruendling, T., Falkenhagen, J. and Weidner, S. Eds., *Mass Spectrometry in Polymer Chemistry*. Wiley-VCH Verlag & Co., Weinheim, pp. 5–32, 20–21, 86–87.)

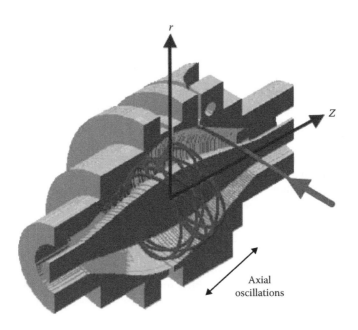

FIGURE 5.9 Section from an Orbitrap instrument. (Reprinted from Perry, R.H. et al., 2008, *Mass Spectrom Rev*, 27:661–699.)

ion trajectories between two opposite rods. Stable ion trajectories turn in an orbital motion around the central electrode and at the same time oscillate on the z-axis. The *m/z* values are derived from the axial harmonic oscillation frequencies of the trapped ions as they progress across the spindle, which is independent of the initial properties of the ions. This independence provides the high resolution and mass accuracy of the orbitrap. The electric current produced by the movement of ions between the rods is measured. The magnitude of the current is proportional to the number of ions. As the ions axially oscillate back and forth, an image current is produced and this is measured using one or both parts of the outer electrode. The image current acquired in the outer electrodes (induced by an axial motion of ions) is fast Fourier-transformed to produce a frequency spectrum. Frequencies are converted to *m/z*. Orbitrap is used to determine low-abundance ions and samples with high complexity. The image current detection technique is used in conventional ion traps, FT and Orbitrap [39]. The high resolution of Orbitrap is achieved by injecting the ions in a very short time or in bursts, so that the ions that belong to a given *m/z* ratio move together. The curved linear ion trap is one of the most efficient ways of injecting ions to the Orbitrap.

A targeted-untargeted high-resolution tandem MS method for the identification of 24 cannabinoid-like compounds, including 20 synthetic analogs and 4 natural species, is developed for the identification of illegal substances and adulterated products [22]. To do this, an LTQ Orbitrap XL instrument coupled with an HPLC system was used to identify all analytes in one run. The compounds were identified in eight minutes using the accurate mass, retention time, and fragmentation pattern (MS^2) features of the hybrid instrument. The researchers applied their method to 11 seized samples. These included four incense sticks, two cigarettes, three herbal samples, one cannabis sample, and one tablet. Seven of these were found as positive for cannabinoids. Two synthetic cannabinoids were present in a single incense stick. They reported that the method produced neither false positives nor false negatives.

5.2.3.2 TOF-MS

One of the most popular MS systems in the detection of NPSs is the TOF instrument. TOF analyzers work according to the principle of accelerating ions that are produced by an ion source, using a fixed potential along a drift region of a certain length [36], so that all the ions with different masses, but of the same charge, moving in the same direction with a (more-or-less) constant kinetic energy, will have a corresponding distribution of velocities [36,41]. The kinetic energy formula is important here; $KE = 1/2\ mv^2$, where m is the mass and v is the velocity of the ion. In other words, TOF is based on accelerating a set of ions to a detector, where all of the ions are given the same amount of energy [42]. Lower-mass ions obtain a greater velocity and travel the distance of the drift region in a shorter time, so the ions are separated according to their *m/z* (Figure 5.10) [36]. The instrument calculates the ion velocities using the length of the drift region and the time (that the ions reach the detector) measured ($v = d/t$). Then, using this velocity, the *m/z* values of the ions are determined [42]. The analyzer is called TOF because the *m/z* values are determined according to the "time of flight" of the ions. In screenings, employing the TOF-MS identification of compounds is based on the accurate mass of the protonated/deprotonated molecular

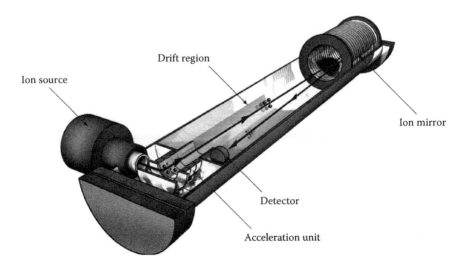

FIGURE 5.10 Diagram of the basic components of an orthogonal acceleration TOF mass analysis system featuring an ion mirror, and the means by which it achieves *m/z*-based ion separation. (Reprinted from Hart-Smith, G. and Blanksby, S. J. 2012. Mass analysis. In Barner-Kowollik, C., Gruendling, T., Falkenhagen, J. and Weidner, S. Eds., *Mass Spectrometry in Polymer Chemistry*. Wiley-VCH Verlag & Co., Weinheim, pp. 5–32, 20–21, 86–87.)

ion, isotopic pattern, and retention time [43]. The isotopic pattern is important in reducing the false-positive results, but not up to 100%. Combining a quadrupole and a TOF instrument (QTOF) provides a precursor selection, which can afterwards be fragmented in a collision cell. A high-resolution product ion spectrum is obtained as the fragments are measured in TOF.

The selection of a characteristic precursor is important and it is performed in the light of the user's predefined criteria (information-dependent acquisition, IDA). Using the optimal settings for all situations seems impossible, because there is always a risk of missing the characteristic ions in IDA mode, especially in complex samples. In data acquisition with the SWATH mode, a survey scan with low collision energy in the user-defined mass range is performed. In this stage, Q_1 is set to full transmission. Then a Q_1 range (typically 20 Da) is defined and, consecutively, the mass range is scanned, applying a series of collision energies to obtain a fragmentation ion spectra (for the product ion). An untargeted acquisition of data combined with a targeted or untargeted data analysis is possible with the SWATH mode. It is suitable for qualitative and quantitative analysis of even complex samples (e.g., whole blood, serum, etc.), without the need to reinject the sample. All types of postrun experiments (e.g., neutral loss experiments, mass defect filtering, etc.) can be applied for the research purposes. It also proved to be more suitable for the screening purposes than the IDA methods. The IDA mode has been reported to miss analytes that were easily caught by the SWATH mode.

Because of the possibility of fast analyses and sharp peaks, UHPLC is becoming the most widespread LC approach, where small particle sizes (lower than 2 µm) and ultrahigh-pressures (up to 1300 bars) are used. It requires MS instruments with higher acquisition speed that can provide enough sampling points for a reliable peak integration [35,44]. The most suitable mass analyzer for UPLC is obviously the TOF analyzer, with fast acquisition speeds of around 10–50 Hz.

Combination of IT technology with TOF provides excellent selectivity for product ions. The QqQ mass analyzer provides better linear dynamic range, higher precision, less matrix interferences, and better robustness in quantitative analysis [35]. With the high-MA in both full-scan and MS/MS modes, hybrid QTOF is very suitable for structural elucidation. Hybrid FT tandem mass analyzers provide full-scan high-RP mass spectra at scan speeds that are appropriate for UHPLC together and routine measurements with high mass accuracy (<5 ppm). A QTOF mass analyzer can implement IDA with the TOF survey scan and the product-ion scan as dependent scan and MS^n fragmentation and SWATH in which sequential precursor ions are used to collect the same spectrum of precursor and fragment ions using a collision energy range.

Identification of unknown NPSs using the LC-TOF/MS and ion trap consists of four steps [45]: First, the sample is analyzed with LC-TOF/MS in full scan, seeking large unknown peaks using a mild CID fragmentation. Second, the generated empirical formulas and any A^{+2} isotopes, such as Cl, Br, or S (if present), are searched in the Merck Index or ChemIndex. In the third step, MS^2 or MS^3 analysis is applied with the ion trap MS/MS, regarding the proposed structures. A chemical structure-drawing program will help to identify the ion fragments and their accurate masses. Finally, the LC-TOF/MS data of the fragment ions (empirical formula of fragment ions), are combined, if available and identification is made. After the identification, the identified molecule should be confirmed with a standard, if possible.

Forensic scientists often use TOF or QTOF MS to obtain higher resolutions [12]. Separation of the ions in QTOF, according to the time that they travel a defined distance in a vacuum, provides better resolution than the classical separation of the ions by the m/z ratios. While the traditional quadrupole-MS provides mass resolution up to the last digit or one decimal after comma, QTOF measures up to four decimal places; for example, 238.0999. As a whole instrument, LC-ESI-QTOF-MS is a very good choice in forensic and clinical analysis, for its high sensitivity and ability to identify the unknowns [46]. The QTOF-MS/MS offers precise mass, isotopic pattern, and MS/MS fragmentation pattern at the one analysis result. By the help of accurate mass measurements, the assessment of empirical formulas of unknown molecules is very useful in designer drug analysis. Using only MS spectra to identify unknown substances is not sufficient, as different substances could have the same empirical formula, but a different fragmentation pattern; therefore, using the MS/MS mode is obligatory. Enhancement of the mass resolving and mass determining capabilities of MS technology gives the chance of fast analysis through coupling in tandem with chromatographic separation techniques as UPLC or HPLC with monolithic column. Accurate mass molecular ion is used in identification and the software assists in accurate mass MS-MS spectral matching and structural elucidation in TOF-MS.

5.2.4 Which Matrix, for Which Aim?

Urine is the most common matrix for the NPS testing, due to the longer detection window than blood and oral fluid, a higher concentration of metabolites for several days after intake, and the chance of obtaining a larger sample volume [47]. However, blood (plasma, serum) screening may be a requirement; for example, if urine cannot be provided because of reasons such as drug-induced urinary retention or acute renal failure [48]. Also in the quantitation of NPSs, normalization to creatinine would be of much help against the risk of adulteration or intentional dilution of the samples by the users (through drinking a large amount of water or pouring water into urine). For some NPSs, e.g., many synthetic cannabinoids, the parent compound is not detectable in biological samples and, in such cases, the only way to understand the drug use is through the detection of metabolites [47]. Incubation with HLM is the most common *in vitro* model for metabolite profiling due to low cost, availability, and simplicity of use. However, this may not show *in vivo* metabolites or the relative abundance of metabolites, as in the case of 5F-AKB48 and AM-2201. Authentic urine synthetic cannabinoid metabolites can be predicted through *in vitro* studies with human hepatocytes, because they are complete functioning cells and contain comprehensive phase I and II metabolic enzymes and cofactors, proteins that bind to the drugs and the uptake and efflux transporters.

5.2.5 Metabolism Studies with LC-MS/MS and Isomer Analysis

LC-MS/MS is also preferred in the investigation of the metabolism and determination of the metabolites of new designer drugs, on most of which little information is available. ElSohly et al. [49] investigated the metabolism of JWH-018 in HLMs and identified the metabolites of JWH-018 in the urine specimen of a user. Three of its urinary metabolites, which are the 6-OH-, the N-alkyl-OH (terminal hydroxyl)-, and the N-alkyl terminal carboxy metabolites, were determined using LC-MS/MS. The 2-mL urine sample, hydrolyzed using 1.1 M sodium acetate buffer and 100 μL of β-glucuronidase through incubation at 37°C for four hours in an oven, was vortexed and alkalified with 1 mL 1.5-M carbonate buffer. Then it was extracted with 8 mL $CHCl_3$/isopropanol (9:1) and vortexed for one minute. After the evaporation of the organic phase to 1 mL, the IS was added, vortex mixed, and the mixture transferred to LC-MS vial where the solvent was evaporated to dryness. The residue was reconstituted with 150 μL MeOH and analyzed using a Synergi Hydro column (150 × 3.0 mm, 4 μ; 80A) and a gradient mobile phase program using acetonitrile/water (50:50) containing 0.02% formic acid and acetonitrile/water (80:20) containing 0.05% formic acid. A software program for metabolite identification was used. A full-scan mass spectrum was generated for each peak based on the ions trapped in the Qtrap. This allowed for the generation of a TIC for each group of metabolites sharing the same molecular ion.

Drug metabolism studies focus on understanding the biotransformation and the *in vitro* or *in vivo* production of the drug metabolites and characterization thereof [50]. *In silico* methods have started to be used recently for a fast and rough prediction to drug metabolism in order to quickly detect some potential sources of warnings. In the *in vitro* assays, monitoring of biotransformation, microsomes, hepatocytes, tissues, etc. is frequently preferred in the studies on new designer drugs, to obtain a first insight to the metabolite formation. *In vivo* assays provide more comprehensive and accurate information; however, *in vitro* assays provide faster results and are easier to perform. Once metabolites have been generated, they need to be identified and quantified. Different LC-MS strategies can be chosen for the identification. The full-scan MS analysis may help to predict the biotransformation of the analyte and ionizable metabolites, although it is rarely sufficient to define the exact position of the structural changes. Mass shifts pertaining to formation of the metabolites through a gain or loss of one or several atoms may give an idea of the formed metabolites. Table 5.1 gives mass shifts and Table 5.2 shows typical fragments used in metabolite characterization by neutral loss and product ion scan modes. The MS/MS spectra can help to predict the structure of the metabolite or, at least, to reduce the number of possibilities. Liquid chromatography nuclear magnetic resonance spectroscopy (LC-NMR) can be used as a complementary technique for the characterization of the metabolites. Metabolomics and chemometrics may also be used.

TABLE 5.1
Mass Shifts in General, Associated to the Gain or Loss of One or Several Atoms with Respect to the Parent Drug

Mass Shift	Metabolic Reaction	Example
	Phase I	
+15.9949	Hydroxylation, epoxidations, oxidations to N-oxides, S-oxides and sulfones, oxidation of aldehydes to carboxylic acids	R-CH$_2$ → R-CHOH R-HC=CH-R → R-CHOCH-R R-S-R → R-SO-R
+31.9898	Addition of two oxygen	R-S-R → R-SO$_2$-R
+13.9792	Oxidation of alcohol to carboxylic acid	R-CH$_2$-OH → R-COOH
−14.0156	Oxidative demethylation (N-, 0-, S-dealkylation)	R-O-CH$_3$ → R-OH R-NH-CH$_3$ → R-NH$_2$ R-S-CH$_3$ → R-SH
−2.0156	Formation of an unsaturated bond Oxidation of alcohols to aldehides	R-CH$_2$-OH → R-COH
−29.9741	N reduction	R-NO$_2$ → R-NH$_2$
+18.0105	Epoxide hydratation	R-CH(0)CH-R → R-CH(OH)-CH(OH)-R
	Phase II	
+176.0320	Glucuronidation	R-OH → R-O-Glu R-NH$_2$ → R-NH-GIu
	Addition of glutathione (GSH)	
+305.0681/307.0837	- Glutathione	R-CH=CH$_2$ → R-CH$_2$-CH$_2$-SG
+161.0147/163.0303	- Mercapturic acid	R-CH=CH$_2$ → R-CH$_2$-CH$_2$-S- mercapturic acid
+79.9568	Sulfation	R-OH → R-OSO$_3$H
+14.0156	Methylation	R-OH → R-OCH$_3$
+42.0105	Acetylation	R-NH$_2$ → R-NH-CO-CH$_3$
	Amino add conjugation	
+57.0214	- Glycine	R-COOH → R-CO-Gly
+71.0371	- Alanine	R-COOH → R-CO-AIa
+114.0793	- Ornithine	R-COOH → R-CO-Orn

Source: Reprinted from Saurina, J. and Sentellas, S., 2017, *J. Chromatogr. B.* 1044–5: 103–111.

TABLE 5.2

Typical Fragments Used in Metabolite Characterization by Neutral Loss (LN) and Product Ion (PI) Scan Modes

Metabolite	Mode	Scan
Glucuronide	+/−	NL 176 (-$C_6H_8O_6$)
Phenolic sulfates	+	NL 80 (-SO_3)
Aliphatic sulfates	−	PI 97 (HSO_4)
Aliphatic GSH adducts	+	NL 129 (-$C_5H_7NO_3$)
Aryl GSH adducts	+	NL 275 (-$C_{10}H_{17}N_3O_6$)
GSH adducts	−	PI 272 ($C_{10}H_{14}N_3O_6$)

Source: Reprinted from Saurina, J. and Sentellas, S., 2017, *J Chromatogr B*, 1044–5:103–111.

Since the precursor and product ions might be identical in isobaric compounds, chromatography can be a good solution, especially for the ones in the sample matrix. Isobaric compounds are also sometimes encountered in metabolite profiling. With the accurate mass LC-TOF/MS technique, including a retention time parameter and/ or more MRM transitions for absolute identification, the need for long columns in GC-MS may be eliminated. However, the method may fail for some isomeric compounds unless special conditions are used [27]. The MS^n fragmentation or a second or third product ion can help in identification, but may still fail in allowing correct quantification because of isotopic crosstalk. Monitoring more than one ion pair can help, but it is not always successful. More sophisticated hybrid instruments with high-resolution Q_3 scanning and data-dependent high-resolution precursor ion scans might reveal the main isotopic peak of the high-concentration interferent.

One of the newest LC-MS/MS studies carried out to identify the metabolites, among which isobaric compounds existed, includes a brand new synthetic cannabinoid labeled as ADB-CHMINACA (MAB-CHMINACA), which has high potency and from which many adverse events and fatalities have been reported [47]. For ADB-CHMINACA and its analogue MDMB-CHMICA, some metabolites were detected in urine specimens with no detectable parent compound. Human hepatocytes incubated for 3h with 10 µmol/L ADB-CHMINACA were analyzed with LC-HRMS/MS (Orbitrap) on a 100 × 2.1 mm, 3-µm biphenyl column and metabolite identification software. A gradient elution was performed at 30°C with 0.1% formic acid in water (A) and 0.1% formic acid in acetonitrile (B) at a 0.5-mL/min flow rate. The MS parameters were as follows: spray voltage, 4 kV; sheath gas flow rate, 40 a.u.; auxiliary gas flow rate, 5 a.u.; sweep gas flow rate, 2 a.u.; S-lens RF level, 50 a.u.; auxiliary gas heater temperature, 400°C; capillary temperature, 300°C. Each sample was injected twice, first in the full-scan MS/data-dependent MS/MS mode with a list of expected metabolites, second in the full-scan MS/all-ion fragmentation/data-dependent MS/MS mode with a list of neutral losses based on ADB-CHMINACA fragmentation pattern. Sulfate and glucuronide losses were included in the case of phase II metabolism. The authors have identified major metabolites and recommend two ADB-CHMINACA hydroxycyclohexylmethyl isomers and ADB-CHMINACA 4″-hydroxycyclohexyl metabolite to detect ADB-CHMINACA intake. Precise location of hydroxylation cannot be predicted with MS/MS analysis and required reference standards that are not available. At this point, NMR use is suggested. The same authors determined the human metabolic fate of ADB-PINACA and 5F-PINACA using human hepatocytes in another study [51]. The molecular structures of ADB-CHMINACA and MDMB-CHMICA, the ADB-FUBINACA and PINACA analogs are given in Figure 5.11. DB-PINACA and 5F-PINACA-containing human hepatocytes were incubated and analyzed in LC-HRMS. Nineteen and twelve major ADB-PINACA and 5F-ADB-PINACA metabolites, respectively, were successfully identified. For ADB-PINACA, the major metabolic reactions included pentyl hydroxylation, hydroxylation followed by oxidation (ketone formation), and glucuronidation. Oxidative defluorination followed by carboxylation was found as the main metabolism for 5F-ADB-PINACA. The authors recommend the ADB-PINACA ketopentyl and hydroxypentyl, and the ADB-PINACA 5-hydroxypentyl and pentanoic acid, as

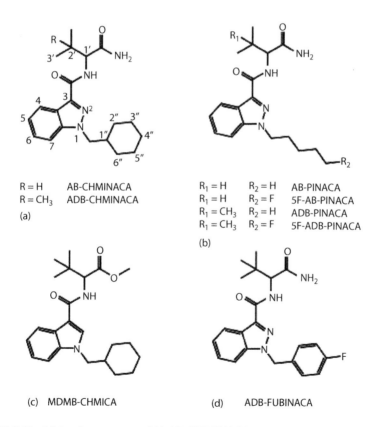

R = H   AB-CHMINACA
R = CH₃   ADB-CHMINACA
(a)

R₁ = H   R₂ = H   AB-PINACA
R₁ = H   R₂ = F   5F-AB-PINACA
R₁ = CH₃   R₂ = H   ADB-PINACA
R₁ = CH₃   R₂ = F   5F-ADB-PINACA
(b)

(c) MDMB-CHMICA

(d) ADB-FUBINACA

FIGURE 5.11   Molecular structures of (a) AB-CHMINACA, ADB-CHMINACA, (b) PINACA analogues, (c) MDMB-CHMICA and (d) ADB-FUBINACA analog. (Reprinted from Carlier, J. et al., 2017, AAPS J, 19:568–577.)

biomarkers for the ADB-PINACA and 5F-ADB-PINACA intake, respectively. Since the primary metabolites of the two compounds display positional isomers, unique product ions and optimized chromatographic conditions should be found to make a clear distinction between the ADB-PINACA and the 5F-ADB-PINACA intake. The results of such metabolism studies are also useful in guiding analytical standard manufacturers to more efficiently provide suitable references for further studies on the metabolism of new designer drugs.

Fourteen tryptamine analogues with combinations of structural isomers that need to be distinguished from each other, such as 5-MeO-DIPT and 5-methoxy-N,N-dipropyltryptamine (5-MeO-DPT); N,N-diisopropyltryptamine (DIPT) and N,N-dipropyltryptamine (DPT); and 5-methoxy-N,N-diethyltryptamine (5-MeO-DET) and 5-methoxy-N-methyl-N-isopropyltryptamine (5-MeO-MIPT), were analyzed using GC-MS and LC-MS/MS, as complementary to each other [52]. Powder samples (~1 mg) + 1 mL distilled water extracted for three minutes by means of basic LLE (using 1 mL ethyl acetate) were analyzed with GC-MS, using a DB-1ms column, within 15 minutes after trimethylsilyl derivatization. The LC separation was performed using a C_{18} analytical column (150 × 2.1 mm, 3 μm) (40°C). 10 mM ammonium formate (pH 3.5, adjusted with formic acid) and acetonitrile (80:20) were used as the mobile phases (flow rate: 0.2 mL/min). In the MS analysis, the nebulizer gas was 1.5 L/min; interface voltage was 4.5 kV; and collision energy (tandem MS) was –20 eV for EIPT, DIPT, 4-OH-DIPT, and 5-MeO-DIPT, and –15 eV for the other analytes. The structural isomers were discriminated through the EI mass spectra. Structural isomers of the tryptamines were also differentiated by means of LC-MS/MS, except for the combination of 5-MeO-DET and 5-MeO-MIPT. Different product ion spectra for each structural isomer was provided at the higher collision energy.

In addition to structural analogs, there is a new trend arising; positional isomers started to emerge in the illegal drug market, and this brings serious problems just as with determination of the substitution in ortho, meta, or para position on the phenyl ring of some synthetic cathinones [46]. The same is the case for some synthetic cannabinoids. For example, after THJ-2201 as a relatively new synthetic cannabinoid analogue had been scheduled, its positional isomer FUBIMINA (BIM-2201) was quickly introduced to the market and became popular [53]. The metabolic patterns of FUBIMINA and THJ-2201 are the same, and their main metabolites are also pairs of isomers, which complicates the discrimination of the two drugs. This brings a new challenge regarding new designer drugs. All synthetic cannabinoids investigated to date are excreted as metabolites in urine, and to catch them in the analyses is a problem, since the metabolites are initially unknown. Their pharmacokinetics are unknown as well as their pharmacology, toxicity, and safety data. Many of the NPSs contain a stereogenic center [54]. As non-enantioselective synthesis is easier and cheaper, new designer drugs are mainly manufactured as racemic mixtures.

Chiral analysis and using chiral columns in LC-MS/MS can be useful in the methods in the near future because of the rising trend of positional isomers in the illegal drug market. Polysaccharide esters and phenylcarbamate chiral stationary phases are among the most utilized columns for chiral analytical and preparative separation of enantiomers. A wide range of selectors, such as polysaccharide, brush, macrocyclic antibiotic, cyclodextrin, crown ether, protein, etc., are available for chiral separation.

Many of these phases are covalently bonded, resulting in a wide range of column and eluent combinations [55]. Enantioseparation is performed through various types of interactions between the analytes and the immobilized chiral stationary phase [54]. Basically, the enantiomers were separated due to their interaction with the chiral cavities of the polysaccharide. The other interactions which play a role in the enantioseparation are steric, dipole–dipole, π–π interactions, and hydrogen bondings. Polysaccharide-based stationary phases that can also be used in reversed-phase LC may be utilized for chiral separation of basic and acidic compounds. Using these types of stationary phases, acidic mobile phase additives in combination with basic modifiers may improve enantioselectivity, especially with basic compounds. The influence of 0.1% formic acid, which is frequently used in LC-MS/MS, may be suitable for use. The development of reliable chiral separation methods is one of the greatest challenges for analytical chemists [55]. Although SFC has gained popularity over the last decade, HPLC is still the most used technique in enantioseparation. Immobilized polysaccharide CSPs have recently been introduced to the chiral analysis market as a powerful technique. In this technique, there is less limitation on solvent choice. Especially in preparative chromatography, analytes in a wide polarity scale can be dissolved in compatible diluents. Furthermore, samples from asymmetric syntheses can often be injected directly into these columns without sample preparation. The immobilized columns are rugged. The performance can be reversed with specialized reconditioning steps that return the phase to its original state. In screening, the additional midpolarity solvents (e.g., ethyl acetate, methyl tert-butyl ether, tetrahydrofuran, acetonitrile, etc.) can be used to obtain an optimal enantioseparation.

A simple and rapid method for the enantiomer analysis of methamphetamine, amphetamine, cathinone, and methcathinone in plasma and urine using LC-MS/MS with chiral stationary phase was published in 2015 [56]. A sample of only 1 mL plasma and 100 μL urine was enough for a fast liquid–liquid extraction procedure with a >70% recovery and analysis. The extraction was conducted after mixing the sample with 1 mL water. The mixture was alkalified with 50 μL 1 M KOH and extracted with 5 mL MTBE, shaken for ten minutes, and reconstituted with 100 μL solvent (water:methanol, 1:1 (v/v)). The analyte separation was performed on a Chirobiotic V2 column (2.1 × 150 mm, 5 μm) at a temperature of 25°C for six minutes. The mobile phase was methanol and 50 mM ammonium formate/0.01% formic acid in water (95/5, v/v). The isocratic flow rate was 0.45 mL/min. The method was validated. The lower limits of detection for all of the compounds studied were at a low pg/mL level for both plasma and urine. The routine application was demonstrated for the (S)- and (R)-enantiomer differentiation.

In general, the C_{18} stationary phases are widely used for metabolism studies [57]. Other stationary phases have also been introduced (for example, the pentafluorophenyl, hydro-RP modifications) to establish more efficient interactions with polar compounds for better separations. Recently, HILIC has gained popularity for more polar compounds which retain hardly in the reversed phase columns. Two-dimensional LC (2D-LC) is a more powerful technique with an enhanced separation capability. In 2D-LC, two analytical columns of different nature (e.g., the reversed-phase and HILIC) are coupled for the determination of analytes in a broad range of physicochemical characteristics in the highly complex samples. Poorly resolved compounds eluted from the first column are injected into the second column for a better resolution. Furthermore, the optical isomers coeluting in the first column can be separated on the chiral column. The new trend in LC is the use of the capillary columns (<1 mm ID) [35] or even separations on chips. The capillary columns and chips work with the nL/min flow rates, which are suitable for nanoelectrospray ionization. The 2D-LC–ESI-IMS-MS coupling has also a great potential in the biomarker discovery, determination of analytes in complex samples and separation of isobaric compounds. Ion mobility spectrometry (IMS) enables the differentiation of ions by size, shape, charge as well as mass, in addition to the chromatographic separation of the molecules and mass spectrometric separation of ions. In principle, separation of the gas-phase ions at the atmospheric pressure is based on their different mobilities in the low or high electric fields. After the molecules in a liquid-phase are separated in UHPLC, their protonated molecules are separated using the ultrahigh-RP MS in a gas phase, according to their accurate mass-to-charge ratios, while IMS separates them according to their size-to-charge ratio.

Different types of columns are preferred by the different research groups, depending on the needs of the studies. Coreshell columns, which are a relatively new technology, have emerged in the last decade and have been found applicable to the different analyte groups, and have also started to be used for the metabolite profiling [58,59]. Coreshell columns provide sharper peaks with high efficiency and they can also be a choice for the enantioseparation, with the right chromatographic method. A sensitive and selective LC-MS/MS method has been developed to separate the ortho, meta, and para isomers of methylmethcathinone (MMC) and methylethcathinone (MEC) using a coreshell biphenyl analytical column and was applied to real samples in a study published in 2017 [60]. The biphenyl analytical column (100 mm × 2.1 mm, 2.7 μm) was used in 50°C temperature. A gradient program was applied using mobile phases of 0.1% formic acid in water/methanol (95:5, v/v) and 0.1% formic acid in methanol. The Q-trap mass spectrometer was used. The MS parameters were, as follows: curtain gas, 20 psi; ion source gas 1, 40 psi; ion source gas 2, 60 psi; ion spray voltage, 5500 V and temperature, 425°C. The whole blood was centrifuged at 1248 g for ten minutes and the serum was immediately separated from the red blood cells. The 200 μL serum sample fortified with 10 μL butylone-d$_3$ (1 μg/mL) was subjected to protein precipitation using 200 μL methanol. After vortexing and centrifugation (8 minutes, 1625 g), 50 μL of the supernatant was diluted with 150 μL water and analyzed. LLOQ was 5 ng/mL and LOD was <2 ng/mL. The recoveries were generally higher than 74%. The ion chromatogram of the successful separation of the isomers is given in Figure 5.12.

Diao et al. [53] have used LC-MS/MS for the metabolite profiling of NM-2201, in their detailed study. They investigated *in vitro* human metabolism (using HLMs) of NM-2201, to confirm marker metabolites in the authentic urine specimens. For the metabolite identification of hepatocytes, they used protein precipitation and centrifugation. The authentic urine samples were extracted using SLE cartridges and eluted twice with 3 mL ethyl acetate. Extracts were enriched before LC-MS/MS analysis. An ultra-biphenyl column (100 mm × 2.1 mm ID, 3 μm, temperature: 30°C) and gradient elution were used with 0.1% formic acid in water (A) and 0.1% formic acid in acetonitrile (B) at the mobile phase flow rate of 0.5 mL/min. An ESI(+)-triple quadrupole TOF mass spectrometer was used and the MS data were acquired in the IDA mode in combination with

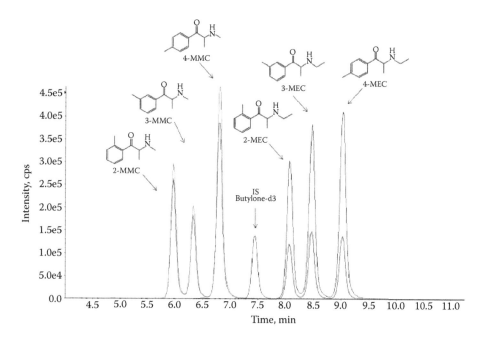

FIGURE 5.12 Ion chromatogram of the successful separation of ortho, meta, and para isomers of MMC and MEC (100 ng/mL in serum) in a coreshell-type biphenyl column. (Reprinted from Maas, A. et al., 2017, *J Chromatogr B*, 1051:118–125.)

the multiple MDFs and dynamic background subtraction. The ESI source temperature and the ion spray voltage were 650°C and 4000 V, respectively; gas 1 and gas 2 were 60 and 75 psi, respectively, and the curtain gas was 45 psi. The collision energy range was 35 ± 15 eV. NM-2201 as [M+H]$^+$ (*m/z* 376.1717) was eluted at 9.60 minutes and its product ions were *m/z* 144.0446, 171.0445, 206.1343, 232.1140, and 358.1611 (transitions were monitored for NM-2201: *m/z* 376.2 > 232.2; *m/z* 376.2 > 144.2). The parent drug was not detected in the real urine samples. The primary metabolites were 5F-PI-COOH and its glucuronide. The run time for 5F-PI-COOH was 6.06 minutes and its protonated molecular ion was *m/z* 250.1250. The characteristic product ions of 5F-PI-COOH were *m/z* 118.0662, 130.0659/132.0816, 144.0450, 174.0552, 206.1344, and 232.1139. The primary metabolites were 5F-PI-COOH and its glucuronide. The glucuronide of 5F-PI-COOH was completely converted to 5F-PI-COOH after the β-glucuronidase hydrolysis. Thus, they proposed 5F-PI-COOH as the best urinary marker for confirming the consumption of NM-2201. However, since 5F-PI-COOH is also a major metabolite of 5F-PB-22, which is an analog of NM-2201, discrimination of NM-2201 from 5F-PB-22 requires detection of the parent compound in blood or oral fluid.

In 2016, mass intoxication of 33 persons occurred because of a synthetic cannabinoid in one New York City neighborhood, in an event described as a "zombie" outbreak because of the appearance of the intoxicated people [61]. Serum, whole blood, and urine of eight patients who were brought to local hospitals were tested using LC-QTOF-MS, as well as a sample of the herbal "incense" product "AK-47 24 Karat Gold" responsible for the outbreak. The synthetic cannabinoid, methyl 2-(1-(4-fluorobenzyl)-1*H*-indazole-3-carboxamido)-3-methylbutanoate (AMB-FUBINACA, also known as MMB-FUBINACA or FUB-AMB), was identified in AK-47 24 Karat Gold at a mean (±SD) concentration of 16.0±3.9 mg/g. No AMB-FUBINACA parent compound was detected in the blood or urine of the patients, but its de-esterified acid metabolite, 2-(1-(4-fluorobenzyl)-1*H*-indazole-3-carboxamido)-3-methylbutanoic acid, was detected in each patient, with serum concentrations of between 77 and 636 ng/mL. In another study, synthetic cannabinoids were found in some urine samples collected from 5956 US athletes [62]. LC-MS/MS was used to test for the presence of JWH-018, JWH-073, and their metabolites. In 4.5% of the samples, metabolites of both compounds were detected; metabolites of JWH-018 and JWH-073 (50%), JWH-018 (49%), and only JWH-073 (1%) were detected in positive samples.

5.2.6   HAIR ANALYSIS IN THE DETERMINATION OF ACUTE AND CHRONIC CONSUMPTION

Hair is one of the most used matrices for analysis in prediction of retrospective use. This matrix can also display the chronology of the use of the drug through segment analysis, if the hair is long enough. Though its importance in the detection of one-time use or the chronic use is known, few data are available on the concentration of some new designer drugs in this matrix. In 2017, an LC-ESI(+)-MS/MS method was developed in order to measure the synthetic cathinones, 4-methylethcathinone (4-MEC) and 3,4-methylenedioxypyrovalerone (MDPV), using the hair of a patient [63]. After decontamination with dichloromethane (immersion for two minutes) and warm water (immersion for two minutes), 20 mg of hair was grinded and incubated for ten minutes in 1 mL of phosphate buffer at pH 5.0 at 95°C, in the presence of 2 ng of MDMA-d_5. The double basic LLE was performed using 4 mL hexane/ethyl acetate (v/v: 1/1) after addition of 2 mL of carbonate buffer (pH 9.7), then organic phase was evaporated to dryness and reconstituted in 80 μL of mobile phase. Samples were separated on a 1.9-μm Hypersil GOLD PFP column (100 × 2.1 mm) at 30°C using gradient elution in 12 minutes. The mobile phase is a gradient of acetonitrile and the formate buffer (2 mM formate in 0.1% formic acid). Compounds were detected by a LCQ TSQ Vantage XP triple-quadrupole mass spectrometer. Data were collected in SRM mode, with the two *m/z* transitions per analyte. The SRM transitions of *m/z* 192.1 > 146.1 and 174.2, *m/z* 276.1 > 175.0 and 205.1, and *m/z* 199.1 > 165.1 were used for 4-MEC, MDPV, and the IS, respectively. The heated capillary temperature was set at 350°C. Nitrogen was employed as sheath and auxiliary gas at a pressure of 50 and 20 arbitrary units, respectively. The method was validated and LOQ was 1.0 pg/mg for both compounds and accuracy was within the range of 85–115% of the nominal values. The method was applied to a 30-year-old man who regularly consumed cathinones for six months, administered intravenously, and who was admitted to a general. hospital for delirium and tachycardia after consumption of 10 g of a powder sold as 4-MEC and 5 g of MDPV. Both 4-MEC (30 ng/mg) and MDPV (1 ng/mg) were identified in the hair at high concentrations, showing regular consumption of these drugs. Many other compounds were also identified in hair (mephedrone, MDMA, MDA, cocaine and metabolites, tramadol, hydroxyzine, aripiprazole, haloperidol).

Hutter et al. [64] reported a validated method for the hair testing of 22 synthetic cannabinoids. JWH-007, JWH-015, JWH-018, JWH-019, JWH-020, JWH-073, JWH-081, JWH-122, JWH-200, JWH-203, JWH-210, JWH-250, JWH-251, JWH-398, AM-694, AM-2201, methanandamide, RCS-4, RCS-4 ortho isomer, RCS-8, WIN 48,098, and WIN 55,212-2 were extracted from 50 mg of hair by 3-h ultrasonification in ethanol. The extracts were analyzed on a triple-quadrupole linear ion trap mass spectrometer in sMRM. The methodology is simple and it has a limit of quantification (LOQ) of 0.5 pg mg^{-1}. The authentic hair samples from chronic consumers showed the presence of two to six synthetic cannabinoids in the same segment. In the first segment, concentrations of up to 78 pg/mg JWH-081 were present. In segmented hair, the concentrations of most substances increased from the first (proximal) to the third segment. The highest concentration was ca. 1100 pg/mg JWH-081.

5.2.7 ORAL FLUID AS AN EASILY COLLECTED AND CONFIDENT SAMPLE

Oral fluid is becoming an important matrix for the analysis of NPSs [65]. The increasing number of produced NPSs is a challenge for the applications to this matrix [66–68]. Since the collection of oral fluid is easy and noninvasive, there is an increasing interest in the use of oral fluid in drug screening [68]. Furthermore, oral fluid can be sampled in the accompaniment of the legal stuff so there is no chance for the users to dilute, substitute, or adulterate their own specimen. There is a wide-scope UHPLC(ESI+)-triple-quadrupole MS method described in the literature that contains a selection of 21 drugs of abuse, including amphetamines, cocaine, cannabis, opioids, and benzodiazepines [69]. The method was fully validated for 200 μL oral fluid/buffer mix using an oral fluid sampling kit. A fully automated SLE method has been developed and validated. An extraction time of 1.5 hours was enough for 96 samples combined with a relatively short UHPLC-MS/MS run time of 7.1 minutes. The MRM mode was used with two transitions for both analytes and an IS. The capillary voltage was 1.0 kV and the source block temperature was 150°C. Desolvation gas was 650°C and its flow rate was 1100 L/h. The cone gas (N$_2$) flow rate was 60 L/h. The method is suited for handling large amounts of the samples. The method seems to be for the detection of 7-AN, 7-AC, alprazolam, nitrazepam, THC, N-desmethyldiazepam, and diazepam, which have recoveries that are far from analytically acceptable range. For the rest of the analytes, it can be used for both detection and quantification. In studies with oral fluids, the stability of the analytes in the oral fluid kits should be known or tested in storage conditions after sampling: preferably at 20°C, 4°C, and –20°C, due to different storage times. Great losses of some compounds were observed in some kits in the previous studies, so that the choice of oral fluid kit is very important.

5.2.8 SELECTED STUDIES IN DIFFERENT MATRICES
AND WITH DIFFERENT EXTRACTION TECHNIQUES

In 2015, an analysis method was developed for the screening of different classes of NPSs (a total of 78 analytes, including cathinones, synthetic cannabinoids, phenethylamines, piperazines, ketamine and its analogs, benzofurans, and tryptamines) from blood samples [24]. The simultaneous extraction of the analytes, including 4-FA, 4-MA, 4-MEC, butylone, CB13, buphedrone, butylone/ethylone, MBDB, MBZP, MTA, MDAI, MDPV, mephedrone, methylone, AM2201, AM2233, AM 694, and RCS4, was performed by DLLME, a very rapid, cheap, and efficient technique that employs microliter amounts of organic solvents. The samples were deproteinized with 500 μL methanol. Supernatant was transferred into a tube containing 1 mL water +0.2 g of NaCl +100 μL carbonate buffer (pH 9). In order to obtain a cloudy solution, 350 μL of a mixture of chloroform/methanol 1:2.5, the extractant and the disperser solvent, were rapidly added to obtain the formation of a turbid mixture. The sample was sonicated for two minutes and then it was centrifuged. The extractant phase that sedimented at the bottom of the tube phase (about 50 ± 5 μL) was transferred into a vial, evaporated, and reconstituted in 20 μL methanol. The analyses were performed in the MRM mode using a superficially porous Kinetex C_{18} column (2.6 μm, 100 × 2.1 mm) at 40°C. Because of varying structures in the analyte scale, two different chromatographic methods were developed. Gradient elution was used with the mobile phases A: 5 mM ammonium formate containing 0.1% formic acid and B: methanol with 0.1% of formic acid. Capillary voltage of the MS was set to 4000 V, the ion source was heated up to 350°C and nitrogen was used as a nebulizing and collision gas at 12 L/min and 40 psi, respectively; the EM voltage was set to +1000 V and the nozzle voltage at 2000 V. The LOD values ranged from 0.2 to 2.0 ng/mL. The method was then applied to 60 authentic real samples from forensic cases, demonstrating its suitability for the screening of a wide number of NPSs. This is one of the prominent methods in the literature, with a wide screening capability and sufficient sensitivity. Although most of the recoveries were found lower than an analytically acceptable range, the LOD values were adequate for the detection, which allows it to be used as a screening method with LC-MS/MS in two chromatographic runs. However, maximum care should be taken to obtain good repeatability with such small amounts of extractants.

Some new studies in wastewater are emerging to analyze the extent of the NPS problem. The analysis of wastewater to estimate illicit drug consumption based on biomarkers has traditionally focused on the most common illicit drugs, i.e., cocaine, cannabis, amphetamine, methamphetamine, and 3,4-methylenedioxymethamphetamine (ecstasy, MDMA), leaving a large information gap on other illicit drugs and NPSs [70,71]. LC-MS/MS is the technique of choice for the quantitative determination of illicit drugs in wastewater, due to the low concentrations involved and the high sensitivity of the instruments [71]. In a recent study, a sensitive LC-MS/MS method was developed for the quantitative determination of a number of NPSs of the synthetic cathinone and the phenethylamine families: butylone, ethylone, methylone, naphyrone, methedrone, methylenedioxypyrovalerone (MDPV), mephedrone, 25-I-NBOMe, 25-CNBOMe, and 25-B-NBOMe in wastewater, which are frequently encountered in recent reports. Different influent wastewater samples were utilized in the development and validation of the present method, originating from Zurich, Copenhagen, and Castellon. The 24-h composite samples were taken from Zurich, Copenhagen, Oslo, Castellon, Milan, Brussels, Utrecht, and Bristol. Since it has been shown that the metabolic patterns of the selected phenethylamine-based designer drugs show rather slow metabolism rates with parent drugs being the primary biomarkers of consumption, the authors focused on the parent drugs. SPE was performed at pH 2 using 100 mL samples, then the eluates were enriched to 1 mL through evaporation and reconstitution. UHPLC-ESI(+)-MS/MS with the C_{18} column, 1.7 μm, 50 mm ×

2.1 mm (i.d.) was used along with a gradient elution. The MS/MS parameters were optimized, as 3.0 kV for capillary voltage, 150°C for source temperature, and 650°C for desolvation temperature (precursor ion was: [M+H]⁺). Nonspecific transitions, such as the loss of water, were avoided as much as possible (except for methylone, mephedrone, and N-ethylcathinone, where they were used as confirmation transitions), in order to reduce the likelihood of the false positives. For the three 25-X-NBOMe compounds, the same product ions were seen (m/z 121 and 91), corresponding to the cleavage of the methoxybenzoyl moiety (m/z 121) and the further loss of the methoxy group producing the tropylium ion (m/z 91). These transitions were not very specific and could come from any compound with a methoxybenzoyl moiety; however, no other transitions were available for these compounds. The LOD and LOQ values were between 50 and 200 ng/mL, and 1 to 5 μg/mL. The stability results for the selected drugs showed that no significant degradation occurred in non-acidified samples after refrigeration at 4°C or freezing at –20°C within the seven days. When samples were acidified to pH 2, all investigated compounds were more stable and could even be maintained at room temperature for up to seven days. Only three types of drugs could be detected. When the real samples from all over Europe were investigated, the results were found to be comparable with EMCDDA seizure data; for example, mephedrone (20%), methylone (7%) and MDPV (9%). Bade et al. [70] could detect three types of designer drugs in the wastewater of eight European cities. An estimation of NPS consumption within the population is claimed to be achieved with such studies [71]. However, these kinds of studies should be performed very carefully. The sampling regions, sampling sites, and sampling periods are important, because in the certain cities or regions where the drug use ratio is high, some of the frequently used designer drugs may be found; however, in the regions where the ratio is low, some designer drugs probably remain nondetected. When the concentrations of the drugs are calculated theoretically, regarding the total daily flow of the produced wastewater, the percentage of drug users and the solid parts of the wastewater that may absorb some of the analytes, the results may give low pg/mL or even lower concentrations amounts that require very sensitive techniques and very strong enrichment methods. Stability and chemical interaction of some drugs in other chemicals in wastewater are also another issue to be considered.

Dried blood spot (DBS) analysis provides chance for easy sample collection and detection, which may also be a very helpful technique for the blood samples found in crime scenes or in traffic controls. A validated QTRAP LC/MS-MS method was reported in 2014, upon the dried blood spots (DBS) for 64 NPSs, including some amphetamine derivatives, 2C drugs, aminoindanes, tryptamines, desoxypipradol, ephedrine, pseudoephedrine, ketamine, norephedrine, norpseudoephedrine, and PCP [72]. Ten venous blood samples were used for application on the blood spot cards and dried for at least three hours at room temperature. A DBS of 1-cm diameter was punched out and placed in an Eppendorf tube. The extraction of the analytes was performed by adding 500 μL methanol and 10 μL of the IS solution (10 ng/mL) and vortexed for 15 minutes. The methanolic solution was acidified and dried, and reconstituted with 100 μL of water/formic acid (99.9/0.1; v/v) solution and eventually analyzed. Gradient elution was performed with water and acetonitrile, both containing 0.1% formic acid, using a Synergi Polar-RP column (100 × 2.0 mm, 2.5 μm) at 50°C, the ESI(+) mode at the ion spray voltage of 5000 V, and the ion source temperature of 700°C.

Curtain gas, collision gas, gas 1 and gas 2 were 30, 6, 40, and 60 psi, respectively. For the analytes with the isobaric precursor and fragments, three transitions were monitored. The LODs and LLOQs were within the 1–10 and 2.5–10 ng/mL range, respectively. All isobaric compounds could be separated except flephedrone and 3-FMC, and MDDMA and MDEA. Another method using LC-MS/MS was developed in 2015, for an unambiguous identification and quantification of 56 NPSs in the blood and urine samples of clinical and forensic cases, e.g., intoxications or driving under the influence of drugs [73]. The method was validated to quantify 56 NPSs in blood and urine, including amphetamine derivatives, 2C compounds, aminoindanes, cathinones, piperazines, tryptamines, dissociatives, and others. Again, a Synergi Polar-RP column and a QTrap mass spectrometer were used in the analyses, with 20-minute run times.

The study from Scheidweiler et al. demonstrated the utility of LC-QTOF in the simultaneous identification of 47 common synthetic cannabinoid metabolites (derived from compounds from 21 synthetic cannabinoid families) in urine [5]. SWATH-MS was used, which is a nontargeted data acquisition method that has been utilized in the other systemic toxicological studies. In that study, ultra-high performance supercritical fluid chromatography (UHPSFC) demonstrated its potential in providing an improved separation of synthetic cannabinoids, particularly positional isomers and diastereomers.

For faster and repeatable analysis results, automated sample preparation systems can be used, making an analysis process much easier, decreasing the need for analytical skills, and saving time. An automated robotic SPE and UHPLC with TOF-MS detection was used to develop a screening method for 256 illicit compounds in blood, and 95 of these compounds were validated [74]. The LOD ranged from 0.001 to 0.1 mg kg^{-1}. The method was applied to 1335 forensic traffic cases and 74% of them were positive. Commonly abused drugs such as amphetamine, cocaine, and frequent types of benzodiazepines were the major findings. Nineteen less-frequently encountered drugs were detected, i.e., buprenorphine, butylone, cathine, fentanyl, lysergic acid diethylamide, m-chlorophenylpiperazine, MDPV, mephedrone, 4′-methylamphetamine, p-fluoroamphetamine, and p-methoxy-N-methylamphetamine.

5.2.9   Sample Studies with Multitechniques in the Screening, Identification, and Confirmation of NPSs

Tandem MS often combines sequential quadrupole devices (triple-quadrupole; QQQ) or quadrupole with TOF (QTOF) [75]. In some cases, the analysis is targeted, looking for the detection of a particular analyte in a given sample. To do this, the analytical hardware and the acquisition software should be optimized to avoid false positives. Triple-quadrupole LC/MS systems with triggered MRM (tMRM) acquisition software are suitable for this goal, as this system produces quantitative data and a searchable library spectrum in order to avoid the false positives. Alternatively, the purpose may be the identification of a variety of compounds, including NPSs, in a complex mixture, which is the case in the forensic routine laboratories. For this, the appropriate hardware must be combined with a comprehensive library. Here, one might choose LC-Q-TOF-MS for an accurate mass analysis feature combined with an ability to search the library data to match with the new compounds without reinjection. The manufacturers provide libraries but the users are recommended to develop their own libraries with their experimental data as well. Of course, LC-Q-TOF-MS is nowadays one of the most used LC-MS/MS techniques; however, in order to catch and identify the unknown new designer drugs or the isomeric NPSs/metabolites with high certainty, supportive identifying techniques have been required in recent years.

Synthetic cannabinoids and cathinones cause a challenge for MS analysis, as many isomers have identical masses and cannot be distinguished by MS or MS/MS. One example is the analysis of flephedrone (4-fluoromethcathinone, 4-FMC) and 3-FMC, which are, respectively, para- and ortho-substituted isomers of a cathinone that may be found in the bath salt-type powders. In such cases, a FTIR spectrometry can be coupled with GC for a high-resolution solid-phase transmission spectrum for each component of a sample. A spectral program can be obtained for coupling these two instruments. The IR spectroscopy can distinguish among the ortho-, meta-, and para-substituted isomers; even the diastereomers can be distinguished by infrared spectroscopy.

The impurity profiling in forensic analysis provides the chance to predict the synthetic route of the target drug and the source that it comes from. There are impurity profiling studies in the literature, such as the one that Ko et al. carried out in 2007 [76]. They describe a method of classifying the seized methamphetamine samples according to the types of impurities analyzed by GC/MS. A very good example of impurity profiling with the LC/MS-IT-TOF method was published in 2016 [77]. The impurities of the identified drug 2C-E and its synthetic route were deduced with this method. The fragment pathway of 2C-E was identified; some impurities were separated and isolated by the preparative HPLC and then identified by MS and NMR. Details of this study are given in Chapter 15, "2C Derivatives of Phenylethylamines and Their Analysis."

Recently, the use of NMR gained importance in elucidating the structure of new compounds that are not on the screening lists of laboratories. For example, in 2014, new designer drugs were analyzed using UPLC-ESI-MS, GC-MS, and NMR spectroscopy [78]. A total of 33 designer drugs, including FUB-PB-22, 5-fluoro-MN-18, THJ-2201, XLR-12, AB-CHMINACA, DL-4662, a-PHP, 4-methoxy-a-POP, 4-methoxy-a-PHPP, 4-fluoro-a-PHPP, and acetylfentanyl, were detected in 104 illegal products, in 60 different combination patterns. Portions of 10 mg, 2 mg, and 2 μL of an herbal product, powder, and a liquid product, respectively, were extracted with 1 mL methanol. After centrifugation of each extract, the supernatant solution was passed through a centrifugal filter and analyzed. The 0.1% solution of formic acid in water and 0.1% formic acid in acetonitrile were used as mobile phases. Three different elution programs were used for the analysis. The accurate mass numbers of the target compounds were measured by LC-ESI-QTOF-MS. Two programs used for the GC-MS analysis and the obtained GC mass spectra were compared with those of an EI-MS library of the instrument as well as with an in-house library of designer drugs obtained by the authors. The LC-UV-PDA spectra were also obtained. Structural assignments were made based on interpretation of the ^{1}H NMR and ^{13}C NMR spectra, HMBC, HMQC, ^{15}N HMBC, and double quantum-filtered correlation spectroscopy (DQF-COSY) spectra. The number of the compounds detected in each product varied between one and seven. In addition, some products contained three different types of the drugs, such as synthetic cannabinoids, cathinone derivatives, and phenethylamine derivatives.

Another study regarding the use of NMR to catch new designer drugs, which are out of the routine screening scale, was published in 2016 [79]. Besides the reports of EMCDDA, numerous NPSs have been identified in Japan and assessed for their risks, and subsequently regulated as scheduled substances. Using HPLC-UV during an investigation carried out in 2014 in the case of an herbal drug product commercially available in the Tokyo metropolitan area, the authors detected a small unknown peak along with an intense peak of FUB-144. They identified the compound present in this small peak. Toward this aim, the unknown compound was isolated using a silica gel column, which was then assessed with use of LC-QTOF/MS and GC-MS to have a molecular weight of 241 Da. The accurate mass measurement suggested an elementary composition of $C_{16}H_{19}NO$. Using these mass data together with those obtained by the NMR analysis, the compound was finally identified as (1H-indol-3-yl)(2,2,3,3-tetramethylcyclopropyl) methanone (despentyl-UR-144; DP-UR-144), an analog of FUB-144. In addition, this compound was revealed to have affinities for the cannabinoid receptors CB1 and CB2, with EC50s of 2.36×10^{-6} and 2.79×10^{-8} M, respectively. Their results suggest that components present in small amounts can contribute to the effects of a major component in their mother product, if they have sufficient pharmacological activities. This result suggests that even small amounts of characterized and well-evaluated components can make it easier to catch new marketed designer drugs in the samples.

The herbal product obtained from the market was extracted with acetonitrile under ball-milling for three minutes and then the supernatant was filtered. The PDA-sliced UV spectrum of the unknown peak showed the maximal absorption at 215.8, 240.7, and 296.4 nm, which seemed similar to the spectrum of FUB-144. The mass spectrum of the unknown peak showed the base peak at m/z 242.0 [M+H]$^+$ in the positive scan mode. The UPLC HTTS T3 column (50 mm × 2.1 mm i.d., particle size 1.8 μm) was used at 40°C. Composition of mobile phase, gradient elution, and PDA conditions were identical to the conditions applied in their previous study [80]. In the ESI(+)-MS analysis, the temperatures of the desolvation gas and ion source were 400°C and 150°C, respectively, and the cone voltage was 20 V. The total ion chromatogram by GC-MS showed the unknown peak and the peak of FUB-144 at the electron ionization mode. In GC-MS analysis, HP-5MS (30 m × 0.25 mm i.d., 0.25-μm film thickness) GC column was used. The MS scan range was m/z 20–600 and the split ratio was 1:4. The molecular-related ions of the above-mentioned compounds were at m/z 241 and 349, respectively, and a major fragment ion of the unknown peak appeared at m/z 144. High-resolution mass spectrum of the isolated target compound was obtained by LC-QTOF-MS. The accurate mass spectrum gave an ion signal at m/z 242.1547 in the positive scan mode, suggesting that the protonated molecular formula of the target compound was $C_{16}H_{20}NO$ (calcd. 242.1545). Additionally, the accurate mass spectrum gave an ion signal at m/z 144.0450, suggesting that the molecular formula of the fragment ion was C_9H_6NO (calcd. 144.0449). The NMR assignments were made by ^1H and ^{13}C NMR, with DEPT, HMBC, HMQC, and the COSY spectra. Finally, they identified (1H-indol-3-yl)(2,2,3,3-tetramethylcyclopropyl) methanone (named as DP-UR-144) as a new analogue of UR-144, where the n-pentyl chain is absent.

In a paper dated 2017, 20 powdered samples purchased via the Internet, which police officials submitted, were also analyzed with the multitechniques [81]. Identifications were based on LC-QTOF-MS, GC-MS, and NMR spectroscopy. The identified compounds consisted of 14 synthetic cathinones and 6 tryptamines, and the authors reported the analytical properties of the four synthetic cathinone derivatives; (1) iso-4-BMC or iso-brephedrone, (2) β-TH-naphyrone, (3) mexedrone, and (4) 4-MDMC.

5.2.10 REMARKS ON VALIDATION WITH LC-MS/MS

As in each quantitative technique, standard validation protocols from the FDA (Food and Drug Administration), ICH (International Conference on Harmonization), AOAC (Association of Official Analytical Chemists; now AOAC International), and EURACHEM can be used. However, it may not be satisfactory to adhere to one validation standard because different LOQ and LOD calculation methods may better fit to different methods. The most reliable and strict technique in LOQ calculation is the Eurachem method. However, the analyst should be able to choose the calculation that is the most suitable for the developed analytical method. Linear range should be as comprehensive as possible, to allow samples containing a wide range of the analyte concentrations to be analyzed without further manipulation (dilution), and the method's precision and accuracy should also be high [31].

In methods with internal or external standards, the calibration graphs are constructed using standard solutions of the analytes in pure solvents; however, it must be considered that if the signal intensities obtained reflect the real condition of the analytes in the matrix (interaction of the analyte and internal standard with the matrix) or the matrix effect on the performance of the mass spectrometer should be investigated [82]. A way to minimize these effects is to prepare the calibration solutions in a matrix in which the analytes are found (matrix-matched calibration). Matrix effects are minimized in the matrix-matched calibrations. Besides minimizing the matrix effects, matrix-matched calibration also has a limitation that the composition of the matrix may vary widely and thus matrix effects may differ from that of the matrix used in the preparation of the calibration solutions [83]. This may also cause variations in ion suppression or ion enhancement between different matrices and should be tested. Matrix effects may affect accuracy, precision, and robustness of the analytical measurements in LC-MS/MS. Lack of selectivity may occur due to ion suppression or enhancement caused by the sample matrix and interferences from the metabolites [83]. Because of this, the matrix effect should be investigated as one of the validation parameters.

Moreover, system suitability tests should be applied during the study, in addition to the classical validation steps. Stability should also be investigated in the study conditions, in order to be sure that the studied analytes do not degrade in the used matrix, if no previous study has been performed for the same analytes in the same matrix, in the study conditions. If the samples are stored in a certain temperature (refrigerated or frozen, etc.) for a period before analysis, stability studies gain importance.

In the multianalyte studies, where the recoveries change in a certain range, the recovery of some analytes may be far from the recovery of internal standard and this case may cause a miscalculation if area of analyte peak/area of internal standard peak ratio is used. In such cases, areas of the analyte peaks are recommended to be used only for the recovery calculations, to avoid misleading results in this parameter, and the ratio to be used for the rest of the parameters.

5.3 CONCLUSIONS AND FUTURE PROSPECTS

In recent years, the identification and separation of new designer drugs has gained enormous importance worldwide. Today, old conventional drug screens do not detect newer compounds. For the constantly changing and growing illicit drug market, analytical determination methods, especially with the different LC-MS/MS techniques, are becoming rapidly more common in the literature. However, constant modifications of chemical structures by black science laboratories keeps the NPS manufacturers one step ahead of the legal process, causing a "cat-and-mouse" situation [84]. A considerable challenge of these NPSs is that, due to their novelty, the available analytical methods for their identification and separation are unsatisfactory or limited in number [54]. Quantitative analysis is generally difficult or even impossible since there are limited reference standards, or the prices of these standards are high. Furthermore, there is a lack of knowledge on their pharmacology and toxicology data as well as the long-term damage they cause. Since NPSs have been introduced to the illegal drug market, a great forensic and medical interest has arisen in developing chromatographic methods for their analysis.

High-resolution and high-accuracy LC/HRMS techniques, such as TOF-MS, offer the advantage of accurate molecular formula determination of the unknown compounds. TOF-MS also can search the TOF data sets again for information on unexpected and unknown observed species [85]. These advantages, including selectivity and sensitivity, are among the reasons why high-resolution LC-MS/MS techniques have started to take over the role of GC-MS in recent years.

Recent advances in software allow combinatorial chemistry to be used along with the high sample capacity of the automated LC/MS systems [86]. There are systems that use an automated SLE, SPE, and SPME. In particular, the automated systems where reactions are carried out in well plates with 96 samples (or greater) capacity provide fast analysis in routine applications. Plates are placed on the autosampler; the method is specified; the instrument analyzes the samples in each well in the plate; and the data system prints a report showing whether the compounds detected are of the expected molecular weight. There is no mass range limitation for the samples analyzed by the LC but there are limitations for an MS analyzer. Finally, LC can employ inorganic buffers and, for MS, the volatile buffers are preferred. Recent developments in the APCI sources have expanded the molecular weight and sample polarity and have overcome the flow-rate limitation of the older LC/MS techniques. In many cases, analysts can use HPLC methods without any modification. LC-MS/MS is superior to GC-MS from the point of view that it can determine all volatile and nonvolatile compounds, while GC-MS allows determination of only the volatile, semivolatile, and derivatized compounds that can be volatilized. No derivatization is used in LC-MS/MS. Since it has lower detection and quantification limits, in most cases small amounts of a sample are enough also in this technique. LC-MS/MS seems to offer all the advantages of GC-MS without the disadvantages of a compromised detection sensitivity and long chromatography run times [27]. However, without proper attention to the sample cleanup in LC-MS/MS, its sensitivity promise cannot be realized fully; and the same is true for its specificity.

Besides all the advantages of LC-MS/MS, the results of the analysis of new designer drugs would be more confident if confirmed with both GC/MS and LC-MS/MS [87]. If the sample preparations are automated, the repeatability and reproducibility will get higher. For example, when testing hair, the most important factor is the sensitivity of the technique and the kind of molecules attempting to be identified in a single run. So LC-MS/MS is the most adapted technique due to its high sensitivity. In the future, high resolution will become a standard practice in catching the emergence of new designer drugs, especially in the analysis of seized products. It will also be necessary to work on the compound libraries in order to identify these new drugs. In such cases, high-resolution MS (or TOF-MS) and NMR spectroscopy are especially useful to clarify the detailed chemical structure of such compounds [7].

If confirmation of the results or structural elucidation with the other methods is necessary, a good background knowledge as well as a multidisciplinary approach is needed instead of a decision only regarding the outputs of the devices. For example, it should be known that there could be unconfirmed positives in LC-MS/MS or false negatives in GC-MS. It is also a point of consideration, especially in the analysis of the new designer drugs, whether the analyte is volatile (i.e., whether its boiling point is low) or not, or whether it decomposes at the temperature used in the GC-MS method, its half-life, pK_a (if known), excretion in the body, its stability in the corpse in postmortem conditions, and whether its concentration is close to the LOD or LOQ values of the method, etc. Basic knowledge, laboratory skills, and interpretation power of the analyst and the experts defines what should be written in the report.

Detection limits should be as low as possible, especially in such matrices as urine, serum, hair, etc. Although the concentration is influenced by the sampling time after drug intake and by the intake amount, concentrations of these drugs in such matrices are generally in the ng/mL amounts, and may be even lower than 1 ng/mL. In this respect, LC-MS/MS screening seems to be the best choice. Kneisel and Auwärter [88] demonstrated the simultaneous detection of 30 synthetic cannabinoids in serum and the LOD and LOQ values were within the range of 0.01–2.0 and 0.1–2.0 ng/mL, respectively. In most of the LC-MS/MS methods, the C_{18} columns were used as analytical columns and analyses were performed in the gradient mode.

Urine and blood are first-choice matrices for drug testing [7]. However, many of these drugs, especially synthetic cannabinoids, are cleared from biofluids after a short period. Hair analysis may be helpful in such cases. Other matrices, such as oral fluid, sweat, postmortem liver, vitreous humor, etc., can also prove the consumption of these drugs and will receive increased attention with time. Stability of some of the new designer drugs in blood samples and in final extracts is affected by pH. The phenomena of the decline of the found concentrations in biological matrices kept after addition of NaF/potassium oxalate as preservative should also be investigated.

Alpha-PVP, 25I-NBOMe, furanyl fentanyl, U-47700, 4,4-DMAR, and MDMAR are some of the new compounds with significant forensic implications resulting in deaths. For example, in Sweden, 20 people recently died after an intake of the morphine-like substance acetylfentanyl [89]. In cases with or without death, the users or their relatives may be unable to provide accurate information on what was used. Clinical management may also be complicated by the poly-drug combinations. Identification of the responsible agents in serious NPS intoxications requires collaboration among analytical chemists, clinical laboratories, health professionals, law enforcement agencies, and synthetic organic chemists, so that information about the causative agent can be obtained on time [61]. Nowadays, the analysis of NPS requires more than the typical targeted drug panels used in the emergency departments, forensic laboratories, etc. and it requires more sophisticated analytical platforms that can rapidly identify previously unreported compounds. Also, clinical history and information received from healthcare professionals helps chemists in toxicological analysis by eliminating the possibility of conventional drugs such as cocaine, heroin, and methamphetamine. The ability to predict and rapidly generate the reference standards for new drugs and their metabolites is also important for the identification of previously unknown NPSs for which commercial reference standards are not available after identification of the compound. Such multidisciplinary coordinations gains importance for the timely resolution of future NPS outbreaks.

To determine a new unknown drug in a matrix is difficult without information, even with the simple LC-MS/MS. High-resolution MS or TOF-MS allow estimation of the parent drugs and their metabolites with synthetic cannabinoids. For structural elucidation of an unknown drug in an herbal product; GC-MS, LC-MS/MS, and HRMS (or TOF-MS) can be used. The target compound may then be purified by means of the preparative LC or preparative TLC, to obtain more than several milligrams of the pure compound. NMR spectroscopy is then used to help in elucidating the chemical structure in detail. NMR and LC-MS/MS have frequently been used together in recent studies, sometimes along with other techniques.

Characterization of an unknown molecule is not always possible using NMR alone [90]. Common functional groups such as carboxylic acid, phenol, and amino groups are NMR-silent in many solvents because of proton–deuterium exchange. Nitro groups and sulfate conjugates do not have protons and they cannot be directly detected in the ^{1}H NMR spectra; however, they can be detected by MS. Molecular weight, empirical formula, and fragmentation information can be obtained via MS, but generally these are insufficient to elucidate the structure of an unknown drug. In the literature, HPLC is reported to be coupled to NMR and MS detectors to form LC-NMR-MS. Such an instrument can be used in natural products, combinatorial chemistry, and drug metabolism studies. There are also recent improvements in lowering the NMR detection limits, use of CryoFlow Probes, and online solid-phase extraction combined with an NMR (LC-SPE-NMR) system. These techniques might provide faster identification of unknown NPSs, their analogs, and positional isomers, and combat the fast production of new designer drugs. In the near future, SPE-LC-NMR-MS might be the choice of the forensic chemists and will gather sample preparation, chromatographic separation, and identification in one step, with use of one instrument.

Future advances expected for the analysis of new designer drugs are the manufacturing of portable in-field detection systems/kits for NPSs in seized products (especially in bulk samples and adulterated products, such as alcoholic drinks) and in various matrices. Moreover, the development of fast, simple, selective, and validated comprehensive hyphenated chromatographic methods with high recovery values is still important for screening and confirmation of common NPSs in the forensically most important biological matrices, including a wide scope of the analogs, isomers, and principle metabolites. Progress in the impurity profiling and/or the source identification methods with common NPSs is also needed in the forensic area [91]. There are comprehensive methods published for a wide scope of the analytes; however, some of them are not validated, some others provide the LOD and/ or LOQ values only, and some can quantitate a number of the analytes; however, in some studies, the rest of the detected species provide very low recovery values, some even below 30%. Apart from many comprehensive methods, semiquantitative methods are available from the literature as well. Also, if more powerful methods with better sample preparation techniques with high recoveries are developed (not lower than 70%), the risk of missing an analyte will decrease and the accuracy of the results will be higher.

LC-MS/MS is one of the most reliable techniques for the detection of designer drugs; however, drug courts need to be aware that this testing is new, and there are still many unsolved global issues, like cutoff levels, which have not yet been defined or standardized for many NPSs [92]. There are no standardized protocols, independent quality control products, or proficiency testing. Detection windows in urine vary among the laboratories and reliability of the tests cannot be controlled from the point of quality assurance. Furthermore, creatinine quantification (which shows if there is an intentional dilution of urine) is not applied during routine drug tests in many laboratories.

To combat the use of designer drugs, drug courts should collaborate with more than one laboratory and/or university research departments working in this area, to take an advantage of the most recent, comprehensive, and reliable detection methods [92]. Drug courts are encouraged to evaluate laboratory services carefully. They should consider which laboratories offer more comprehensive testing, with lower detection limits. Furthermore, the courts should be sure of that the cut-off limit of the laboratory is calculated regarding all consequences related to the excretion and metabolism of such drugs and possible diseases. The most traditional method may not be the best method in every case. Finally, the key issue is probably the quality of scientific expert that a laboratory uses to develop these methods, to carry out the analyses, and to interpret the results of designer drug investigations.

GLOSSARY

AOAC: Association of Official Analytical Chemists, now AOAC International
APCI: atmospheric-pressure chemical ionization
CE: capillary electrophoresis

CEDIA: cloned enzyme donor immunoassay
CI: chemical ionization
CID: collision-induced ionization
COSY: correlation spectroscopy
CSP: chiral stationary phases
DALT: N,N-diallyl substituted tryptamine
DART: direct analysis in real time
DBS: dried blood spot
DC: direct current
DEPT: distortionless enhancement by polarization transfer
2D-LC: two-dimensional LC
DLLME: dispersive liquid/liquid microextraction
DQF-COSY: double quantum-filtered correlation spectroscopy
EI: electron impact
ELISA: enzyme-linked immunosorbent assay
EMCDDA: European Monitoring Center for Drugs and Drug Addiction
EMIT: enzyme multiplied immunoassay
ESI: electrospray ionization
FAB: fast atom bombardment
FDA: Food and Drug Administration
FT: Fourier transform
FTIR: Fourier-transform infrared
GC-EI-IT-MS: gas chromatography electron impact ion trap mass spectrometry
GC-MS: gas chromatography–mass spectrometry
GHB: γ-hydroxybutyric acid
HILIC: hydrophilic interaction liquid chromatography
HLM: human liver microsomes
HMBC: heteronuclear multiple-bond correlation
HMQC: heteronuclear multiple-quantum coherence
HRMS: high-resolution mass spectrometer
ICH: International Conference on Harmonization
ICR: ion cyclotron resonance
IDA: information-dependent acquisition
IMS: ion mobility spectrometry
IR: infrared
IT: ion trap
LC-MS/MS: liquid chromatography–mass spectrometry
LC-NMR: liquid chromatography nuclear magnetic resonance spectroscopy
LC-PDA: liquid chromatography photodiode array detector
LC-QqQ-MS: liquid chromatography, triple quadrupole mass spectrometry
LC-SPE-NMR: online solid-phase extraction combined with NMR
LTQ Orbitrap XL: a hybrid linear ion trap-orbitrap mass spectrometer
LIT: linear ion traps
LOD: limit of detection
LQT: linear quadrupole ion trap
MA: mass accuracy

MALDI HRMS: matrix-assisted laser desorption ionization high-resolution mass spectrometry
MDF: mass defect filter
MRM: multiple reaction monitoring
MS: mass spectrometry
MS^n: multistage MS
NMR: nuclear magnetic resonance spectroscopy
NPS: new psychoactive substances
PCP: phencyclidine
RF: radio frequency
RP: resolving power
SFC: supercritical and subcritical fluid chromatography
SIM: selected ion monitoring
SLE: solid–liquid extraction
SPME: solid-phase liquid microextraction
SRM: selected reaction monitoring
sMRM: scheduled multiple-reaction monitoring mode
SWATH: sequential window acquisition of all theoretical fragment ion spectra
TIC: total ion chromatogram
tMRM: triggered MRM
TOF-MS: time-of-flight mass spectrometer
TSP: thermospray
UHPLC: ultra-high-performance liquid chromatography
UHPSFC: ultra-high-performance supercritical fluid chromatography

REFERENCES

1. Wohlfarth, A. and Weinmann, W. 2010. Bioanalysis of new designer drugs, *Bioanalysis*, 2:965–979.
2. Chavant, F., Boucher, A., Le Boisselier, R., Deheul, S., and Debruyne, D. 2015. New synthetic drugs in addictovigilance, *Thérapie*, 70:179–189.
3. Australian Drug Foundation. 2016. *New psychoactive substances (synthetics)—Alcohol and Drug Information Factsheet.*
4. Dunn, T. N. 2016. Prison Drugs Scandal: One in 10 prisoners are high on dangerous designer drugs as deadly new epidemic sweeps jail, https://www.thesun.co.uk/news/1828789/one-in-10-prisoners-are-high-on-dangerous-designer-drugs-as-deadly-new-epidemic-sweeps-jail/, accessed 20 April 2017, Copyright: City of Edinburgh Council.
5. Lin, D. 2016. Designer drugs—A brief overview, *Therapeutics and Toxins News. Newsletter for the TDM and Toxicology Division of AACC*, 2:1–10
6. Hill, S. L. and Thomas, S. H. 2011. Clinical toxicology of newer recreational drugs, *Clin Toxicol*, 49:705–719.
7. Namera, A., Kawamura, M., Nakamoto, A., Saito, T., and Nagao, M. 2015. Comprehensive review of the detection methods for synthetic cannabinoids and cathinones, *Forensic Toxicol*, 33:175–194.
8. Krasowski, M. D. and Ekins, S. 2014. Using cheminformatics to predict cross reactivity of "designer drugs" to their currently available immunoassays, *J Cheminform*, 6:22.

9. Abbott, R., and Smith, D. E. 2015. The new designer drug wave: A clinical, toxicological, and legal analysis, *J Psychoact Drugs*, 47:5, 368–371.

10. Petrie, M., Lynch, K. L., Ekins, S., Chang, J. S., Goetz, R. J., Wu, A. H. B. and Krasowski, M. D. 2013. Cross-reactivity studies and predictive modeling of "bath salts" and other amphetamine-type stimulants with amphetamine screening immunoassays, *Clin Toxicol*, 51:83–91.

11. Musselman, M. E. and Hampton, J. P. 2014. "Not for human consumption": A review of emerging designer drugs, *Pharmacotherapy*, 34:745–757.

12. May, M. 2016. *Forensic Mass Spectrometry*, Labcompare, 24 May 2016. http://www.labcompare.com/10-Featured-Articles/186874-Forensic-Mass-Spectrometry/Forensic Mass Spectrometry, accessed 1 May 2017.

13. Harris, D. N., Hokanson, S., Miller, V. and Jackson, G. P. 2014. Fragmentation differences in the EI spectra of three synthetic cannabinoid positional isomers: JWH-250, JWH-302, and JWH-201, *Int J Mass Spectrom*, 368:23–29.

14. Kusano, M., Zaitsu, K., Nakayama, H., Nakajima, J., Hisatsune, K., Moriyasu, T., Matsuta, S., Katagi, M., Tsuchihashi, H., and Ishii, A. 2015. Positional isomer differentiation of synthetic cannabinoid JWH-081 by GC-MS/MS, *J Mass Spectrom*, 50:586–591.

15. György, V. and Vekey, K. 2004. Solid-phase microextraction: A powerful sample preparation tool prior to mass spectrometric analysis. *J Mass Spectrom*, 39:233–254.

16. Boatto, G., Nieddu, M., Pirisi, M. A. and Dessì, G. 2007. Simultaneous determination of new thioamphetamine designer drugs in plasma by capillary electrophoresis coupled with mass spectrometry, *Rapid Commun Mass Spectrom*, 21:3716–3720.

17. De Jong, G. 2016. *Capillary Electrophoresis–Mass Spectrometry (CE-MS): Principles and Applications*, John Wiley & Sons, Weinheim, p. 316

18. Týčová, A., Ledvina, V., and Klepárník, K. 2016. Recent advances in CE-MS coupling: Instrumentation, methodology, and applications, *Electrophoresis*, 38:115–134.

19. Ostermann, K. M., Luf, A., Lutsch, N. M., Dieplinger, R., Mechtler, T. P., Metz, T. F., Schmid, R. and Kasper, D. C. 2014. MALDI Orbitrap mass spectrometry for fast and simplified analysis of novel street and designer drugs, *Clin Chim Acta*, 433:254–258.

20. Gwak, S. and Almirall, J. R. 2015. Rapid screening of 35 new psychoactive substances by ion mobility spectrometry (IMS) and direct analysis in real time (DART) coupled to quadrupole time-of-flight mass spectrometry (QTOF-MS), *Drug Test Anal*, 7:884–893.

21. Brandt, S. D., Kavanagh, P. V., Dowling, G., Talbot, B., Westphal, F., Meyer, M. R., Maurer, H. H. and Halberstadt, A. L. 2017. Analytical characterization of N,N diallyltryptamine (DALT) and 16 ring substituted derivatives, *Drug Test Anal*, 9:115–126.

22. Mayer, M. J. 2015. Designer Drugs: Monitoring Synthetic Cannabinoids, Thermo Fisher Scientific, https://www.thermofisher.com/blog/proteomics/designer-drugs-monitoring -synthetic-cannabinoids/, accessed 13 November 2017.

23. Roškar, R. and Lušin, T.T. 2012. Analytical Methods for Quantification of Drug Metabolites in Biological Samples, in *Chromatography—The Most Versatile Method of Chemical Analysis*, Ed. Calderon, L., InTech, available from https://www.intechopen.com/books /chromatography-the-most-versatile-method-of-chemical-analysis/analytical-methods-for -quantification-of-drug-metabolites-in-biological-samples (accessed 13 November 2017).

24. Odoardi, S., Fisichella, M., Romolo, F. S., and Strano-Rossi, S. 2015. High-throughput screening for new psychoactive substances (NPS) in whole blood by DLLME extraction and UHPLC–MS/MS analysis, *J Chromatogr B*, 1000:57–68.

25. Want, E. J., O'Maille, G., Smith, C. A., Brandon, T. R., Uritboonthai, W., Qin, C., Trauger, S. A., and Siuzdak, G. 2006. Solvent-dependent metabolite distribution, clustering, and protein extraction for serum profiling with mass spectrometry, *Anal Chem*, 78:743–752.

26. Mizuno, K. and Kataoka, H. 2015. Analysis of urinary 8-isoprostane as an oxidative stress biomarker by stable isotope dilution using automated online in-tube solid-phase microextraction coupled with liquid chromatography–tandem mass spectrometry, *J Pharm Biomed Anal*, 112:36–42.

27. Grebe, S. K. G. and Singh, R. J. 2011. LC-MS/MS in the clinical laboratory—Where to from here?, *Clin Biochem Rev*, 32:5–31.

28. Anilanmert, B., Çavuş, F., Narin, I., Cengiz, S., Sertler, Ş., Özdemir, A. A. and Acikkol, M. 2016. Simultaneous analysis method for GHB, ketamine, norketamine, phenobarbital, thiopental, zolpidem, zopiclone and phenytoin in urine, using C18 poroshell column, *J Chromatogr B*, 1022:230–241.

29. Liquid Chromatography–Mass Spectrometry. 2017. EAG Laboratories, CA. http://www.eag.com/liquid-chromatography-mass-spectometry-lc-ms/, accessed May 2017.

30. Leeds, S. M. 1999. *Chapter 3—Experimental in Characterisation of the Gas-Phase Environment in a Microwave Plasma Enhanced Diamond Chemical Vapour Deposition Reactor Using Molecular Beam Mass Spectrometry*, PhD thesis, Bristol.

31. Ardrey, R. E. 2003. *Liquid Chromatography–Mass Spectrometry: An Introduction*, John Wiley & Sons, Ltd., Chichester, pp. 2, 3, 46, 100, 101, 123, 185.

32. *Basics of LC-MS*. 2001. Agilent Technologies, http://ccc.chem.pitt.edu/wipf/Agilent%20 LC-MS%20primer.pdf, accessed 13 November 2017.

33. Tiller, P. R. and Drexler, D. M. 1999. *Quadrupole MS Technology Evaluating the Differences between Source CID or Real MS/MS*, Thermo Finnigan LC/MS Technical Report, Thermo Finnigan. Printed in USA 4/99 A0895-776 #719 9M, http://www.thermo.com/eThermo/CMA/PDFs/Articles/articlesFile_11351.pdf, accessed May 2017.

34. Błażewicz, A., Bednarek, E., Sitkowski, J., Popławska, M., Stypułkowska, K., Bocian, W., and Kozerski, L. 2017. Identification and structural characterization of four novel synthetic cathinones: α-methylaminohexanophenone (hexedrone, HEX), 4-bromoethcathinone (4-BEC), 4-chloro-α-pyrrolidinopropiophenone (4-Cl-PPP), and 4-bromo-α-pyrrolidinopentiophenone (4-Br-PVP) after their seizures, *Forensic Toxicol*, 35:317–332.

35. Holcapek, M., Jirásko, R., and Lísa, M. 2012. Recent developments in liquid chromatography–mass spectrometry and related techniques, *J Chromatogr A*, 1259:3–15.

36. Hart-Smith, G. and Blanksby, S. J. 2012. Mass analysis. In Barner-Kowollik, C., Gruendling, T., Falkenhagen, J. and Weidner, S. Eds., *Mass Spectrometry in Polymer Chemistry*. Wiley-VCH Verlag & Co., Weinheim, pp. 5–32, 20–21, 86–87.

37. Hutter, M., Broecker, S., Kneisel, S., and Auwärter, V. 2012. Identification of the major urinary metabolites in man of seven synthetic cannabinoids of the aminoalkylindole type present as adulterants in 'herbal mixtures' using LC-MS/MS techniques, *J Mass Spectrom*, 47:54–65.

38. Baugh, P. J. 2004. Analytical Mass Spectrometry, Quay Pharma Training Programme, Day 4, 28 September 2004.

39. Crawford Scientific. 2017. *Mass Spectrometry, Fundamental LC-MS, Orbitrap Mass Analyzers*, Chromacademy, e-learning for the analytical chemistry committee, www .chromacademy.com (accessed 14 April 2017).

40. Perry, R. H., Cooks, R. G., and Noll, R. J. 2008. Orbitrap mass spectrometry: Instrumentation, ion motion and applications, *Mass Spectrom Rev*, 27:661–699.

41. Guilhaus, M. 1995. Special feature: Tutorial, principles and instrumentation in time-of-flight mass spectrometry, physical and instrumental concepts, *J Mass Spectrom*, 30:1519–1532.

42. Lewis, J. K., Wei, J., and Siuzdak, G. 2000. Matrix-assisted laser desorption/ionization mass spectrometry in peptide and protein analysis. In *Encyclopedia of Analytical Chemistry*, Meyers, R.A. Ed., John Wiley & Sons Ltd, Chichester, pp. 5880–5894.

43. Roemmelt, A. T., Steuer, A. E., Poetzsch, M., and Kraemer, T. 2014. Liquid chromatography, in combination with a quadrupole time-of-flight instrument (LC QTOF), with sequential window acquisition of all theoretical fragment-ion spectra (SWATH) acquisition: Systematic studies on its use for screenings in clinical and forensic toxicology and comparison with information-dependent acquisition (IDA), *Anal Chem,* 86:11742–11749.

44. Guillarme, D. and Veuthey, J. L. Eds. 2012. *UHPLC in Life Sciences,* The Royal Society of Chemistry, Cambridge.

45. Thurman, E. M., Ferrer, I., and Fernández-Alba, A. R. 2005. Matching unknown empirical formulas to chemical structure using LC/MS TOF accurate mass and database searching: Example of unknown pesticides on tomato skins, *J Chromatogr A,* 1067:127–134.

46. Błażewicz, A., Bednarek, E., Sitkowski, J., Popławska, M., Stypułkowska, K., Bocian, W., and Kozerski, L. 2017. Identification and structural characterization of four novel synthetic cathinones: a-methylaminohexanophenone (hexedrone, HEX), 4-bromoethcathinone (4-BEC), 4-chloro-apyrrolidinopropiophenone (4-Cl-PPP), and 4-bromo-apyrrolidinopentiophenone (4-Br-PVP) after their seizures, *Forensic Toxicol.* Published online, 6 March 2017.

47. Carlier, J., Diao, X., Sempio, C., and Huestis, M. A. 2017. Identification of new synthetic cannabinoid ADB-CHMINACA (MAB-CHMINACA) metabolites in human hepatocytes, *AAPS J,* 19:568–577.

48. Helfer, A. G., Michely, J. A., Weber, A. A., Meyer, M. R., and Maurer, H. H. 2017. Liquid chromatography–high resolution–tandem mass spectrometry using Orbitrap technology for comprehensive screening to detect drugs and their metabolites in blood plasma, *Anal Chim Acta,* 965:83–95.

49. ElSohly, M. A., Gul, W., ElSohly, K. M., Murphy, T. P., Madgula, V. L., and Khan, S. I. 2011. Liquid chromatography–tandem mass spectrometry analysis of urine specimens for K2 (JWH-018) metabolites, *J Anal Toxicol,* 35:487–495.

50. Saurina, J. and Sentellas, S. 2017. Strategies for metabolite profiling based on liquid chromatography, *J Chromatogr B,* 1044–1045:103–111.

51. Carlier, J., Diao, X., Scheidweiler, K. B., and Huestis, M. A. 2017. Distinguishing intake of new synthetic cannabinoids ADB-PINACA and 5F-ADB-PINACA with human hepatocyte metabolites and high-resolution mass spectrometry, *Clin Chem,* 63:1008–1021.

52. Nakazono, Y., Tsujikawa, K., Kuwayama, K., Kanamori, T., Iwata, Y. T., Miyamoto, K., Kasuya, F., and Inoue, H. 2014. Simultaneous determination of tryptamine analogues in designer drugs using gas chromatography–mass spectrometry and liquid chromatography–tandem mass spectrometry, *Forensic Toxicol,* 32:154–161.

53. Diao, X., Carlier, J., Zhu, M., Pang, S., Kronstrand, R., Scheidweiler, K. B., and Huestis, M. A. 2017. In vitro and in vivo human metabolism of a new synthetic cannabinoid NM-2201 (CBL-2201), *Forensic Toxicol,* 35:20–32.

54. Taschwer, M., Grascher, J., and Schmid, M. G. 2017. Development of an enantioseparation method for novel psychoactive drugs by HPLC using a Lux1 Cellulose-2 column in polar organic phase mode, *Forensic Sci Int,* 270:232–240.

55. Sharp, V. S., Gokey, M. A., Wolfe, C. N., Rener, G. A., and Cooper, M. R. 2015. High performance liquid chromatographic enantioseparation development and analytical method characterization of the carboxylate ester of evacetrapib using an immobilized chiral stationary phase with a non-conventional eluent system, *J Chromatogr A,* 1416:83–93.

56. Wang, C. C., Hartmann-Fischbach, P., Krueger, T. R., Lester, A., Simonson, A., Wells, T. L., Wolk, M. O. and Hidlay, N. J. 2015. Fast and sensitive chiral analysis of amphetamines and cathinones in equine urine and plasma using liquid chromatography tandem mass spectrometry, *Am J Anal Chem,* 6:995.

57. Saurina, J. and Sentellas, S. 2017. Strategies for metabolite profiling based on liquid chromatography, *J Chromatogr B*, 1044–1045: 103–111.
58. Kloos, D. P., Lingeman, H., Niessen, W. M., Deelder, A. M., Giera, M. and Mayboroda, O. A. 2013. Evaluation of different column chemistries for fast urinary metabolic profiling, *J Chromatogr B Anal Technol Biomed Life Sci*, 927:90–96.
59. Dubbelman, A. C., Cuyckens, F., Dillen, L., Gross, G., Hankemeier, T. and Vreeken, R. J. 2014. Systematic evaluation of commercially available ultra-high performance liquid chromatography columns for drug metabolite profiling: Optimization of chromatographic peak capacity, *J Chromatogr A*, 1374:122–133.
60. Maas, A., Sydow, K., Madea, B., and Hess, C. 2017. Separation of ortho, meta and para isomers of methylmethcathinone (MMC) and methylethcathinone (MEC) using LC-ESI-MS/MS: Application to forensic serum samples, *J Chromatogr B*, 1051:118–125.
61. Adams, A. J., Banister, S. D., Irizarry, L., Trecki, J., Schwartz, M., and Gerona, R. 2017. "Zombie" outbreak caused by the synthetic cannabinoid AMB-FUBINACA in New York, *N Engl J Med*, 376:235–242.
62. Heltsley, R., Shelby, M. K., Crouch, D. J., Black, D. L., Robert, T. A., Marshall, L., Bender, C. L., DePriest A. Z., and Colello, M. A. 2012. Prevalence of synthetic cannabinoids in US athletes: Initial findings, *J Anal Toxicol*, 36:588–593.
63. Alvarez, J. C., Etting, I., Abe, E., Villa, A., and Fabresse, N. 2017. Identification and quantification of 4-methylethcathinone (4-MEC) and 3,4-methylenedioxypyrovalerone (MDPV) in hair by LC–MS/MS after chronic administration, *Forensic Sci Int*, 270:39–45.
64. Hutter, M., Kneisel, S., Auwärter, V., and Neukamm, M. A. 2012. Determination of 22 synthetic cannabinoids in human hair by liquid chromatography–tandem mass spectrometry, *J Chromatogr B*, 903:95–101.
65. Øiestad, E. L., Øiestad, Å. M. L., Gjelstad, A. and Karinen, R. 2016. Oral fluid drug analysis in the age of new psychoactive substances, *Bioanalysis*, 8:691–710.
66. Strano-Rossi, S., Anzillotti, L., Castrignanò, E., Romolo, F. S., and Chiarotti, M. 2012. Ultra high performance liquid chromatography–electrospray ionization–tandem mass spectrometry screening method for direct analysis of designer drugs, spice and stimulants in oral fluid, *J Chromatogr A*, 1258:37–42.
67. Rodrigues, W. C., Catbagan, P., Rana, S., Wang, G., and Moore, C. 2013. Detection of synthetic cannabinoids in oral fluid using ELISA and LC–MS–MS, *J Anal Toxicol*, 37:526–533.
68. Øiestad, E. L., Johansen, U., Christophersen, A. S., and Karinen, R. 2013. Screening of synthetic cannabinoids in preserved oral fluid by UPLC–MS/MS, *Bioanalysis*, 5:2257–2268.
69. Valen, A., Leere Øiestad, Å. M., Strand, D. H., Skari, R., and Berg, T. 2016. Determination of 21 drugs in oral fluid using fully automated supported liquid extraction and UHPLC-MS/MS, *Drug Test Anal*, 9:808–823.
70. Bade, R., Bijlsma, L., Sancho, J. V., Baz Lomba, J. A., Castiglioni, S., Castrignano, E., Causanilles, A., Gracia-Lor, E., Kasprzyk-Hordern, B., Kinyua, J., McCall, A. K., van Nuijs, A. L. N., Ort, C., Plosz, B., Ramin, P., Rousis, N. I., Ryu, Y., Thomas, K. V., de Voogt, P., Zuccato, E., and Hernández, F. 2017. Liquid chromatography–tandem mass spectrometry determination of synthetic cathinones and phenethylamines in influent wastewater of eight European cities, *Chemosph*, 168:1032–104.
71. Thomas, K. V., Bijlsma, L., Castiglioni, S., Covaci, A., Emke, E., Grabic, R., Hernandez, F., Karolak, S., Kasprzyk-Hordern, B., Lindberg, R. H., Lopez de Alda, M., Meierjohann, A., Ort, C., Pico, Y., Quintana, J. B., Reid, M., Rieckermann, J., Terzic, S., van Nuijs, A. L. N., and de Voogt, P. 2012. Comparing illicit drug use in 19 European cities through sewage analysis, *Sci Total Environ*, 432:432–439.

72. Ambach, L., Redondo, A. H., König, S., and Weinmann, W. 2014. Rapid and simple LC-MS/MS screening of 64 novel psychoactive substances using dried blood spots, *Drug Test Anal*, 6:367–375.

73. Ambach, L., Redondo, A. H., König, S., Angerer, V., Schürch, S., and Weinmann, W. 2015. Detection and quantification of 56 new psychoactive substances in whole blood and urine by LC–MS/MS, *Bioanalysis*, 7:1119–1136.

74. Pedersen, A. J., Dalsgaard, P. W., Rode, A. J., Rasmussen, B. S., Muller, I. B., Johansen, S. S., and Linet, K. 2013. Screening for illicit and medicinal drugs in whole blood using fully automated SPE and ultra-high-performance liquid chromatography with TOF-MS with data-independent acquisition, *J Sep Sci*, 36, 2081–2089.

75. Brock, T.G. 2001. Analysis of synthetic cannabinoids and designer drugs, Cayman Chemical, Michigan. www.caymanchem.com/article/2199, accessed 4 December 2017.

76. Ko, B. J., Suh, S., Suh, Y. J., In, M. K., and Kim, S. H. 2007. The impurity characteristics of methamphetamine synthesized by Emde and Nagai method, *Forensic Sci Int*, 170:142–147.

77. Li, Y., Wang, M., Li, A., Zheng, H., and Wei, Y. 2016. Identification of the impurities in 2, 5-dimethoxy-4-ethylphenethylamine tablets by high performance liquid chromatography mass spectrometry-ion trap-time of flight, *Anal Methods*, 8:8179–8187.

78. Uchiyama, N., Shimokawa, Y., Kawamura, M., Kikura-Hanajiri, R., and Hakamatsuka, T. 2014. Chemical analysis of a benzofuran derivative, 2-(2-ethylaminopropyl)benzofuran (2-EAPB), eight synthetic cannabinoids, five cathinone derivatives, and five other designer drugs newly detected in illegal products, *Forensic Toxicol*, 32:266–281.

79. Ichikawa, Y., Nakajima, J. I., Takahashi, M., Uemura, N., Yoshida, M., Suzuki, A., Suzuki, J., Nakae, D., Moriyasu, T. and Hosaka, M. 2017. Identification of (1H-indol-3-yl)(2,2,3,3-tetramethylcyclopropyl) methanone (DP-UR-144) in a herbal drug product that was commercially available in the Tokyo metropolitan area, *Forensic Toxicol*, 35:146–152.

80. Nakajima, J., Takahashi, M., Seto, T., Yoshida, M., Kanai, C., Suzuki, J., and Hamano, T. 2012. Identification and quantitation of two new naphthoylindole drugs-of-abuse, (1-(5-hydroxypentyl)-1H-indol-3-yl)(naphthalen-1-yl)methanone (AM-2202) and (1-(4-pentenyl)-1H-indol-3-yl)(naphthalen-1-yl)methanone, with other synthetic cannabinoids in unregulated "herbal" products circulated in the Tokyo area, *Forensic Toxicol*, 30:33–44.

81. Qian, Z., Jia, W., Li, T., Liu, C., and Hua, Z. 2017. Identification and analytical characterization of four synthetic cathinone derivatives iso-4-BMC, β-TH-naphyrone, mexedrone, and 4-MDMC, *Drug Test Anal*, 9:274–281.

82. Chambers, E., Wagrowski-Diehl, D. M., Lu, Z., and Mazzeo, J. R. 2007. Systematic and comprehensive strategy for reducing matrix effects in LC/MS/MS analyses, *J Chromatogr B Analyt Technol Biomed Life Sci*, 852:22–34.

83. Matuszewski, B. K., Constanzer, M. L., and Chavez-Eng, C. M. 2003. Strategies for the assessment of matrix effect in quantitative bioanalytical methods based on HPLC–MS/MS, *Anal Chem*, 75:3019–3030.

84. Zawilska, J. B. and Andrzejczak, D. 2015. Next generation of novel psychoactive substances on the horizon—A complex problem to face, *Drug Alcohol Depend*, 157:1–17.

85. Lurie, I. S., Marginean, I., and Rowe, W. 2015–2016. *Analysis of Synthetic Cannabinoids in Seized Drugs by High-Resolution UHPLC/MS and GC/MS*, Application Note, 012433A_01, PerkinElmer, Inc., MA.

86. Agilent Technologies. 1998. *Basics of LC-MS*. www.agilent.com/cs/library/support/documents/a05296.pdf (accessed 11 April 2017).

87. Vass, L. 2016. Editorial Article: Webinar Highlights: The Future of Designer Drug Analysis, *Select Science, News and Advice.* http://www.selectscience.net/editorial-articles/webinar-highlights-the-future-of-designer-drug-analysis/?artID=42208, 12 December 2016 (accessed 4 December 2017).

88. Kneisel, S. and Auwärter, V. 2012. Analysis of 30 synthetic cannabinoids in serum by liquid chromatography-electrospray ionization tandem mass spectrometry after liquid–liquid extraction, *J Mass Spectrom*, 47:825–835.

89. Tuv, S. S., Krabseth, H. M., Strand, M. C., Karinen, R. A., Wiik, E., Vevelstad, M. S., Westin, A. A., Øiestad, E. L., and Vindenes, V. 2016. New designer drugs from the web, *Tidsskr Nor Legeforen*, 136:721–723.

90. Corcoran, O. and Spraul, M. 2003. LC–NMR–MS in drug discovery, *Drug Discov Today*, 8:624–631.

91. Smith, J. P., Sutcliffe, O. B., and Banks, C. E. 2015. An overview of recent developments in the analytical detection of new psychoactive substances (NPSs), *Analyst*, 140:4932–4948.

92. Cary, P. L. 2014. *Designer Drugs: What Drug Court Practitioners Need To Know, Drug Court Practitioner*, Fact Sheet, National Drug Court Institute (NADCP), The Professional Services Branch of NADCP, Alexandria, VA, 9: 1–13. www.ndcrc.org/sites/default/files/designer_drugs_2.pdf, accessed 4 December 2017.

6 Ambient Plasma Ionization Techniques in Forensic Analysis

Przemysław Mielczarek and Marek Smoluch

CONTENTS

6.1 INTRODUCTION

Ambient plasma ionization mass spectrometry techniques are based on electrical discharges in a flowing gas. The ionization takes place at ambient conditions, allowing for direct sample analysis without or with minor sample preparation. This feature makes ambient plasma ionization methods suitable in forensic analyses, where high speed is often required in combination with high specificity of mass spectrometry data.

6.2 APPLICATIONS

The oldest ambient plasma ionization method is DART—direct analysis in real time [1]. Alongside ASAP (atmospheric solids analysis probe) [2], DART is the only commercially available plasma-based ADI (ambient desorption ionization) source. This fact, next to the specific features of this method, have resulted in DART being used in the broadest range of applications compared to other plasma-based ADI methods. This includes analysis of cannabinoids, cathinones, explosives, plant material, counterfeit pharmaceuticals, ink, and others. A full list of all plasma-based ambient ionization techniques, with descriptions and mechanisms of ionization, can be found elsewhere [3] and will not be discussed in this book. Nevertheless, the main advantages of these methods are no or only minor sample preparation, rapid analysis, and a possibility to analyze samples in solid, liquid, or gaseous forms.

6.3 DART—DIRECT ANALYSIS IN REAL TIME

The most widespread application of DART is in the analysis of drugs and pharmaceuticals, although this field is still reserved mainly for GC-MS, LC-MS, NMR, or IR (see the list of abbreviations in Chapter 5). The biggest advantage of DART over already-established methods is simple sample preparation and the short time required to obtain results. This technique was validated for the rapid screening of forensic evidence for drugs of abuse [4]. In this study the determination of LOD (limit of detection), selectivity, and a comparison of this technique to the established analytical protocols were conducted. The LOD was calculated at 0.05 mg/mL. The results were sufficient to incorporate DART into the Virginia Department of Forensic Science analysis scheme for the screening of solid dosage forms of drugs of abuse.

Semiquantitative analysis of synthetic cathinones and their metabolites in urine was performed by LaPointe et al. [5]. Three different cathinones and three metabolites were directly, without sample preparation, detected down to subclinical levels. The authors also demonstrated the use of solid-phase microextraction to increase the detectability by more than one order of magnitude for both drugs and metabolites. The improvement was due to the removal of a majority of MS signals associated with urine that usually dominate spectra obtained from such complex biological samples.

Another study [6] analyzed methadone in untreated urine. Such studies traditionally utilize enzyme immunoassays (EIA) and other techniques applied to confirm the results (GC-MS, LC-MS/MS). In such cases the full procedure of sample analysis requires a lot of time and effort. DART can be used as an alternative approach for screening and confirming the presence of methadone in untreated urine samples. In this case the total time for screening and confirmation of methadone was less than five minutes per sample, compared to the three to five days required by EIA, GC-MS, or LC-MS/MS techniques. DART was proved to be highly specific and selective; allowing for detection of methadone at 250 ng/mL, which is comparable with the EIA technique (300 ng/mL).

DART is often used for the analysis of cannabinoids. Synthetic cannabinoids are widely used as components of designer drugs and cause problems due to the difficulties connected with their identification in complex mixtures. The standard analysis utilizing LC-MS requires extraction of the cannabinoid from plant material and a further sample cleanup step. Habala et al. [7] describes the sampling techniques for the provided herbal material containing the cannabinoids, directly as plant parts or as a methanol extract. Both techniques can be used, depending on the user's requirements. In case of quantitative or semiquantitative analyses, application of the sample in solution seems to be a better choice. Quantitative analysis directly from the surface is problematic and requires homogeneous deposition of the internal standard on the surface. DART can also be supported by the NMR technique, serving as a complementary approach to conventional GC-MS or LC-MS methods. DART circumvents time-consuming sample extraction, derivatization, chromatography, and other preparative steps required for analysis by the more conventional mass spectrometric methods.

Marino et al. [8] applied proton nuclear magnetic resonance (NMR) spectroscopy to promptly screen synthetic cannabinoids after their direct detection in herbs and powders by DART-MS. A simplified sample preparation protocol was employed utilizing 50 mg of herbal matrices for quick NMR detection. Ten synthetic cannabinoids were found in 15 herbal incenses.

DART can also be coupled to thin-layer chromatography (TLC), which is often employed in the forensic laboratories [9]. Usually, the detection from the TLC plate is accomplished by using visualization reagents or is performed under UV light. DART enables direct mass spectrometry detection from the TLC plate, without time-consuming scratching of the spot from the plate followed by the extraction of the separated compound and its subsequent analysis by GC-MS or LC-MS.

DART has proved to be an excellent tool for the characterization and comparison of printing ink evidence. Ink analysis is crucial for examination of a questionable document. Trejos et al. [10] designed and validated a searchable printing database to gather chemical information from the analyses of ink evidence. The database contains more than 300 samples from printing sources (toner, inkjet, offset, and intaglio inks). The database was built not only for DART, but also for Fourier-transform infrared (FTIR), scanning electron microscopy–energy-dispersive x-ray (SEM-EDS), laser ablation-inductively coupled plasma-mass spectrometry (LA-ICP-MS), and pyrolysis-gas chromatography-mass spectrometry (Py-GC-MS) methods. It is worth noticing that the amount of time since an ink has been deposited on paper has a large influence on its spectrum [11]. The biggest changes are observed during the first few months after ink deposition, when the most volatile compounds evaporate. After that time spectra stabilize.

A very interesting approach of DART is the possibility for scanning for drugs of abuse directly in hair [12]. Analysis of hair segments can be used to obtain retrospective timelines of drug use. Traditionally such analysis is a laborious process, in contrary to application of DART where analysis of locks of hair takes a few moments. The DART hair-scan method was validated for cocaine and compared to an accredited liquid chromatography/tandem mass spectrometry (LC/MS/MS) method. The detection limit for cocaine in hair was found to be below the cutoff value of 0.5 ng/mg recommended by the Society of Hair Testing, meaning that the DART hair-scan method is amenable to forensic cases. DART was also successfully applied in direct analysis of tetrahydrocannabinol (THC) from hair [13]. This application shows also that DART not only scans the surface of a hair, but also penetrates deep enough to measure incorporated THC. The schematic diagram of the setup is shown in Figure 6.1.

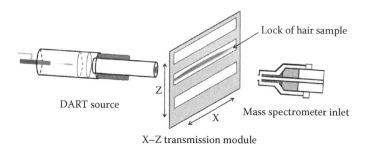

FIGURE 6.1 Experimental setup used for hair analysis by DART-orbitrap MS using the X–Z transmission module. (Based on Duvivier, W.F. et al., 2014. *Rapid Commun Mass Spectrom.* 28: 682–690.)

DART can also be helpful in cases of sexual assault crimes. Routine methods in such cases are those implementing DNA analysis, which is not possible with DART. However, non-DNA-related trace evidences (like spermicide nonoxynol-9, compounds used in condom manufacturing, or fatty acids identified in latent fingerprints) are also important as probative evidences and can often be analyzed by DART [14].

DART has also been shown to be an excellent tool for identification of alprazolam in counterfeit and routine drug identification cases [15]. The method allows for analysis of alprazolam tablets directly without laborious sample preparation, and the procedures also fulfil forensic standards.

The trace detection and analysis of potential, partially nitrated, and dimerized byproducts of homemade nitrate ester explosive synthesis has also been demonstrated [16]. Five compounds related to the synthesis of nitroglycerin and pentaerythritol tetranitrate (PETN) were successfully examined at nanogram to subnanogam levels. This application also proves the applicability of DART in this field of forensic research.

6.4   OTHER AMBIENT PLASMA IONIZATION TECHNIQUES

DART remains the main ambient plasma ionization technique used in forensic analysis, but other methods, such as flowing atmospheric pressure afterglow (FAPA-MS) [17] or dielectric barrier discharge ionization (DBDI-MS) [18], are gaining increased attention.

Several designer drugs (JWH-122, 4BMC, pentedrone, 3,4-DNNC, and ETH-CAT) were analyzed by FAPA-MS [19]. Two approaches of sample introduction to the plasma stream have been shown (Figures 6.2 and 6.3). In one case, the samples were

FIGURE 6.2   FAPA MS with analyte applied as methanolic aerosol from the nebulizer. (From Smoluch, M. et al., 2016. *Talanta*. 146: 29–33. With permission.)

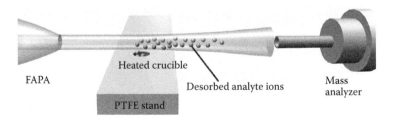

FIGURE 6.3   FAPA MS with the analyte applied by thermal desorption from the heated crucible. (From Smoluch, M. et al., 2016. *Talanta*. 146: 29–33. With permission.)

applied in the form of methanolic aerosol, and, in the other case, they were analyzed by desorption supported by sample heating. Both sample application methods can be used in any ambient plasma ionization source. FAPA offers rapid and reliable sample identification, without any pre-separation.

FAPA was also a useful technique to analyze methcathinone synthesized by a protocol found on the Internet [20]. The method was able to identify methcathinone directly from the crude reaction mixture without any sample preparation. The ability for rapid analysis of such mixtures can also be a useful supporting tool for initial identification of psychoactive compounds.

For toxicology studies, it is important to consider not only intact psychoactive compounds but also their metabolites generated in the human body. Metabolism of psychoactive substances is of utmost importance for the diagnosis and treatment of often unconscious persons under the influence of drugs who are brought into the emergency room. In the era of extremely popular so-called legal highs among adolescents, or different kinds of products containing psychoactive substances not scheduled or included on drug enforcement lists controlled by the law, the problem of their rapid identification is extremely important. Determination of psychostimulants and their metabolites generated by electrochemical techniques linked online to FAPA has been demonstrated [21]. In this work, an electrochemical flow cell was used to generate potential drug metabolites in order to study simulated metabolic pathways. The obtained products were identified by FAPA-MS. The schematic diagram of the coupling is shown in Figure 6.4.

A similar approach was used for the electrochemical generation and detection of selegiline metabolites by the electrochemistry-liquid chromatography-dielectric barrier discharge ionization-mass spectrometry system (EC-LC-DBDI-MS) [22]. Selegiline—a drug used for the treatment of Parkinson's disease—metabolizes in the human body to, among others, amphetamine and methamphetamine, compounds

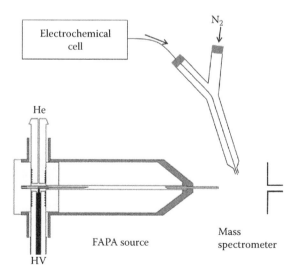

FIGURE 6.4 Schematic diagram of EC-FAPA-MS coupling. (From Smoluch, M. et al., 2014. *Analyst.* 139: 4350–4355.)

with well-known psychoactive properties. Dielectric barrier discharge ionization (DBDI), similar to DART, can be useful for the analysis of cannabinoids. For this purpose, a sample heater was employed for more efficient sample desorption/ionization [23]. This approach is similar to the one shown in Figure 6.3, but the FAPA source was replaced by DBDI and allowed for direct identification of eight seized synthetic cannabinoids in botanical matrices. In summary, a combination of FAPA/DBDI with electrochemical generation of potential metabolites is a useful approach to identification of the forthcoming threats brought by this novel wave of psychoactive compounds. This may also help, in the near future, in the application of a predictive toxicology approach to saving the lives of intoxicated patients.

6.5 CONCLUSIONS

The application of ambient plasma ionization methods minimizes the time necessary to deliver analytical evidence, as these methods circumvent laborious sample preparation. As a consequence, these methods can reduce the backlog of analyses and accelerate criminal prosecution. These methods have huge potential to be among the main methods used for drug screening outside of the laboratory, as they can be miniaturized and can work in ambient conditions. However, one of the biggest disadvantages of the described methods is their problematic quantitation capability. However, this feature may be improved in the near future as extensive work is done in this field.

REFERENCES

1. Cody, R.B., Laramee, J.A., and Durst, H.D. 2005. Versatile new ion source for the analysis of materials in open air under ambient conditions, *Anal Chem.* 77: 2297–2302.
2. McEwen, C.N., McKay, R.G., and Larsen, B.S. 2005. Analysis of solids, liquids, and biological tissues using solids probe introduction at atmospheric pressure on commercial LC/MS instruments, *Anal Chem.* 77: 7826–7831.
3. Smoluch, M., Mielczarek, P., and Silberring J. 2016. Plasma-based ambient ionization mass spectrometry in bioanalytical sciences, *Mass Spectrom Rev.* 35: 22–34.
4. Steiner, R.R. and Larson, R.L. 2009. Validation of the direct analysis in real time source for use in forensic drug screening, *J Forensic Sci.* 54: 617–622.
5. LaPointe, J., Musselman, B., O'Neill, T., and Shepard, J.R. 2015. Detection of "bath salt" synthetic cathinones and metabolites in urine via DART-MS and solid phase microextraction, *J Am Soc Mass Spectrom.* 26: 159–165.
6. Beck, R., Carter, P., Shonsey, E., and Graves, D. 2016. Tandem DART™ MS Methods for Methadone Analysis in Unprocessed Urine, *J Anal Toxicol.* 40: 140–147.
7. Habala, L., Valentová, J., Pechová, I., Fuknová, M., and Devínsky F. 2016. DART–LTQ ORBITRAP as an expedient tool for the identification of synthetic cannabinoids, *Leg Med (Tokyo).* 20: 27–31.
8. Marino, M.A., Voyer, B., Cody, R.B., Dane, A.J., Veltri, M., and Huang, L. 2016. Rapid identification of synthetic cannabinoids in herbal incenses with DART-MS and NMR, *J Forensic Sci.* 61(Suppl 1): 82–91.
9. Howlett, S.E. and Steiner, R.R. 2011. Validation of thin layer chromatography with AccuTOF-DART™ detection for forensic drug analysis, *J Forensic Sci.* 56: 1261–1267.

10. Trejos, T., Torrione, P., Corzo, R., Raeva, A., Subedi, K., Williamson, R., Yoo, J., and Almirall, J. 2016. A novel forensic tool for the characterization and comparison of printing ink evidence: Development and evaluation of a searchable database using data fusion of spectrochemical methods, *J Forensic Sci*. 61: 715–724.
11. Jones, R.W. and McClelland, J.F. 2013. Analysis of writing inks on paper using direct analysis in real time mass spectrometry, *Forensic Sci Int*. 231: 73–81.
12. Duvivier, W.F., van Putten, M.R., van Beek, T.A., and Nielen, M.W. 2016. (Un)targeted scanning of locks of hair for drugs of abuse by direct analysis in real-time, high-resolution mass spectrometry, *Anal Chem*. 88: 2489–2496.
13. Duvivier, W.F., van Beek, T.A., Pennings, E.J., and Nielen, M.W. 2014. Rapid analysis of Δ-9-tetrahydrocannabinol in hair using direct analysis in real time ambient ionization orbitrap mass spectrometry, *Rapid Commun Mass Spectrom*. 28: 682–690.
14. Musah, R.A., Cody, R.B., Dane, A.J., Vuong, A.L., and Shepard, J.R. 2012. Direct analysis in real time mass spectrometry for analysis of sexual assault evidence, *Rapid Commun Mass Spectrom*. 26: 1039–1046.
15. Samms, W.C., Jiang, Y.J., Dixon, M.D., Houck, S.S., and Mozayani, A.J. 2011. Analysis of alprazolam by DART–TOF mass spectrometry in counterfeit and routine drug identification cases, *Forensic Sci*. 56: 993–998.
16. Sisco, E. and Forbes, T.P. 2016. Direct analysis in real time mass spectrometry of potential by-products from homemade nitrate ester explosive synthesis, *Talanta*. 150: 177–183.
17. Andrade, F.J., Shelley, J.T., Wetzel, W.C., Webb, M.R., Gamez, G., Ray, S.J., and Hieftje, G.M. 2008. Atmospheric pressure chemical ionization source. 1. Ionization of compounds in the gas phase, *Anal Chem*. 80: 2646–2653.
18. Na, N., Zhao, M., Zhang, S., Yang, C., and Zhang, X. 2007. Development of a dielectric barrier discharge ion source for ambient mass spectrometry, *J Am Soc Mass Spectrom*. 18: 1859–1862.
19. Smoluch, M., Gierczyk, B., Reszke, E., Babij, M., Gotszalk, T., Schroeder, G., and Silberring, J. 2016. FAPA mass spectrometry of designer drugs, *Talanta*. 146: 29–33.
20. Smoluch, M., Reszke, E., Ramsza, A., Labuz, K., and Silberring, J. 2012. Direct analysis of methcathinone from crude reaction mixture by flowing atmospheric-pressure afterglow mass spectrometry, *Rapid Commun Mass Spectrom*. 26: 1577–1580.
21. Smoluch, M., Mielczarek, P., Reszke, E., Hieftje, G.M., and Silberring, J. 2014. Determination of psychostimulants and their metabolites by electrochemistry linked on-line to flowing atmospheric pressure afterglow mass spectrometry, *Analyst*. 139: 4350–4355.
22. Mielczarek, P., Smoluch, M., Kotlinska, J.H., Labuz, K., Gotszalk, T., Babij, M., Suder, P., and Silberring J. 2015. Electrochemical generation of selegiline metabolites coupled to mass spectrometry, *J Chromatogr A*. 1389: 96–103.
23. Smoluch, M., Babij, M., Zuba, D., Schroeder, G., Gotszalk, T., and Silberring, J. 2015. Heat assisted sample introduction and determination of cannabinoids by dielectric barrier discharge ionization mass spectrometry, *Int. J. Mass Spectrom*. 386: 32–36.

7 Application of LC-QTOFMS for the Identification of the Structure of New Psychoactive Substances

Karolina Sekuła and Dariusz Zuba

CONTENTS

7.1 CHARACTERIZATION OF QUADRUPOLE TIME-OF-FLIGHT MASS SPECTROMETRY

A quadrupole time-of-flight tandem mass spectrometer (QTOFMS) is a hybrid of a quadrupole (Q) mass filter and a time-of-flight (TOF) analyzer. While originally targeted at the analysis of peptides and other large biological polymers, this device is now applied to environmental and food samples, and analysis of pharmaceutical products and psychoactive substances in seized samples and biological material.

The greatest advantages of QTOF mass spectrometers are good mass accuracy and high mass resolution. Mass accuracy (Δm) is defined as a difference between measured (experimental) mass (m_{exp}) and theoretical mass (m_{theor}) calculated from the elemental composition of the ion. It can be expressed in atomic mass units (Da) or, more often, in parts per million (ppm).

$$\Delta m = m_{exp} - m_{theor} \ (\text{Da}) \tag{7.1}$$

$$\Delta m = \frac{m_{exp} - m_{theor}}{m_{theor}} \times 10^6 \ (\text{ppm}) \tag{7.2}$$

To realize how the difference between the masses affects the value of mass accuracy, an exemplary correlation is shown in Table 7.1.

The QTOF mass spectrometers obtain 2–5 ppm mass accuracy. To minimize mass error, the QTOF system is continuously controlled by measuring the reference ions. Based on the difference between the measured mass-to-charge (m/z) values and theoretical mass of the reference ions, the accurate mass of the analyte ions is corrected and determined.

Mass resolution is closely connected with mass accuracy and it defines the ability of the analyzer to separate adjacent ions. In a mass spectrum, mass resolution is the observed m/z value divided by the smallest difference $\Delta(m/z)$ for two ions that can be separated: $(m/z)/\Delta(m/z)$. For a single peak corresponding to singly charged ions, the following equation, which determines resolution (R), is commonly used:

$$R = \frac{m}{\Delta m} \tag{7.3}$$

The value Δm can be expressed by the width of the peak measured at a specified fraction of the peak height, generally 50%, which is called the full width at half maximum (FWHM). For an isolated peak, resolution determines the mass accuracy of the analyte, i.e., correspondence between the measurement value and the theoretical mass.

In mass spectrometry, the term "resolving power" is also used. Mass resolving power is a measure of the ability of a mass spectrometer to provide a specified value of mass resolution.

The mass resolution values obtained by QTOF mass spectrometers are typically above 10,000. As in the case of the mass accuracy, the change in mass of the

TABLE 7.1

Correlation between Mass of a Compound and Mass Error

Mass (Da)	Mass Error (Da)	Mass Error (ppm)
300	±0.3000	1000
300	±0.0300	100
300	±0.0030	10
300	±0.0003	1
100	±0.0010	10
250	±0.0010	4
500	±0.0010	2
1000	±0.0010	1

TABLE 7.2

Correlation between Mass of Investigated Compounds and Possible Mass Resolution

Mass (m)	FWHM (Δm)	Resolution (R)
100	0.10	1000
500	0.10	5000
1000	0.10	10,000
2000	0.10	20,000
100	0.01	10,000
500	0.05	10,000
1500	0.15	10,000

Note: FWHM, full width at half maximum.

analyzed compounds also changes the resolution of the spectrometer. The higher the value of m/z of the investigated compound, the easier it is to get higher resolution. These correlations are shown in Table 7.2.

The characteristics of the QTOF mass spectrometer are the result of its construction. The QTOF system is complex, but the most important elements are, obviously, quadrupole, time-of-flight analyzer, and ionization source (Figure 7.1). The mass analyzer represents the heart of the mass spectrometer, i.e., the device able to measure the m/z ratios of gas-phase ions. To allow a free transition of the ions through the analyzer towards the detector, the analyzer must be operated under high vacuum conditions. The lower the pressure (typically in the range 10^{-4} to 10^{-7} torr), the longer the mean free path of the gas-phase ions and consequently the better sensitivity and mass resolution [1].

Before the analysis of a test sample, it must be introduced into the mass spectrometer first. The basic requirement is that a substance is able to ionize in the

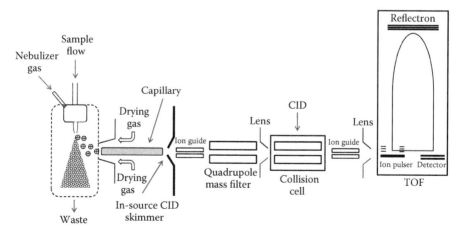

FIGURE 7.1 Schematic of the ESI-QTOFMS system.

applied conditions. The sample components separated in a liquid chromatograph are first entered into the ionization source. The LC-MS interface has a double task of eliminating the solvent from the LC eluent and producing gas-phase ions from the analyte [1]. The QTOF mass spectrometer with atmospheric pressure ionization (i.e., electrospray and nanoelectrospray) is the most frequent connection. An electrospray (ESI) is produced by applying a strong electric field under atmospheric pressure to a liquid passing through a capillary tube with a weak flux (1–10 µl/min). A capillary outlet is located in a stream of gas (usually nitrogen). Thanks to the use of gas and the potential difference (typically 3–5 kV) in the ionization chamber, the flow of liquid increases in the so-called Taylor cone [2]. The applied field induces a charge accumulation at the liquid surface located at the end of the capillary, which then breaks up to form highly charged droplets. As the solvent contained in these droplets evaporates, they shrink to the point where the repelling coulombic forces come close to their cohesion forces, thereby causing their explosion [3]. This process occurs repeatedly until gas-phase ions are produced by direct emission from the microdroplets.

If a compound can be charged at multiple sites in the solution, it will carry multiple charges under ESI conditions. This allows the use of mass analyzers with limited mass-to-charge (m/z) range (e.g., quadrupoles) to identify the high-molecular-weight compounds. The multiple charged ions are common in analysis of peptides or other large biological polymers by an ESI-QTOF mass spectrometer [1]. Electrospray can also be used in the case of molecules without any ionizable sites, by the formation of sodium, potassium, ammonium, or other adducts [3]. Moreover, dimers in combination with sodium or potassium ions ([2M+Na$^+$] or [2M+K$^+$]) are formed under ESI conditions.

Electrospray is classified as a soft ionization method. This means that during ionization there is practically no ion fragmentation of the analyte molecules (as opposed to, for example, electron impact ionization). Typical ions produced under ESI conditions are protonated molecular ions [M+H]$^+$, sodium or potassium adducts, or solvent adducts (in the positive mode), and deprotonated molecular ions [M–H]$^-$, or formate or acetate adducts (in the negative mode) [1].

Typical liquids used as solvents in electrospray ionization are alcohols (e.g., methanol, ethanol), nitriles (acetonitrile), and water (typically mixed with methanol). Depending on the planned mode of detection of ions (positive or negative), a small amount of selected acids or alkalis is added (typically 0.01–0.1% v/v) to the solution of analytes. In the case of analysis of ions with a positive charge, acid (usually formic acid) is used as a proton donor. If there is an intention to record the negative ions, a small addition of a base (generally ammonia) should be applied [2].

As mentioned, electrospray ionization produces very little fragmentation. However, fragmentation can be forced by adding kinetic energy to the ions of an analyte within the intermediate vacuum region placed between the ion source and the mass analyzer. By applying a difference in potential at two ends of this region (this parameter is called fragmentor voltage), the ions are accelerated and forced to collide with other molecular species (gas molecules, residual solvent, coeluting compound, etc.) and the energy gained as a result of these collisions is dissipated by fragmentation. This technique is called in-source fragmentation (or in-source collision-induced dissociation/

in-source CID) [1]. The application of in-source fragmentation to obtain full-scan mass spectra with main fragments of investigated compound is an important step in the identification process of a substance. Therefore, fragmentor voltage is one of the parameters to be optimized in the QTOFMS analysis to determine the degree of in-source fragmentation of the molecule.

Another part of the mass spectrometer is the skimmer. This is the separation element of regions at different pressures in a mass spectrometer, because, in the subsequent elements of the device, there is a growing vacuum. The additional task of the skimmer is to separate the analyte ions from the carrier gas, impurities, and residual solvent. Further, the ions are transferred through a quadrupole or an octopole ion guide (depending on manufacturer) followed by the lenses and go to the quadrupole mass filter. This additional quadrupole or octopole ion guide is used for collisional cooling and focusing of the ions entering the instrument.

The quadrupole consists of four parallel rods or poles equally spaced around a central axis. An electrical potential is applied to each rod so that each two adjacent poles have opposite polarities. By applying a precisely controlled combination of two electrostatic fields—one direct current (DC) and one at varying radiofrequency (RF)—a resonance frequency for a specific m/z ratio is obtained so that ions at that m/z ratio can reach the detector, whereas ions with lower and higher m/z ratios are discarded.

The quadrupole can work in two different modes. For single MS (or TOFMS) measurements, the quadrupole is operated in the total transmission ion mode so that it serves merely as a transmission element, while the TOF analyzer is used to record spectra. For MS/MS (or QTOFMS) measurements, the quadrupole is operated in the ion isolation mode to transmit only the specific parent ion, typically selecting in a mass window from 1 to 3 m/z wide depending on the desire to transmit the full isotopic cluster [4].

The essential fragmentation process of ions occurs in the collision cell. This phenomenon is the essence of tandem mass spectrometry, where the parent (precursor) ion m_p^+ isolated in a first analyzer is then subjected to a fragmentation giving daughter (product) ions m_d^+ and neutral fragments m_n.

$$m_p^+ \rightarrow m_d^+ + m_n \qquad (7.4)$$

The fragmentation process involves collision-induced dissociation (CID), also called collisionally activated dissociation (CAD). During this phenomenon, the precursor ion is accelerated and fragmented into product ions by collision with an inert gas. As a collision gas nitrogen or argon is most commonly used. As a result of collisions with gas, dissociation of ions occurs. Depending on the amount of collision energy, the fragmentation rate of ions changes. The greater the mass of a molecule, the more energy is needed to dissociate the ions. However, the degree of ion fragmentation also depends on the strength of each bond in a molecule. Therefore, the collision energy is the parameter that is optimized in the QTOFMS analysis.

It should be noted that single mass spectra can be acquired either with or without collision gas in the collision cell. In the former case, all parameters in the quadrupole section are set as for MS/MS mode, but the collision energy should be kept closed to 0 eV to avoid fragmentation.

Fragmented ions pass on to the time-of-flight analyzer. However, before ions enter this part of the spectrometer, the ion beam has to be flattened so that ions entering the analyzer have the same path length to cross. In this purpose, the quadrupole ion guide and ion focus lenses are commonly used.

The principle of operation of a TOF analyzer is based on measuring the time in which the ions take to cover a defined distance (i.e., the path in a flight tube to reach a detector) depending on the m/z value. Ions are accelerated by an electric field pulse. This acceleration results in ions of the same charges having equal kinetic energy before entering a field free-drift region, also known as the flight tube. Thus, the velocity of the ion depends on the mass-to-charge ratio (the lower the ion's mass, the greater the velocity and the ion reaches a detector faster). Typical flight times are 5–100 µs. In order to obtain an accurate measure of the flight time, the starting time of ions (i.e., the time at which ions enter the flight tube) needs to be precisely defined. Therefore, coupling with LC requires rapid electric field switching ("gating") to allow the ions produced in the ion source to enter the flight tube in a short and defined time. The TOF analyzers can operate in two ways: in a linear mode (the ions move in a straight line) and in a reflectron mode (using mirrors curving an ion stream). However, the reflectron-type TOF is currently the most common TOF analyzer in analytical toxicology applications because of much better mass resolution of up to 20,000–30,000 [1,2].

The QTOF detection system usually uses a microchannel-plate detector. The ion beam strikes the microchannel plate and generates a stream of electrons which are changed to photons. The photons strike a photomultiplier tube, generating an amplified signal proportional to ion flux [5].

7.2 CHALLENGES IN IDENTIFICATION OF STRUCTURES OF NEW PSYCHOACTIVE SUBSTANCES

In forensic laboratories, gas chromatography coupled to mass spectrometry with electron impact ionization (GC-EI-MS) is routinely applied to the identification of drugs. However, when new psychoactive substances began to appear on the drug market, this method was often insufficient. This was due to several reasons. First of all, the use of databases was often limited because the mass spectra of novel substances were obviously not added to libraries yet. Second, the problem is the structural similarity of designer drugs. Many positional isomers, homologs, analogs, and other types of derivatives that have been marketed have relatively simple and similar chemical structures (Figure 7.2).

Therefore, the risk of incorrect identification based only on an EI-MS spectrum is relatively high. The use of more sophisticated techniques is highly recommended. The very effective spectrometric technique is tandem mass spectrometry. It is based on controlled dissociation of the analyte ions into fragments and analysis of their masses. It is used to obtain detailed information about the molecular structure of

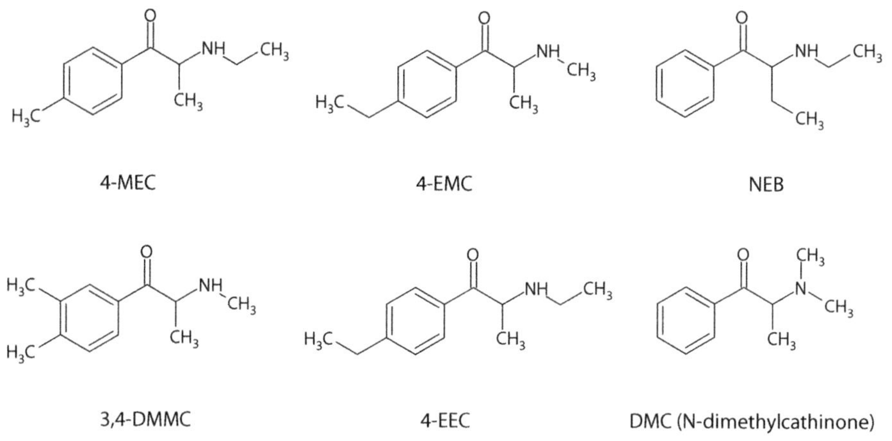

FIGURE 7.2 Structural similarity of compounds illustrated by the example of cathinone derivatives.

the investigated substance. Thanks to the process of fragmentation, we can find out which chemical groups are present in the sample and how they are bonded.

One of the methods used for identification of the structures of new psychoactive substances is liquid chromatography coupled to quadrupole time-of-flight mass spectrometry (LC-QTOFMS). This technique can be applied to both analyze seized materials containing novel compounds as well as to confirm the structures of designer drugs identified in biological materials.

7.3 PROCEDURES USED IN IDENTIFICATION OF NEW PSYCHOACTIVE SUBSTANCES BY LC-QTOFMS

The identification process of new psychoactive substances will be presented in this chapter in the example of quadrupole time-of-flight mass spectrometry (QTOFMS) under electrospray ionization (ESI). Analyzing the substances in various conditions, we can obtain different information about the investigated compounds.

The first great advantage of analysis by ESI-QTOFMS is the ability to determine the molecular mass of the compound. Because the electrospray is a soft ionization method, the pseudomolecular ions are obtained during the analysis. These ions are formed by addition or removal of one or more protons. Therefore, depending on the ionization mode, positive or negative, the [M+H]⁺ or [M–H]⁻ ions are observed for singly charged molecules. It should be noted that most new designer drugs are generally positive-ionized.

To obtain information about molecular mass, the analysis should be performed in MS mode (without collision-induced dissociation of ions in the collision cell) at a low fragmentor voltage to prevent an in-source fragmentation. The masses of most currently known psychoactive substances are in the range of 150–450 Da. Therefore, to avoid the fragmentation of pseudomolecular ion, a fragmentor voltage is maximally 100–150 V. An exemplary LC-ESI-QTOFMS spectrum (JWH-122) is presented in Figure 7.3.

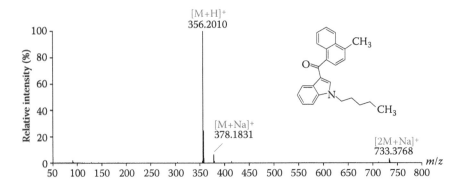

FIGURE 7.3 LC-ESI-QTOFMS spectrum of JWH-122 recorded in MS mode (fragmentor voltage 100 V).

Using an additive, e.g., formic acid to a LC mobile phase, we expect to get the ions formed by attachment of H^+ ions. Therefore, to calculate the molecular mass of an investigated compound on the basis of the pseudomolecular ion, $[M+H]^+$, the mass of hydrogen cation should be subtracted. The mass of a hydrogen atom is 1.0078 Da, the mass of an electron is about 0.0005 Da, thus the mass of the H^+ ion is 1.0073 Da. As a consequence, subtracting the value of 1.0073 from the value of m/z for pseudomolecular ion observed in the spectrum. For example, for the ion observed in Figure 7.3, we get the mass of the compound JWH-122 equal to 355.1937 Da. Comparison of experimental mass with theoretical mass of JWH-122 (m_{theor} = 355.1936 Da) indicates that the mass error, Δm, is 0.3 ppm.

In the case of an electrospray ion source, besides $[M+H]^+$ ions, sodium or potassium adducts are often observed in the spectra. For an inexperienced operator, it can be confusing if the m/z values observed in the mass spectrum come from the pseudomolecular ion and some fragment ions or perhaps from the pseudomolecular ion and some adducts. If we observe that the difference between these ions is about 22 Da or 38 Da, then we know that ions $[M+H]^+$ and $[M+Na]^+$ or $[M+H]^+$ and $[M+K]^+$ were probably registered. This assumption may be confirmed by calculating the exact difference in the masses of ions and compared to the mass of Na^+ or K^+ cations. Furthermore, in an electrospray chamber the ions of the tested compound can sometimes form dimers and they can also combine with sodium or potassium ions, resulting in the ions $[2M+Na]^+$ or $[2M+K]^+$ being observed in the ESI-QTOFMS spectrum. The $[M+H]^+$, $[M+Na]^+$, and $[2M+Na]^+$ ions of JWH-122 are marked in Figure 7.3.

Moreover, information on whether an investigated molecule of designer drug contains a halogen atom may be derived from the characteristic isotopic distribution. It is known that most elements occur in nature as a mixture of isotopes. Isotopes are atom species of the same chemical element that have different masses. They have the same number of protons and electrons, but a different number of neutrons. For example, the main elements and their naturally occurring isotopes are:

For hydrogen:	^{1}H (mass 1.0078 Da, abundance 99.99%),
	^{2}H (mass 2.0141 Da, abundance 0.01%);
For carbon:	^{12}C (mass 12.0000 Da, abundance 98.90%),
	^{13}C (mass 13.0034 Da, abundance 1.10%);
For oxygen:	^{16}O (mass 15.9949 Da, abundance 99.76%),
	^{17}O (mass 16.9991 Da, abundance 0.04%),
	^{18}O (mass 17.9992 Da, abundance 0.20%).

It can be noted that the lightest isotope is also the most abundant for these elements. In mass spectrometry, a monoisotopic mass is used. The monoisotopic mass is the sum of the masses of the atoms in a molecule using the principal isotopic mass of each atom instead of the isotope-averaged atomic mass (for example, the monoisotopic mass of mephedrone is 177.1154 Da). The average atomic mass of an element is defined as the weighted average of the masses of all its naturally occurring stable isotopes (the average mass of mephedrone is 177.2429 Da). On the other hand, in common language a nominal mass is often used. The nominal mass of an ion or a molecule is calculated using the integer mass (ignoring the mass defect) of the most abundant isotope of each element (the nominal mass of mephedrone is 177 Da). If we consider the elements that are components of the molecules of designer drugs, we can notice a characteristic isotopic composition for two halogens, i.e., chlorine and bromine atoms:

For chlorine:	^{35}Cl (mass 34.9689 Da, abundance 75.78%),
	^{37}Cl (mass 36.9659 Da, abundance 24.22%);
For bromine:	^{79}Br (mass 78.9183 Da, abundance 50.69%),
	^{81}Br (mass 80.9163 Da, abundance 49.31%).

Therefore, compounds that include these elements in their structures will have a characteristic isotope pattern. Figure 7.4 presents the isotope distributions of the molecules of the phenethylamine derivatives 25C-NBOMe (a) and 2C-B (c), containing, respectively, chlorine and bromine atoms, and they are compared with predicted relative isotopic abundances for the pseudomolecular ions of 25C-NBOMe (b) and 2C-B (d).

The ability to determine the mass of a compound is very useful in the identification of new psychoactive substances. In the case of the GC-MS method, we have to deal very often with a situation where the ions corresponding to small molecule fragments are observed in the spectrum. This is due to the fact that the electron impact technique is a hard ionization method. This makes the identification of novel substances difficult. Determination of the accurate molecular mass reduces the list of possible candidate formulae. For example, let us consider the situation when the pseudomolecular ion [M+H$^+$] for an unknown compound is observed at m/z value equal to 276.1592 Da. If we want to match the molecular formula to the mass of this

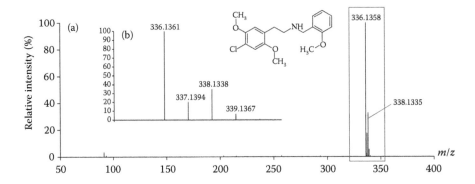

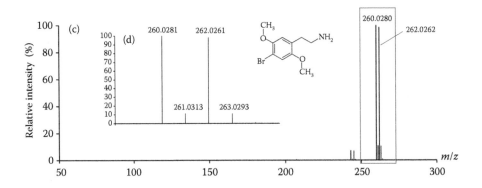

FIGURE 7.4 Isotopic pattern for the pseudomolecular ions of 25C-NBOMe (a) and 2C-B (c) compared with the predicted relative isotopic abundance for 25C-NBOMe (b) and 2C-B (d).

molecule, i.e., 275.1519 Da (neutral species), we can assume that the carbon, hydrogen, oxygen, nitrogen, chlorine, and bromine atoms are available in the following amounts: C: 0–50 atoms, H: 0–50 atoms, O: 0–5 atoms, N: 0–5 atoms, Cl: 0–2 atoms, and Br: 0–2 atoms. Using formula-finding software we can match four compounds with a maximum tolerance of 5 ppm, i.e.:

1. $C_{16}H_{21}NO_3$ $m_{theor} = 275.1521$ Da $\Delta m = 0.9$ ppm
2. $C_{11}H_{22}ClN_5O$ $m_{theor} = 275.1513$ Da $\Delta m = 2.2$ ppm
3. $C_{13}H_{24}ClN_2O_2$ $m_{theor} = 275.1526$ Da $\Delta m = 2.6$ ppm
4. $C_{14}H_{19}N_4O_2$ $m_{theor} = 275.1508$ Da $\Delta m = 4.0$ ppm

For comparison, if a high-resolution spectrometer is not available, we can match 38 compounds with a maximum tolerance up to 100 ppm. That is why the mass analyzer providing high mass accuracy allows the reduction of the number of possible molecular formulae matching to observed mass in the spectrum. Moreover, if the characteristic isotope pattern for chlorine is not observed in the spectrum, we can eliminate compounds having this atom in their structure. Thus, in this case, the compound was identified as a derivative of cathinone, i.e., MDPV with the chemical formula $C_{16}H_{21}NO_3$.

Therefore, sometimes only the determination of molecular mass by the QTOFMS method is sufficient to obtain information on which compound is present in the test sample. To consider this situation, the EI-GC-MS spectra of substances potentially present in a seized sample of plant material are shown in Figure 7.5.

It was suspected that, in the sample, compound A-834,735 or an N-heptyl analog of UR-144 was found. Both substances belong to a group of synthetic cannabinoids. They have the same nominal mass, i.e., 339 Da, but differ in the chemical formula ($C_{22}H_{29}NO_2$ for A-834,735 and $C_{23}H_{33}NO$ for heptyl derivative of UR-144). The EI-GC-MS spectra of these substances are very similar. If a laboratory does not have reference standards of these substances and therefore their retention times are not known, it is difficult to decide which compound is present in the sample. In this situation QTOFMS method is very useful. The difference between the monoisotopic masses for tested compounds is 107 ppm (m_{theor} = 339.2198 Da for A-834,735 and m_{theor} = 339.2562 Da for N-heptyl analog of UR-144). Because the mass accuracy measured by the QTOF mass spectrometer is 2–5 ppm, the mass of the synthetic cannabinoid can be easily determined with an accuracy of four decimal places and thus the correct compound can be identified.

However, to ascertain the structure of the tested compounds a fragmentation is needed. As mentioned before, fragmentation can occur in the source or in the collision cell. In-source fragmentation requires the use of higher fragmentor voltage. For currently known molecules of designer drugs, the fragmentor voltage in the range of 200–300 V is used to cause the dissociation of ions. The higher the voltage, the more significant the fragmentation. The exemplary QTOFMS spectra of mephedrone and buphedrone recorded in the MS mode using fragmentor voltage of 200 V are presented in Figure 7.6. The observed m/z values correspond to proposed structures of ions. It can be seen that the fragmentation of derivatives of cathinone is based on loss of water and further loss of methyl or ethyl radical attached to α-carbon adjacent to the nitrogen atom. It was deduced because the ions with m/z 160.1120 for mephedrone and m/z 160.1125 for buphedrone correspond to $C_{11}H_{14}N^+$ formula. In turn, the ions with m/z 145.0887 for mephedrone and m/z 131.0745 for buphedrone are formed by dissociation of fragments with m/z 15.0233 and 29.0380, respectively. Thus, these fragments originate from methyl and ethyl radicals.

Nevertheless, sometimes the analysis of fragment ions obtained by the in-source fragmentation is not sufficient to distinguish between two compounds. Figure 7.7 shows the spectra of compounds JWH-007 and JWH-019 registered in the MS mode using fragmentor voltage 300 V. These compounds have the same molecular formula, and obviously the same monoisotopic mass, but their structures differ. Therefore, they cannot be distinguished by analyzing them in the MS mode with low fragmentor voltage because the same pseudomolecular ions would be observed. As shown in Figure 7.7, the spectra recorded at a higher fragmentor voltage (300 V) are also very similar. The ions with m/z in the range 127.0540–127.0541 and 155.0490–155.0491 correspond to naphthyl and naphthoyl cations, whereas the ion with m/z in the range 228.1378–228.1381 originates from a substituted indole ring bonded to a carbonyl

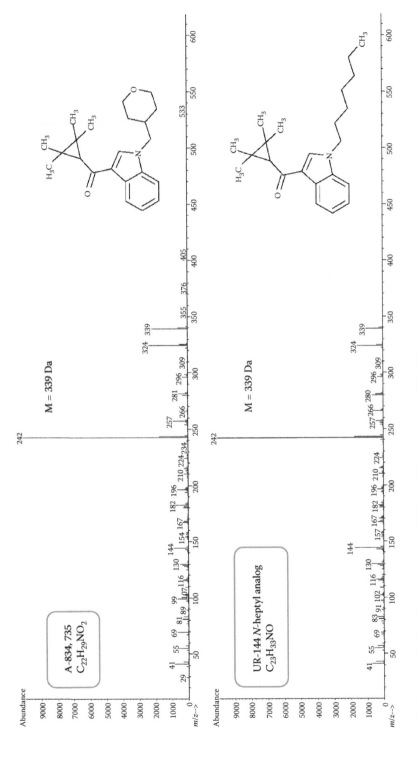

FIGURE 7.5 EI-GC-MS spectra of A-834,735 and *N*-heptyl analog of UR-144.

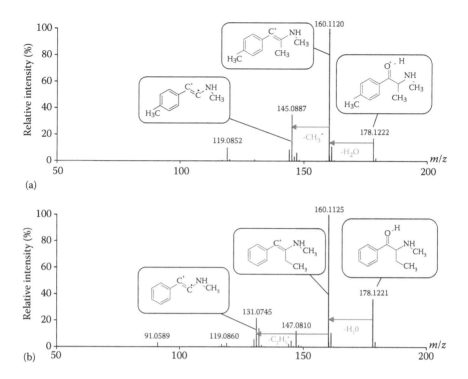

FIGURE 7.6 LC-ESI-QTOFMS spectra of mephedrone (a) and buphedrone (b) recorded in MS mode (fragmentor voltage 200 V).

group. Therefore, in order to distinguish compounds, the analysis in the MS/MS mode should be performed. The MS/MS experiments are based on the fragmentation of the selected ion in the collision cell. In this case, the ion with theoretical m/z value equal to 228.1383 should be chosen, because it corresponds to the fragments of the molecules, which are different for JWH-007 and JWH-019. As a consequence, the in-source fragmentation is needed to obtain a fragment with m/z 228.1383 and then, this ion is dissociated in the collision cell using collision energy CE = 30 eV. It can be seen that MS/MS spectra are useful to distinguish between these two synthetic cannabinoids, because the ion with m/z 158.0596 for JWH-007 and the ion with m/z 144.0440 for JWH-019 were observed. Structures for the obtained fragments are proposed and illustrated in Figure 7.7b and d.

Similarly to the in-source CID, the higher the collision energy used, the stronger the fragmentation of the parent ion. This correlation for synthetic cannabinoid NM-2201 is shown in Figure 7.8. The spectra were registered in MS/MS mode and a fragment ion with m/z 232.1127 was submitted to dissociation in the collision cell using collision energy in the range of 10–50 eV. It can be noticed that with increasing collision energy, the intensity of the parent ion m/z 232.1127 decreases, while the intensities of the ions with lower m/z values are higher.

It should be emphasized that determination of the structure of a compound is sometimes not obvious. For example, let us consider the analysis of a

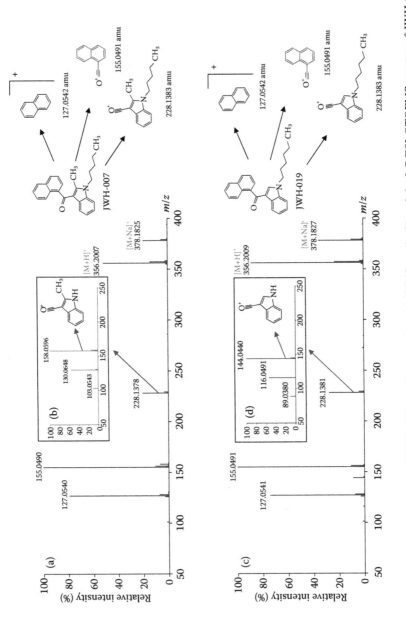

FIGURE 7.7 LC-ESI-QTOFMS spectra of JWH-007 in MS mode (a) and MS/MS mode (b) and the LC-ESI-QTOFMS spectra of JWH-019 in MS mode (c) and MS/MS mode (d).

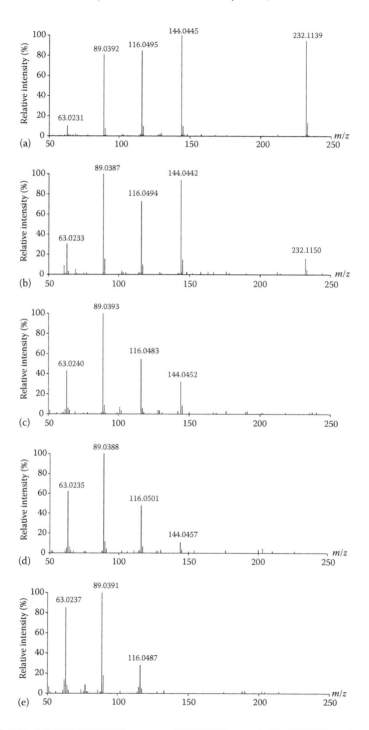

FIGURE 7.8 LC-ESI-QTOFMS spectra of NM-2201 recorded in MS/MS mode using the collision energy 10 eV (a), 20 eV (b), 30 eV (c), 40 eV (d), and 50 eV (e).

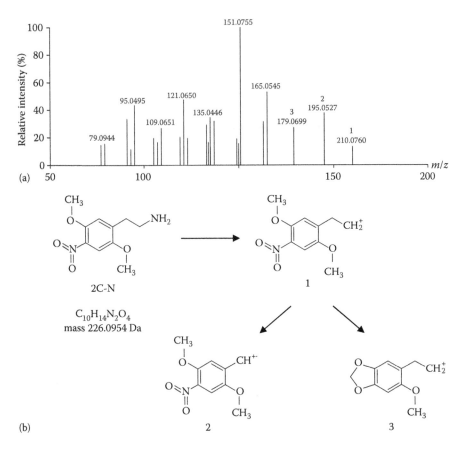

FIGURE 7.9 LC-ESI-QTOFMS spectrum of 2C-N recorded in MS/MS mode (a) and proposed structures of the obtained product ions (b).

phenethylamine derivative with acronym 2C-N. Figure 7.9a shows the spectrum of this compound registered in the MS/MS mode. The proposed structures of the obtained ions are illustrated in Figure 7.9b. The ion with m/z 210.0760 (1) is formed by loss of ammonium moiety, which is characteristic for compounds from the 2C family. The determined formula for this fragment is $C_{10}H_{12}NO_4^+$ ($\Delta m = -0.5$ ppm). The ion observed at m/z 195.0527 (2) may be formed by the removal of methyl radical, because this mass corresponds to $C_9H_9NO_4^+$ formula ($\Delta m = 0.5$ ppm). The next characteristic ion was recorded at m/z 179.0699 (3). Based on this exact mass, it was assumed that its chemical formula is $C_{10}H_{11}O_3^+$ ($\Delta m = -2.2$ ppm). In this case, this cation (3) seems to be formed by the dissociation of HNO moiety (NO from the nitro group and H from the OCH_3 group substituted at position 5 in the ring), and further rearrangement, which leads to the ring closure between oxygen atom from the nitro group and carbon atoms (first one from methoxy group and the second from benzene ring). This process leads to the formation of a methylenedioxy group, which is characteristic for MDMA molecule. Further fragmentation of

2C-N occurs in two ways through the cleavage of fragments 2 and 3. As described above, the atom rearrangement processes should also be considered when determining the structure of fragmented ions.

The presented cases relate to the analysis carried out in the so-called "targeted MS/MS" mode, where the parent ion is selected by the operator. Nevertheless, the experiments can also be conducted in "auto MS/MS" mode, where precursor ions are chosen automatically by a software algorithm. However, analysis of the spectra, in the latter case, seems to be faster, but ultimately is often more difficult, because the ions of non-analyte can be considered and may hinder the identification process.

It should be mentioned that the analysis of the obtained QTOFMS spectra can obviously be performed using available databases. In the case of MS experiments, the search algorithm is based on finding the mass of the compound with the minimum mass error. Thanks to the high mass accuracy of the QTOF mass spectrometer, the list of possible candidates is significantly reduced. For MS/MS experiments, the mass spectra of investigated compounds are compared with MS/MS spectra included in the libraries. For obvious reasons, the test sample must be analyzed at the same or very similar conditions (the fragmentor voltage and collision energy) in which spectra from mass libraries were registered. Otherwise, it may lead to misidentification of the investigated compound.

7.4 LC-QTOFMS FOR ANALYSIS OF SEIZED SAMPLES CONTAINING NEW PSYCHOACTIVE SUBSTANCES

Analysis of psychoactive substances by QTOF mass spectrometry using databases seems to be an easy and accurate method of identifying these compounds. The measurements of the exact masses of the parent compounds and their fragments allow confirmation of the presence of the component in the sample. Nevertheless, in case of such dynamic growth in the number of new substances on the drug market, the novel compound is often not added yet to mass spectra library. Therefore, the knowledge on fragmentation patterns of particular groups of designer drugs is necessary to predict the structures of the novel compounds. Recently, a growing number of papers in the field of forensic toxicology have been related to the identification of new psychoactive substances by means of liquid chromatography coupled to quadrupole time-of-flight mass spectrometry.

For example, this method was used to establish rules of fragmentation of naphthoylindoles under electrospray ionization. The developed procedure allowed the determination of substituent groups of the core naphthoylindole structure and distinction between positional isomers. The obtained results were used for the prediction of the ESI-MS spectra for many compounds from the naphthoylindoles family with a high affinity to cannabinoid receptors [6].

With regard to the synthetic cannabinoids, the analysis of six herbal samples containing JWH-019, JWH-081, JWH-203, and JWH-250 was performed using UHPLC-QTOFMS by Ibáñez et al. Based on accurate masses of ions observed in the spectra, the structures of designer drugs and their products of fragmentation were proposed [7].

In another study, the QTOF mass spectra were useful in determining the formula of a cannabinoid that was new at the time, i.e., compound A-834,735, which has a tetramethylcyclopropyl moiety instead of a common aromatic substituent (benzyl or naphthyl) found in compounds from the JWH family [8].

LC-QTOFMS was also applied for the identification of new derivatives of phenethylamine. After publication of the book *Phenethylamines I Have Known and Loved: A Chemical Love Story* by A. Shulgin, the compounds from the 2C family became very popular. At the beginning, derivatives with alkylthio and halogen group (e.g., 2C-T-2, 2C-T-7, 2C-B, 2C-I) were introduced to the drug market. Several years later, the first 2C-series members with an alkyl substituent at position 4 were reported to the European Monitoring Centre for Drugs and Drug Addiction (EMCDDA): 2C-D and 2C-E were identified in 2004, 2C-P in 2005, and 2C-G and 2C-N in 2011. In one paper, a comparison of fragmentation pattern of isomers 2C-E and 2C-G was presented and formulae of the product ions obtained in QTOFMS experiments were proposed. An identical set of ions was recorded for these substances using the aforementioned method in both MS and tandem mass spectrometry (MS/MS) mode. However, the distinction of the isomers was possible based on differences in the selected ion intensities [9]. Similar studies were carried out for compound 2C-N with a nitro group, which was identified in a light-yellow powder [10]. After extensive research on affinity for serotonin receptors, *N*-(2-methoxy)benzyl derivatives of the 2C-series of phenethylamines became popular on the drug market. This new class of hallucinogens was named 25-NBOMe, or simply NBOMe. The studies concerning analytical characterization of three new compounds, i.e., 25D-NBOMe, 25E-NBOMe, and 25G-NBOMe using, inter alia, LC-QTOFMS method were performed in 2012. Because of the chemical similarity of these hallucinogens to the compounds from the 2C group, i.e., 2C-D, 2C-E and 2C-G, respectively, analogous dissociation of molecules was reported [11]. As a result of similar identification of active substance applied on blotter paper, the presence of 25C-NBOMe was also confirmed. The characteristic isotope distribution of the molecules containing a chlorine atom was observed and it was compared with the predicted relative isotopic abundance for the pseudomolecular ion [12]. Six *N*-benzylphenethylamines based on the 2,5-dimethoxy-4-iodophenethylamine structure (including the positional isomers of 25I-NBOMe and 25I-NBBr) were also investigated by Brandt et al. These substances were extensively characterized using techniques commonly employed in forensic toxicology, inter alia, electrospray high mass accuracy quadrupole time-of-flight tandem mass spectrometry. The implementation of mass spectral techniques was helpful for differentiation between isomers [13].

The appearance on the drug market of another compound from a group of phenethylamines named 25I-NBMD was in response to the results of structure–affinity relationship (SAR) studies. In one paper, the analysis of fragments obtained during QTOFMS experiments was performed. The authors proposed the chemical formulae of 22 fragment ions and suggested the structures for them. The mass accuracy of *m/z* values observed in the spectra was calculated too. Additionally, the fission mechanism of 25I-NBMD and compound 2C-I from 2C-series members was compared based on the analysis of the same precursor ion present in both hallucinogens. Moreover, the fragmentation of the *N*-(2,3-methylenedioxy)benzyl cathion from 25I-NBMD and the *N*-(3,4-methylenedioxy)benzyl ion present in MDMA molecule was compared and discussed [14].

Studies of cathinone derivatives were also carried out by LC-QTOFMS. Analysis of samples containing the compounds from this group causes many problems because these designer drugs are very similar in their structure and they have many isomers. In the publication written by Zuba, the mass spectrometric methods were used to identify cathinones and other active components of "legal highs." It was found that as a result of QTOFMS fragmentation, characteristic ions were produced via loss of a water molecule in cathinones, forming secondary amines. The targeted MS/MS mode allowed identification of the structures of many unknown substances [15].

The analysis of cathinone derivatives was also conducted by Fornal et al. [16–18]. Six 3,4-methylenedioxy derivatives, methylone, butylone, pentylone, MDPBP, MDPV, and BMDP, were detected and identified by liquid chromatography coupled to a hybrid quadrupole time-of-flight mass spectrometer with nanoelectrospray ionization (nanoESI). Knowledge of the fragmentation pattern of these compounds in collision-induced dissociation (CID) could facilitate future screening and identification of new synthetic cathinones. For 3,4-methylenodioxy-derivative cathinones the loss of neutral groups CH_4O_2, H_2O, amines, and imines was observed [16]. In another paper, collision-induced fragmentation of 38 protonated cathinone derivatives (including six groups: phenyl, 4-methylphenyl, 4-methoxyphenyl, 4-ethylphenyl, 3,4-dimethylphenyl, and 3,4-dimethoxyphenyl derivatives) generated by electrospray ionization was examined by LC-QTOFMS. This method revealed that odd-electron product ions were often formed by collision-induced dissociation of protonated aryl α-primary amino ketones, contradicting the even-electron rule. Radical cations were among the most characteristic and most abundant ions in the CID-MS/MS product spectra, and were usually represented by basic peaks in the spectra [18].

Another investigated group were tryptamine derivatives [13]. The authors studied twelve substituted *N*-benzyl-5-methoxytryptamines. Due to the problems associated with the identification of newly emerging substances, analytical data based on mass spectral techniques were presented. The suggested structures of key ions formed during analysis by QTOF-MS/MS were presented.

Some papers presented very innovative approaches to the identification of new psychoactive substances. In one article, a rapid screening and determination of 11 new psychoactive substances by direct analysis in real-time mass spectrometry (DART-MS) and liquid chromatography quadrupole time-of-flight mass spectrometry were described and compared. The analysis included four cathinones, one phenylethylamine, and six synthetic cannabinoids. DART is a representative ion source of ambient mass spectrometry that has been commercialized and used for the analysis of various small molecules without sample preparation. Samples could be analyzed in various states including gases, liquids, and solids. Preparation procedures in both methods were simple and only need a small amount of solvent. However, compared to DART-MS, the LC-QTOFMS method provided a more satisfactory LOD, linearity, recovery, and repeatability. Nevertheless, the high analytical speed of the DART-MS method allowed quick screening of the confiscated samples that contained novel substances in high concentration [19].

7.5 LC-QTOFMS IDENTIFICATION OF NOVEL SUBSTANCES IN BIOLOGICAL MATERIAL

Quadrupole time-of-flight tandem mass spectrometry is increasingly being used for the analysis of biological material on the content of the new designer drugs. This method is slightly more difficult to use for routine analysis, but due to high accuracy of mass determination of the compound, it is well suited for identifying new substances and their metabolites in biological material.

Studies on metabolites of the designer drug methylenedioxypyrovalerone (MDPV) were performed by Strano-Rossi et al. Phase I and phase II metabolism of MDPV was evaluated *in vitro*. The resulting metabolites were subsequently liquid/liquid extracted and analyzed using gas chromatography coupled to mass spectrometry (GC-MS) as trimethylsilyl (TMS) derivatives. Structures of the metabolites were further confirmed by accurate mass measurements using a liquid chromatography/ quadrupole time-of-flight (LC-QTOF) mass spectrometer. The studies demonstrated that the metabolic pathway of MDPV is not complex. Similar to the designer drugs MDMA and MDA, the metabolism occurs mainly on the methylenedioxy functional group [20].

The next publication, written by Adamowicz et al., concerned the case of suspected intoxication by cannabinoids. The GC-MS analysis of seized traces of powder allowed the detection of synthetic cannabinoid named UR-144 and its pyrolysis product. The concentration of UR-144 in blood collected from a patient was determined using liquid chromatography–triple quadrupole tandem mass spectrometry (LC-QQQMS). Additionally, a urine sample was analyzed by liquid chromatography– quadrupole time-of-flight tandem mass spectrometry (LC-QTOFMS). The parent substance and its pyrolysis products were not detected in the urine sample, while five metabolites of UR-144 were found. The accurate mass experiments allowed the elucidation of the structures of the metabolites [21].

The objective of another study was to compare the performance of an immunoassay screening for synthetic cannabinoids with a newly developed confirmation method using liquid chromatography quadrupole time-of-flight mass spectrometry. The screening included metabolites of JWH-018, JWH-073, and AM-2201. The confirmation concerned metabolites of AM-2201, JWH-018, JWH-019, JWH-073, JWH-081, JWH-122, JWH-210, JWH-250, JWH-398, MAM-2201, RCS-4, and UR-144. The authors concluded that the rapid exchange of one cannabinoid to another may pose problems for any method, but immunoassays are generally becoming less useful for the screening of newly emerging drugs. They believed that the strategy of using a qualitative method and further a sensitive and selective technique such as quadrupole time-of-flight mass spectrometry yielded a flexible approach that can be adjusted to the analytes available on the drug market [22].

Similarly, in the next publication, due to limitations in immunoassay-based screening techniques, the studies relied on more laborious chromatographically based screening approaches in order to detect synthetic cathinones in biological material. Solid-phase extraction (SPE) and liquid chromatography quadrupole time-of-flight (LC-QTOF) mass spectrometry were used to identify 22 compounds from the aforementioned group in urine and blood. Using traditional reversed-phase chromatography, all positional isomers, including 3-FMC and 4-FMC, were separated in 12 minutes. The method was validated in accordance with the Scientific Working Group for Forensic Toxicology (SWGTOX) Standard Practices for Method Validation and provided the necessary sensitivity and specificity for use with authentic casework samples [23].

In another paper, the development of a liquid chromatography high-resolution mass spectrometry quadrupole-time-of-flight (LC-HRMS-QTOF) method for the analysis of new stimulant designer drugs (e.g., phenethylamine, amphetamine, cathinone, and piperazine derivatives) and common drugs of abuse (e.g., ketamine and ritalinic acid) in urine was reported by Paul et al. Sample preparation was carried out by a fast salting-out assisted liquid–liquid extraction (SALLE) procedure. The data was generated by a preferred target list combined with untargeted data-dependent acquisition recording additional sample information (i.e., not listed metabolites of target compounds or database-stored drugs). The identification was realized by a fully automated data-extraction algorithm, taking into account accurate mass spectra, fragment masses, and retention times [24].

A qualitative screening workflow based on data-independent acquisition mode (all-ions MS/MS) on liquid chromatography coupled to quadrupole-time-of-flight mass spectrometry for the detection and identification of new psychoactive substances in biological matrices was presented by Kinyua et al. According to the authors, developing targeted methods for analysis of novel compounds can be difficult and costly due to their variability on the drug market. A proposed workflow was applied to the two case studies involving drug intoxications. The method was able to confirm the presence of the parent compounds ketamine, 25B-NBOMe, 25C-NBOMe, and several predicted phase I and II metabolites that were not previously reported in urine and serum samples [25].

7.6 SUMMARY

Quadrupole time-of-flight tandem mass spectrometry (QTOFMS) was originally designed for the analysis of high-molecular-weight molecules, but in recent years has also been applied to environmental and food samples, pharmaceutical products, and forensic analysis. Due to the rapid growth in the number of new psychoactive substances on the drug scene, this method is increasingly being used to identify designer drugs. The main advantages of this device result from a hybrid combination of quadrupole and time-of-flight mass analyzer. The QTOF mass spectrometer offers high resolution over 10,000 and mass accuracy of 2–5 ppm. The use of electrospray ionization allows protonated $[M+H]^+$ or deprotonated $[M-H]^-$ molecules to be obtained and the mass of the investigated compound to be determined. Moreover, the characteristic isotope pattern may provide us with the information on whether the molecule contains a chlorine or bromine atom. In turn, based on the fragmentation, the structure of the test compound is ascertained. The fragmentation of the molecule can occur within the intermediate vacuum region placed between ion source and the mass analyzer (in-source collision-induced dissociation) and/or in the collision cell (collision-induced dissociation). By optimizing a fragmentor voltage and collision energy, we decide on the degree of the molecule fragmentation. Based on the exact masses of product ions, the structural formula of the investigated compound can be determined. High mass accuracy allows a reduction in the list of potential candidates for the analyte and the identification of which atoms form the molecule.

In the case of the analysis of new psychoactive substances, the use of spectral databases is often limited because the mass spectra of novel compounds are not added to the libraries yet. Therefore, the interpretation of QTOF mass spectra of designer drugs is often not performed automatically. This situation forces the need for knowledge of the fragmentation pattern of particular groups of new psychoactive substances in order to predict the mass spectra of a novel compound.

The LC-ESI-QTOFMS could certainly be treated as supplementary methods to the GC-EI-MS, which is routinely used in forensic laboratories. The mechanisms of fragmentation under different ionization techniques lead to different spectra and consequently, the identification of designer drugs can be based on independent analytical data.

ACRONYMS OF CHEMICALS USED IN THE CHAPTER

25B-NBOMe: 2-(4-bromo-2,5-dimethoxyphenyl)-*N*-(2-methoxybenzyl)ethanamine
25C-NBOMe: 2-(4-chloro-2,5-dimethoxyphenyl)-*N*-(2-methoxybenzyl)ethanamine
25D-NBOMe: 2-(2,5-dimethoxy-4-methylphenyl)-*N*-(2-methoxybenzyl)ethanamine
25E-NBOMe: 2-(4-ethyl-2,5-dimethoxyphenyl)-*N*-(2-methoxybenzyl)ethanamine
25G-NBOMe: 2-(2,5-dimethoxy-3,4-dimethylphenyl)-*N*-(2-methoxybenzyl)
          ethanamine
25I-NBBr: *N*-[(2-bromophenyl)methyl]-2-(4-iodo-2,5-dimethoxyphenyl)
          ethanamine
25I-NBMD: *N*-(1,3-benzodioxol-4-ylmethyl)-2-(4-iodo-2,5-dimethoxyphenyl)
          ethanamine
25I-NBOMe: 2-(4-iodo-2,5-dimethoxyphenyl)-*N*-(2-methoxybenzyl)ethanamine
2C-B: 2-(4-bromo-2,5-dimethoxyphenyl)ethanamine
2C-D: 2-(2,5-dimethoxy-4-methylphenyl)ethanamine
2C-E: 2-(4-ethyl-2,5-dimethoxyphenyl)ethanamine
2C-G: 2-(2,5-dimethoxy-3,4-dimethylphenyl)ethanamine
2C-I: 2-(4-iodo-2,5-dimethoxyphenyl)ethanamine
2C-N: 2-(2,5-dimethoxy-4-nitrophenyl)ethanamine
2C-P: 2-(2,5-dimethoxy-4-propylphenyl)ethanamine
2C-T-2: 2-[4-(ethylsulfanyl)-2,5-dimethoxyphenyl]ethanamine
2C-T-7: 2-[2,5-dimethoxy-4-(propylsulfanyl)phenyl]ethanamine
3-FMC: 1-(3-fluorophenyl)-2-(methylamino)-1-propanone
4-FMC: 1-(4-fluorophenyl)-2-(methylamino)-1-propanone
A-834,735: [1-(tetrahydro-2*H*-pyran-4-ylmethyl)-1*H*-indol-3-yl](2,2,3,3
          -tetramethylcyclopropyl)methanone
AM-2201: [1-(5-fluoropentyl)-1*H*-indol-3-yl](1-naphthyl)methanone
BMDP: 1-(1,3-benzodioxol-5-yl)-2-[(phenylmethyl)amino]-1-propanone
JWH-007: (2-methyl-1-pentyl-1*H*-indol-3-yl)(1-naphthyl)methanone
JWH-018: 1-naphthyl(1-pentyl-1*H*-indol-3-yl)methanone
JWH-019: (1-hexyl-1*H*-indol-3-yl)(1-naphthyl)methanone

JWH-073: (1-butyl-1*H*-indol-3-yl)(1-naphthyl)methanone
JWH-081: (4-methoxy-1-naphthyl)(1-pentyl-1*H*-indol-3-yl)methanone
JWH-122: (4-methyl-1-naphthyl)(1-pentyl-1*H*-indol-3-yl)methanone
JWH-201: 2-(4-methoxyphenyl)-1-(1-pentyl-1*H*-indol-3-yl)ethanone
JWH-203: 2-(2-chlorophenyl)-1-(1-pentyl-1*H*-indol-3-yl)ethanone
JWH-250: 2-(2-methoxyphenyl)-1-(1-pentyl-1*H*-indol-3-yl)ethanone
JWH-398: (4-chloro-1-naphthyl)(1-pentyl-1*H*-indol-3-yl)methanone
MAM-2201: [1-(5-fluoropentyl)-1H-indol-3-yl](4-methyl-1-naphthyl)methanone
MDA: 1-(1,3-benzodioxol-5-yl)-2-propanamine
MDMA: 1-(1,3-benzodioxol-5-yl)-N-methyl-2-propanamine
MDPBP: 1-(1,3-benzodioxol-5-yl)-2-(pyrrolidin-1-yl)butan-1-one
MDPV: 1-(1,3-benzodioxol-5-yl)-2-(1-pyrrolidinyl)-1-pentanone
NM-2201: 1-naphthyl 1-(5-fluoropentyl)-1H-indole-3-carboxylate
RCS-4: (4-methoxyphenyl)(1-pentyl-1H-indol-3-yl)methanone
UR-144: (1-pentyl-1H-indol-3-yl)(2,2,3,3-tetramethylcyclopropyl)methanone

REFERENCES

1. Politi, L., Groppi, A., and Polettini, A. 2006. Ionisation, ion separation and ion detection in LC-MS. In *Application of LC-MS in Toxicology*, ed. A. Polettini, 1–22. London: Pharmaceutical Press.
2. Suder, P. 2006. Jonizacja pod ciśnieniem atmosferycznym (API). In *Spektrometria mas*, eds. P. Suder and J. Silberring, 51–65. Krakow: Wydawnictwo Uniwersytetu Jagiellonskiego.
3. Barker, J. 1999. *Mass Spectrometry*. Chichester: John Willey & Sons.
4. Chernushevich, I. V., Loboda, A. V., and Thomson, B. A. 2001. An introduction to quadrupole-time-of-flight mass spectrometry, *J Mass Spectrom*, 36: 849–865.
5. Agilent Technologies. 2015. Technical Overview. Ion optics innovations for increased sensitivity in hybrid MS systems. https://www.agilent.com/cs/library/technicaloverviews /Public/5989-7408EN_HI.pdf (accessed 16 February 2017).
6. Sekuła, K., Zuba, D., and Stanaszek, R. 2012. Identification of naphthoylindoles acting on cannabinoid receptors based on their fragmentation patterns under ESI-QTOFMS, *J Mass Spectrom*, 47: 632–643.
7. Ibáñez, M., Bijlsma, L., van Nuijs, A. L., Sancho, J. V., Haro, G., Covaci, A., and Hernández, F. 2013. Quadrupole-time-of-flight mass spectrometry screening for synthetic cannabinoids in herbal blends, *J Mass Spectrom*, 48: 685–694.
8. Zuba, D., Geppert, B., Sekuła, K., and Żaba, C. 2013. 1[-(2,2,3,3tetramethylcyclopropyl)]-methanone: A new synthetic cannabinoid identified on the drug market, *Forensic Toxicol*, 31: 281–291.
9. Zuba, D. and Sekuła, K. 2013. Identification and characterization of 2,5-dimethoxy-3,4-dimethyl-β-phenethylamine (2C-G)—A new designer drug, *Drug Test Anal*, 5: 549–559.
10. Zuba, D., Sekuła, K., and Buczek, A. 2012. Identification and characterization of 2,5-dimethoxy-4-nitro-β-phenethylamine (2C-N)—A new member of 2C-series of designer drug, *Forensic Sci Int*, 222: 298–305.
11. Zuba, D. and Sekuła, K. 2013. Analytical characterisation of three hallucinogenic *N*-(2-methoxy)benzyl derivatives of the 2C-series of phenethylamine drugs, *Drug Test Anal*, 5: 634–645.
12. Zuba, D., Sekuła, K., and Buczek, A. 2013. 25C-NBOMe—New potent hallucinogenic substance identified on the drug market, *Forensic Sci Int*, 227: 7–14.

13. Brandt, S. D., Elliott, S. P., Kavanagh, P. V., Dempster, N. M., Meyer, M. R., Maurer, H. H., and Nichols, D. E. 2015. Analytical characterization of bioactive *N*-benzyl-substituted phenethylamines and 5-methoxytryptamines, *Rapid Commun Mass Spectrom*, 29: 573–584.
14. Sekuła, K. and Zuba, D. 2013. Structural elucidation and identification of a new derivative of phenethylamine using quadrupole time-of-flight mass spectrometry, *Rapid Commun Mass Spectrom*, 27: 2081–2090.
15. Zuba, D. 2012. Identification of cathinones and other active components of 'legal highs' by mass spectrometric methods, *Trends Analyt Chem*, 32: 15–30.
16. Fornal, E. 2013. Identification of substituted cathinones: 3,4-Methylenedioxy derivatives by high performance liquid chromatography-quadrupole time of flight mass spectrometry, *J Pharm Biomed Anal*, 81–82: 13–19.
17. Fornal, E., Stachniuk, A., and Wojtyla, A. 2013. LC-Q/TOF mass spectrometry data driven identification and spectroscopic characterisation of a new 3,4-methylenedioxy-*N*-benzyl cathinone (BMDP), *J Pharm Biomed Anal*, 72: 139–144.
18. Fornal, E. 2013. Formation of odd-electron product ions in collision-induced fragmentation of electrospray-generated protonated cathinone derivatives: Aryl α-primary amino ketones, *Rapid Commun Mass Spectrom*, 27: 1858–1866.
19. Nie, H., Li, X., Hua, Z., Pan, W., Bai, Y., and Fu, X. 2016. Rapid screening and determination of 11 new psychoactive substances by direct analysis in real time mass spectrometry and liquid chromatography/quadrupole time-of-flight mass spectrometry, *Rapid Commun Mass Spectrom*, 30 (Suppl 1): 141–146.
20. Strano-Rossi, S., Cadwallader, A. B., de la Torre, X., and Botre, F. 2010. Toxicological determination and in vitro metabolism of the designer drug methylenedioxypyrovalerone (MPDV) by gas chromatography/mass spectrometry and liquid chromatography/quadrupole time-of-flight mass spectrometry, *Rapid Commun Mass Spectrom*, 24: 2706–2714.
21. Adamowicz P., Zuba, D., and Sekuła, K. 2013. Analysis of UR-144 and its pyrolysis product in blood and their metabolites in urine, *Forensic Sci Int*, 233: 320–327.
22. Kronstrand, R., Brinkhagen, L., Birath-Karlsson, C., Roman, M., and Josefsson, M. 2014. LC-QTOF-MS as a superior strategy to immunoassay for the comprehensive analysis of synthetic cannabinoids in urine, *Anal Bioanal Chem*, 406: 3599–3609.
23. Glicksberg, L., Bryand, K., and Kerrigan, S. 2016. Identification and quantification of synthetic cathinones in blood and urine using liquid chromatography–quadrupole/time of flight (LC-Q/TOF) mass spectrometry, *J Chromatogr B*, 1035: 91–103.
24. Paul, M., Ippisch, J., Herrmann, C., Guber, S., and Schultis, W. 2014. Analysis of new designer drugs and common drugs of abuse in urine by a combined targeted and untargeted LC-HR-QTOFMS approach, *Anal Bioanal Chem*, 406: 4425–4441.
25. Kinyua, J., Negreira, N., Ibáñez, M., Bijlsma, L., Hernández, F., Covaci, A., and van Nuijs, A.L. 2015. A data-independent acquisition workflow for qualitative screening of new psychoactive substances in biological samples, *Anal Bioanal Chem*, 407: 8773–8785.

8 Rapid Generation of Metabolites by Electrochemical Method

Przemysław Mielczarek and Marek Smoluch

CONTENTS

8.1 INTRODUCTION

Addiction to drugs and psychoactive substances can cause severe problems, not only associated with the health of abusers but also social misery and economical aspects. Addiction has a high impact on the addict's family, but also on the society that is responsible for the financial expenses of therapy, treatment, infectious diseases, increase of robbery, and delinquency. Knowledge about addiction is constantly increasing, but to get the complete view about changes that occur in the body, new research is essential, especially for novel substances like legal highs and designer drugs. In 2014 alone, over 100 new psychoactive compounds were introduced to the market. Explanation of the biological pathways and particular mechanisms of each drug can help to create new therapies for addicted people.

Currently, the metabolism of xenobiotics is mainly studied *in vivo* using laboratory animals or *in vitro* by incubation of the examined substance with liver microsomes. Both strategies require application of advanced separation techniques to analyze metabolites in complex biological matrices obtained from the animals' tissues. What is more, this type of analysis is time-consuming and may cause ethical problems concerning experiments involving laboratory animals.

Cytochromes P450 (also called CYPs) are the main enzymes present in liver microsomes responsible for metabolic processes. Metabolism present in living cells can be divided into two phases. Phase I metabolism is mainly conducted by cytochromes P450 and relies on the oxidation and hydrolysis of xenobiotics. Metabolites of Phase I can also undergo Phase II metabolism, which draws up metabolism in conjugation with other molecules, such as glutathione or glucuronic acid [1]. Products of Phase II metabolism have an increased molecular mass and contain polar groups as a result of such reactions. They are less active and can easily be removed from the organism.

It was shown that electrochemical oxidation of several drugs like acetaminophen (paracetamol), amodiaquine, cocaine, selegiline, and other drugs may successively simulate metabolic processes [2,3]. Additionally, Phase II metabolism based on conjugation with glutathione can also be mimicked by electrochemical oxidation, as demonstrated for acetaminophen and amodiaquine [4].

The main function of cytochromes P450 is oxidation of compounds, which helps to eliminate them from the organism. CYPs are also responsible for the metabolism of molecules such as narcotics including cocaine, codeine, amphetamine, methamphetamine, MDMA (called ecstasy) and others [5]. Even knowing the molecular structure of a psychoactive compound, it is extremely difficult to predict its properties, such as toxicity and its influence on the human body. It is also important to determine which metabolic pathways are involved in the processing of this molecule and its metabolites. The best option to analyze metabolism would be *in vivo* tests of blood or urine collected from addicts; yet, for obvious reasons, such a solution would be unattainable. As a result, *in vitro* tests are the most important, fast, and noninvasive way to study metabolism. Electrochemical oxidation is an excellent screening test to determine the metabolites of novel compounds such as synthetic cannabinoids, legal highs, and designer drugs. A practical example of such an approach is presented in the forthcoming sections.

8.2 MATERIALS AND METHODS

8.2.1 Chemicals

XLR-11 ((1-[5′-fluoropentyl]indol-3-yl)-(2,2,3,3-tetramethylcyclopropyl)methanone) was obtained by courtesy of Professor Dariusz Zuba (Institute of Forensic Research, Krakow, Poland) under permission No. FAKR-I.857.3.31.2014.

Ammonium bicarbonate and ammonium formate were purchased from Fluka. Ammonium hydroxide, ethylenediaminetetraacetate (EDTA), formic acid, glycerol, β-nicotinamide adenine dinucleotide 2′-phosphate-reduced tetrasodium salt hydrate (NADPH), potassium chloride, pyrophosphate, and tris(hydroxymethyl)aminomethane (TRIS) were obtained from Sigma-Aldrich. Acetonitrile high-performance liquid chromatography (HPLC) (far UV/gradient grade) and methanol HPLC (gradient grade) were from J.T. Baker. Milli-Q water was purified using the Simplicity UV Water Purification System (conductivity 18.2 MΩ cm^{-1}) provided by Millipore.

A stock solution of XLR-11 was prepared in methanol at a concentration of 10 mM. The stock solution (stored at −20°C) was used to prepare samples shortly before each experiment.

8.2.2 INCUBATION OF XLR-11 WITH MICROSOMES

Microsomes were prepared as described earlier [5]. Incubation of XLR-11 with the rat liver microsomes was performed as follows. 100 mM phosphate buffer (pH 7.4) was incubated at 37°C and the freshly prepared NADPH solution was added to reach the final concentration of NADPH equal to 1 mM. Then the XLR-11 stock solution was added to obtain the final concentration of XLR-11 equal to 10 μM and the mixture was preincubated at 37°C. Microsomes were added to initiate the reaction (0.5 mg/mL final protein concentration) and further incubated for one hour at 37°C. The total volume of the solution was equal to 500 μL. The reaction was quenched by addition of an ice-cold methanol to reach 33% (v/v). The samples were centrifuged at 10,000 g for 10 minutes at 4°C and the supernatant was stored at −20°C for further processing. For the negative control, the experiment was performed as described above with the addition of a phosphate buffer instead of the NADPH solution.

To purify the produced metabolites, solid phase extraction (SPE) was carried out using UltraMicroSpin Columns, C18 beads (The Nest Group, Inc., Southborough, Massachusetts, U.S.), according to the manufacturer's protocol. After cleanup, the samples were dried and redissolved in mobile phase A used for chromatography.

To determine the metabolites of XLR-11, LC-MS analyses were performed. The experiments were performed with the application of the Nexera UHPLC system (Shimadzu, Tokyo, Japan). For all separations a Kinetex 2.6 μm, C18, 100 A column (Phenomenex, Torrance, California, U.S.) was used. The flow rate was equal to 400 μL/min. Separation under gradient conditions was used. Mobile phase A consisted of 0.1% formic acid in water, and mobile phase B's composition was 0.1% formic acid in acetonitrile. The gradient was 2–80% mobile phase B in 20 minutes operated under ambient temperature. The detection was performed in a positive-ion mode using the single quadrupole mass spectrometer LCMS-2020 (Shimadzu, Tokyo, Japan). Table 8.1 collects the applied settings of the mass spectrometer.

TABLE 8.1
Mass Spectrometer Settings Used for Detection during Liquid Chromatography

Parameter	Value
DL temperature	300°C
DL voltage	Default
Heat block	400°C
Detector voltage	1.7 kV
Interface voltage	4.5 kV
Qarray DC voltage	Default
Drying gas	15 L/min
Nebulizing gas	1.5 L/min

8.2.3 ELECTROCHEMICAL OXIDATION OF **XLR-11**

The electrochemical oxidation of XLR-11 as a simulation of the Phase I metabolism was performed using a potentiostat ROXY™ with a dual piston syringe pump (Antec, Zoeterwoude, The Netherlands) with an electrochemical ReactorCell™ (see Figure 8.1) that were controlled by the Dialog software (Antec, Zoeterwoude, The Netherlands).

The XLR-11 stock solution was diluted with three electrolytes at various pH presented in Table 8.2. All solutions were introduced to the electrochemical cell (the effective volume of the electrochemical cell was equal to 0.75 µL) using a flow rate of 10 µL/min. The electrochemical oxidation was performed using the boron-doped diamond working electrode (BDD), which is an ultrathin crystalline diamond layer deposited on top of a silicon substrate [6]. The experimental potential of the working electrode was linearly ramped up from 0 V to 3 V, measured versus Pd/H_2 reference electrode, at the rate of 10 mV/s. The surface of the working electrode was electrochemically activated before each experiment, as described in [7].

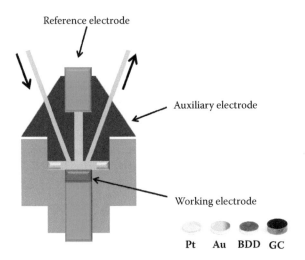

FIGURE 8.1   Construction of the electrochemical cell.

TABLE 8.2

Composition of Electrolytes Used during Electrochemical Oxidation of XLR-11

Nr	Electrolyte	Organic modifier	pH
1	0.1% (v/v) formic acid	50% (v/v) acetonitrile	2.0
2	20 mM ammonium formate	50% (v/v) acetonitrile	7.4
3	50 mM ammonium bicarbonate	50% (v/v) acetonitrile	9.0

The electrochemical products of XLR-11 collected during three cycles of electrochemical oxidation using scan mode were further analyzed by liquid chromatography linked to mass spectrometry. During the experiments, a homemade capillary column 10 cm × 75-μm i.d. packed with the ReproSil-Pur C-18-AQ, 3-μm stationary phase (Dr. Maisch GmbH, Germany) was used with a 2 cm × 100-μm precolumn packed with the 5-μm beads. Analyses were performed using the Proxeon EASY-nLC II (Bruker Daltonics, Germany) and the total flow rate was maintained at 300 nL/min. Gradient separation was applied using two solvents. Mobile phase A consisted of 0.1% formic acid in water, and mobile phase B was 0.1% formic acid in acetonitrile. The gradient profile 2–80% phase B in 50 minutes was used.

The electrochemical oxidation products were directly analyzed online by the amaZon ETD mass spectrometer (Bruker Daltonics, Germany) with an ion trap analyzer and electrospray (ESI) ion source operated in a positive-ion mode. For determination of the molecular structure, fragment mass spectra of the compounds were recorded.

8.3 RESULTS

8.3.1 NATURAL METABOLITES OF XLR-11

Natural metabolites of XLR-11 (peak 1) were determined based on the incubation with the rat liver microsomes. There are two main oxidation pathways of this synthetic cannabinoid, as shown in Figure 8.2. The first one is hydroxylation of the molecule. Three main metabolites of this process (isomers) are presented in Table 8.3 (peaks 2–4), all with the same molecular mass equal to 345.2 Da. The second metabolic pathway is N-dealkylation and the structure of this metabolite is shown in Table 8.3 (peak 5). The generated molecule has molecular mass equal to 241.3 Da.

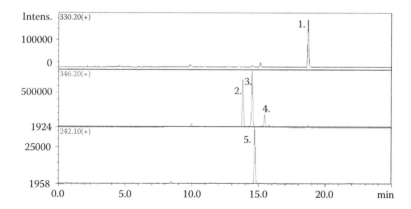

FIGURE 8.2 Chromatograms presenting separation of XLR-11 metabolites acquired after incubation with rat liver microsomes. (Peak numbers correspond to the structures presented in Table 8.3.)

TABLE 8.3

Overview of XLR-11 Metabolites and Electrochemical Oxidation Products

Peak Number	Formula	Structure	Monoisotopic Mass (Da)
1.	$C_{21}H_{28}FNO$	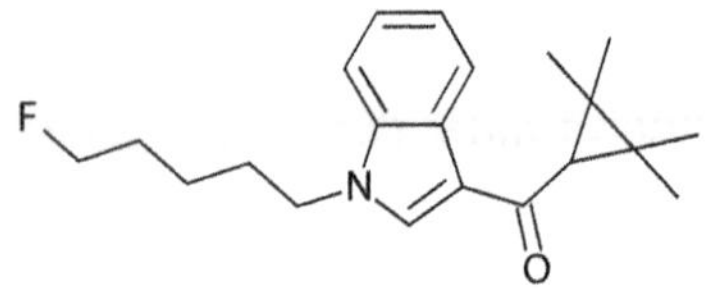	329.2155
2.	$C_{21}H_{28}FNO_2$	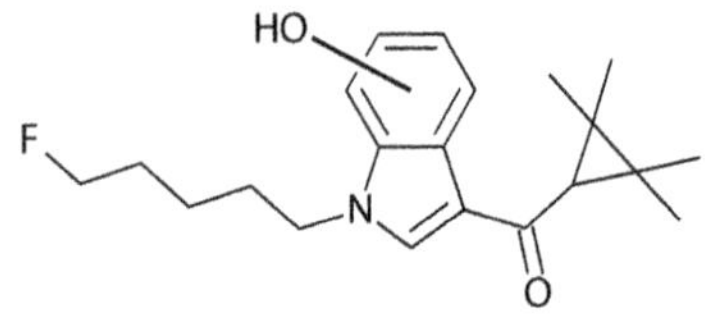	345.2104
3.	$C_{21}H_{28}FNO_2$	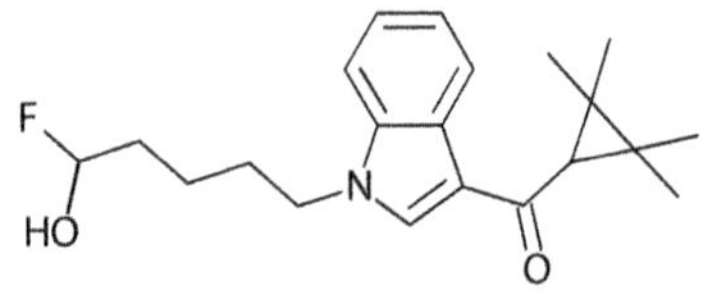	345.2104
4.	$C_{21}H_{28}FNO_2$	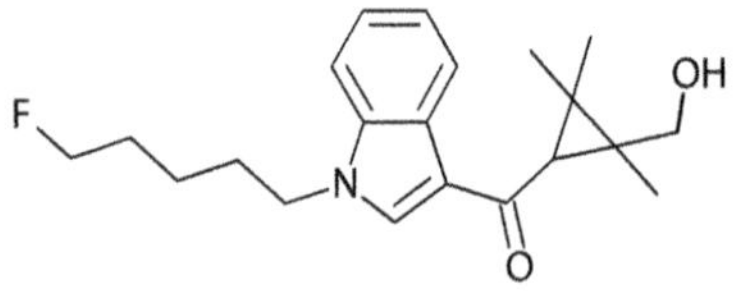	345.2104
5.	$C_{16}H_{19}NO$	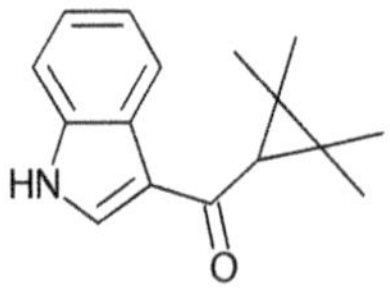	241.1467
6.	$C_{16}H_{19}NO_2$	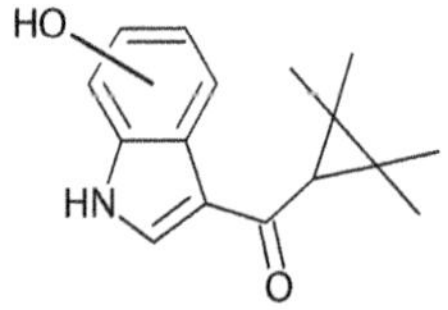	257.1416
7–8.	$C_{21}H_{28}FNO_3$	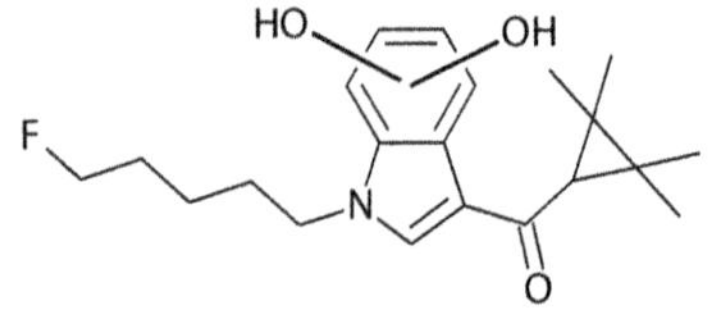	361.2053

(Continued)

TABLE 8.3 (CONTINUED)
Overview of XLR-11 Metabolites and Electrochemical Oxidation Products

Peak Number	Formula	Structure	Monoisotopic Mass (Da)
9.	$C_{21}H_{28}FNO_3$		361.2053
10.	$C_{21}H_{28}FNO_3$		361.2053
11.	$C_{21}H_{28}FNO_3$		361.2053

Note: Numbers correspond to the peaks present in the chromatograms.

Finally, there are several metabolites formed by a combination of these two processes, oxidation and N-dealkylation [8].

8.3.2 OXIDATION PRODUCTS OF XLR-11

Products of electrochemical oxidation of XLR-11 are presented in Figures 8.3 and 8.4. The first metabolite (peak 6) is a product of hydroxylation of the aromatic structure conjugated with N-dealkylation. The monoisotopic molecular mass of this product equals 257.1 Da and it was also identified as a natural metabolite after incubation with the rat liver microsomes.

The second group contains doubly hydroxylated forms of XLR-11. The main products generated electrochemically under acidic and basic conditions contain two hydroxyl groups substituted in the aromatic structure of the molecule (peaks 7 and 8). Both of them are natural metabolites of XLR-11 [8]. The second oxidation product (peak 9) is N-oxide, mainly formed at neutral pH and not identified after incubation of XLR-11 with the rat liver microsomes. Other metabolites (peaks 10 and 11) are XLR-11 derivatives containing hydroxyl groups substituted in the aliphatic structure of the molecule. This type of metabolite could not be generated electrochemically using the applied conditions and the BDD working electrode.

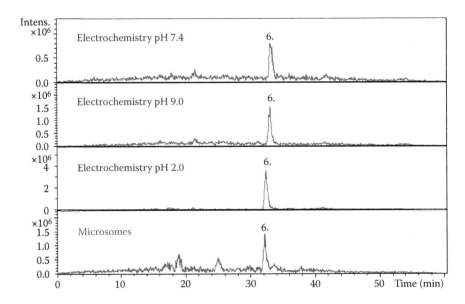

FIGURE 8.3 Extracted ion chromatograms of m/z 258.1 for electrochemical oxidation products of XLR-11 and a natural metabolite obtained by incubation with rat liver microsomes.

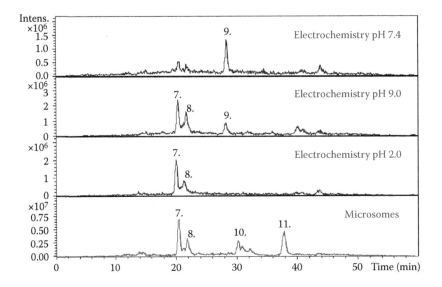

FIGURE 8.4 Extracted ion chromatograms of 362.2 m/z for electrochemical oxidation products of XLR-11 and natural metabolite obtained by incubation with rat liver microsomes.

8.4 CONCLUSIONS

Electrochemical oxidation of XLR-11 validates the possibility to simulate liver metabolism of drugs and psychoactive compounds. Reactions such as hydroxylation and N-dealkylation, both catalyzed by CYPs in the human body, can be effectively mimicked. However, hydroxylation of aliphatic structures could not be achieved.

8.5 SUMMARY

The electrochemical system shows new possibilities to rapidly simulate metabolism and fast analyses of the properties of new drugs and psychoactive compounds. Electrochemical oxidation products might be directly analyzed by mass spectrometry or separated by HPLC prior to mass spectrometry analysis. The use of animals can be avoided, similarly to tissue samples or their homogenates. This methodology is less time-consuming and does not require application of the advanced separation techniques, especially important for handling of biological samples.

The biggest problem pertains to proper optimization of analytical conditions, such as an appropriate electrolyte composition, the best potential of working electrode and the type of material it is constructed of. However, even the best optimization may not lead to the generation of all natural metabolites.

The described electrochemical oxidation is complementary to other routinely used techniques for studying metabolism, particularly for screening new, unknown substances with uncharacterized psychoactive properties like legal highs and designer drugs. These substances are popular on the illegal market and are potentially dangerous for adolescents that are unaware about the risks of their use. Finally, this technique can help to create new clinical tests, which might save the lives of abusers.

8.6 ACKNOWLEDGMENTS

The authors acknowledge SHIM-POL A. M. Borzymowski Company for granting the possibility to perform experiments on the Nexera UHPLC system with a single quadrupole mass spectrometer LCMS-2020 (Shimadzu, Tokyo, Japan).

REFERENCES

1. Meunier, B., de Visser, S.P., and Shaik, S. 2004. Mechanism of oxidation reactions catalyzed by cytochrome p450 enzymes, *Chem. Rev.*, 104: 3947.
2. Lohmann, W., Karst, U., and Baumann, A. 2010. Electrochemistry and LC–MS for metabolite generation and identification: Tools, technologies and trends, *LCGC Europe*, 23: 8.
3. Jahn, S. and Karst, U. 2012. Electrochemistry coupled to (liquid chromatography/) mass spectrometry—Current state and future perspectives, *J. Chromatogr. A*, 1259: 16.
4. Lohmann, W. and Karst, U. 2006. Simulation of the detoxification of paracetamol using on-line electrochemistry/liquid chromatography/mass spectrometry, *Anal. Bioanal. Chem.*, 386: 1701.

5. Mielczarek, P., Raoof, H., Kotlinska, J.H., Stefanowicz, P., Szewczuk, Z., Suder, P., and Silberring, J. 2014. Electrochemical simulation of cocaine metabolism—A step toward predictive toxicology for drugs of abuse, *Eur. J. Mass Spectrom. (Chichester)*, 20(4): 279–85.

6. Kraft, A., Stadelmann, M., and Blaschke, M. 2003. Anodic oxidation with doped diamond electrodes: A new advanced oxidation process, *J. Hazard. Mater.*, 103: 247–61.

7. Mielczarek, P., Smoluch, M., Kotlinska, J.H., Labuz, K., Gotszalk, T., Babij, M., Suder, P., and Silberring J. 2015. Electrochemical generation of selegiline metabolites coupled to mass spectrometry, *J. Chromatogr. A*, 1389: 96–103.

8. Jang, M., Kim, I.S., Park, Y.N., Kim, J., Han, I., Baeck, S., Yang, W., and Yoo, H.H. 2016. Determination of urinary metabolites of XLR-11 by liquid chromatography-quadrupole time-of-flight mass spectrometry, *Anal. Bioanal. Chem.*, 408(2): 503–16.

9 Gas Chromatography– Mass Spectrometry in the Analysis of Designer Drugs

Bogumiła Byrska and Roman Stanaszek

CONTENTS

9.1 INTRODUCTION

Analytical procedures used for the identification and detection of drugs of abuse seized from the drug market have undergone big changes and developments and faced great challenges in recent years. It is caused by the fast growth of a number of new psychoactive substances introduced onto the drug market, and their structural similarity. This is why the application of such popular analytical techniques like gas chromatography with flame ionization detection (GC-FID) or high-performance liquid chromatography with diode array detection (HPLC-DAD), which are commonly used in forensic laboratories, turned out to be insufficient. Even the use of gas chromatography coupled to mass spectrometry (GC-MS), which is still a basic tool for the forensic analysis of drugs, is limited while analyzing new psychoactive compounds. Many new psychoactive substances exhibit similar or even identical structural, chemical, chromatographic, and spectral properties, and thus the process of detection and identification of them becomes more complex, almost impossible to achieve by the means of routine analysis. This chapter deals with the application of GC-MS in the identification of designer drugs and problems that may arise when using it.

9.2   GAS CHROMATOGRAPHY–MASS SPECTROMETRY (GC-MS) IN FORENSIC DRUG ANALYSIS

Gas chromatography (GC) is a perfect separating technique but provides no molecular identification information about the chromatographed compound. On the contrary, mass spectrometry (MS) alone can give molecular identification information but, having no separating ability, will not be usable to analyze a mixture of chemical species at a time. One of the major drawbacks of the EI mass spectra is that molecular or pseudomolecular ions are frequently very small or missing. For designer drugs, there are many compounds showing very similar patterns of fragment peaks. In such cases, the appearance of a distinct molecular peak would be very useful for differentiating compounds with similar structures except for regioisomers. The coupling of the two instrumental devices has a very broad analytical power. Gas chromatography combined to mass spectrometry (GC-MS) is one of the major analytical tools of the forensic laboratory. It is capable of rapid routine separation and identification of many components in complex matrices. GC-MS has become a well-established analytical technique in a large variety of forensic applications and is now a method of choice used in many working analytical procedures in forensic laboratories.

The popularity of GC-MS is due to the fact that this technique provides reproducible mass spectra, thus allowing the building of databases, such as spectral libraries. One routine, user-friendly, fast, and simple method for identification of an unknown compound is to acquire the electron ionization (EI) mass spectrum of this compound by GC-MS at 70 eV and to perform a data search using commercially available mass spectral libraries. These databases are largely limited to EI mass spectra at 70 eV obtained by GC-MS, because of their excellent reproducibility. A collected spectrum of the analyzed compound can also be used to search in an in-house user's libraries. While probability base-matching (PBM) systems are very powerful, the best practice is to compare spectra of analytes with standards analyzed on the laboratory's equipment and under the same conditions. A commercial mass spectral library like the NIST (National Institute of Standards and Technology) Mass Spectral Library is the world-standard database for EI-MS, covering more than 220,000 spectra for a wide range of organic compounds and their derivatives. There are also specialized libraries, e.g., Maurer/Pfleger/Weber MS Library, which covers mass spectral data of 8,000 drugs, poisons, and their metabolites. Unfortunately, most active components of "legal highs" are not included in the NIST or Maurer/Pfleger/Weber libraries, whose application is limited. There are also commercial libraries available on the market that are dedicated for specific applications, such as the base for designer drugs created by Rosner et al. [1], but in the case of completely novel compounds, they are unavailable in such databases.

Moreover because of the close similarity in chemical structure of new compounds to those which are already banned, there is a great risk of an incorrect identification in seized material or during detection in biological specimens, especially by an unexperienced analyst. Therefore, noncommercial libraries created by specialists who are engaged in the work of scientific groups dealing with new psychoactive substances, such as the Drugs Working Group of the European Network of Forensic Science Institutes (ENFSI DWG), the European Monitoring Centre for Drugs and Drug Addiction (EMCDDA), or the American Scientific Working Group for the Analysis of Seized Drugs (SWGDRUG), play a valuable role. Their libraries are frequently updated and redistributed among their members, or published openly. The case of "legal highs" indicates clearly that international collaborative efforts are very valuable and supportive in the identification of previously unknown psychoactive compounds that are newly synthetized and introduced onto the drug market. There is also a recent trend and idea among specialists and institutions dealing with new psychoactive substances to create open, publicly available, easy-to-use, worldwide analytical databases of new psychoactive substances (NPS). These trends are already followed by scientific projects aimed at creating such libraries as the European Union (EU)-cofounded RESPONSE (Collect, Analyse, Organise, Evaluate, Share—A Response to Challenges in Forensic Drugs Analyses) project with an already-existing and very useful detailed analytical and spectral database, the European Commission Joint Research Centre's (JRC) electronic cloud repository, which enables the sharing of data, or the German Bundeskriminalamt—American NIST NPS data hub for nuclear magnetic resonance (NMR) and other analytical data with a Wikipedia-like approach to future public access.

Another problem exists in the availability of reference standards where there is a need to update quickly. The cost of all new certified standards is another issue the forensic laboratories must deal with. Continuous introduction of novel designer drugs onto the drug market requires continuous investment in purchasing new standards. It takes money but it also takes time. Commercial purchasing of new compounds that are sometimes already controlled in other countries may take up to several months, possibly too long for the laboratory's customers.

Because of the similarity of the designer drugs molecules, their structures, retention times, and mass spectra, the use of selected-ion monitoring (SIM) acquisition in GC-MS identification of such compounds seems to be strongly limited. Unlike full-scan spectral acquisition, assays utilizing SIM will generally not detect the presence of unsuspected and unexpected drugs that may be of forensic significance. Under several conditions it can be applied to the targeted analysis of certain drugs in biological specimens.

Tandem mass spectrometry (MS/MS) can provide an additional dimension of structural information for pure compounds and can serve as a tool for the detection of trace components in complex mixtures. More often, tandem mass spectroscopy is used in case of liquid chromatography–tandem mass spectrometry (LC-MS-MS), which is described in detail in other chapters.

Although chemical ionization mass spectrometry (CI-MS) may be used to achieve a gentler, less energetic fragmentation and improve the probability of obtaining a molecular ion or more precisely pseudomolecular ion, the accuracy with which such a mass is obtained is too small for differentiation between compounds that may possess similar molecular masses. Designer drugs, in many cases, are isomeric compounds that possess identical molecular mass and cannot be differentiated with

GC-MS chemical ionization. Therefore, CI-MS is quite limited in the analysis of legal highs. In such cases, high-resolution mass spectrometry systems (for example, time-of-flight mass detectors—TOF-MS) may find great application.

In cases where a certain compound is unavailable as a reference standard there is a strong need to apply more sophisticated techniques, such as liquid chromatography coupled to hybrid quadrupole–time-of-flight mass spectrometry (LC-QTOF-MS), NMR, or even Fourier transform infrared spectroscopy (FT-IR) in order to provide an unequivocal identification to a customer. These techniques give an analyst deeper structural information about a compound, like exact molecular mass, fragmentation pattern, or characteristic bonds. It is sometimes necessary to synthesize the suggested substances and make their analytical profile as broad as possible, but few forensic laboratories are equipped to perform organic synthesis.

9.3 APPLICATION OF GC-MS IN THE ANALYSIS OF DESIGNER DRUGS

9.3.1 ANALYSIS OF SYNTHETIC CATHINONES

Synthetic cathinones, like other designer legal highs, are entering the drug market apparently faster than they can be restricted. Analysis of such NPS and the identification of individual compounds may help in banning their production and their abuse. By the time enough information can be gathered about a drug in order to place it under temporary or permanent scheduling, replacement compounds are already created and made ready for distribution. Quick and efficient extraction methods will surely allow for rapid analysis of these compounds and increase the overall laboratory throughput. A comprehensive analytical method could potentially lead to reducing the time a new drug is available on the market. Based on the structure of the unsubstituted cathinone molecule, the designer modifications are possible in three distinct regions of the molecule: the aromatic ring, the alkyl side chain, and the amino group. Based on currently visible trends in the market, all three of the mentioned areas of cathinone structure are under exploration by illicit researchers in order to introduce designer-type modifications.

Synthetic cathinones are functionally similar and they fragment under electron ionization (EI) conditions to fragments with similar m/z. Therefore, they produce relatively similar spectra using EI. This is the reason that the analysis of the drugs utilizing only GC-MS becomes difficult, especially in the case when the compounds appear to have the same base peak and similar fragmentation patterns. Following a general rule, mass spectra show base peaks in the low mass and small or absent intensities for the molecular ions (M+). Zuba introduced the systematic identification of cathinones using the mass spectra obtained. First, it should be checked whether the molecular ion is observed. The immonium ion ($m/z = 16 + 14 n, n = 1, 2, 3, ...$) is then checked in the EI spectrum. If the immonium ion is found in the spectrum, the substance could be a straight-chained cathinone. If not, it is checked whether the ion for a pyrrolidine ring is observed ($m/z = 70 + 14 n, n = 1, 2, 3, ...$). If this ion is found in the spectrum, the substance could be a cathinone with a pyrrolidine ring in the molecule [2]. The mass-spectral fragmentation pattern of cathinone derivatives under electron impact conditions includes the formation of different iminium ions at m/z 44, 58, 72, 86, and 100. For cathinones with a pyrrolidine ring in the side chain, fragmentation leads to the formation of characteristic ions at m/z 70, 55, 42, and 41, results of the pyrrolidine ring degradation [3–5]. The alternative cleavage reaction, typical for aromatic ketones, results in the formation of the product of decarboxylation, the phenyl cation (m/z 77) [2]. All isomers of fluoromethcathinone evidently show significant fragments at m/z 95 and m/z 123; these correspond to the fluorophenyl cation and fluorobenzoyloxy cation, respectively [6].

Another work concerned the analysis of a series of homologs and regioisomers of 3,4-methylenedioxypyrovalerone (MDPV). The analyzed aminoketones show major fragments in their mass spectra corresponding to the regioisomeric and homologous immonium cation fragments primarily from the loss of the methylenedioxybenzoyl radical species. All ten analyzed compounds show equivalent EI MS fragments for the 3,4-methylenedioxybenzoyl fragments (m/z 149) and the methylenedioxybenzene fragment (m/z 121). The m/z 149 results from ionization of the carbonyl oxygen followed by an alpha-cleavage fragmentation. The loss of CO from this ion yields the m/z 121 fragments common to all spectra. The regioisomeric aminoketones produce equivalent mass spectra including mass equivalent regioisomeric immonium cation base peaks. Comparing the elution order of the regioisomers makes clear that the ring size has a significant effect on the elution order, thus the five-membered pyrrolidine ring (eluted before the piperidine six-membered ring and the seven-membered azepane ring) has the highest retention time [7].

An efficient extraction of the active compounds from the samples is crucial, especially because the products are not soluble due to adulterants, cutting agents, or other added materials. Leffler et al. tested numerous solvents, such as methanol, acetonitrile, toluene, hexane, acetone, ethyl acetate, carbon disulfide, methylene chloride, and distilled water. Internal and external quantification of the compounds showed that methanol removed the majority of the active substance in the first wash and was found to be the best extraction solvent [8].

Zuba et al. describes the first two cases elaborated in the Institute of Forensic Research (IFR) in which buphedrone was detected and quantitated in biological and seized materials [9]. Qualitative analysis of powders was performed by using GC-MS and quantitative analysis of the active ingredient was carried out using HPLC-DAD. Analyses of blood specimens were carried out using liquid chromatography–tandem mass spectrometry (LC-MS/MS). Target compounds in the GC-MS method were identified by matching their retention times and spectra against reference libraries, including ENFSI Drugs Working Group Mass Spectral Library and the authors' own library [10]. The GC-MS method has been successfully applied for the analysis of buphedrone and other common ingredients of "legal highs" in products seized on the market.

Daeid et al. proposed a screening GC-MS method for the seized samples, enabling the separation and identification of 16 cathinone derivatives. All the considered samples were dissolved in methanol with no derivatization and analyzed individually, as well as in combination with three commonly founded adulterants: benzocaine, lidocaine, and procaine. All analyzed cathinones were resolved from each other and from the three adulterants. In the cited paper, authors presented also the mass spectral data for each compound. The rapid single step extraction of samples in methanol with no necessity for derivatization made this method very useful for screening analysis of designer drugs [11].

9.3.2 ANALYSIS OF SYNTHETIC CANNABINOIDS

Some synthetic cannabinoids have never been described in the available literature, thus making their identification significantly difficult. Moreover, some substances have disappeared from the market after several months, bearing the risk that by the time reference standards are available, the product is no longer available, and people have already switched to other products. This apparent "creativity" of producers poses real problems for analytical laboratories to implement new substances into their screening methods. New analogs appear in the drug market right after the preceding drug comes under regulation [12].

When the structure of a side chain or substitution is slightly different from the one of the scheduled drug, the analog is perceived as being beyond the scope of the regulation in the countries with nongeneric legislation legal systems. These emerging drugs always display psychoactive actions because their chemical structures are very similar to those of the drugs being controlled.

Methanol and ethanol are often used to extract synthetic cannabinoids from herbal products. For example, Uchiyama and colleagues used methanol [13,14], and Moosmann et al. [15] recommended ethanol because of the high solubility of synthetic cannabinoids in this solvent. However, sometimes alcoholic solutions (methanol and ethanol) caused remarkable degradation of synthetic cannabinoids and gave degradation products different than the pattern compounds [16]. Molecular (M+) and/or fragment ions observed by full scan data acquisition of GC-MS reflect the structures of the synthetic cannabinoids. The fragmentation pathways of naphthoylindoles have been well studied for the identification of synthetic cannabinoids by GC-MS [17]. In naphthoylindoles (e.g., JWH-018), the carbonyl group fragment ions caused by α-cleavage of the alkylamino group of the indole and naphthoyl, and $[M-17]^+$ are usually observed. For benzoylindoles (e.g., RCS-4), the fragment ions being caused by α-cleavage of the alkylamino group of the indole and carbonyl groups are observed, although $[M-17]^+$ is not present on the spectra. In naphthoylindoles with methylpiperidine moiety (e.g., AM-1220), the ion of this group is observed as the base peak. For phenylacetyl (e.g., JWH-250), cyclopropyl, or adamantyl (e.g., APICA) indoles, the base peak of the fragment ion caused by the N-alkylindole 3-carbonyl moiety is observed solely in full scan spectrum [18,19]. In analogs, where the indole skeleton is changed to an indazole, molecular and N-dealkylated ions are typically observed in the spectrum (e.g., THJ-018, THJ-2201) [20]. In amide- or ester-type analogs bonded with an N-alkylindole or N-alkylindazole 3-carbonyl moiety (e.g., APICA, NNEI, ADBICA, QUPIC (PB-22), ADB-PINACA, AB-CHMINACA), the abundance of the molecular ion is low, and the fragment ion caused by the indoyl (or indazolyl) moiety is observed as a base peak. Although the fragment ion caused by elimination of the terminal $CO-NH_2$ is lower than that of the cleavage of the amide moiety in the indole analogs, such as ADBICA [14], the fragment ion caused by elimination of terminal $CO-NH_2$ is as intense as that of the cleavage of the amide moiety in indazole analogs such as ADB-PINACA and AB-CHMINACA [21].

9.4 DERIVATIZATION IN THE IDENTIFICATION OF DESIGNER DRUGS

A compound to be analyzed by gas chromatography should be stable in the vapor phase and have a minimum of about 1 torr vapor pressure at a maximum temperature near 300°C. Many compounds that do not have sufficient volatility can be derivatized to modify their structure by converting polar substituent groups to less polar entities that will be more volatile and thermally stable. Derivatization serves several important functions in the GC-MS analysis. It can greatly affect the volatility of a compound, improve its chromatographic behavior, and enhance the specificity of a compound's mass spectrum. Large derivatives yield molecular and fragment ions with higher masses, thus they may be more useful for smaller molecules (amphetamines, cathinones), while small derivatives may also be used for large-mass compounds (cannabinoids, opioids). Common derivatization methods are silylation (with bis-(trimethylsilyl)-trifluoroaceticamide BSTFA) and tifluoroacetylation (with trifluoroacetic acid anhydrate TFAA). They can be applied in many assays, which is why they are popular in many laboratories. However, one must be aware that the upper limit of the mass range of most benchtop spectrometers is approximately 650 amu and the molecular mass of the derivatives should not exceed this limit. Several extensive reviews of derivatization procedures and techniques have been published in the past [22].

Application of trifluoroacetylation in the analysis of three hallucinogenic N-(2-methoxy)-benzyl derivatives of the 2C-series of phenethylamine drugs is described by Zuba and Sekuła [23]. This type of derivatization made it possible to determine the molecular masses of the investigated substances that were absent in the GC-MS analysis without derivatization. The relatively intense ion in the 25D-NBOMe (2-(2,5-dimethoxy-4-methylphenyl)-N-(2-methoxybenzyl)ethanamine) spectrum was recorded at $m/z = 411$, whereas in the spectra of 25E-NBOMe (2-(4-ethyl-2,5-dimethoxyphenyl)-N-(2-methoxybenzyl)ethan-1-amine) and 25G-NBOMe (2,5-dimethoxy-N-[(2-methoxyphenyl)methyl]-3,4-dimethyl-benzeneethanamine) a signal was observed at $m/z = 425$. These masses correspond to the mono-substituted molecular ions (one hydrogen atom was substituted by the TFAA group). The dominant ion in the spectrum of 25D-NBOMe was observed at $m/z = 178$, whereas $m/z = 192$ was observed for the other two molecules. This was useful in the preliminary process of identification, although not sufficient because the latter two compounds had the same mass spectra and molecular and fragment ions.

Similar application of TFA derivatization was carried out successfully in the study that dealt with the identification of 25C-NBOMe (2-(4-chloro-2,5-dimethoxyphenyl)-N-(2-methoxybenzyl)ethanamine) in blotter papers [24]. EI-MS spectra of derivatized and underivatized 25C-NBOMe are shown in Figure 9.1. The characteristic chlorine isotope distribution can be observed in the underivatized compound's spectrum, while no molecular ion mass can be seen. Mono substitution of hydrogen by the TFA group allows for the detection of the $m/z = 431$ ion, which corresponds to the molecular mass of this derivative and thus to the parent compound. Cleavage of the C–N bond produces $m/z = 198$, which represents the remaining part after the $NCOCF_3C_9H_{11}O$ moiety loss.

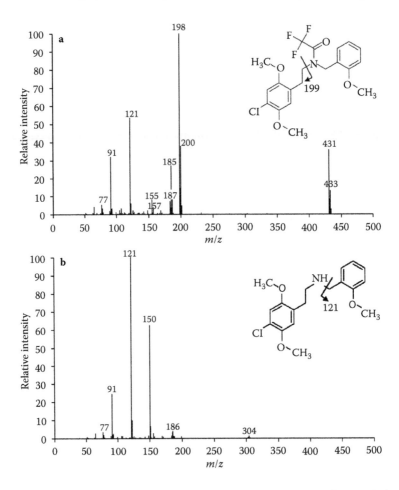

FIGURE 9.1   EI-MS spectra of TFA derivative of (a) 25C-NBoMe and (b) underivatized 25C-NBoMe.

Trifluoroacetic acid anhydrate may also be used as a derivatizing agent in the identification study of some phenethylamines, such as 2C-G [25] and 2C-N [26].

Derivatization may also prevent the phenomenon of transesterification, which has been observed when injections of the underivatized drugs have been made with methanol used as a solvent or even spontaneously when stocked in a methanolic solution.

9.5   RETENTION TIME LOCKING (RTL) AS AN ANALYTICAL TOOL IN IDENTIFICATION OF ISOMERS

Identification of designer drugs using GC-MS is based on mass spectra and retention times of the analyzed compounds. It is important to obtain repeatable results in the subsequent analyses. The mass spectra of regio- and ring-substituted analogs of designer drugs present on the drug market are very similar, therefore the identification of isomers by GC-MS is a challenge. Relying only upon the mass spectra of isomers may lead to misidentification, therefore retention time complements the information obtained with MS by comparing the retention time of the unknown peak with that of a reference standard. Electron impact mass spectra are reproducible in the subsequent analyses performed at the standard operating parameters of the mass spectrometer (70 eV). At the same time, slight changes of the GC parameters might affect the retention time. Retention time is directly dependent on the linear velocity of the carrier gas and its flow rate. The parameters are controlled by the appropriately adjusted inlet pressure of the carrier gas and the column's temperature. Contemporary gas chromatographs are equipped with electronic pressure control (EPC) and an oven temperature control system that allows keeping constant linear velocity or flow rate of the carrier gas in the course of the whole temperature program and reaching to retention time precisions better than 0.005 min [27]. Moreover, the type of the column also affects the precision of the retention time. The use of capillary columns increases the repeatability of the retention times of the analyzed compounds. Despite this, the retention time shifts frequently between subsequent analyses, because of, for example, elution of the stationary phase from the column. Routine maintenance procedures such as column trimming alter the retention times as well.

In a laboratory working on several instruments utilizing the same methods, comparison of retention time among chromatographs using the same system configuration and running under nominal conditions can be problematic. Differences between similarly configured GC systems may be caused by several reasons:

1. Columns of the same part number can vary slightly in length, diameter, and film thickness.
2. GC pneumatics can have small variations in the actual inlet pressure applied at a given set point.
3. The actual temperature of the GC oven has temporary deviations from the indicated value [28].

In order to compare the retention times between the analyses or among the instruments, retention time locking (RTL) is introduced. The background theory of the RTL method is thoroughly described in the literature [29]. The fundamental principle of RTL is that the retention times on a given GC setup can be configured by making the required adjustments to the inlet pressure.

RTL is realized using a standard (target analyte) that should express certain features, i.e., it should be chemically stable in the conditions of the analysis, it should have good chromatographic features, and its peak should be symmetric. Its retention time should be placed in the most critical area of the temperature program (between 1/3 and 2/3 of the whole analysis range). Moreover, it should be easily available for other potential users of this method. This standard is added to each sample. In order to lock a given method, either on one instrument or in order to repeat it on another system, it is necessary to: set the conditions of the method and then, after choosing an appropriate standard and choosing the target ions to be monitored, run the analysis. The next phase is the calibration of the retention time of the standard vs. the inlet pressure. In order to do this, the target analyte is analyzed five times using different pressures (the nominal method pressure and at ±10% and ±20%). The retention time of the standard is determined for each run. Then, after automatic software-based data analysis, the program computes the retention time. It is possible to alter this value—if one wants to obtain identical retention times on different apparatuses one should enter the desired value and save the methods. Five pairs of retention time and flow intensity are entered to the system, and the results of these measurements are shown as a calibration curve, with an appropriate equation shown. One should here turn attention to two values: measured time and blocked time of the standard. If one has an already blocked method on one apparatus and would like to achieve repeatable outcomes on another one, instead of blocked retention time one should enter another value; in this case the apparatus would recompute the new pressure, which, when used in the new method, would achieve the desired retention time of the standard.

In our routine work we applied the RTL method to differentiate JWH-122 and six of its regioisomers. The overlaid chromatograms and the structure of JWH-122 and its regioisomers was shown in Figure 9.2. As can be seen, retention time of JWH-122 and its 5-methylnaphthyl isomer is very similar. The mass spectra of these compounds show no differences. Reanalyses confirmed that the repeatability of retention times of these isomers is sufficient and differences in retention times are statistically significant (Student's t-distribution t-test value $= 31.3$, $p < 0.0001$). Detailed values of retention times of JWH-122 and its isomer in the 5 position are shown in Table 9.1.

In order to check if the RTL method is robust, 34 NSPs were analyzed, using five analyses per day (for four days in a row), on three different GC/MS instruments. The results of our experiments confirmed a very good retention time precision even among different apparatuses (RSD $< 0.5\%$).

Applying RTL makes work in the laboratory more efficient. Once the locked method can be unlocked and locked again, for example, after shortening or replacing the column, after changing the method for another detector, the same retention times for the analyzed substances can be achieved again. The software developed by

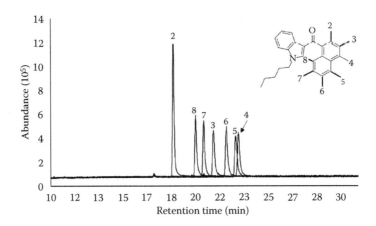

FIGURE 9.2 Overlaid chromatograms of JWH-122 (position 4) and its regioisomers (positions 2, 3, 5, 6, 7, and 8).

TABLE 9.1
Detailed Values of Retention of JWH-122 and Its 5-methylnaphthyl Isomer

Number of Analysis	Retention Time (min)	
	JWH 122 5-methylnaphthyl Isomer	JWH-122 4-methylnaphthyl Isomer
1	21.334	21.534
2	21.323	21.523
3	21.331	21.522
4	21.341	21.513
5	21.314	21.512
Mean ± SD	21.329 ± 0.010	21.521 ± 0.009

Agilent can be also be used for creating the databases of the RTL, making possible automatic identification of the analyzed substances. The databases can be created automatically or by manual adding of the new substances. In this case, defining of the retention time and its acceptable deviation as well as main ion and qualifying ions for each compound, along with acceptable variability ranges, is crucial. The software can also be used for conducting quantitative analyses. After using the same library and the same method, the results may be compared on different instruments using the same or a different configuration. Moreover, one can also compare the results of the analyses among the laboratories. Thanks to this method, the identification of substances is easier and more reliable.

9.6 THERMAL DEGRADATION OF DESIGNER DRUGS AS A COMMON PROBLEM IN THE ANALYSIS OF DESIGNER DRUGS BY GC-MS

The high temperature of the injector during GC-MS analysis causes some of the thermally liable compounds to undergo thermal degradation. Some of the factors that can affect this process are: injector temperature, injection mode (split, splitless), matrix, and the surface activity of the inlet liner. Serah Kerrigan et al. identified *in situ* thermal degradation products for 18 synthetic cathinones [30]. In their study, the loss of two hydrogen atoms, yielding a characteristic 2 Da mass shift, arose from the oxidative degradation reaction. Degradation products were characterized by prominent iminium base peaks with a mass-to-charge ratio 2 Da lower than the parent drug, and, in the case of the pyrrolidine-containing cathinones, predominant molecular ions arising from 2,3-enamine. Alpha-PVP is also suspected to be thermally labile and some other authors describe the thermal degradation of this compound during the injection process in GC-MS. Alpha-PVP is similarly decomposed to form an enamine with the double bond located on the alkyl side chain [31]. As an example from our case work, in Figure 9.3 we present the GC-MS total ion chromatogram, EI-MS spectra, and proposed fragments for 4-CMC and its thermal decomposition artifact shown as enamine. Less degradation was achieved by lowering the injection temperature, eliminating active sites during chromatographic analysis and resident

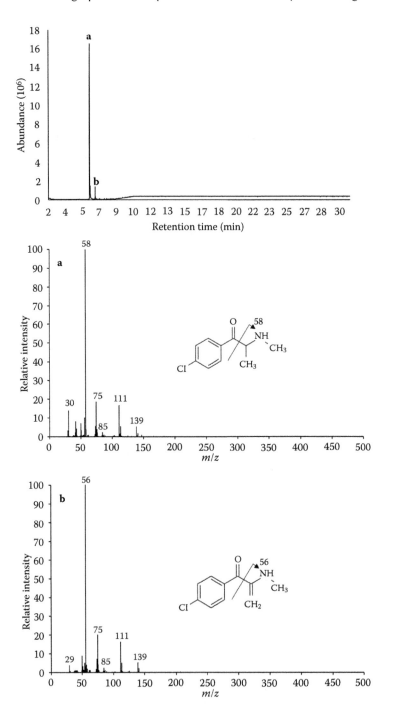

FIGURE 9.3 GC-MS total ion chromatogram and EI-MS spectra and proposed fragments for (a) 4-CMC and (b) its thermal decomposition artifact (shown as the enamine).

time in the inlet. Although derivatization can also improve thermal stability, not all the synthetic cathinones (e.g., alpha-PVP) are readily derivatized.

Another interesting example of thermal degradation is decomposition of QUPIC (quinolin-8-yl 1-pentyl-(1H-indole)-3-carboxylate) in the GC-MS system. When the sample was dissolved in alcoholic solution (methanol and ethanol), QUPIC decomposed to give methyl 1-pentyl-(1H-indole)-3-carboxylate, ethyl 1-pentyl-(1H-indole)-3-carboxylate, and methyl idole-3-carboxylate. Nonalcoholic solvents such as acetone, chloroform, and ethyl acetate gave a major peak of an unchanged QUPIC and a minor peak of the degradation product 2($1H$)-quinoline (which is also observed in alcoholic solution) [32]. As an example of hydrolytic degradation in the presence of methanol, we present transesterification of 5F-NPB-22, which is shown in Figure 9.4, with the corresponding mass spectra of the parent drug, its transesterification and thermal degradation products namely 5F-PB-22 indazole methylester analog and 2($1H$)-quinoline, respectively, shown in Figure 9.5. Also, other synthetic cannabinoid compounds with ester linkages (e.g., PB-22, 5F-SDB-005, FDU-PB-22) are susceptible to this process. To avoid thermal hydrolytic degradation in the presence of methanol, the samples containing such synthetic cannabinoids could be dissolved in a non-alcoholic solvent like acetonitrile or chloroform.

The emergence of new psychoactive substances continues to prove challenging to the forensic scientist. Some of the newly introduced compounds onto the drug market that are not yet controlled or banned may undergo decomposition or transformation into substances that are already controlled by the law. AM-2201 ([1-(5-fluoropentyl)-1H-indol-3-yl]-1-naphthalenyl-methanone) is an example of such a situation. During GC-MS analysis of the plant sample containing AM-2201, we detected JWH-022 ((4-methyl-1-naphthalenyl)(1-pentyl-1H-indol-3-yl)-methanone) besides the parent drug. The analysis by ultrapressure liquid chromatography with photo diode array detection (UPLC-PDA) revealed only the presence of AM-2201 in the sample.

According to Donohue and Steiner, JWH-022 may have occurred from the combustion of AM-2201 that was present in the plant material [33]. In our opinion, a similar process may take place in the GC-MS injector. Figure 9.6 shows GC-MS total ion chromatogram and EI-MS spectra of AM-2201 and its degradation product JWH-022, which was obtained during the GC-MS analysis of a herbal high sample delivered to the Institute of Forensic Research, Kraków, for forensic analysis. Liquid chromatography analysis is performed in much lower temperature and, hence, the thermal degradation process does not occur. This phenomenon may lead to some wrong interpretation and false legal consequences.

Synthetic cannabinoids with a tetramethyl cyclopropyl group (e.g., UR-144, XLR-11) exposed to elevated temperatures may undergo ring-opening conversion of the cyclopropyl ring and form several thermally depredated products, e.g., 3,3,4-trimethyl-1-(1-pentyl-1H-indol-3-yl)pent-4-en-1-one in the case of UR-144 [34]. It is hypothesized that the heat from smoking the synthetic cannabinoid as well as the injection port temperature causes the cyclopropyl ring to open.

Another interesting example of thermal degradation is decomposition of 4-acetylpsilocin (4-AcO-DMT) to psilocin (HO-DMT), shown in Figure 9.7,

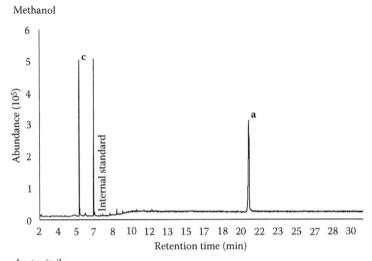

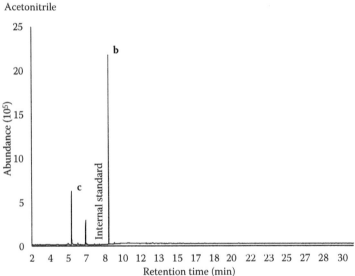

FIGURE 9.4   GC-MS total ion chromatograms obtained for 5F-NPB-22 dissolved in metha-nol and acetonitrile (peaks: (a) 5F-NPB-22, (b) 5F-PB-22 indazole methylester analog, and (c) 2(1*H*)-quinoline).

the latter one being on the list of banned substances; although, in some countries esters of psilocin may also be considered as controlled drugs.

The potential for synthetic cathinones, cannabinoids, and other thermally labile drugs should be taken into consideration during the identification process. It is also likely that users are unknowingly being exposed to novel designer drug structures from thermal decomposition and pyrolysis, which may have different pharmacologi-cal properties from the original chemical entities.

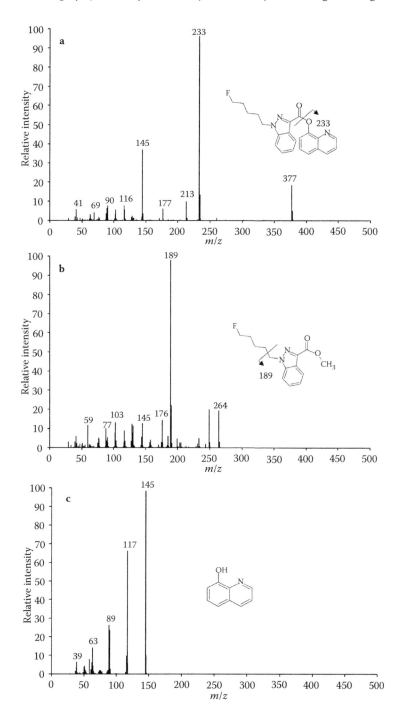

FIGURE 9.5   EI-MS spectra of (a) 5F-NPB-22, its transesterification and thermal degradation products, (b) 5F-PB-22 indazole methylester analog, and (c) 2(1H)-quinoline).

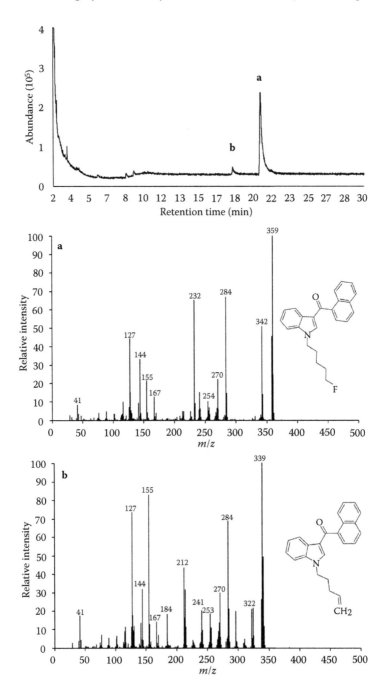

FIGURE 9.6   GC-MS total ion chromatogram and EI-MS spectra of (a) AM-2201 (MW = 359) and (b) its degradation product JWH-022 (MW = 339).

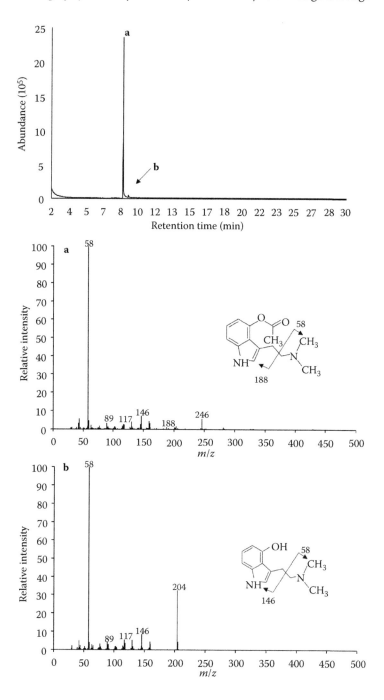

FIGURE 9.7   GC-MS total ion chromatogram and EI-MS spectra of (a) 4-acetylpsilocin and (b) its degradation product, psilocin.

9.7 SUMMARY

Identification of new psychoactive substances found in recreational preparations and detection of these compounds in biological specimens is not trivial. In order to properly follow the law, the forensic laboratory must be able to correctly identify the compounds that are present in the NPS. The process of detection and identification of such new substances is quite difficult to attain because many of them exhibit similar or even identical chemical, chromatographic, and spectral properties. It should be noted that not all analytical problems in the identification of designer drugs can be or should be addressed by such a robust and popular technique as gas chromatography coupled to mass spectrometry. The results of GC-MS analyses have to be interpreted carefully by experienced analysts. Modern analytical techniques, when used alone, may be unreliable. Thus, sometimes other sophisticated methods, such as LC-QTOF-MS, NMR, FT-IR, and GC-IR, must be applied in order to support unequivocal identification of an unknown compound. Furthermore, in the absence of good quality standards that might be available at a reasonable price and reference spectral data, identification becomes more complex, almost impossible to achieve by the means of routine analysis.

REFERENCES

1. Rosner, P., Junge, Th., Westphal, F., Fritschi, G. 2007. *Mass Spectra of Designer Drugs*, Wiley VCH, Weinheim.
2. Zuba, D. 2012. Identification of cathinones and other active components of 'legal highs' by mass spectrometric methods. *TrAC*. 32:15–30.
3. Paillet-Loilier, M., Cesbron, A., Le Boisselier, R., Bourgine J., Debruyne, D. 2014. Emerging drugs of abuse: Current perspectives on substituted cathinones. *Subst Abuse Rehabil*. 26(5):37–52. DOI: 10.2147/SAR.S37257.
4. Westphal, F., Junge, T., Girreser, U., Greibl, W., Doering, C. 2012. Mass, NMR and IR spectroscopic characterization of pentedrone and pentylone and identification of their isocathinone by-products. *Forensic Sci Int*. 217(1–3):157–167. DOI: 10.1016/j .forsciint.2011.10.045.
5. Westphal, F., Junge, T. 2012. Ring positional differentiation of isomeric N-alkylated fluorocathinones by gas chromatography/tandem mass spectrometry. *Forensic Sci Int*. 223:97–105.
6. Archer, R. P. 2009. Fluoromethcathinone, a new substance of abuse. *Forensic Sci Int*. 185(1–3):10–20.
7. Hamad Abiedalla, Y. F., Abdel-Hay, K., De Ruiter, J., Randall Clark, C. 2012. Synthesis and GC-MS analysis of a series of homologs and regioisomers of 3,4-methylenedioxypyrovalerone (MDPV). *Forensic Sci Int*. 223:189–197.
8. Leffler, A. M., Smith, P. B., De Armas, A., Dorman, F. L. 2014. The analytical investigation of synthetic street drugs containing cathinone analogs. *Forensic Sci Int*. 234:50–56. DOI: http://dx.doi.org/10.1016/j.forsciint.2013.08.021.
9. Zuba, D., Adamowicz, P., Byrska, B. 2012. Detection of buphedrone in biological and non-biological material—Two case reports. *Forensic Sci Int*. 227(1–3):15–20. DOI: 10.1016/j.forsciint.2012.08.034.
10. Zuba, D., Byrska, B., Pytka, P., Sekuła, K., Stanaszek, R. 2011. *Widma masowe składników aktywnych preparatów typu dopalacze*, Institute of Forensic Research Press, Kraków.

11. Daeid, N. N., Savag, K. A., Ramsay, D., Holland, C., Sutcliffe, O. B. 2014. Development of gas chromatography–mass spectrometry (GC-MS) and other rapid screening methods for the analysis of 16 'legal high' cathinone derivatives. *Sci Just.* 54(1):22–31.

12. Zuba, D., Byrska, B. 2013. Analysis of the prevalence and coexistence of synthetic cannabinoids in 'herbal high' products in Poland. *Forensic Toxicol.* 31(1):21–30.

13. Kikura-Hanajiri, R., Uchiyama, N., Kawamura, M., Goda, Y. 2013. Changes in the prevalence of synthetic cannabinoids and cathinone derivatives in Japan until early 2012. *Forensic Toxicol.* 31:44–53.

14. Uchiyama, N., Matsuda, S., Kawamura, M., Kikura-Hanajiri, R., Goda, Y. 2013. Two new-type cannabimimetic quinolinyl carboxylates, QUPIC and QUCHIC, two new cannabimimetic carboxamide derivatives, ADB-FUBINACA and ADBICA, and five synthetic cannabinoids detected with a thiophene derivative a-PVT and an opioid receptor agonist AH-7921 identified in illegal products. *Forensic Toxicol.* 31:223–240.

15. Moosmann, B., Kneisel, S., Wohlfarth, A., Brecht, V., Auwärter, V. 2013. A fast and inexpensive procedure for the isolation of synthetic cannabinoids from 'Spice' products using a flash chromatography system. *Anal Bioanal Chem.* 405:3929–3935.

16. Tsujikawa, K., Yamamuro, T., Kuwayama, K., Kanamori, T., Iwata, Y. T., Inoue, H. 2014. Thermal degradation of a new synthetic cannabinoid QUPIC during analysis by gas chromatography–mass spectrometry. *Forensic Toxicol.* 32:201–207. DOI: 10.1007 /s11419-013-0221-6.

17. Hudson, S., Ramsey, J. 2011. The emergence and analysis of synthetic cannabinoids. *Drug Test Anal.* 3:466–478.

18. Namera, A., Kawamura, M., Nakamoto, A., Saito, T., Nagao, M. 2015. Comprehensive review of the detection methods for synthetic cannabinoids and cathinones. *Forensic Toxicol.* 33:175–194. DOI: 10.1007/s11419-015-0270-0.

19. Akutsu, M., Sugie, K., Saito, K. 2017. Analysis of 62 synthetic cannabinoids by gas chromatography–mass spectrometry with photoionization. *Forensic Toxicol.* 35:94–103. DOI:10.1007/s11419-016-0342-9.

20. Uchiyama, N., Shimokawa, Y., Kawamura, M., Kikura-Hanajiri, R., Hakamatsuka, T. 2014. Chemical analysis of a benzofuran derivative, 2-(2-ethylaminopropyl)benzofuran (2-EAPB), eight synthetic cannabinoids, five cathinone derivatives, and five other designer drugs newly detected in illegal products. *Forensic Toxicol.* 32:266–281.

21. Uchiyama, N., Shimokawa, Y., Matsuda, S., Kawamura, M., Kikura-Hanajiri, R., Goda, Y. 2014. Two new synthetic cannabinoids, AM-2201 benzimidazole analog (FUBIMINA) and (4-methylpiperazin-1-yl)(1-pentyl-1H-indol-3-yl)methanone (MEPIRAPIM), and three phenethylamine derivatives, 25H-NBOMe 3,4,5-trimethoxybenzyl analog, 25B-NBOMe, and 2CN-NBOMe, identified in illegal products. *Forensic Toxicol.* 32:105–115.

22. Blau, K. and Halket, J., Eds. 1993. *Handbook of Derivatives for Chromatography*, John Wiley & Sons, Chichester.

23. Zuba, D., Sekuła, K. 2012. Analytical characterisation of three hallucinogenic N-(2-methoxy)-benzyl derivatives of the 2C-series of phenethylamine drugs. *Drug Test Anal.* 5(8):634–645. DOI: http://dx.doi.org/10.1002/dta.1397.

24. Zuba, D., Sekuła, K., Buczek, A. 2012. 25C-NBOMe—New potent hallucinogenic substance identified on the drug market. *Forensic Sci Int.* 227(1–3): 7–17. DOI: http://dx.doi.org/10.1016/j.forsciint. 2012.08.027.

25. Zuba, D., Sekuła, K. 2013. Identification and characterization of 2,5-dimethoxy-3,4-dimethyl-β-phenethylamine (2C-G)—A new designer drug. *Drug Test Anal.* 5(7):549–559. DOI: http://dx.doi.org/10.1002/dta.1396.

26. Zuba, D., Sekuła, K., Buczek, A. 2012, Identification and characterization of 2,5-dimethoxy-4-nitro-β-phenethylamine (2C-N)—A new member of 2C-series of designer drug. *Forensic Sci Inter.* 222:298–305.

27. Etxebarria, N., Zuloaga, O., Olivares, M., Bartolomé, L. J., Navarro, P. 2009. Retention-time locked methods in gas chromatography. *J Chromatogr A*. 1216(10):1624–1629. DOI: 10.1016/j.chroma.2008.12.038.
28. Giarocco, V., Quimby, B., Klee, M. 1997. *Retention Time Locking: Concepts and Applications*. Agilent Technologies, Application Note 228-392, Publication (23) 5966-2469E.
29. Blumberg, L. M., Klee, M. S. 1998. Method translation and retention time locking in partition GC. *Anal. Chem.* 70(18):3828–3839. DOI: 10.1021/ac971141v.
30. Kerrigan, S., Savage M., Cavazos, C., Bella, P. 2016. Thermal, degradation of synthetic cathinones: Implications for forensic toxicology. *J Anal Toxicol*. 40(1):1–11. DOI: 10.1093/jat/bkv099.
31. Tsujikawa, K., Kuwayama, K., Kanamori, T., Iwata, Y. T., Inoue, H. 2013. Thermal degradation of α-pyrrolidinopentiophenone during injection in gas chromatography/mass spectrometry. *Forensic Sci Int*. 231(1–3): 296–299. DOI: 10.1016/j.forsciint.2013.06.006.
32. Tsujikawa, K., Yamamuro, T., Kuwayama, K., Kanamori, T., Iwata, Y. T., Inoue, H. 2014. Thermal degradation of a new synthetic cannabinoid QUPIC during analysis by gas chromatography–mass spectrometry. *Forensic Toxicol*. 32:201–207. DOI:10.1007/s11419-013-0221-6.
33. Donohue, K. M., Steiner, R. R. 2012. JWH-018 and JWH-022 as combustion products of AM2201. *Microgr J*. 9(2):52–65.
34. Kavanagh, P., Grigoryev, A., Savchuk, S., Mikhura, I., Formanowsky, A. 2013. UR-144 in products sold via Internet: Identification of related compounds and characterization of pyrolysis products. *Drug Test Anal*. 5:683–692.

10 NMR Spectroscopy in the Analysis of Illegal Drugs

Vlatka Vajs, Iris Djordjević, Ljubodrag Vujisić, and Slobodan M. Milosavljević

CONTENTS

10.1 INTRODUCTION

The detection of illegal drugs in various samples has been among the most fascinating objects for the practicing analytical chemist involved in forensic analysis. For centuries (until the end of the nineteenth century), drugs were almost entirely of natural origin, mostly herbal, and in some cases slightly chemically modified (e.g., through acetylation of morphine to heroin). By the development of synthetic organic chemistry (commencing at the beginning of the twentieth century), quite a few synthetic drugs designed to mimic the pharmacological effects of drugs originating from natural sources have been designed (e.g., cannabimimetic indazole derivatives or amphetamines, etc.). Detection of such synthetic drugs, named "designer drugs," due to a large number of new substances continuously emerging on the market, is crucial. The efficacy of this analysis has become a tremendous challenge in the analytical forensic field. A development of fast identification and structural characterization is greatly facilitated by the advent of modern spectroscopic techniques, namely nuclear magnetic resonance spectroscopy (NMR) and modern mass spectrometry (MS) as well as separation techniques, such as gas and liquid chromatography (GC and LC). Nowadays, rapid strategies for chemical characterizations of illicit drugs are available.

10.2 HYPHENATED TECHNIQUES

The coupling of spectrometers to chromatographs resulted in very powerful so-called hyphenated (coupled) techniques enabling online direct analysis of complex matrices containing illegal drugs. The GC-MS, developed during the 1950s, after being originated by James and Martin in 1952,[1] was practically the only heavily used hyphenated method for half a century, but its application is rather limited to the thermally stable nonpolar volatiles. In the second half of the twentieth century, the development of the MS ionic sources compatible with LC, such as APCI (atmospheric pressure chemical ionization) by Henion[2,3] and ESI (electrospray ionization) by Fenn[4], enabled combination of LC with MS as a tandem LC-MS technique applicable to the variety of polar nonvolatiles. Today, various hyphenated methods involving coupling LC and MS are becoming routine, e.g., the LC-UV-photodiode array detection–MS (LC-UV-MS), the LC-tandem MS (LC-MS-MS), and the LC–multiple stage MS (LC-MS[n]). Although GC-MS (and nowadays LC-MS) screening in combination with a mass spectral library search is a very efficient technique, enjoying "gold standard" status in most forensic laboratories worldwide (see Chapters 4 and 8), the similarity in chemical structures of the interesting compounds often results in similar chromatographic and spectral properties complicating the identification of new compounds, not only due to the lack of the reference mass spectra in common analytical libraries and the limited availability of certified reference material, but also due to the need for an analytical method that can discriminate between (stereo)isomers. In such cases, nuclear magnetic resonance (NMR) spectroscopy, which is one of the most potent structural elucidation techniques, could be the method of choice enabling unambiguous determination of structure and stereochemistry of the compounds in question. In many cases, direct NMR spectroscopy (without chromatographic separation) could be used for identification and also quantification of drugs in different matrices, as demonstrated in a recent report regarding quantification of synthetic cannabinoids in herbal smoking blends.[5]

The LC-NMR technique is considered as the direct LC-NMR technique, whereas the LC solid phase extraction NMR (LC-SPE-NMR) and the microflow NMR are considered as indirect LC-NMR techniques. LC-SPE-NMR is the most used NMR hyphenated technique for natural product chemistry and metabolomics studies, but the lack of publications in forensics means that LC-NMR is still not routinely used. The main reasons for this are complicated automatization, hard interpretation of results, and time-consuming analysis. Nevertheless, this technique could be very useful for quantitative 1D or 2D NMR analysis, and, in that case, it could be a time-saving analysis.[6–8]

10.3 1D AND 2D NMR SPECTROSCOPY

Since the first experiment in 1945, NMR has become one of the foremost methods for molecular identification, for evaluating detailed molecular structures, for understanding conformations, and for probing molecular dynamics. NMR spectroscopy can also be used for quantitative analysis. Using NMR (sometimes in conjunction with the other types of spectroscopy, but often by itself), chemists can usually define a complete molecular structure in a very short time.

NMR spectroscopy is a special branch of spectroscopy that uses magnetic properties of the atomic nucleus. The method is based on the fact that some atomic nuclei have magnetic properties, namely the nuclear spin. Nuclei that possess spin (for example, a proton) placed in the magnetic field behave like bar magnets and respond to the influence of an external field, tending to align themselves in the manner of a compass needle in the earth's magnetic field. Because of the quantum nature of the phenomena, a proton can adopt only two orientations with respect to an external magnetic field, aligned with the field (the lower-energy state), or opposed to the field (the higher-energy state). If a proton is in the aligned orientation, it can absorb energy, and pass into opposed orientation.

Upon the irradiation of the precessing nuclei with the beam of the radiofrequency (Rf) energy of the correct frequency, the low-energy nuclei may absorb this energy and move to the higher-energy state. The precessing nuclei will only absorb energy from the Rf source if the precessing frequency is the same as that of the radiofrequency beam. When this occurs, the nucleus and radiofrequency are said to be in resonance, hence the term *nuclear magnetic resonance*. Many good textbooks describe the theoretical background of NMR spectroscopy.[9,10]

The simplest NMR experiment consists in exposing the protons in a molecule to a powerful external magnetic field. If protons have different chemical surroundings, not all of them will precess at the same frequency. When these protons are irradiated with an appropriate energy, it will promote the protons from the low-energy state (aligned) to the high-energy (opposed) state. The NMR analyst records this absorption of energy in the form of an NMR spectrum. During the first three decades of NMR spectroscopy, all measurements relied on one-dimensional modes of observation, which resulted in the spectra having just one frequency axis, the second axis being used to display signal intensities. The development of two-dimensional experiments started a new era in NMR spectroscopy. Spectra recorded by these methods have two frequency axes, intensities being displayed in the third dimension.

The idea of two-dimensional (2D) NMR, proposed by Jeener in 1971,[11] and its subsequent spectacular developments by Ernst and coworkers triggered acceleration in NMR activity, with new possibilities being further opened up by an avalanche of novel ideas from many laboratories. In 1976, Ernst's paper about the application of 2D NMR[12] utilized Jeener's idea to produce spectra. Manufacturers also played a major role by providing the pulse programmers and probe hardware that allowed everyone to join in the fun. Introduction of routine 2D NMR measurements in the 1980s afforded much more efficient and more reliable interpretation of the NMR data of organic molecules. Application of this methodology led to full structural assignments for many compounds, and the subsequent development of 3D and 4D NMR techniques (Wütrich)[13] in the 1980s enabled the 3D structure determination of a small protein in solution using only NMR data. In 1991, Ernst won the Nobel Prize in chemistry for his contributions to Fourier transform NMR. Wütrich also earned the Nobel Prize for chemistry in 2002 for his development of nuclear magnetic resonance spectroscopy for determining the three-dimensional structure of biological macromolecules in solution, i.e., the development of 3D and 4D NMR techniques.

10.4   ROUTINE METHODS OF 2D NMR SPECTROSCOPY

2D NMR spectroscopy methods used routinely nowadays are listed in Table 10.1. Since it was discovered more than 60 years ago that even "simple," one-dimensional NMR spectra contained a wealth of information about a molecule and because the extraction of the NMR parameters was not always straightforward, there were many attempts to develop methods for analyzing the spectra. To understand the complexity of measured spectra, computer programs evolved to perform quantum mechanical calculations of NMR spectra. As a help in analyzing NMR spectra, quite a lot of web resources for structural elucidation are available; some of them are summarized in recent publications in the *Journal of Chemical Education*[14] and in *Magnetic Resonance in Chemistry*.[15] Nowadays, visualization, processing, and analysis of the NMR data can be done even on smartphones and tablets.[16]

TABLE 10.1

The 2D NMR Methods Used Routinely Nowadays

Homonuclear Correlation

H,H-COSY (COrrelated SpectroscopY), connects signals of protons via scalar coupling (J_{HH}).

TOCSY (TOtal Correlated SpectroscopY), or **HOHAHA** (HOmonuclear HArtman HAhn spectroscopy), connects all spins within a *J*-coupled spin system (A and B).

NOESY (Nuclear Overhauser Effect SpectroscopY), connectivity of close spins (r < 4 - 5 Å) via dipolar couplings (through space).

EXSY (EXchange SpectroscopY), connectivity via chemical or conformational exchange.

Heteronuclear Correlation

HSQC (Heteronuclear Single Quantum Coherence) and **HMQC** (Heteronuclear Multiple Quantum Coherence), connect heteronuclei (e.g., ^{13}C and H or ^{15}N and H) via direct scalar coupling ($^{1}J_{CH}$ or $^{1}J_{NH}$).

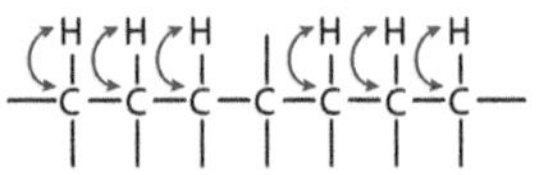

HMBC (Heteronuclear Multiple Bond Correlation), connects heteronuclei (e.g., ^{13}C and H) via long-range ($^{2}J_{CH}$, $^{3}J_{CH}$); used to establish connectivity between two separate spin systems (A and B).

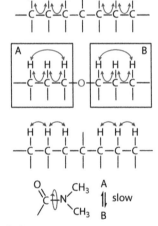

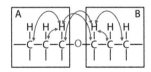

10.5 SELECTED EXAMPLES OF THE NMR ANALYSIS OF ILLEGAL DRUGS

Since its foundation (1966), the Laboratory of Instrumental Analysis, at the Faculty of Chemistry, and the Centre of Chemistry (ICTM), University of Belgrade is involved in different areas of instrumental organic analysis.[17–19] The research focuses on isolation and structural determination of natural products and in that respect NMR analysis is the most valuable source of information.[20–22]

In addition, the Laboratory (accredited according to ISO 17025) is engaged in solving different practical problems, such as quality control of raw materials and final products in chemical industry, analysis of chemical weapons, forensic analyses involving identification of illegal compounds (see below), etc.

Application of NMR spectroscopy in the detection of illegal drugs carried out in our laboratory is demonstrated through the selection of six examples given in Sections 10.5.1–10.5.6.

10.5.1 ACETIC ACID ANHYDRIDE (AA) AS A PRECURSOR OF ILLICIT NARCOTIC AND PSYCHOTROPIC DRUGS

The analysis of an unknown liquid taken from a tank truck held at the border crossing at Horgoš (between Serbia and Hungary) was carried out on the demand of the Ministry of Health in 2003. The content of the tank truck was officially declared as "anhydrous acetic acid." The NMR and IR spectra of the suspicious sample taken from the tank truck unambiguously revealed acetic acid anhydride and not anhydrous acetic acid, as declared. Obviously, this was an attempt at illegal importation of acetic acid anhydride, a chemical whose traffic is strongly controlled, because it is on the list of precursors used for heroin production (by acetylation of morphine), and also for the synthesis of benzyl-methyl ketone, the basic precursor for Leuckart synthesis of amphetamine (Figure 10.1).

10.5.2 ILLEGAL PRODUCTION OF AMPHETAMINE

The amphetamine (AM) group and related aromatic ring-substituted substances are generally synthetic compounds, belonging to one of the most heavily abused drug groups in recent years. Some compounds in this class also originate from plants. Amphetamine, its N-methyl derivative (meth), and their analogs are among the many illicit psychotropic drugs stimulating the central nervous system. They can induce numerous unwanted effects, such as hypertension and long-term addiction leading to violent destructive behavior and acute psychosis similar to paranoid schizophrenia. International attention is increasingly focusing on the growing issue of amphetamine-type stimulants (ATSs). Particularly over the last 10 to 15 years, abuse of the ATSs, involving amphetamines (amphetamine and methamphetamine) and substances of the ecstasy group (MDMA, MDA, MDEA, etc.), has become a global problem. There are regional differences, but today no country is spared of one of the many facets of ATS manufacture, trafficking, or abuse. This was the reason that the UN Office on Drugs and Crime in Vienna published a very comprehensive "Manual for Use by

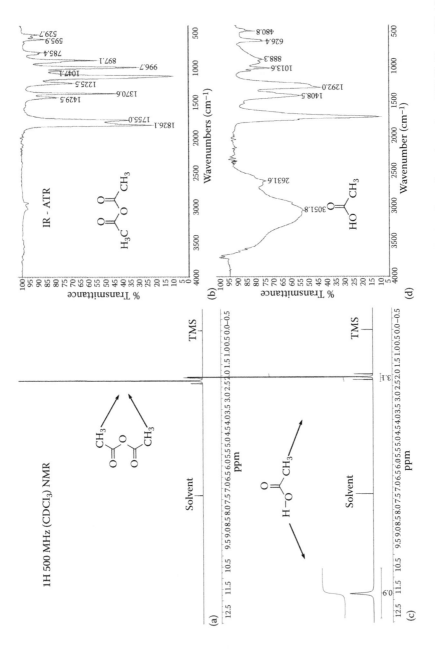

FIGURE 10.1   Proton NMR (a) and IR spectra (b) of the sample of "anhydrous acetic acid" taken from the tank truck and NMR (c) and IR (d) spectra of the glacial acetic acid.

National Drug Testing National Laboratories."[23] In this manual, the following instrumental analytical techniques for the analysis of ATS are presented: presumptive tests (color tests, anion tests, microcrystal tests), thin-layer chromatography (TLC), gas chromatography–flame ionization detector (GC-FID), gas chromatography–mass spectrometry (GC-MS), high-performance liquid chromatography (HPLC), Fourier-transform infrared (FTIR) spectroscopy, ^{1}H nuclear magnetic resonance (NMR) techniques, capillary electrophoresis (CE), solid-phase microextraction–gas chromatography (SPME-GC), and gas chromatography–Fourier-transform infrared spectroscopy (GC-FTIR).

In 2003, our laboratory received *ca.* 100 samples from the Agency for Medicines, Serbia, collected during an action of the police in the "pharmaceutical" plant "Lenal Pharm," suspected of illegal production of illicit chemical substances (Figure 10.2).

The samples were analyzed using the FT infrared (IR) spectroscopy and nuclear magnetic resonance (^{1}H NMR; Figures 10.3–10.5). The ^{1}H NMR spectra were measured directly by dissolving the crystalline samples in D_2O. It should be noted that a similar procedure has been subsequently recommended in the above Manual from the UN[23]: "Dissolve about 20 mg of the drug sample in 1 mL D_2O. If insoluble materials are present, centrifuge, otherwise transfer directly the supernatant into an NMR tube. Record the spectrum of this solution containing the salt form of the ATS."

The assignments of proton signals, based on a simple first-order analysis, using chemical shifts and coupling patterns, proved the basic structure of amphetamine in the form of salts, such as sulfate and tartrate. The epilogue was that the firm owner and the remaining persons involved in illegal production were sentenced to long-term imprisonment.

FIGURE 10.2 Part of the "Lenal Pharm" pharmaceutical plant, in the vicinity of Belgrade, where the analyzed samples were collected.

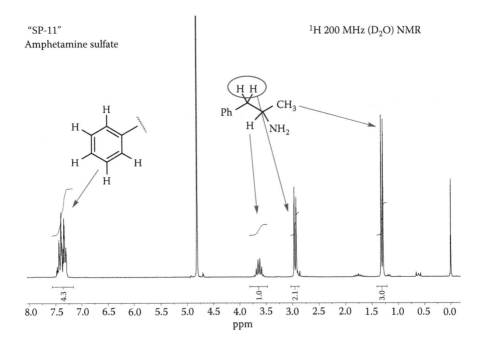

FIGURE 10.3 ¹H NMR spectrum of sample "SP-11" collected in the "Lenal Pharm" pharmaceutical plant.

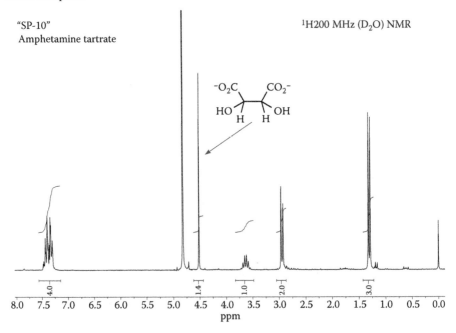

FIGURE 10.4 ¹H NMR spectrum of sample "SP-10" collected in the "Lenal Pharm" pharmaceutical plant.

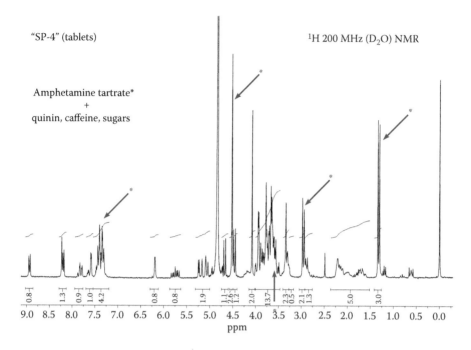

"SP-4" (tablets) ^{1}H 200 MHz (D$_2$O) NMR

Amphetamine tartrate*
+
quinin, caffeine, sugars

FIGURE 10.5 ^{1}H NMR spectrum of sample "SP-4" (tablets) collected in the "Lenal Pharm" pharmaceutical plant.

10.5.3 SYNTHETIC CANNABINOIDS

The detection of synthetic cannabinoids within the recreational drug market was first reported in Germany in 2008,[24] and has since spread to become a worldwide phenomenon. The identification of new substances of abuse (SOA) continuously emerging on the market has become a tremendous challenge in the forensic field. About 100 SOAs were reported to the European Monitoring Centre for Drugs and Drug Addiction (EMCDDA) in 2015, a 25% increase from 2014.[25] A development of fast identification and structural characterization of newly emerging SOA is crucial.

Various products available through the Internet contain synthetic cannabinoids as psychoactive adulterants. Identification of the new SOA, with a slightly altered chemical structure compared with an original illegal drug, usually implies isolation of the compound, and complete structure elucidation. Since the first identification of synthetic cannabinoids in the "herbal mixture," the market has grown continuously. Owing to legal restrictions, initially added substances have meanwhile nearly vanished from the market. As the "herbal highs" market continues to boom, the added synthetic cannabinoids are exchanged with new ones, trying to stay at least one step ahead of legal restrictions.

Most of the added substances belong to the chemical group of the aminoalkylindoles (see Scheme 10.1 for compounds **1** and **2**, isolated in 2012), and for many of them the synthetic procedures have been described in the literature. During regular market surveillance, Moosmann and coworkers[26] identified two new synthetic

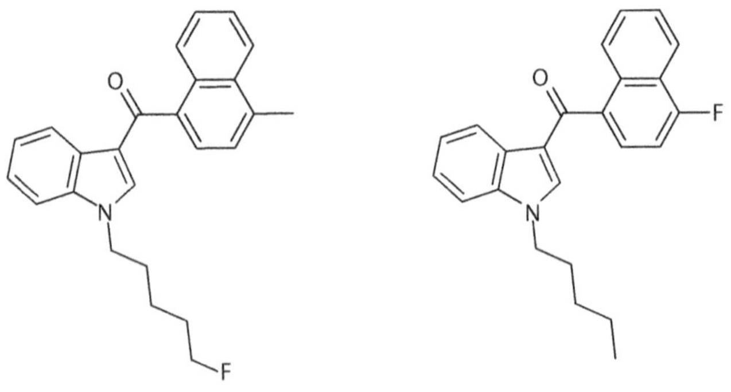

SCHEME 10.1 Aminoalkylindoles identified by Moosmann and coworkers.

cannabinoids, one in the "herbal mixture" labeled XoXo (**1**), and the other (**2**) obtained from a seizure.

The methods recommended by the Laboratory and Scientific Section United Nations Office on Drugs and Crime, Vienna for the identification and analysis of synthetic cannabinoids are presented in the manual.[27] They comprise the following techniques: presumptive tests, TLC, GC-FID, GC-MS, ultra-high-performance liquid chromatography (UHPLC), liquid chromatography–tandem mass spectrometry (LC-MS-MS), ATR-IR and FTIR, gas chromatography–infrared detection (GC-IRD), ambient ionization mass spectrometry, high-resolution mass spectrometry (HRMS), matrix-assisted laser desorption ionization time-of-flight mass spectrometry (MALDI-TOF-MS), and nuclear magnetic resonance (NMR) spectroscopy.

In August 2014, the composition of an "air freshener" was tested in our laboratory as a preparation of a lab exercise for the summer school of chromatography in the Petnica Science Center (Valjevo, Serbia). The sample was supplied from a nearby shop, under the name "BAD MAD marshmallow leaf vanilla," a product of Maya World Trading (EU). The extract of the above sample obtained by the combined acid/base extraction procedure described by Agilent[28] was directly analyzed using GC-MS, as well as 1D and 2D NMR techniques (Figures 10.6–10.10).

As it could be seen from Figure 10.6, the sample contained a single compound with a molecular ion m/z 324 (t_R = 17.8 minutes). However, the best match obtained by the NIST 11 library was rather low, *ca.* 58%, so it was obvious that the spectrum of an unknown compound was not available in the library. This prompted us to measure the ^{1}H and ^{13}C NMR spectra of the unknown substance (Figures 10.7 and 10.8). The NMR spectra, assigned using the 2D NMR techniques COSY and HSQC (Figures 10.9 and 10.10, respectively), together with a comparison with the published data[29] revealed the structure of the indazole derivative named (AB-FUBINACA). Thus, it was obvious that the analyzed sample contained an illicit cannabimimetic drug.

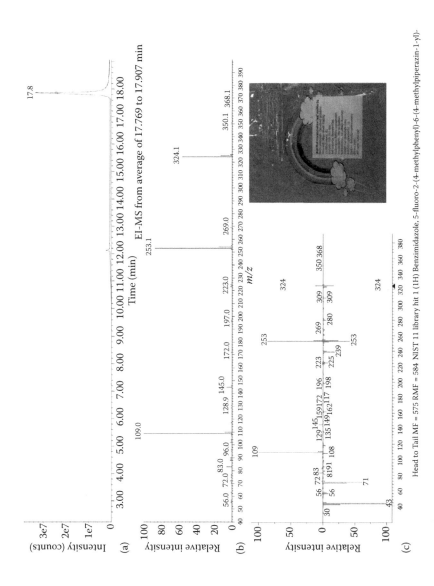

FIGURE 10.6 GC-MS analysis of CH$_2$Cl$_2$ extract of the "air freshener;" (a) total ion current (TIC) chromatogram, (b) EI MS of the peak at t_R = 17.8 min, (c) EI MS of the unknown component (upper trace), and EI MS from the NIST 11 library (lower trace) (match factor *ca.* 58%).

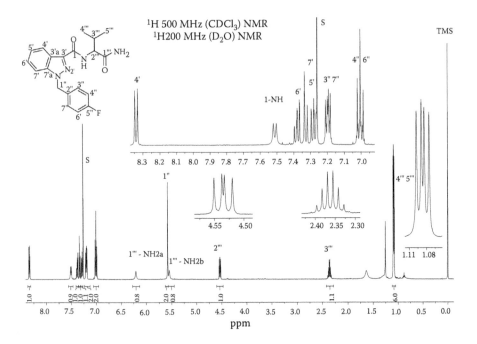

FIGURE 10.7   ¹H (500 MHz) NMR spectrum of the crude extract of "air freshener."

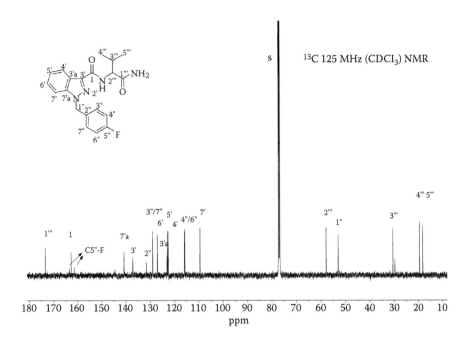

FIGURE 10.8   ¹³C (125 MHz) NMR spectrum of the crude extract of "air freshener."

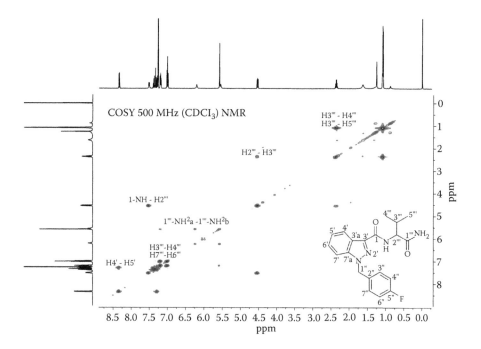

FIGURE 10.9   H,H COSY NMR of the crude extract of "air freshener."

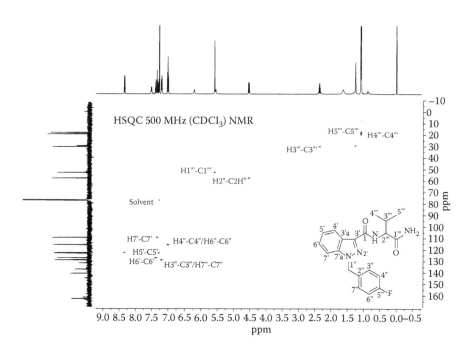

FIGURE 10.10   C,H HSQC NMR spectrum of the crude extract of "air freshener."

10.5.4 FAKE ANTITETANUS VACCINE

As it was pointed out recently,[30] falsified medicines are not controlled and it is therefore not known what these products really contain. Neither the quality nor the efficacy of these products can be assured. Analytical techniques used for identifying the counterfeit drugs have been reviewed by Martino et al.[31] The techniques discussed were colorimetry, thin-layer chromatography, gas chromatography (GC), high-performance liquid chromatography (HPLC), mass spectrometry (MS), and different vibrational spectroscopic methods. Also ^{1}H NMR has been proven very useful for detecting and quantifying the illegal compounds.[32] The following example involving the analysis of a false antitetanus vaccine was carried out in our laboratory, mostly using the ^{1}H NMR spectroscopy (Figures 10.11 through 10.13). From the Belgrade Police Department, our laboratory received a request to carry out the analysis of the imported antitetanus vaccine *Tetaglobuline* allegedly produced in the Institute Pasteur Mérrieux Connaught (it was confirmed that serial numbers printed on the labels of vaccine were never issued by the Institute Pasteur).

A big difference between these two spectra (Figures 10.11 and 10.13) was obvious at a first glance. The spectrum (broad signals) of Tetagam P was typical of a protein molecule that should be the active constituent in the proper antitetanus vaccine, whereas the spectrum of the sample under suspicion unambiguously indicated the presence of gentamicin, the broad-spectrum antibiotic, *ca.* ten times cheaper than the antitetanus vaccine. The latter was proved by the identity of the above NMR spectrum with that of the sample of gentamicin (not shown). Thus, it could be concluded

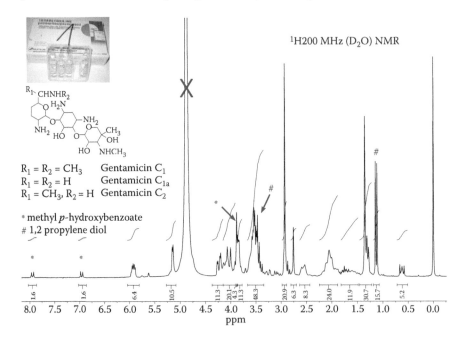

FIGURE 10.11 ^{1}H NMR spectrum of a fake vaccine, distributed under the name Tetaglobuline.

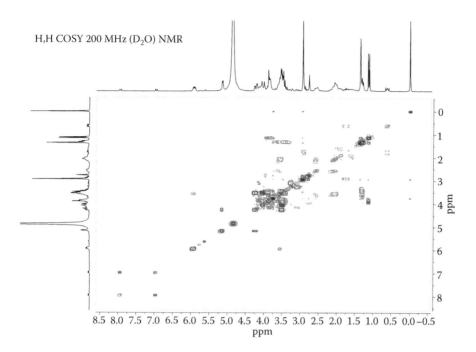

FIGURE 10.12 H,H COSY spectrum of a fake vaccine, distributed under the name Tetaglobuline.

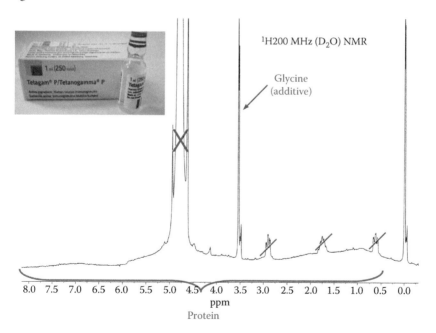

FIGURE 10.13 ^{1}H NMR spectrum of the correct vaccine, bought in a pharmacy under the name Tetagam P.

that the so-called *Tetaglobuline* was the fake vaccine and the Institute Pasteur Mérrieux Connaught has never issued serial numbers printed on the labels of vaccine.

10.5.5   Satibo™ Capsules for the Treatment of Erectile Dysfunction

From the Municipal Office for Health Protection, Belgrade, we received to analyze a sample of Satibo (Fuji, Chinese product) for the treatment of erectile dysfunction, declared as a 100% natural herbal preparation, safe and without any side effects, based on Chinese traditional medicine (Figure 10.14). As a health food, Satibo is claimed to be refined from the Chinese herbal medicines having the function of regulation of humoral immunity, enhancing the physical strength and improving immunity and declared as "*Rhizoma dioscorea* (Chinese Yam), *Fructus Lycium barbarium* (Wolfberry), *Radix Glycyrrhiza glabra* (Licorice), *Semen Coix lacryma* (Jobi seed), *Semen Euryale ferox* (Gorgon fruit), and *Bulbus Lilium lanciforum* (Lily)."

Analytical procedure: The content of a Satibo capsule was sonicated with CD_3OD (0.5 mL) *ca.* one hour, and transferred to a standard NMR tube after filtration. A low-field part (δ 6.5–8.5 ppm) of ¹H and TOCSY 500 MHz NMR is shown in Figure 10.15.

This part of the spectrum contained two sets of aromatic protons (denoted as **V** and **C**, Figure 10.15) indicating two aromatic compounds in the ratio of **V/C** *ca.* 6:1, based on the relative intensities (integral per proton) of these signals. Two separate aromatic spin systems are also supported by cross-peaks occurring in TOCSY. The first-order analysis of the spectrum confirmed two aromatic compounds that could be easily assigned by first-order analysis, according to characteristic coupling patterns and chemical shifts as synthetic compounds Viagra (main component **V**) and Cialis (**C**) (Scheme 10.2). The remaining resonances at a higher field (not shown) also fit this conclusion. The presence of the above adulterants is also supported by LC-ESI MS TOF analysis showing [M+H]⁺ ions at *m/z* 475.21168 ($C_{22}H_{30}N_6O_4S$) and 390.14350 ($C_{22}H_{19}N_3O_4$) corresponding to Viagra and Cialis, respectively.

FIGURE 10.14   Satibo (Fuji, Chinese product) capsules for the treatment of erectile dysfunction.

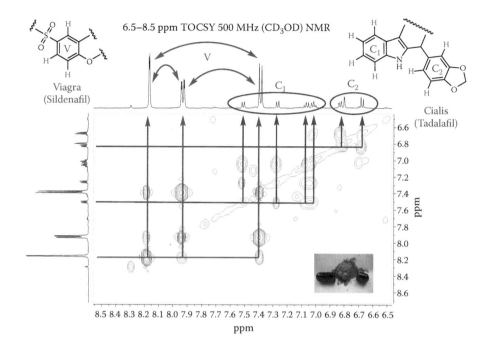

FIGURE 10.15 Aromatic portion (δ 6.5–8.5 ppm) of ¹H and TOCSY 500-MHz NMR spectra of a crude CD₃OD extract of a Satibo capsule.

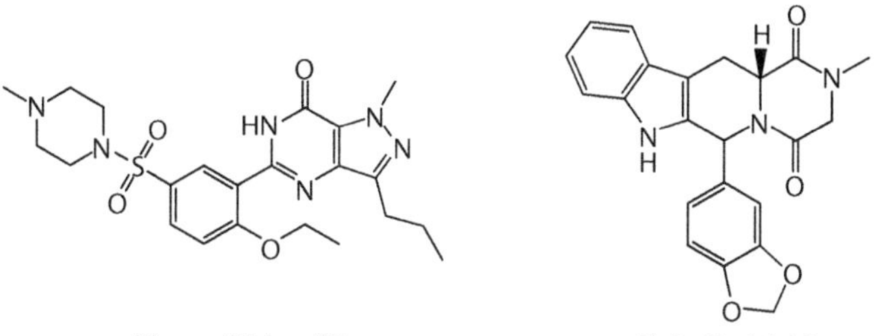

Viagra (Sildenafil)

Cialis (Tadalafil)

SCHEME 10.2 Identified adulterants in Satibo capsules.

10.5.6 GREEN COFFEE FOR SLIMMING

This rather popular dietary supplement, sold under the name "Green coffee for weight loss," a product of Elephant Co.–Strong Nature (Belgrade, Serbia), was analyzed in our laboratory about seven years ago. The analyzed sample was supplied from a nearby drugstore. The package contained 14 aluminum bags (two weeks' dose), each of them declared to contain only natural ingredients, such as 5.87 g coffee, 2.920 g powdered milk, 17.5 mg dried fructus of unripe orange, 12.5 mg Chinese rose essence, 300 mg Chinese lotus leaves essence, 12.5 mg wild rose essence, 10.75 mg juniper essence, 600 mg chitin, 20 mg vitamin B3, 5 mg vitamin B6, and 250 mg vitamin C.

Analytical procedure: The content of one bag (10 g), after the combined acid/base extraction, was extracted with CH_2Cl_2 (25 mL) at ambient temperature. After filtration and evaporation *in vacuo*, the extract was transferred to a standard NMR tube with $CDCl_3$. The ^1H 200 MHz NMR spectrum of the extract is shown in Figure 10.16.

The NMR spectrum unambiguously revealed the presence of one main component in this extract whose resonances were typical of the synthetic anorexiant named sibutramine (or reductil).[33] The structure of this compound was also supported by LC-ESI MS TOF analysis showing the $[M+H]^+$ and $[M+H+2]^+$ ions at m/z 280.1836 and 282.1809, corresponding to the molecular formula $C_{17}H_{26}NCl$ of sibutramine.

Sibutramine (usually in the form of the hydrochloride monohydrate salt) is an oral anorexiant. Until 2010, it was marketed and prescribed as an adjunct in the treatment of exogenous obesity, along with diet and exercise. It has been associated with increased cardiovascular events and strokes, and has been withdrawn from the market in several countries and regions including Australia, Canada, China, the European Union (EU), Hong Kong, India, Mexico, New Zealand, the Philippines, Thailand, the United Kingdom, and the United States. Unfortunately, in parallel with its withdrawal from the legal market, sibutramine and some other anorexiants reappeared in illegally adulterated products, generally claimed to contain only herbal drugs or extracts; however, their efficacy is in fact based on the presence of the marked amounts of synthetic compounds.

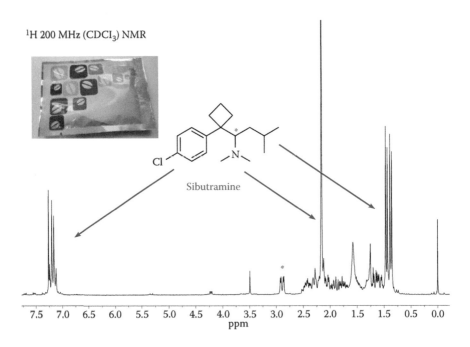

^1H 200 MHz ($CDCl_3$) NMR

Sibutramine

7.5 7.0 6.5 6.0 5.5 5.0 4.5 4.0 3.5 3.0 2.5 2.0 1.5 1.5 0.5 0.0
ppm

FIGURE 10.16 ^1H (200 MHz) NMR spectrum of a CH_2Cl_2 extract of "Green coffee for weight loss."

10.6 CONCLUSION

NMR spectroscopy provides the forensic analyst with an extremely powerful tool for the detection and quantification of drugs. A whole range of one-dimensional (1D) and two-dimensional (2D) NMR techniques is available for performing the required analyses. These NMR methods may be used for routine purposes, such as to confirm the identity of a drug, or to quantify the amount of an illicit substance.

In many cases, direct NMR spectroscopy (without chromatographic separation) could be used for the identification and also quantification of drugs in different matrices. Although NMR is one of the essential spectroscopic techniques for the structure elucidation of unknown compounds, it is still far less common than mass spectrometry in routine usage in forensic laboratories. Basic weaknesses of NMR are its relatively low sensitivity, and the high price of the instruments and the deuterated solvents needed for the analysis. However, there are a couple of ways to lower these expenses: using probes with smaller diameters (if the substances of interest are soluble in small amounts of solvent), or using nondeuterated solvents in the standard sample tube with a reusable internal capillary filled with deuterated solvent with solvent suppression.[34]

For the more complex samples requiring chromatographic separation, a hyphenated LC-NMR technique is now available. In recently published books[6–8] NMR was considered a potential universal detector for liquid chromatography and this hyphenated technique was discussed in detail.

ACKNOWLEDGMENTS

This work was supported by the Ministry of Technological Development of Serbia (grant No. 172053) as well as the Serbian Academy of Science and Art (SASA grant F-188).

REFERENCES

1. James, A.T. and Martin, A.J.P. 1952. Gas-liquid partition chromatography: The separation and micro-estimation of volatile fatty acids from formic acid to dodecanoic acid, *Biochem. J.*, 50: 679–680.
2. Henion, J.D., Thomson, B.A., and Dawson, P.H. 1982. Determination of sulfa drugs in biological fluids by liquid chromatography/mass spectrometry/mass spectrometry, *Anal. Chem.*, 54: 451–456.
3. Covey, T.R., Lee, E.D., and Henion, J.D. 1986. High-speed liquid chromatography/tandem mass spectrometry for the determination of drugs in biological samples, *Anal. Chem.*, 58: 2453–2460.
4. Fenn, J.B., Mann, M., Meng, C.K., Wong, S.F., and Whitehouse, C.M. 1989. Electrospray ionization for mass spectrometry of large biomolecules, *Science*, 246(4926): 64–71.
5. Dunne, S.J. and Rosengren-Holmberg, J.P. 2017. Quantification of synthetic cannabinoids in herbal smoking blends using NMR, *Drug Test. Anal.*, 9: 734–743.
6. Elipe, M.V.S. 2011. *LC-NMR and Other Hyphenated NMR Techniques: Overview and Applications*, John Wiley & Sons, Inc., Hoboken, NJ.
7. Gonnella, N. 2013. *LC-NMR Expanding the Limits of Structure Elucidation*, CRC Press Taylor & Francis Group, Boca Raton, FL.

8. Bohni, N., Queiroz, F.E., and Wolfender, J.L. 2014. On-line and at-line LC-NMR and related micro-NMR methods, Chapter 14, in *Encyclopedia of Analytical Chemistry*, Online ©2006–2014. John Wiley & Sons, Ltd., Chichester, UK.

9. Williams, D.H. and Fleming, I. 2007. *Spectroscopic Methods in Organic Chemistry*, sixth edition, Amazon Co., UK.

10. Friebolin, H. 1998. *Basic One- and Two-Dimensional NMR Spectroscopy*, third revised edition, Wiley-VCH, Verlag GmbH, Weinheim, Germany.

11. Jeener, J. 1971. *Lecture at Ampere Summer School*, Baško Polje, Yugoslavia.

12. Aue, W.P., Bartholdi, E., and Ernst, R.R. 1976. Two-dimensional spectroscopy. Application to nuclear magnetic resonance, *J. Chem. Phys.*, 64: 2229–2246.

13. Braun, W., Wider, G., Lee, K.H., and Wüthrich, K. 1983. Conformation of glucagon in a lipid-water interphase by ^{1}H nuclear magnetic resonance, *J. Mol. Biol.*, 169: 921–948.

14. Graham, K.J., McIntee, E.J., and Schaller, C.P. 2016. Web-based 2D NMR spectroscopy practice problems, *J. Chem. Educ.*, 93: 1483–1485.

15. Jeannerat, D. 2017. Human- and computer-accessible 2D correlation data for a more reliable structure determination of organic compounds. Future roles of researchers, software developers, spectrometer managers, journal editors, reviewers, publisher and database managers toward artificial-intelligence analysis of NMR spectra, *Magn. Reson. Chem.*, 55: 7–14.

16. Cobas, C., Iglesias, I., and Seoane, F. 2015. NMR data visualization, processing, and analysis on mobile devices, *Magn. Reson. Chem.*, 53: 558–564.

17. Vajs, V., Jokić, A., and Milosavljević, S. 2017. Artemisinin story from the Balkans, *Nat. Prod. Commun.*, 12: 1157–1160.

18. Makarov, S., Vujisić, Lj., Ćurčić, B., Ilić, B., Tešević, V., Vajs, V., Vučković, I., Mitić, B., Lučić, L., and Djordjević, I. 2012. Chemical defense in the cave-dwelling millipede *Brachydesmus troglobius* Daday, 1889 (Diplopoda, Polydesmidae), *Int. J. Speleol.*, 41: 95–100.

19. Andjelković, B., Vujisić, Lj., Vučković, I., Tešević, V., Vajs, V., and Godjevac, D. 2017. Metabolomics study of Populus type propolis, *J. Pharm. Biomed. Anal.*, 135: 217–226.

20. Djordjević, I., Vajs, V., Bulatović, V., Menković, N., Tešević, V., Macura, S., Janaćković, P. and Milosavljević, S. 2004. Guaianolides from two subspecies of *Amphoricarpos neumayeri* from Montenegro, *Phytochemistry*, 65: 2337–2345.

21. Vujisić, Lj., Vučković, I., Makarov, S., Ilić, B., Antić, D., Jadranin, M., Todorović, N., Mrkić, I., Vajs, V., Lučić, L., Ćurčić, B. and Mitić, B. 2013. Chemistry of the sternal gland secretion of the Mediterranean centipede *Himantarium gabrielis* (Linnaeus, 1767) (Chilopoda: Geophilomorpha: Himantariidae), *Sci. Nat. (Naturwissenschaften)*, 100: 861–870.

22. Aljančić, I., Vučković, I., Jadranin, M., Pešić, M., Djordjević, I., Podolski-Renić, A., Stojković, S., Menković, N., Vajs, V., and Milosavljević, S. 2014. Two structurally distinct chalcone dimers from *Helichrysum zivojinii* and their activities in cancer cell lines, *Phytochemistry*, 98: 190–196.

23. Laboratory and Scientific Section United Nations Office on Drugs and Crime Vienna, *Recommended Methods for the Identification and Analysis of Amphetamine, Methamphetamine and Their Ring-Substituted Analogues in Seized Materials*, New York 2006, available on https://www.unodc.org/pdf/scientific/stnar34.pdf.

24. Lindigkeit, R., Boehme, A., Eiserloh, I., Lubbecke M., Wiggerman, M., Ernst L., and Beuerle, T. 2009. Spice: A never ending story?, *Forensic Sci. Int.*, 191: 58–63.

25. EMCDDA, European Drug Report—Trends and Development, available on http://www.emcdda.europa.eu/edr2015.

26. Moosmann, B., Kneisel, S., Girreser, U., Brecht, V., Westphal, F., and Auwarter, V. 2012. Separation and structural characterization of the synthetic cannabinoids JWH-412 and 1-[(5-fluoropentyl)-1H-indol-3yl]-(4-methylnaphthalen-1-yl)methanone using GC-MS, NMR analysis and a flash chromatography system, *Forensic Sci. Int.*, 220: e17–e22.

27. Laboratory and Scientific Section United Nations Office on Drugs and Crime Vienna, *Recommended Methods for the Identification and Analysis of Synthetic Cannabinoid Receptor Agonists in Seized Materials 2013*, available at https://www.unodc.org/documents/scientific/STNAR48_Synthetic_Cannabinoids_ENG.pdf.

28. Identification of Synthetic Cannabinoids in Herbal Incense Blends by GC/MS Application Compendium, Agilent, USA, 2016.

29. Uchiyma, N., Matsuda, S., Wakana, D., Kikura-Hanjairi, R., and Goda, Y. 2013. New cannabimmimetic indayole derivatives, N-(1-amino-3-methyl-1-oxobutan-2-yl)-1-pentyl-1H-indazole-3-carboxamide (AB-PINACA) and N-(1-amino-3-methyl-1-oxobutan-2-yl)-1-(4-fluorobenzyl)-1H-indazole-3-carboxamide (AB-FUBINACA) identified as designer drugs in illegal products, *Forensic Toxicol.*, 3: 93–100.

30. Johansson, M., Fransson, D., Rundlöf, T., Huynh, N.H., and Arvidsson, T. 2014. A general analytical platform and strategy in search for illegal drugs, *J. Pharm. Biomed.*, 100: 215–229.

31. Martino, R., Malet-Martino, M., Gilard, V., and Balayssac, S. 2010. Counterfeit drugs: Analytical techniques for their identification, *Anal. Bioanal. Chem.*, 398: 77–92.

32. Holzgrabe, U. and Malet-Martino, M. 2011 Analytical challenges in drug counterfeiting and falsification—The NMR approach, *J. Pharm. Biomed. Anal.*, 55: 679–687.

33. Csupor, D., Boros, K., Dankó, B., Veres, K., Szendrei, K., and Hohmann, J. 2013. Rapid identification of sibutramine in dietary supplements using a stepwise approach, *Pharmazie*, 68: 15–18.

34. Gama, L.A., Merlo, B.B., Lacerda, V. Jr, Romao, W., and Neto, A.C. 2015. No-deuterium proton NMR (No-D NMR): A simple, fast and powerful method for analyses of illegal drugs, *Microchem. J.*, 118: 12–18.

11 Preparation of Biological Material for Toxicological Analysis

Milena Majchrzak and Rafał Celiński

CONTENTS

11.1 INTRODUCTION

The toxicological analysis of biological material consists of the qualitative and quantitative determination of xenobiotics present in a human organism. From the standpoint of judicial toxicological analysis (which is an important tool in judicial endorsement), the most important task is to establish a cause–effect relationship between the presence of a toxic factor in the system and its biological response. In order to reach this goal, the judicial toxicologists have to select proper material for the investigation and determine adequate analytical techniques and methods. From the perspective of jurisdiction, toxicological analysis of biological material focuses on identification and quantification of substances affecting the central nervous system, such as psychotropic and stupefying agents, ethyl alcohol, and also new psychoactive substances (NPSs), popularly known as designer drugs. Based on analytical results targeting the presence of psychoactive substances, judicial opinions are issued that refer to the perpetrators of traffic accidents or victims of sexual assaults, and with mortal cases, identification of psychoactive substances is connected with a judgment on the cause of death related to the history and the mode of a drug intake.

In toxicological analyses targeting the presence of substances affecting the central nervous system, the most frequently utilized biological materials collected from the living persons are body fluids (i.e., blood and urine) and hair samples. In postmortem analyses, blood and urine are analyzed, but also liquid collected from the eyeball and the tissues of inner organs (mostly tissues of the liver, kidney, brain, and stomach, with its inner content). Less frequently the tissues of lungs, muscles, and guts are investigated. From the standpoint of instrumental analysis, proper preparation of a given sample most probably is a crucial and even decisive step for an overall evaluation outcome. Chemical analysis of biological material in the form of blood, urine, or tissue samples is difficult, due to the presence of numerous metabolites and biochemical factors therein, which may obscure the final result. Proper extraction of a biological matrix can ensure the correct evaluation of toxic agents, avoiding contamination of very sensitive analytical instruments. In this chapter, we present techniques of extraction of new psychoactive substances (i.e., designer drugs) from biological material collected from living and deceased subjects.

11.2 APPLICATIONS

11.2.1 Screening Analysis of Biological Material

Each toxicological analysis of biological material for the presence of psychotropic and stupefying agents, and also new psychoactive substances (NPSs), has to start from a preliminary screening analysis. For this purpose, both specific techniques can be used (e.g., liquid chromatography coupled with mass spectrometry in the multiple reaction monitoring (MRM) mode and optimized for the identification of a predefined group of compounds), and nonspecific techniques (e.g., the immunoenzymatic ELISA [enzyme-linked immunosorbent assay] test). In toxicological analysis, ELISA is the most popular and the most frequently used screening method, which allows detecting certain proteins in the investigated material with use of enzyme-coupled mono- or polyclonal antibodies [1]. In such an investigation, a mixture of a standardized and enzyme-tagged antigen (a drug-enzyme conjugate) and an antigen contained in the investigated sample is used, both of them competing for the specific monoclonal antibodies immobilized of the surface of a solid phase (microplate). Thus, the antigen and the antibody form an immunological complex that is permanently connected with this solid surface. Unconnected antigens are washed out and the enzyme conjugated with the standardized antigen catalyzes the reaction of the substrate added to the reaction medium. The resulting colored product of the reaction can be spectrophotometrically quantified and the absorbance value remains in reverse proportion to the analyte concentration in the sample. Manufacturers sell the items necessary to perform the aforementioned ELISA tests in kits, which consist of a plate with holes coated with antibodies, a solution of the standardized and enzyme-tagged antigen, negative and cutoff controls, a washing out buffer, a buffer used for sample preparation, a reagent being the substrate of the reaction performed, and a reagent used to stop this reaction. Preparation of biological material (blood and urine) for this analysis is quite simple and consists in adding of a dedicated buffer solution to an analyzed sample in a proper volume ratio. The prepared sample is directly introduced to the antibody-coated holes and further analysis is carried out in accordance with the manufacturer's instructions [1,2]. Apart from commercially available ELISA tests with antibodies enabling detection of such psychotropic and stupefying agents as, e.g., amphetamine derivatives, delta-9-tetrahydrocannabinol, benzodiazepine and methamphetamine derivatives, cocaine, and opiates, many test kits are also available with antibodies targeting the presence of the designer drugs (like, e.g., cathinone derivatives and synthetic cannabinoids). An advantageous quality of the ELISA test is its simplicity and the rapidity of the biological sample preparation, whereas its main drawback is a possible occurrence of cross-reactions [2].

It needs to be remembered that the immunoenzymatic screening analyses are but a preliminary step aiming to assess the presence of psychoactive substances in biological samples. Their advantage is that they drive the further analytical procedure (pointing out to the most favorable extraction procedure) toward a predefined, narrower group of compounds. Due to the lack of specificity of the screening tests, the reference methods need, however, to be the decisive ones.

11.2.2 BLOOD AND URINE LIQUID–LIQUID EXTRACTION

Liquid–liquid extraction (LLE) is the most frequently used technique of isolating psychoactive substances from body fluids (such as blood, urine, and the fluid from the eyeball). Detailed procedures are selected, based on the type of tissue analyzed and the nature of the identified and quantified compound (in the first instance, on its acidity or basicity).

Liquid–liquid extraction of basic NPSs (such as cathinone derivatives) from blood or urine can be carried out in a number of different ways. Useful methodology is presented in a paper by Sorensen [3]. A 300-µL aliquot of blood or urine collected from a living or deceased subject is mixed with 100 µL methanol solution of an inner standard containing 500 µg L^{-1} of the deuterated methcathinone or mephedrone analog in the 2-mL test tube. Then 600 µL methanol is added to the test tube and the entity is vortexed for several seconds. After a pause lasting *ca.* ten minutes, the whole mixture is centrifuged and 300 µL of a pure supernatant is transferred to a test tube equipped with a filter (in the discussed study, the Millipore Amicon Ultra unit with a 0.5-mL reservoir and a 30-kDa membrane of regenerated cellulose was applied). Then the 10-µL aliquot of formic acid is added to remove the soluble high-molecular-weight components and to obtain stable extracts, and the entity is vortexed for ten minutes. The 100-µL aliquot of the filtrate is collected, 100 µL water is added to lower the sample elution strength, and the entity is placed in the autosampler vial. Extracts are analyzed by means of liquid chromatography with mass spectrometric detection, using the electrospray ionization mode (LC-ESI-MS). The author demonstrated that acidification of the obtained extracts significantly influenced the stability of the quantified compounds. For example, storage for 24 hour at 10°C of the non-acidified samples of the cathinone derivatives (e.g., flephedrone) considerably lowered their concentration, which was not the case with the samples acidified with formic acid.

Another variant of isolating psychoactive substances of basic character from body fluids by means of LLE can be the process carried out in the basic environment. Golasik et al. [4] describe preparation of the urine sample by adding 200 µL 0.5 M NaOH and 1.2 mL ethyl acetate to 500 µL urine and shaking the entity for three minutes. After centrifugation, the organic layer was collected and 50 µL 0.02 M HCl was added to it. Then the solvent was evaporated to dryness in a stream of nitrogen at 40°C and the residue was dissolved in 300 μL of a mixture composed of 500 mL deionized water with an addition of 100 μL conc. phosphoric acid–acetonitrile, 1:1 (v/v). The resulting extracts were analyzed by means of high-performance liquid chromatography (HPLC). The LLE variant most frequently employed in the laboratories of judicial toxicology to isolate the basic NPSs is most often carried out at a pH higher than 8 [4,5].

Kneisel and Auwärter investigated several methods of isolation followed by identification of NPSs from a group of synthetic cannabinoids [6]. In their studies, they focused on isolation of more than 30 different NPS compounds contained in 833 blood serum samples. As a result, the most effective extraction mixture proposed was composed of 990 mL *n*-hexane and 10 mL ethyl acetate (99:1, v/v). After adding to each biological material sample 10 μL of the deuterated inner standard and 0.5 mL carbonate buffer, 1.5 mL extraction mixture was added and the entity was centrifuged. Finally, 1 mL of the organic supernatant was evaporated to dryness in a stream of nitrogen at 40°C. The dry residue was dissolved in 100 μL mobile phase A+B (50:50, v/v), where A: 0.2% formic acid with an addition of 2 mmol L^{-1} ammonium formate and B: pure methanol. Extracts were analyzed by means of liquid chromatography coupled with the electrospray ionization-type mass spectrometer. Owing to its high efficiency, the proposed method of isolating synthetic cannabinoids is frequently utilized in toxicological laboratories [6].

An increasingly popular method of isolating synthetic cannabinoids from biological material (basically, from blood and urine) becomes precipitation with ice-cold acetonitrile, owing to the rapidity of this method combined with its high efficiency [7,8]. In this case, a 200-μL sample is treated with 20 μL of an inner standard (its concentration 10 ng mL^{-1}), and then 600 μL of the ice-cold acetonitrile is added to the entity contained in the Eppendorf tube, with constant shaking of the tube. Upon precipitation, the sample has to be centrifuged, the supernatant is transferred to a 2-mL glass vial, and the entity is evaporated to dryness in order to remove acetonitrile. The dry residue has to be dissolved in 100 μL 0.1% aqueous formic acid, transferred to the autosampler vial, and analyzed by means of liquid chromatography coupled with mass spectrometer equipped with the triple quadrupole operating in positive ionization, in the multiple reaction monitoring (MRM) mode [7,8]. The described method is well suited for the extraction of the incriminated substances from the samples of inner organs, provided the tissues are mechanically homogenated with the addition of water.

In the literature on NPSs from the group of opioid derivatives (such, as acetylfentanyl, butyrfentanyl, or methylfentanyl), numerous descriptions of time-consuming procedures are encountered, with use of a wide spectrum of extractants. Ojanperä et al. [9] prepared a 1-mL urine sample originating from an autopsy by its 16-hour lasting hydrolysis with β-glucuronidase at the temperature of 46°C (the sample was incubated in a water bath). Then the external standard (i.e., deuterated fentanyl, 20 μL, 1 μg μL^{-1}) and 400 μL Na$_2$HPO$_4$ buffer (pH = 9) was added to 1 mL hydrolyzed body fluid. Eventually, the mixture was extracted with 600 μL butyl acetate by shaking the sample for two minutes. After centrifugation and separation of the organic phase, it was evaporated to dryness in a stream of air at 40°C and the solid residue was dissolved in 150 μL of ammonium acetate buffer (10 mmol L^{-1},

0.1% formic acid, pH 3.2), then sonicated and centrifuged. After centrifugation, the extract underwent the analysis by means of LC-MS/MS.

In the opinion of the authors of paper [9], the proposed LLE procedure applied to the analysis of NPSs from the group of opioids—although highly effective—at the same time is excessively time-consuming and for this reason applicable basically to scientific investigations. In the case of toxicological analyses carried out in criminalistic and judicial laboratories and in hospital wards treating heavy intoxications, the same authors suggest usage of the extraction procedures presented in the earlier parts of this section, which are considerably faster and still plausibly effective.

It must be added that in the course of many autopsies, no urine is found in the bladder of a deceased person. Such cases are typical of traffic accident victims (with extensive body damages) and of suicide victims (both those who jump from an elevation or who hang themselves). When collection of urine proves impossible, an alternative material is the liquid from the eyeball and the vitreous humor, which can be extracted with use of the methods earlier described for the extraction of blood and urine samples.

11.2.3 Blood and Urine Solid Phase Extraction

In terms of the frequency of usage, solid phase extraction (SPE) appears in second position in the laboratories of judicial toxicology, just after liquid–liquid extraction, and is used to prepare biological material for analyses aiming to detect psychotropic and stupefying agents, and NPSs.

One possibility of applying the SPE procedure to blood and urine samples originating from autopsy and targeting basic compounds is presented in a paper by Dickson et al. [10]. The assumed procedure can be performed in the following way: to 1 or 2 mL of body fluid, 3 mL 0.1 M phosphate buffer (at pH = 6) and the internal standard (e.g., 0.5 mg L^{-1} mepivacaine or ethylmorphine) is added, and then the obtained mixture is sonicated for 15 minutes and ultimately centrifuged. Then the sample is introduced on top of the mixed-mode silica-based SPE column, earlier conditioned with 3 mL methanol, 3 mL deionized water, and 2 mL phosphate buffer. After filtration of the sample, the column is rinsed with 2 mL deionized water, 2 mL 20% aqueous acetonitrile, and 2 mL 0.1 M acetic acid, and then it is vacuum dried for three minutes, rinsed with 2 mL n-hexane and 3 mL methanol, and again vacuum dried for ten minutes. Finally, the analytes adsorbed on solid phase are eluted from the column with 3 mL of the dichloromethane/isopropanol/ammonia mixture (78:20:2, v/v/v) and after evaporation of the eluate in the stream of nitrogen, the dry residue is transferred to 50 µL acetonitrile and analyzed by means of gas chromatography with mass spectrometric detection (GC-MS). The obtained chromatogram coupled with the mass spectrum confirms an efficiency of extraction of basic NPS compounds (such as, e.g., cathinone derivatives) originating from biological material.

An effective, although in toxicological laboratories less frequently applied, modification of solid phase extraction is solid phase microextraction (SPME). Usage of this particular technique in the extraction of NPSs, and more specifically, to the extraction of α-PVP (a cathinone derivative) was described by Saito et al. [11].

Prior to the first usage and also prior to each consecutive day of running the analyses, the SPME fibers (i.e., 100-μm polydimethylsiloxane (PDMS), 75-μm Carboxen (CAR)/PDMS, 65-μm PDMS/divinylbenzene (DVB), and 50/30-μm DVB/CAR/PDMS fibers) were conditioned in the GC injector port, in order to avoid any contamination. Then 0.05-mL aliquots of the blood sample were placed in 4-mL vials sealed with silicon septa and the following solutions were added: 5 μL of a standard solution (1 μg mL^{-1}), 10 μL of an inner standard solution (α-PVP-d_8, 1 μg mL^{-1}), and from 0 to 300 μL of a 1N NaOH solution. Each fiber was introduced to the vial and heated to the temperature 60, 70, or 80°C. In the course of extraction lasting from 5 to 50 minutes, the fiber was inserted in the needle and placed for two minutes at the temperature of 250°C in the GC injector port. From keen observation of the parameters, which considerably influence the extraction process (such as the temperature, the effect of alkalization, and the time needed to obtain thermodynamic equilibrium), it can be deduced that the extraction yields are affected mostly by time and temperature, and the type of the fiber used. An efficiency of extraction of the compounds structurally resembling pyrovalerone (i.e., methylenodioxypyrovalerone [MDPV] and α-pyrrolidynovalerophenone [α-PVP]) was rather insignificant, yet the best results were obtained at 80°C with the 65-μm-long PDMS/DVB fibers, with an extraction process lasting 30 minutes, and adding 200 μL of 1N NaOH solution.

11.2.4 EXTRACTION OF INTERNAL ORGANS

Inner organs collected in the course of judicial autopsy are a frequent investigation material in toxicological analysis. The most frequently collected samples are those of liver, kidney, and stomach, including its inner contents (when a suspicion exists of a suicidal attempt executed with huge amounts of psychoactive substances), the brain (with lethal cases, due to a considerable intake of compounds with high affinity to the brain receptors; e.g., the NBOMe derivatives of phenylethylamine), and of the putrefied liquids from the organs of a strongly decomposed cadaver. Less frequently used are samples of guts, the heart muscle, lungs, spleen, and pancreas, and also fatty tissue, in the case of the highly lipophilic compounds.

The impact of a solid biological matrix on the results of an investigation can be very high, so that proper preparation of a sample is of a paramount importance. It happens quite often that the solid biological matrix is putrefied, thus additionally jeopardizing extraction of psychoactive substances. In fact, each case needs an individual approach and a preliminary visual inspection of the inner organs in question is needed, although effective methods of preparation of solid tissues for toxicological analysis can successfully eliminate the influence of the matrix.

Samples of inner organs can undergo the extraction of body fluids contained therein (such as blood, urine, or eyeball fluid), yet the procedure has to be preceded by morcelation and homogenization of the sample, followed by filtration. An example of isolation of psychoactive substances from solid tissues following the aforementioned approach is given in a paper by Hasegawa et al. [12,13]. The authors describe the procedure of isolating NPSs from a group of cathinone derivatives from nine different types of tissue obtained from autopsy, such as brain, lung, spleen, the heart muscle, liver, kidney, pancreas, skeletal muscle, and fatty tissue, applying the modified QuEChERS method. One hundred grams of each tissue was morcelated with the use of surgical scissors and placed in 5-mL centrifugal tubes containing 4.9 L acetonitrile and 100 ng inner standard compound dissolved in 10 µL acetonitrile. Then five stainless steel balls were added to each tube (to crush the examined tissues), and the tubes were stoppered and mechanically shaken for a period of five minutes. The contents of the test tube containing fatty tissue were additionally heated for ten minutes at 80°C prior to shaking, which was unnecessary with the remaining eight samples. After homogenization of the tissues and removal of the stainless steel balls, the respective suspensions were transferred to larger test tubes, then the 5-mL aliquots of acetonitrile were added and the entity was delicately mixed. From each obtained mixture (its total volume equal to 10 mL), six 1-mL portions were collected and spiked (or not) with the 10-µL aliquots of the standard solution, in order to perform the calibration curves. The contents of the tubes were energetically vortexed for 30 seconds and then centrifuged for two minutes. The obtained supernatant was decanted into a QuEChERS dispersive-SPE centrifuge tube containing 25 mg of primary amine (PSA), 25 mg of end-capped octadecylsilane, and 150 mg of magnesium sulfate, and the test tubes were again vortexed for 30 minutes and centrifuged for two minutes. The upper acetonitrile layer was filtered through a Captiva ND Lipids cartridge. Eventually, the 3.5-µL aliquots of the eluates were analyzed by LC-MS-MS.

The method of adding standards to several portions of the analyzed extracts was assumed to be one which allows circumventing matrix effects and differences in recovery levels. Moreover, there is no longer the need to use reference materials (such as blood free from the identified substances), which is in conformity with ethical recommendations that say not to collect unimportant human tissues [12–15].

11.2.5 HAIR PREPARATION FOR TOXICOLOGICAL ANALYSIS

Hair is opined as a valuable biological material from a medical and judicial standpoint. Most xenobiotics (including psychotropic and stupefying agents) are able to build in the hair structure and segment analysis of the hair of addicts enables the retrospective assessment of an intake of psychoactive substances. Although some reports are available on the possibility to identify psychoactive compounds in the hair of people even after a single contact with a given drug, a prevailing number of experts agree that toxicological analysis of hair makes sense in cases of deadly intoxications, chronic exposure to a given substance, or at least several contacts with this substance in relatively short time intervals, because given xenobiotics can build in the hair structure only based on a high and/or steady level thereof in an organism [16].

Let us assume that hair grows with an average rate of 1 cm per month. The most appropriate hair to be cropped for analysis grows on the occipital part of the head close to the skin, where hair growth rates are most stable. The minimum hair sample needed for toxicological analysis corresponds with pencil thickness [17]. It must be added that stability of an inbuilt compound (and the result of toxicological analysis) can be negatively affected by such hairdressing actions as hair dying or bleaching with hydrogen peroxide.

A preliminary step after collecting the hair sample is its washing and removal of grease, with an aim to remove possible hair cosmetics, sebum, and sweat, but also environmental pollutants. The most frequently used washing liquids are methyl chloride, methanol, acetone, and n-hexane [18]. After washing and drying the hair sample, the next step is its micronization with use of surgical scissors or a special mill. Once the material is micronized, it can undergo the extraction process under optimal working conditions.

Namera et al. describe a technique of isolating some NPSs from the group of cathinone derivatives known as MonoSpin extraction [19]. MonoSpin® C_{18} silica columns have been specially developed for the preparation of biological material for the analysis and in paper [19], the MonoSpin® C_{18} octadecylsilylated silica monolith columns were applied. The hair sample was washed three times with a 0.1% water solution of sodium dodecylsulfate, three times with distilled water, and once with ethanol, and then dried at ambient temperature. Then it was divided into 1-cm-long segments, and each segment was weighed, and then treated with 0.1 mL 1 M NaOH solution in a test tube stoppered with a Teflon septa. The contents of the vials were heated at 70°C for 20 minutes in order to digest the hair tissue and then 0.8 mL buffer (pH = 13) and 3 µL of the inner standard solution (α-PVP-d_8, its concentration equal to 1 µg mL^{-1}) were added. The MonoSpin® C_{18} columns were conditioned with methanol and buffer, and then the contents of the vials were transferred to them, centrifuged, and washed with 0.4 mL buffer. The analytes adsorbed on the columns were eluted with 0.1 mL of 50% aqueous methanol, centrifuged and analyzed by means of the LC-MS technique.

An example of a faster yet equally efficient procedure of isolating cathinone derivatives from a hair matrix is liquid–liquid extraction described by Alvarez et al. [20] in their most recent publication on toxicological analysis of hair. Firstly, the biological material was decontaminated by soaking for two minutes in dichloromethane, and then in warm water. The purified samples were morcelated and homogenized in the ball mill, then the 20-mg samples were collected and incubated for ten minutes at a temperature of 95°C in the presence of MDMA-d_5 (as an inner standard), with an addition of 1 mL phosphate buffer (pH = 5). Finally, liquid extraction was carried out with 4 mL of n-hexane–ethyl acetate (1:1, v/v) in the carbonate buffer medium (pH = 9.7). After 15 minutes vortexing followed by two minutes centrifugation, the organic phase was collected and evaporated to dryness, and the residue was transferred to 80 µL mobile phase (a mixture of acetonitrile and the 2 mmol L^{-1} solution of ammonium formate in 1% aqueous formic acid), and analyzed by means of LC-MS/MS.

Isolation of NPSs from hair is a time-consuming procedure and for this reason, it cannot be utilized for screening analyses. On the other hand, screening of the other biological materials (like blood or urine) also needs an ultimate confirmation with use of the reference methodology. It has to be emphasized that hair as a biological material is very valuable when it comes to formulating medical and judicial opinions due to the reasons explained in the paragraph opening this section.

11.2.6 QuEChERS Technique in the Preparation of Biological Material for Toxicological Analysis

In the most recent reports on the identification of new cathinone derivatives (making a recognized group of NPSs) from the autopsy materials, applications of the QuEChERS methodology (Quick, Easy, Cheap, Effective, Rugged, and Safe) are increasingly more frequently encountered, which is a kind of novelty in toxicological analysis. This novelty consists in the fact that QuEChERS has been developed and initially utilized for the analysis of food samples only, targeting pesticides contained in food. Although the liquid–liquid and liquid–solid extraction methods are simple and therefore convenient to use, risks of an insufficient sample cleansing, contamination of the instrumental equipment, and imprecise final results are always there.

In 2012, Usui et al. [21] elaborated a rapid method of extracting psychotropic substances from human blood with the use of a modified QuEChERS technique, which can be as selective as SPE, as simple as LLE, and yet much faster, cheaper, and safer for the equipment than the other two. A two-step procedure was elaborated and at step one (extraction/partitioning), liquid samples (e.g., blood or urine) underwent triple dilution with distilled water, and then they were placed in plastic test tubes containing 0.5 g of a commercial preparation (i.e., magnesium sulfate, sodium acetate, stainless steel balls, and 1 mL inner standard). Then the entity of the test tube was vigorously vortexed and centrifuged. From the obtained supernatant, compounds of the acidic character could be analyzed by means of LC-MS/MS. To extract basic compounds, step two had to be performed (the so-called dispersive/solid phase extraction step, dSPE). To this effect, 600 µL supernatant had to be transferred to the test tube with the commercially prepared mixture of a primary amine, the end-capped octadecylsilane and magnesium sulfate, in order to cleanse it. Then the test tube was vigorously vortexed and centrifuged, and the upper layer made an extract that could be analyzed at the next step.

Advantages of the proposed method are its rapidity, a suppressed risk of polluting the equipment, and cost effectiveness. This has resulted in a growing number of the QuEChERS applications (performed either directly, or with certain modifications) to the analysis of NPSs contained in biological material [21,22].

REFERENCES

1. Ellefsen, K.N., Anizan, S., Castaneto, M.S., Desrosiers, N.A., Martin, T.M., Klette, K.L., and M.A. Huestis. 2014. Validation of the only commercially available immunoassay for synthetic cathinones in urine: Randox Drugs of Abuse V Biochip Array Technology. *Drug. Test. Anal.* 6: 728–738.
2. Swortwood, M.J., Hearn, W.L., and A.P. DeCaprio. 2014. Cross-reactivity of designer drugs, including cathinone derivatives, in commercial enzyme-linked immunosorbent assays. *Drug. Test. Anal.* 6: 716–727.
3. Sørensen, L.K. 2011. Determination of cathinones and related ephedrines in forensic whole-blood samples by liquid-chromatography-electrospray tandem mass spectrometry. *J. Chromatogr. B Analyt. Technol. Biomed. Life Sci.* 879: 727–736.

4. Golasik, M., Wodowski, G., Gomółka, E., Herman, M., and W. Piekoszewski. 2014. Urine as a material for evaluation of exposure to manganese in methcathinone users. *J. Tr. Elem. Med. Biol.* 28: 338–343.

5. Grapp, M., Kaufmann, C., and M. Ebbecke. 2017. Toxicological investigation of forensic cases related to the designer drug 3,4-methylenedioxypyrovalerone (MDPV): Detection, quantification and studies on human metabolism by GC-MS. *Forensic Sci. Int.* 273: 1–9.

6. Kneisel, S. and V. Auwärter. 2012. Analysis of 30 synthetic cannabinoids in serum by liquid chromatography-electrospray ionization tandem mass spectrometry after liquid-liquid extraction. *J. Mass. Spectrom.* 47: 825–835.

7. Adamowicz, P. 2016. Fatal intoxication with synthetic cannabinoid MDMB-CHMICA. *Forensic Sci. Int.* 261: e5–e10.

8. Adamowicz, P. and J. Gieroń. 2016. Acute intoxication of four individuals following use of the synthetic cannabinoid MAB-CHMINACA. *Clin. Toxicol. (Phila).* 54: 650–654.

9. Ojanperä, I., Gergov, M., Liiv, M., Riikoja, A., and E. Vuori. 2008. An epidemic of fatal 3-methylfentanyl poisoning in Estonia. *Int. J. Legal Med.* 122: 395–400.

10. Dickson, A.J., Vorce, S.P., Levine, B., and M.R. Past. 2010. Multiple-drug toxicity caused by the coadministration of 4-methylmethcathinone (mephedrone) and heroin. *J. Anal. Toxicol.* 34: 162–168.

11. Saito, T., Namera, A., Osawa, M., Aoki, H., and S. Inokuchi. 2013. SPME–GC–MS analysis of α-pyrrolidinovaleorophenone in blood in a fatal poisoning case. *Forensic Toxicol.* 31: 328–332.

12. Hasegawa, K., Suzuki, O., Wurita, A., Minakata, K., Yamagishi, I., Nozawa, H., Gonmori, K., and K. Watanabe. 2014. Postmortem distribution of α-pyrrolidinovalerophenone and its metabolite in body fluids and solid tissues in a fatal poisoning case measured by LC-MS-MS with the standard addition method. *Forensic Toxicol.* 32: 225–234.

13. Hasegawa, K., Wurita, A., Minakata, K., Gonmori, K., Nozawa, H., Yamagishi, I., Watanabe, K., and O. Suzuki. 2015. Postmortem distribution of PV9, a new cathinone derivative, in human solid tissue in a fatal poisoning case. *Forensic Toxicol.* 33: 141–147.

14. Wurita, A., Suzuki, O., Hasegawa, K., Gonmori, K., Minakata, K., Yamagishi, I., Nozawa, H., and K. Watanabe. 2013. Sensitive determination of ethylene glycol, propylene glycol and diethylene glycol in human whole blood by isotope dilution gas chromatography-mass spectrometry, and the presence of appreciable amounts of the glycols in blood of healthy subjects. *Forensic Toxicol.* 31: 272–280.

15. Wurita, A., Hasegawa, K., Minakata, K., Gonmori, K., Nozawa, H., Yamagishi, I., Suzuki, O., and K. Watanabe. 2014. Postmortem distribution of α-pyrrolidinobutiophenone in body fluids and solid tissues of a human cadaver. *Leg. Med.* 16: 241–246.

16. Montesano, C., Johansen, S.S., and M.K.K. Nielsen. 2014. Validation of a method for the targeted analysis of 96 drugs in hair by UPLC-MS/MS. *J. Pharm. Biomed. Anal.* 88: 295–306.

17. Cooper, G.A.A., Kronstrand, R., and P. Kintz. 2012. Society of Hair Testing guidelines for drug testing in hair. *Forensic Sci. Int.* 218: 20–24.

18. Znaleziona, J., Ginterová, P., Petr, J., Ondra, P., Válka, I., Ševčík, J., Chrastina, J., and V. Maier. 2015. Determination and identification of synthetic cannabinoids and their metabolites in different matrices by modern analytical techniques—A review. *Anal. Chim. Acta.* 874: 11–25.

19. Namera, A., Konuma, K., Saito, T., Ota, S., Oikawa, H., Miyazaki, S., Urabe, S., Shiraishi, H., and M. Nagao. 2013. Simple segmental hair analysis for α-pyrrolidinophenone-type designer drugs by MonoSpin extraction for evaluation of abuse history. *J. Chromatogr. B Anal. Technol. Biomed. Life Sci.* 942–943: 15–20.

20. Alvarez, J.C., Etting, I., Abe, E., Villa, A., and N. Fabresse. 2017. Identification and quantification of 4-methylethcathinone (4-MEC) and 3,4-methylenedioxypyrovalerone (MDPV) in hair by LC-MS/MS after chronic administration. *Forensic Sci. Int.* 270: 39–45.
21. Usui, K., Hayashizaki, Y., Hashiyada, M., and M. Funayama. 2012. Rapid drug extraction from human whole blood using a modified QuEChERS extraction method. *Leg. Med. (Tokyo)*. 14: 286–296.
22. Kudo, K., Usumoto, Y., Usui, K., Hayashida, M., Kurisaki, E., Saka, K., Tsuji, A., and N. Ikeda. 2014. Rapid and simultaneous extraction of acidic and basic drugs from human whole blood for reliable semi-quantitative NAGINATA drug screening by GC–MS. *Forensic Toxicol.* 32: 97–104.

12 Discrimination among Designer Drug Isomers by Chromatographic and Spectrometric Methods

Piotr Adamowicz and Dariusz Zuba

CONTENTS

12.1 ISOMERS AND ISOMERISM

Isomerism is the phenomenon of existence of two or more compounds of the same molecular formula (the same elemental composition and molecular weight), but of a different structure. Such compounds are called isomers. There are two main types of isomerism: constitutional isomerism (structural isomerism) and stereoisomerism.

Constitutional isomers have the same molecular formula, but a different order of connectivity of atoms. They always have different properties, especially when different functional groups are present in the molecules. Examples of structural isomers are n-propyl alcohol, isopropyl alcohol, and methoxyethane (all share the same chemical formula, C_3H_8O). Three main categories of constitutional isomers are skeletal, positional (regioisomers), and functional isomers. Skeletal isomers have variable amounts of branching in hydrocarbon chains (e.g., n-butane and isobutane). Positional isomers have different functional groups at the same carbon skeleton (e.g., above-mentioned n-propyl alcohol and isopropyl alcohol). Functional isomers have various functional groups (e.g., ethyl acetate and butyric acid).

Stereoisomers have the same molecular formula, the same connectivity of atoms, but a different arrangement of atoms in space. This class of isomers includes diastereomers and enantiomers. Diastereomers are not mirror images of each other and often contain chiral centers, but there are some diastereomers that are neither chiral nor contain chiral centers. Diastereomers show differences in physical properties, i.e., melting point, boiling point, or solubility. Their optical properties (optical rotation) may be similar or very different. Example of diastereomers are geometric isomers, e.g., cis-trans diastereomers. Enantiomers (optical isomers) are nonsuperimposable mirror images of each other and always contain chiral centers. The ability of a compound to have an enantiomer depends on the geometrical characteristics called chirality. All physical properties and the vast majority of chemical properties are almost identical for the two enantiomers. A mixture of two enantiomers is called a racemic mixture.

Configuration in stereochemistry is defined in several ways and a few methods of distinguishing and naming of optical isomers exist. An absolute configuration is determined according to the Cahn–Ingold–Prelog priority rules. Each chiral center can be labelled as R (Rectus) or S (Sinister), according to a priority of substituents. In biochemistry, the enantiomers are commonly named by the direction in which they rotate the plane of polarized light. If it rotates the light clockwise, that enantiomer is labelled (+), and if it rotates counterclockwise, it is labelled (−). The (+) and (−) isomers have also been termed d- and l-, respectively (for dextrorotatory and levorotatory). However, labelling enantiomers with d- and l- is discouraged, because it can be easy to confuse with D- and L-naming. The D/L system (named after dexter and laevus; right and left) is the method of classification of optical isomers by analyzing the relation of the molecule to glyceraldehyde. This system is unrelated to the d/l system and, for example, different compounds of D-configuration can be both d (+) and l (−). For example, L- and D-ephedrine are enantiomers, as well as L- and D-pseudoephedrine, but pseudoephedrine and ephedrine are diastereomers.

12.2 SIGNIFICANCE OF DRUG ISOMERISM

Both constitutional isomers and stereoisomers may have different properties and biological activity. Isomerism is especially important in pharmacology and pharmacotherapeutics, because isomers differ in their pharmacokinetic and pharmacodynamic properties.

Structural isomers may have similar (e.g., isoflurane and enflurane) or different (e.g., promazine and promethazine) actions [1]. The pharmacological properties of structural isomers are usually different. For example, para-amino isomers show strong local anesthetic activity, while ortho-amino isomers of procaine-type aminobenzoic acid esters show weak activity. The acetylcholine derivative, methacholine, has about the same affinity for muscarinic receptors as acetylcholine, but its nicotinic effect is 200-fold lower than that of the endogenous agonist [2]. Compounds with similar structure can also show dramatic differences in toxicity. An example may be the three isomers of dichlorobenzene (DCB), of which 4,5-epoxides are the most toxic metabolites. Both 1,2- and 1,3-DCB can be formed from the 4,5-epoxide, but because 1,4-DCB has a chlorine at the 4-position, it cannot form this metabolite.

In this case, the difference of toxicity appears to be a direct result of the different toxicity displayed by the different metabolites of the different isomers [3].

Optical isomers are very widespread in nature and play an important role in biochemistry. Most important substances involved in life processes are optical isomers. Moreover, living organisms usually tolerate only one kind of optical isomer of each substance. Asymmetric surface of the protein distinguishes the enantiomers of chiral compounds. For example, in the human body only L amino acids are active. D-Glucose is fully bioavailable by the body, and L-glucose is not absorbed at all (although its taste is sweet). Therefore, some molecules exist in nature only in the one form, e.g., all amino acids have L configuration and sugars have D configuration. In addition, only one isomer of adrenaline has strong stimulating activity.

The same applies to many drugs and other xenobiotics. Most optically active drugs are chiral and there are many examples of such drugs. They show radically different biological activity depending on which enantiomer is considered. One optical isomer of the same substance can be neutral to the body and does not take part in biochemical processes, while another isomer can be active or even harmful and consequences introducing it to the body can then be disastrous. An example might be thalidomide—a drug used in the 1950s as a sedative or hypnotic, as well as against nausea and to alleviate morning sickness in pregnant women. Prior to registration of the drug, only the R-form of thalidomide was tested (shown to be safe), but the racemic mixture of two enantiomers was introduced to sell. As it turned out, the S-enantiomer of this compound not only did not have the expected healing properties, but it was a potent mutagen, acting especially on the DNA of the fetus (causing severe deformation of the body). A classic example of different actions of stereoisomers is quinine (L isomer) and quinidine (D isomer). Quinine is used as an antimalarial drug, and quinidine as antiarrhythmic. Similarly, R-naproxen is used for arthralgic pain, while S-naproxen is teratogenic. Levomethorphan is a potent opioid analgesic, while dextromethorphan is a cough suppressant. L-Methotrexate is better adsorbed than D-methotrexate, esomeprazole is more bioavailable than racemic omeprazole, levocetirizine has a smaller volume of distribution than its dextroisomer, d-propranolol is more extensively bound to proteins than l-propranolol, S-warfarin is more potent and more extensively bound to albumin than R-warfarin, and they also differ in the way of metabolism and the time of half-life. Numerous other compounds exhibit stereoselective drug metabolism and elimination. Such has been reported for a ketamine, whose R(−)-ketamine inhibits clearing more rapidly than S(+)-ketamine. Moreover, S(+)-ketamine causes fewer psychotic emergence reactions, less agitated behavior, and better inoperative amnesia and analgesia than R(−)-ketamine. The metabolism of R(−)- and S(+)-methadone is different and also dependent on the concentration. S(+)-Methadone antagonizes respiratory depression action of R(−)-methadone. Despite the fact that single enantiomers have more selective action and less adverse reactions, many drugs are sold as racemic mixtures. An example is racemic ibuprofen, while S-ibuprofen is the active form and R-ibuprofen is inactive [4,5].

The activity and toxicity of illegal drugs may also depend on the isomeric form, and the biotransformation reactions in the body demonstrate isomeric preference. It was estimated that over half of illicit compounds possess at least one chiral center [6]. For example, D-cocaine was found to have greater activity, more rapid onset, and shorter duration than L-cocaine. Amphetamine isomers have different physiological effects; the d-isomer shows a stronger effect on the central nervous system and is approximately four times more potent than the l-isomer as a releaser of dopamine (DA).

12.3   ISOMERS OF DESIGNER DRUGS

The evolution in drugs of abuse has been observed for many decades, but this phenomenon has intensified in recent years. The drug market has started to change significantly since 2008. The total number of substances being monitored by the European Monitoring Centre for Drugs and Drug Addiction (EMCDDA) is over 560, with most of them being reported in the last few years [7]. Designer drugs are structural or functional analogs of controlled substances that have been created to mimic the effects of the classic drugs, while avoiding classification as illegal and detection in standard drug tests. The chemical structures of many new psychoactive substances are based on a skeleton of phenethylamine, cathinone, tryptamine, or piperazine. In addition, a number of new families of psychoactive substances have appeared, characterized by an action similar to controlled substances, e.g., synthetic cannabinoids.

It is clear that considerable variability in pharmacology and toxicology is observed for designer drugs. It will be discussed based on a comparison of structural isomers of synthetic cathinones. Simmler et al. presented the monoamine transporter and receptor interaction profiles of ethcathinone (N-ethylcathinone), buphedrone, and dimethylcathinone [8]. Buphedrone and dimethylcathinone exhibited very low potency at the serotonin (5-hydroxytryptamine, 5-HT) uptake transporter (SERT) and the dopamine transporter (DAT)/SERT inhibition ratios were >10. Ethcathinone was ten times less potent at SERT vs. the DAT. The drugs' effects on transporter-mediated monoamine release were also different. Dimethylcathinone did not induce norepinephrine (NE) and DA release and thus was a pure uptake inhibitor. In contrast, ethcathinone released NE and 5-HT but had no effect on DA release, whereas buphedrone released NE but not DA nor 5-HT. Differences were also observed in binding affinities expressed as the potencies of the drugs (Ki) to inhibit radioligand binding to the NET, DAT, and SERT and different monoamine receptors. Submicromolar affinity (<1 μM) DAT interaction was found with the ethcathinone, while this drug showed relevant binding (<10 μM) to 5-HT1A. Similar affinity (<10 μM) was present for dimethylcathinone to the 5-HT2A and 5-HT2C receptors.

Differences in the effects and action of cathinones were also confirmed by the study of other authors [9]. They investigated the effects on basal and electrically evoked DA efflux in rats. Mephedrone (4-methylmethcathinone, 4-MMC) increased, while ethcathinone had no significant effect on basal levels of DA in an absence of electrical stimulation. 4-MMC caused qualitatively different effects on electrically evoked DA efflux when compared to ethcathinone. The latter compound significantly increased the peak DA efflux after electrical stimulation and showed DA reuptake.

Separation and identification of structural isomers is an important issue for several reasons. Despite having similar structures, they differ in pharmacology and toxicology. Unambiguous identification of these compounds is thus very important.

Another issue is the legal status of individual isomers. The matter is easy if the drugs are controlled by generic law and therefore all new emerging isomers are automatically controlled. The situation is much more difficult in countries in which antidrug law is based on individual lists of controlled substances. In that case, each new substance has to be added to the list. Therefore, unequivocal identification of a substance is crucial both in forensic chemistry (analysis of samples seized from the drug market) and forensic toxicology (analysis of biological material).

The analysis of enantiomers is also important. Many designer drugs are chiral (most phenylethylamines and cathinones) and, likely, the pharmacological potency of the enantiomers of these stimulants differs. Stereospecific effects of amphetamine and cathinone analogs are well known. S-Cathinone is three times more potent than R-cathinone in causing DA release. R- and S-methcathinone produce neurotoxicity in DA neurons but only the S-isomer produces 5-HT neurotoxicity. S-Methcathinone exhibits three-fold greater potency as a discriminative stimulus substituting for cocaine compared with R-methcathinone in rats. Both enantiomers of mephedrone display some differences in the action. S-Mephedrone has a greater serotonergic profile and demonstrated mild locomotor activation as well as no rewarding properties. R-mephedrone possesses more of a dopaminergic stimulant-like profile with both locomotor activation and reward. R- and S-isomer display similar effects on DA release but the R-isomer is much weaker in its ability to release 5-HT [10].

Since nonenantioselective synthesis is easier and cheaper, designer drugs are mainly sold as racemic mixtures. Even the pharmaceutical preparation of diethylpropion (amfepramone—a drug used as an appetite suppressant) is a racemate. However, the illicit manufacturing of synthetic cathinones (among others, mephedrone) often involves synthesis procedures that result in one enantiomer being synthesized in larger quantities than the other (e.g., methcathinone is primarily synthesized as S-methcathinone) [10]. Different batches of "research chemical" products containing 4-fluoromethylphenidate appeared to consist of different quantities of the isomers. One product consisted of (±)-threo-4-fluoromethylphenidate, whereas the second sample consisted of a mixture of (±)-threo and (±)-erythro-4-fluoromethylphenidate [11]. Chiral analysis can provide some information on the synthetic route, which, in turn, may assist in the identification of the source of the drug.

Similar cases happen with synthetic cannabinoids. HU-211 (dexanabinol) and its enantiomer HU-210 can be an example. HU-210 is a potent cannabinoid agonist, while HU-211 does not act as a cannabinoid receptor agonist, but instead has NMDA antagonist effects [12]. It does not produce cannabis-like effects, but is anticonvulsant and neuroprotective. HU-211 seems to have great potential for therapeutic application. Therefore, chiral separation of the cannabinoid stereoisomers is important for pharmaceutical development and identification of impurities.

12.4 DISCRIMINATION AMONG DESIGNER DRUG ISOMERS BY VARIOUS TECHNIQUES

Distinction among isomers is often a complex analytical issue. A number of analytical techniques, including gas chromatography–mass spectrometry (GC-MS), high-pressure liquid chromatography with diode array detection (HPLC-DAD), and liquid chromatography–mass spectrometry (LC-MS), as well as capillary electrophoresis (CE), can be used for this purpose.

Screening analyses for designer drugs, especially in illicit products, are performed mainly by GC-MS, HPLC-DAD, and LC-MS. LC-MS gives information about the molecular weight, while the MS spectra obtained both in LC-MS and GC-MS provide information about the structure of a molecule. The separation of constitutional isomers by these techniques is usually not a big issue; however, it requires the development of a dedicated method. The experience of a person who analyses the results is also necessary, because a large number of isomers create a chance of incorrect identification. The reason for this is that mass spectra of isomers are often very similar, and only a limited number of designer drugs is included in popular commercial databases. For a newly developed and marketed designer drug, which is an isomer of a substance existing on the drug market, there is a reasonable risk that its spectrum is not included in a database, and the search for the best-fitted spectrum will indicate the parent substance.

Enantiomers can be separated in several ways by chromatographic techniques. HPLC is used for nonvolatile analytes, while GC is preferred in the analysis of volatile compounds. Sometimes CE is also employed, but the most important is the use of chiral columns. In chiral chromatography, chiral stationary phase composed of silica gel bonded with chiral molecules is used. One of the enantiomers will interact more strongly with the stationary phase than the other one. Cyclodextrins, cyclic oligomers of glucose, are frequently used in chiral chromatography. Most manufacturers of chromatographic columns (for both LC and GC) have in their offer columns used for the separation of enantiomers. It is also possible to use chiral mobile phase. Another common method is to transform the enantiomers to diastereomers by chemical reaction. Diastereomers have different chemical properties, and therefore it is much easier to separate them. Diastereomers can be obtained by derivatization with a specific agent. In this case, standard stationary phase can be used.

12.4.1 ISOMERS DISCRIMINATION BY GC-MS

The most popular technique used in forensic laboratories dealing with drug analysis is GC-MS. This technique can be effectively used for the analysis of constitutional isomers, in particular when tandem mass spectrometers (GC-MS/MS) are applied.

This ability of GC-MS to distinguish isomers can be discussed in detail using an example of three isomeric designer drugs being the derivatives of phenethylamine, that is, 2C-G (2,5-dimethoxy-3,4-dimethylphenethylamine), 2C-E (4-ethyl-2,5-dimethoxyphenethylamine), and DOM (dimethoxymethylamphetamine, 1-(2, 5-dimethoxy-4-methylphenyl)-2-aminopropane). Their chemical structures are presented in Figure 12.1 and mass spectra in Figure 12.2.

DOM contains a methyl group at the 4-position of the benzene ring, and another methyl group attached to the β-carbon atom of the side chain. The presence of a methyl group in the backbone influences significantly the compound's chromatographic and spectrometric features, making distinguishing of DOM from 2C-G and/or 2C-E easy. The dominant ion at m/z 44 is formed by cleavage of the bond between C_α and C_β atoms in the side chain and formation of the iminium ion,

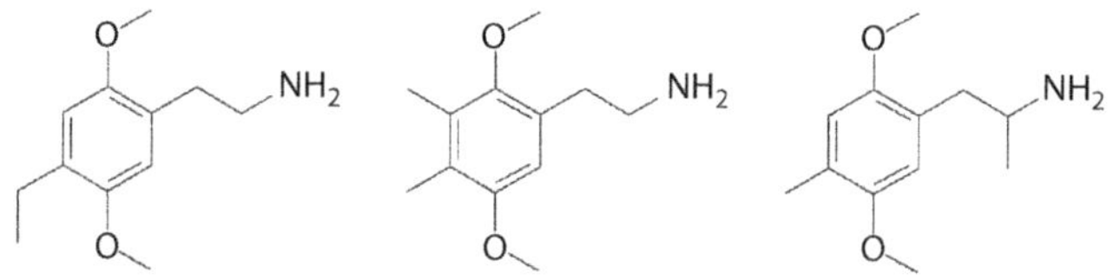

FIGURE 12.1 Chemical structures of three hallucinogenic derivatives of phenethylamine (a: 2C-E, b: 2C-G, c: DOM), which are structural isomers.

which is characteristic for phenethylamines. Iminium ion is observed in the spectra of 2C-G and 2C-E at m/z 30. Slight differences between the spectra of these substances means that they can be wrongly identified. The main differences concern the occurrence of an intense ion at m/z 178 for 2C-G, and the relative abundances of the ions with m/z 165 and 180. The first is dominant in the 2C-G spectrum, while the latter in the 2C-E spectrum. The molecular ion is also more abundant for 2C-E. Further fragmentation of the ion with m/z 165 could lead to the loss of the CH_2 group (appropriate ion was observed at m/z 151). It was clearly seen in 2C-E, while relatively small in 2C-G. The origin of these ions was discussed in [13].

An even more complex set of positional isomers was analyzed by Zuba and Adamowicz [14], which applied GC-MS to distinguish six methyl derivatives of methcathinone (4-MMC; metaphedrone, 3-MMC; 2-methylmethcathinone, 2-MMC; dimethylcathinone, DMC; and ethcathinone, ETC; Figure 12.3). Firstly, the chromatographic features were compared. The differences in the retention times of the isomers turned out to be slight. Dimethylcathinone was eluted as the first at 5.713 minutes, whilst 4-MMC eluted as last at 6.104 minutes (the difference = 0.391 minute). The closest pair was ethcathinone and 2-MMC, for which the retention times differed by only 0.006 minute (less than one second). Although the peak widths of the analyzed isomers were very narrow, the aforementioned pair was not distinguished under the applied conditions. It is theoretically possible to use another column or change the temperature program, but it was clearly seen when analyzing their mass spectra that these data can be effectively used for differentiation. The chromatograms of all isomers for the extracted ions with m/z 58 and 72 are presented in Figure 12.4.

Ethcathinone can be distinguished from the other isomers based on the secondary fragmentation of the immonium ion of m/z 72, that is, the ion with m/z 44. This process occurs efficiently only with substances with an alkyl chain containing two or more carbon atoms substituted at the nitrogen atom. Significant differences for buphedrone and dimethylcathinone are observed for ions with m/z 57 and 42, respectively. On the other hand, the EI-MS spectra of 2-, 3-, and 4-MMC are practically identical and therefore the determination on the location of the methyl group is only possible on the basis of the retention time (or the other analytical methods, e.g., HPLC-DAD). The key solution was the application of retention time locking (RTL), which made the retention time values highly reproducible (relative standard deviations, RSDs, lower than 0.13%). Combining the two types of data, chromatographic and spectrometric, was very useful in distinguishing the isomers as all the analyzed substances could be differentiated unequivocally.

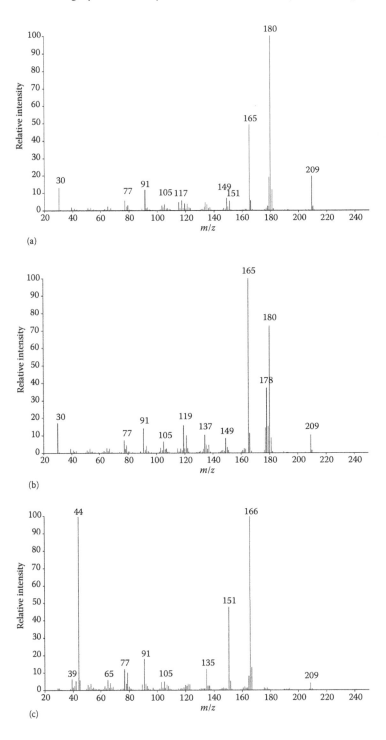

FIGURE 12.2   Mass spectra of (a) 2C-E, (b) 2C-G, and (c) DOM.

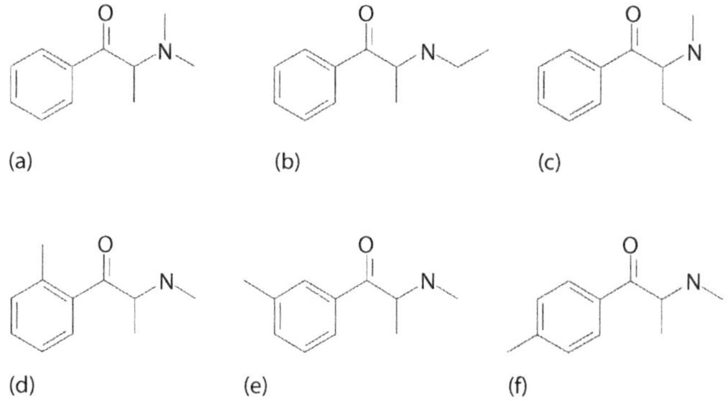

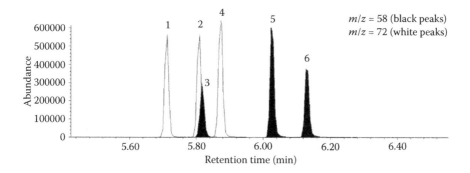

FIGURE 12.3 Chemical structures of six methyl derivatives of methcathinone: (a) N,N-dimethylcathinone, (b) N-ethylcathinone, (c) buphedrone, (d) 2-methylmethcathinone (2-MMC), (e) metaphedrone (3-MMC), and (f) mephedrone (4-MMC).

FIGURE 12.4 Chromatograms of methyl derivatives of methcathinone in the extracted ions mode (m/z 58 and 72). (1) N,N-dimethylcathinone, (2) N-ethylcathinone, (3) 2-MMC, (4) buphedrone, (5) 3-MMC, and (6) 4-MMC.

Other authors [15] reported the attempts to differentiate regioisomers of cathinones with the molecular mass 191 Da, including isomers of ethylmethcathinone (2-, 3-, and 4-), dimethylmethcathinone (2,3-, 2,4-, and 3,4-), 4-methylbuphedrone, 4-methylethcathinone, and pentedrone, using the GC-MS/MS technique. The product ion obtained from the iminium ions allowed for univocal differentiation of cathinones with the same aminoalkyl moiety, and the product ion spectrometry of acylium ions allowed for differentiation of regioisomers resulting from different substitution patterns on the aromatic ring.

Assignment of the location of the positions of the oxygen atom and the double bond in the structure of 6-APB (6-(2-aminopropyl)benzofuran) and 5-APB (5-(2-aminopropyl)benzofuran) was the topic of another paper [16]. It turned out that significant differences in the ratio of the characteristic ions with m/z 131 and 132 allowed for differentiation between these isomers.

Relative ion abundances were also used for differentiation of the 2,5-dimethoxy-N-(2-methoxybenzyl)phenethylamine (NBOMe) series of designer drugs, with its 3- and 4-methoxy isomers [17]. 25H-NBOMe, 25B-NBOMe, 25C-NBOMe, 25D-NBOMe, 25E-NBOMe, 25I-NBOMe, 25N-NBOMe, 25P-NBOMe, 25T2-NBOMe, 25T4-NBOMe, and 25T7-NBOMe were analyzed. The NBOMe methoxybenzyl positional isomers have a base peak ion at m/z 121 and an ion at m/z 150. However, all 2-methoxybenzyl substituted compounds produced a tropylium ion at m/z 91 at a significantly higher relative abundance than their corresponding 3- and 4-methoxybenzyl substituted analogues.

GC-MS was also applied for the discrimination and identification of the six aromatic positional isomers of trimethoxyamphetamine (TMA) [18]. The developed method allowed both mass spectrometric and chromatographic differentiation. The analytes were analyzed without and after trifluoroacetyl derivatization (TFA). However, the mass spectra of the nonderivatized substances, except 2,4,6-trimethoxyamphetamine (TMA-6), exhibited insufficient difference for unambiguous discrimination. The mass spectra of the TFA derivatives of the six isomers showed fragments with significant intensity differences, which allowed unequivocal identification.

Effectiveness of GC-MS/MS in discriminating regioisomers of ring-substituted fluorophenethylamines and N-alkylated fluorocathinones were reported by Westphal et al. [19,20]. The combination of chemical ionization (CI) and product ion spectrometry of the hydrogen fluoride loss ions [M+H-HF](+) allowed a univocal differentiation of a series of N-alkylated ortho-, meta-, and para-fluoroamphetamines and 1-(4-fluorophenyl)butan-2-amines, as well as N-alkylated ortho-, meta-, and para-fluorocathinones without prior derivatization.

Abdel-Hay et al. used GC-MS to evaluate six-ring regioisomeric dimethoxyphenylpiperazines (DOMePPs). GC separation of perfluoroacyl derivatives was achieved on an Rtx-200 column, and their mass spectra showed some differences in relative abundance of ions [21]. The same authors presented the results of their studies on six-ring regioisomeric dimethoxybenzoylpiperazines (DMBzPs) [22]. The obtained mass spectra (without and with derivatization) for these isomers were almost identical and the applied GC-MS method alone did not provide for the confirmation of identity of any one of the isomers to the exclusion of the other compounds. In the next study Abdel-Hay et al. presented analytical profiles of regioisomeric bromodimethoxy piperazines related to designer drug 4-bromo-2,5-dimethoxybenzylpiperazine (2C-B-BZP) using GC-MS [23]. Perfluoroacylation of the secondary amine nitrogen was applied and the derivatives were successfully resolved on a capillary column with stationary phase composed of 100% trifluoropropyl methyl polysiloxane. However, the mass spectra for the seven regioisomeric bromodimethoxy benzyl piperazines also were almost identical with only 2,3-dimethoxy isomers showing unique ions at m/z 214/216. Thus, mass spectrometry alone did not allow for identification of these compounds.

Many synthetic cannabinoids are also isomers. DeRuiter et al. have shown that the electron impact (EI) mass spectra for 1-alkyl-3-acyloindoles can be differentiated from the inverse regioisomeric 1-acyl-3-alkylindoles based on the base peaks [24]. Methods of regioisomeric differentiation of 1-n-pentyl-3-(methoxybenzoyl)indoles and the 1-n-pentyl-3-(methylbenzoyl)indoles were presented in another paper [25]. The meta- and para-isomers displayed fragment ions at equivalent masses with some differences in relative abundance of ions. The ortho-isomers for both methoxybenzoyl and methylbenzoyl showed unique fragment ions. Moreover, these two sets of regioisomers were well resolved by capillary GC and they eluted in the following order: ortho-, meta-, and para-isomers. Six regioisomeric 1-n-pentylbenzoylindoles having the benzoyl group at each of the possible ring substituent positions of the indole ring were also analyzed [26]. The authors used an Rtx-200 column coated with 100% trifluoropropyl methyl polysiloxane in their study. This stationary phase provided excellent resolution of all compounds. The elution order appeared to be related to the relative degree of distance between the two indole substituted groups. 1,3- and 1,5-isomers were the latest eluting. The EI spectra showed the equivalent regioisomeric major fragments while the relative abundance of the ions varied among the six regioisomeric substances. Similar conclusions were achieved in another study concerning six regioisomeric 1-pentyl-3-dimethoxybenzoylindoles [27]. The separation was obtained on an Rtx-1 column.

Discrimination of positional isomers of synthetic cannabinoids by GC-EI-MS was also reported by Asada et al. [28]. Isomeric discrimination of 1-adamantyl and 2-adamantyl isomers of N-adamantyl carboxamides (APINACA, APINACA 2-adamantyl isomer, APICA, APICA 2-adamantyl isomer, 5F-APINACA, 5F-APINACA 2-adamantyl isomer, 5F-APICA, 5F-APICA 2-adamantyl isomer, 5Cl-APINACA, 5Cl-APINACA 2-adamantyl isomer, adamantyl-THPINACA, 2-adamantyl-THPINACA) was presented. Retention times of the isomers were similar, but 1-adamantyl carboxamides were clearly discriminated from their 2-adamantyl isomers based on their different fragmentation patterns in the mass spectra. The EI-MS spectra of adamantylindazole carboxamides showed significant differences when compared to adamantylindole carboxamides.

The same fragment ions were observed in the isomers FUBIMINA (BIM-2201) and THJ-2201, although the relative ion abundances of 177.0467 and 155.0499 were different for these compounds. Moreover, the prominent fragment m/z 273.1041 in FUBIMINA was not observed in THJ-2201 [29].

Kusano et al. [30] analyzed positional isomers of JWH-081 by GC-MS/MS. The authors first investigated isomers in the EI scan mode. EI scan was able to distinguish three of seven isomers: 2-methoxy, 7-methoxy, and 8-methoxy. The remaining isomers exhibited nearly identical spectra and tandem mass spectrometry was applied for differentiation. Precursor ions of m/z 185 and 157 were selected and the obtained product ions enabled the differentiation of 3-methoxy and 5-methoxy isomers. The product ion spectra of 6-methoxy isomer resembled that of JWH-081; however, the relative ion intensities enabled the differentiation from one another. Therefore, the combination of the EI scan and MS/MS allowed for the regioisomeric differentiation of the analyzed compounds.

Fragmentation differences in the EI spectra of three synthetic cannabinoid positional isomers were studied: JWH-250, JWH-302, and JWH-201 [31]. These compounds differ only in the position (ortho-, meta-, or para-) of a methoxy group on one of the aromatic rings (Figure 12.5). Thus, they have almost identical mass spectra. The retention times on HP-5MS capillary column were similar: 3.156, 3.241, and 3.393 minutes for JWH-250, JWH-301, and JWH-201, respectively. However, accurate analysis of the relative abundance of particular product ions was efficiently used to differentiate the isomers. It turned out that the ratios of m/z 121:91 were different

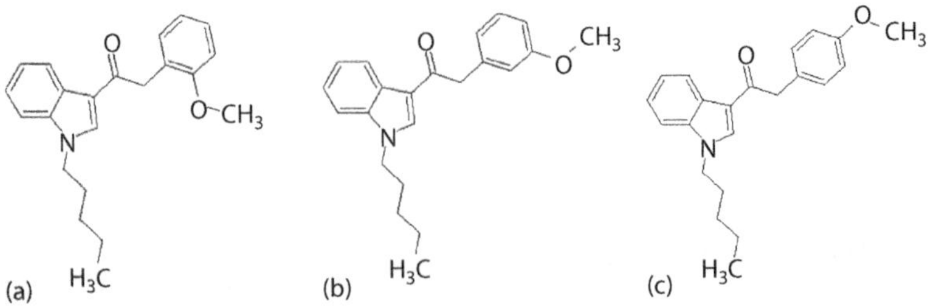

FIGURE 12.5   Chemical structures of (a) JWH-250, (b) JWH-302, and (c) JWH-201.

in the mass spectra of each isomer. The average ion abundance ratios were 0.4 for JWH-250, 1.3 for JWH-302, and 7.2 for JWH-201.

Differentiation of synthetic cannabinoid AB-FUBINACA (para-) and its fluoro positional (ortho- and meta-) isomers in the phenyl ring by electron-ionization-triple quadrupole mass spectrometry was presented [32]. It was found that the three isomers differ in the relative abundance of the ion at m/z 109 and 253 in the product ion spectra.

Stereoisomers of designer drugs were also separated using the GC-MS technique. Eighteen cathinone derivatives were successfully separated by GC-MS using trifluoroacetyl-l-prolyl chloride as chiral derivatization reagent [33]. In the other study, 13 out of 24 compounds were successfully resolved into their enantiomers using the same chiral derivatization agent [34].

12.4.2 Discrimination by HPLC-DAD

The stability of analytical conditions in liquid chromatography is much lower than in gas chromatography. As a consequence, chromatographic data, e.g., the retention times, are not as reproducible as in GC. The use of internal standard(s) and calculation of relative retention times improves the situation; however, identification based on this parameter has to be taken with care.

The latest developments in LC technology have increased the application of this technique in differentiation of isomers, even if other detectors than mass spectrometers are used. These developments include, among others, the use of ultra-high-pressure liquid chromatography (UHPLC) with dedicated monolithic chromatographic columns. Zuba and Adamowicz [14] assessed the usefulness of UHPLC-DAD in the differentiation of six methylmethcathinone isomers. Separation was performed at 30°C on Kinetex C18 (50 × 2.10 mm; 1.7 μm) column. Gradient elution of the mobile phases composed of 0.1% phosphoric acid in water and acetonitryl was used. Methylmethcathinones were eluted in less than two minutes (total analysis time: *ca.* seven minutes). Due to low stability of the retention times (RSD at 0.8–1.5% level), it was impossible to distinguish the isomers based on this parameter and the use of an IS did not improve the situation, but it turned out that analysis of UV/VIS spectra enabled solving the problem occurring in the analysis by GC-MS, that is, distinction among the ring isomers. The UV/VIS spectra of 2-MMC, 3-MMC, and 4-MMC are presented in Figure 12.6.

4-MMC has one intense band at 263 nm and the characteristic shape under 225 nm, in contrast to 2- and 3- isomers that possess two bands, located at 206 nm

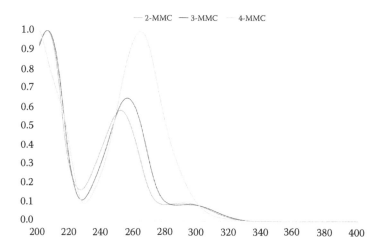

FIGURE 12.6 UV/VIS spectra of 2-MMC, 3-MMC, and 4-MMC.

and 250 or 256 nm (2-and 3-MMC, respectively). This feature is very efficient in distinguishing ring positional isomers of common designer drugs, such as derivatives of cathinone and phenethylamine, including amphetamines. The difference among isomers with the substituents at 4-, 3-, and 2- positions was also proved for other substances. One should remember that a common way to circumvent the law based on an individual drug system by manufacturers of new compounds is to locate the substituent in a different position of a phenyl ring, but the location of a substituent, including the methyl group, in a ring influences the 3D structure of the compound, and therefore the absorption of the UV/VIS light by the molecules of isomers.

A study on different regioisomeric derivatives of cathinone was also performed by Kohyama et al. [15]. The performed experiments enabled differentiation of isomers on the basis of λ_{max} corresponding to aromatic rings observed for 2-ethylmethcathinone (251 nm), 3-ethylmethcathinone (255 nm), and 4-ethylmethcathinone (264 nm), as well as for 2,3-dimethylmethcathinone (254 nm), 2,4-dimethylmethcathinone (263 nm), and 3,4-dimethylmethcathinone (267 nm).

HPLC-DAD was also used for the separation of the positional isomers of chlorophenylpiperazine (m-CPP, o-CPP, and p-CPP) [35]. Analyses were carried out on a Chiracel OJ-RH column (150 × 4.6 mm × 5 μm). The best separation was achieved with isocratic mobile phase of triethyl amine buffer and methanol (70:30, v/v) at pH 9.

Li and Lurie reported the use of a chiral additive in the mobile phase for regioisomeric separations using UHPLC-DAD [36]. They analyzed 24 designer cathinones and phenethylamines and resolved 18 of the regioisomers in eight minutes. Separations were performed on different columns (BEH C18: 2.1 × 100 mm × 1.7 μm; HSS T₃: 2.1 × 100 mm × 1.8 μm; BEH Phenyl: 2.1 × 100 mm × 1.7 μm; CSH Fluoro-Phenyl: 2.1 × 100 mm × 1.7 μm). The mobile phase consisted of 80 mM phosphate buffer at pH 1.8 and either acetonitrile or methanol. Discrimination of compounds was observed in these conditions, but an addition of HP-β-cyclodextrin to the mobile phase improved the separation.

Mohr et al. [37] developed the HPLC-UV method for the chiral separation of 24 cathinone derivatives. A CHIRALPAK AS-H column consisting of amylose tris [(S)-α-methylbenzylcarbamate] coated on 5-μm silica gel was used. The detection was gained at a wavelength of 254 nm. It was found to be suitable to resolve 19 of the tested compounds. The analyses were performed under isocratic conditions with mobile phase consisting of hexane, isopropanol, and triethylamine (97:3:0.1). In the other study [38], chiral mobile phase (methanol: water 2.5:97.5 + 2% sulfated ß-cyclodextrin) and a LiChrospher 100 RP-18e (250 × 4 mm, 5 μm) reversed phase column as the stationary phase were applied for the chiral separation of 25 cathinone and 6 amphetamine derivatives. With the use of such conditions, 17 cathinones were completely or partially chirally separated (only 3 of 25 cathinones were baseline resolved) within 23 minutes.

Enantioseparation of cathinones on chiral strong cation-exchange-type stationary phase was presented by Wolrab et al. [39]. The authors synthesized different chiral phases. Fourteen analytes differed in their substitution on the central aromatic ring as well as in the aliphatic side chains. Either 280- or 254-nm wavelengths were used for detection.

12.4.3 DISCRIMINATION BY LC-MS

LC-MS is a powerful analytical tool, which has a broad application in forensic toxicology and drug analysis. Its advantages include high sensitivity (allowing for efficient analysis of designer drugs in biological fluids) and high specificity, especially when tandem mass spectrometry or high-resolution mass spectrometry is applied.

Zuba and Adamowicz [14] distinguished six methyl derivatives of methcathinone with the use of liquid chromatography coupled to tandem mass spectrometry (LC-MS/MS). Separation was performed on a Kinetex C18 (100 × 4.6 mm, 2.6 μm 100Å) column. Multiple-reaction monitoring (MRM) with positive ion detection was used. The developed and optimized LC-MS/MS method allowed for unequivocal identification of all analyzed isomers. The experiments showed that in order to separate 3- and 4-MMC, it is necessary to apply isocratic phase composition. Six compounds were eluted from the column within six minutes. The use of a deuterated internal standard improved the identification on the basis of retention time. LC-MS/MS chromatograms of the isomers are presented in Figure 12.7.

It would seem that the same molecular masses and very similar chemical structure suggests that the particular MRM transition would be identical, which is not true for all compounds. In this study [14], ten MRM transitions were monitored. For all of them, the precursor ion was m/z 178.1. On the basis of the intensity profiles of product ions it was possible to identify the individual isomers. The profile of dimethylcathinone was different from the others. Intensity profiles for ethcathinone and buphedrone are similar; however, for these two compounds significant differences were observed for transitions (m/z) 178.1→91.1, 178.1→105.1, 178.1→117.1, 178.1→133.1, and 178.1→144.1. The profiles for 2-, 3-, and 4-MMC were practically identical, but the detailed analysis of the relative retention times allowed for unequivocal identification of the isomer.

Rácz et al. developed a liquid chromatographic method for the analysis of JWH-122 and its possible isomers, depending on the position of the methyl group on the

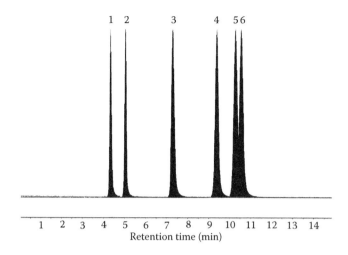

FIGURE 12.7 LC-MS/MS chromatogram of methylmethcathinone isomers. (1) N,N-dimethylcathinone, (2) N-ethylcathinone, (3) buphedrone, (4) 2-MMC, (5) 4-MMC, and (6) 3-MMC.

naphthyl moiety. UV and mass detection was applied in the study. Porous graphitic carbon column and non-aqueous mobile phase were used for separation [40]. The LC-MS analysis was also used for chiral separation of synthetic cannabinoids— (C8)-CP-47,497 (cannabicyclohexanol), its trans-form, and (C7)-CP-47,497 (CP-47,497) [41].

Distinction of isomeric designer drugs can be effectively made with liquid chromatography coupled with high-resolution mass spectrometry. Time-of-flight (TOF) analyzers are very efficient detectors, especially when combined with quadrupole.

Capability of the LC-QTOF method to differentiate isomers may be presented on an example of three synthetic cannabinoids, i.e., JWH-201, JWH-250, and JWH-302 (see Figure 12.5).

The measurements by the LC-QTOFMS instrument can be performed in the MS or the MS/MS mode. The degree of fragmentation depends on the analytical conditions, especially the fragmentor voltage and collision energy, and it is possible to find the optimum conditions. The best distinction among three aforementioned isomers was obtained using the fragmentor voltage 300 V and the collision energy 50 eV, when the fragmentation of the molecules is high [42]. All the cannabinoids had different ions with the highest relative abundance; for JWH-201 the fragment with m/z 121 was dominant, for JWH-250, m/z 91, and for JWH-302, m/z 144. The experiments were repeated at least three times for each substance and the observations were reproducible. Details of LC-QTOFMS experiments are presented in Chapter 7.

12.4.4 DISCRIMINATION BY OTHER METHODS

Fourier-transform infrared (FTIR) spectroscopy is often used as a confirmatory method for the identification of organic compounds. It creates the possibility of distinguishing between positional isomers. The efficiency of FTIR spectroscopy in this field can be discussed based on the example of differentiation between 2C-G and 2C-E, whose structures are presented in Figure 12.1 and spectra in Figure 12.8.

The differences between the spectra of 2C-E and 2C-G are clearly visible. Distinction between the isomers can be made based on different areas. The most significant diversity was recorded in the regions 950–1250 and 1440–1630 cm^{-1}. In the first region, the characteristic doublets (993–1014 and 1099–1124 cm^{-1}) were observed in the spectrum of 2C-G. Such doublets were not recorded in the remaining members of the 2C-series, which had a single band around 1043 cm^{-1}. Another difference concerns the location of the predominant band, which was shifted in 2C-G (1232 cm^{-1}) in comparison to other alkyl-substituted 2Cs (1209–1213 cm^{-1}). This band corresponds to C–O stretching frequency in ethers [13].

The application of FTIR spectroscopy to the analysis of isomeric designer drugs in samples seized from the drug market was also presented for synthetic cathinones.

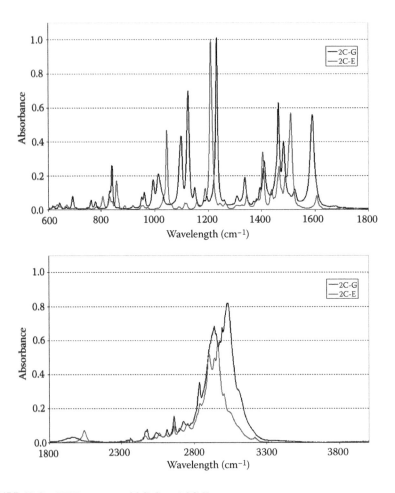

FIGURE 12.8   FTIR spectra of 2C-G and 2C-E.

Isomers of methylmethcathinone (2-, 3-, and 4-) as well as fluoromethcathinone (2-, 3-, and 4-) were characterized by FTIR spectra [43].

Differentiation of the NBOMe methoxybenzyl positional isomers (2-OMe vs. 3-OMe vs. 4-OMe) by FTIR spectroscopy were presented by Casale and Hays [17]. Each compound exhibited characteristic secondary amine HCl ion-pair absorbances between 2500 and 3000 cm^{-1}. The spectra were similar, but characteristic differences allowed each compound to be easily differentiated.

FTIR spectrometers can also be used as detectors in gas chromatography, although there are some technical limitations of such coupling. The GC-IRD method was used, e.g., for the differentiation of the six ring regioisomeric dimethoxybenzoylpiperazines, as well as substituted methoxymethcathinones [22,44].

A comparison of the abilities of GC-MS and FTIR to distinguish 2-MMC, 3-MMC, and 4-MMC, as well as to identify iso-mephedrone and iso-ethcathinone, which are phenylacetone isomers of 4-MMC and ethcathinone formed possibly from contaminated starting material, was presented by McDermott et al. [45]. A comparison of the GC-MS and FTIR properties of six regioisomeric benzoyl substituted-1-n-pentylindoles was also reported [26]. Infrared spectral data showed the carbonyl absorption band for each of the benzoylindoles and provides distinguishing and characteristic information to individualize each of the regioisomers (2-, 3-, 4-, 5-, 6-, and 7-benzoyl-1-n-pentylindoles). FTIR spectra provided also useful data to differentiate six other regioisomeric 1-n-pentyl-3-(dimethoxybenzoyl)-indoles, six-ring regioisomeric dimethoxyphenylpiperazines and seven regioisomeric bromodimethoxy benzyl piperazines [21,23,27].

Christie et al. [46] reported discrimination of cathinone positional isomers by Raman spectroscopy. The authors analyzed, among others, isomers of mephedrone, flephedrone, methylone, butylone, naphyrone, and MDPV. Raman spectra of all the isomers were obtained with far-red excitation (785 nm). The developed method has some advantages, of which the most important is minimal sample preparation.

Capillary electrophoresis on a dynamically coated capillary with PDA detection (CE-PDA) was applied for resolving 24 regioisomers of cathinones and phenethylamines [36]. All regioisomers were separated after modification of the running buffer with 80 mM (2-hydroxypropyl)-β-cyclodextrin (HP-β-CD). The developed chiral CE method was able to discriminate the majority of the enantiomers.

Cyclodextrin-assisted CE with UV and TOF-MS detection was used for chiral separation of 12 cathinone derivatives [47]. The separation was obtained using a buffer system consisting of 10 mM β-cyclodextrin (β-CD) in a 100-mM phosphate buffer for CE-UV, and 0.6% v/v highly sulfated-γ-cyclodextrin (HS-γ-CD) in a 50-mM phosphate buffer for CE-MS. All analytes were separated within 18 minutes in the CE-UV system and substances were identified by TOF-MS.

CE with UV detection at 236 nm was developed for distinction among the chlorophenyl isomers of piperazine, that is, 1-(2-chlorophenyl)piperazine (o-CPP), 1-(3-chlorophenyl)piperazine (m-CPP), and 1-(4-chlorophenyl)piperazine (p-CPP). Optimized background electrolyte contained 20 mmol/L phosphoric acid adjusted to pH 2.5 with triethylamine and 10 mmol/L α-cyclodextrin [48].

Burrai et al. developed CE method for enantiomeric separation of designer drugs from phenethylamine group. Sulfated-β-cyclodextrin was chosen as the best chiral selector. The separation of the analytes was achieved in a fused-silica gel capillary at 20°C using an applied voltage of +25 kV [49].

A chiral capillary zone electrophoresis (CZE) method for the enantioseparation of several cathinone derivatives was developed by Mohr et al. [50]. The following conditions proved to be the best: 20 mg/mL sulfated-β-cyclodextrin in 50 mM ammonium acetate buffer of pH 4.5 containing 10% v/v ACN at the temperature of 40°C and with an applied voltage of 20 kV. The same group of designer drugs was analyzed by capillary electrochromatography (CEC) [51]. Enantiomeric separation of ten cathinone derivatives was carried out on capillary columns packed with amylose tris(5-chloro-2-methylphenylcarbamate) coated on silica. Applying a voltage of 10 kV and a temperature of 20°C, the enantiomers were separated in less than ten minutes.

The last (but not least) analytical technique that can be used for differentiation of isomeric designer drugs is nuclear magnetic resonance (NMR) spectroscopy. It was successfully used to discriminate positional isomers of methylmethcathinones. 4-MMC is a 1,4-para substituted aromatic molecule with a symmetric distribution of protons on the aromatic ring, therefore ^{1}H NMR signals of the aromatic protons exhibit a characteristic splitting pattern. 2-MMC (1,2-ortho) and 3-MMC (1,3-meta) lack the symmetric distribution of aromatic protons and generate more complicated splitting patterns. Thus, 4-MMC can be easily discriminated from its positional isomers; however, it can be difficult to discriminate between 2-MMC and 3-MMC [43].

12.5 CONCLUSIONS

Identification of isomeric forms of the analyzed substances is important not only for seized material, as the penalty can depend on the legal status of the isomer, but also in biological material because the toxicity, action, etc. can differ significantly among substances with the same general formula.

Each of the discussed techniques has advantages and drawbacks in view of its use for distinction among both constitutional isomers and stereoisomers. The most versatile appears to be GC-MS. However, the mass spectra of positional isomers are often similar, causing a risk of misidentification. Moreover, the use of GC-MS in the analysis of biological material is problematic due to the low concentrations of designer drugs, in particular synthetic cannabinoids, in biofluids. This limitation can be eliminated by the application of the selected ion monitoring (SIM) mode of analysis, but then unambiguous identification of isomers may be impossible if they have similar retention times. Therefore, HPLC-DAD and LC-MS are the assays better suited for the analysis of biological material.

Analysis of isomers by a single method, particularly GC-MS or HPLC-DAD, may be insufficient. The use of more comprehensive methods, such as GC-MS/MS or LC-MS/MS, is more desirable. The application of tandem mass spectrometry significantly improves the identification capabilities. But the best solution is to apply several methods based on different principles. The use of multiple analytical procedures for the discrimination of designer drugs is essential and ensures the unequivocal identification of both constitutional isomers as well as stereoisomers.

Enantiomeric separations can be achieved by different chromatographic methods with the use of chiral stationary phases, or an achiral column with a chiral mobile phase or pre-column derivatization with a chiral reagent to create diastereomers. It must be noted that sometimes chiral separation can also be a useful tool for separating positional isomers.

Other methods may also be used for discrimination of isomers, of which nuclear magnetic resonance (NMR) and FTIR spectroscopy are the most commonly used. In turn, capillary electrophoresis (CE) with chiral additives is often applicable for chiral separations.

REFERENCES

1. Steven, M.Y., Nicholas, P.H., James, K.I. 2013. *Anaesthesia and Intensive Care A–Z: An Encyclopaedia of Principles and Practice.* Edinburgh; New York: Churchill Livingston/Elsevier.
2. Mutschler, E., Derendorf, H. 1995. *Drug Actions: Basic Principles and Therapeutic Aspects.* Stuttgart: CRC Press.
3. Sullivan, J.B., Krieger, G.R. 2001. *Clinical Environmental Health and Toxic Exposures.* Philadelphia: Lippincott Williams & Wilkins.
4. Smith, S.W. 2009. Chiral toxicology: It's the same thing... only different. *Toxicol Sci* 110(1):4–30.
5. Chhabra, N., Aseri, M.L., Padmanabhan, D. 2013. A review of drug isomerism and its significance. *Int J Appl Basic Med Res* 3(1):16–8.
6. Mile, B. 2005. Chemistry in court. *Chromatographia* 62:3–9.
7. European Monitoring Centre for Drugs and Drug Addiction. https://emcdda.europa.eu.
8. Simmler, L.D., Rickli, A., Hoener, M.C., Liechti, M.E. 2014. Monoamine transporter and receptor interaction profiles of a new series of designer cathinones. *Neuropharmacology* 79:152–60.
9. Opacka-Juffry, J., Pinnell, T., Patel, N., Bezan, M., Mentel, M., Davidson, C. 2014. Stimulant mechanisms of cathinones—Effects of mephedrone and other cathinones on basal and electrically evoked dopamine efflux in rat accumbens brain slices. *Prog Neuropsychopharmacol Biol Psychiatr* 54:122–130.
10. Gregg, R.A., Baumann, M.H., Partilla, J.S. et al. 2015. Stereochemistry of mephedrone neuropharmacology: Enantiomer-specific behavioural and neurochemical effects in rats. *Br J Pharmacol* 172(3):883–94.
11. McLaughlin, G., Morris, N., Kavanagh, P.V. et al. 2017. Analytical characterization and pharmacological evaluation of the new psychoactive substance 4-fluoromethylphenidate (4 F-MPH) and differentiation between the (±)-threo- and (±)-erythro-diastereomers. *Drug Test Anal* 9(3):347–357.
12. Feigenbaum, J.J., Bergmann, F., Richmond, S.A. et al. 1989. Nonpsychotropic cannabinoid acts as a functional N-methyl-D-aspartate receptor blocker. *Proc Natl Acad Sci U S A* 86(23):9584–7.
13. Zuba, D., Sekuła, K. 2013. Identification and characterization of 2,5-dimethoxy-3,4-dimethyl-β-phenethylamine (2C-G)—A new designer drug. *Drug Test Anal* 5(7):549–59.
14. Zuba, D., Adamowicz, P. 2017. Distinction of constitutional isomers of mephedrone by chromatographic and spectrometric methods. *Aust J Forensic Sci* 49(6):637–649.
15. Kohyama, E., Chikumoto, T., Tada, H., Kitaichi, K., Horiuchi, T., Ito, T. 2016. Differentiation of the isomers of N-alkylated cathinones by GC-EI-MS-MS and LC-PDA. *Anal Sci* 32(8):831–7.

16. Adamowicz, P., Zuba, D., Byrska, B. 2014. Fatal intoxication with 3-methyl-N-methylcathinone (3-MMC) and 5-(2-aminopropyl)benzofuran (5-APB). *Forensic Sci Int* 245:126–32.

17. Casale, J.F., Hays, P.A. 2012. Characterization of eleven 2,5-dimethoxy-N-(2-methoxybenzyl)phenethylamine (NBOMe) derivatives and differentiation from their 3- and 4-methoxybenzyl analogues—Part I. *Microgr J* 9(2):84–109.

18. Zaitsu, K., Katagi, M., Kamata, H. et al. 2008. Discrimination and identification of the six aromatic positional isomers of trimethoxyamphetamine (TMA) by gas chromatography-mass spectrometry (GC-MS). *J Mass Spectrom* 43(4):528–34.

19. Westphal, F., Rösner, P., Junge, T. 2010. Differentiation of regioisomeric ring-substituted fluorophenethylamines with product ion spectrometry. *Forensic Sci Int* 194(1-3):53–9.

20. Westphal, F., Junge, T. 2012. Ring positional differentiation of isomeric N-alkylated fluorocathinones by gas chromatography/tandem mass spectrometry. *Forensic Sci Int* 223(1–3):97–105.

21. Abdel-Hay, K.M., DeRuiter, J., Clark, C.R. 2015. GC-MS and IR studies on the six ring regioisomeric dimethoxyphenylpiperazines (DOMePPs). *J Forensic Sci* 60(2):285–94.

22. Abdel-Hay, K.M., DeRuiter, J., Clark, R.C. 2013. GC-MS and GC-IRD studies on the six ring regioisomeric dimethoxybenzoylpiperazines (DMBzPs). *Forensic Sci Int* 231(1–3):54–60.

23. Abdel-Hay, K.M., DeRuiter, J., Clark, C.R. 2014. Regioisomeric bromodimethoxy benzyl piperazines related to the designer substance 4-bromo-2,5-dimethoxybenzylpiperazine: GC-MS and FTIR analysis. *Forensic Sci Int* 240:126–36.

24. DeRuiter, J., Smith, F.T., Abdel-Hay, K., Clark, C.R. 2014. Analytical differentiation of 1-alkyl-3-acylindoles and 1-acyl-3-alkylindoles: Isomeric synthetic cannabinoids. *Anal Chem* 86(8):3801–8.

25. Abdel-Hay, K.M., DeRuiter, J., Smith, F., Belal, T.S., Clark, C.R. 2015. GC-MS analysis of the regioisomeric methoxy- and methyl-benzoyl-1-pentylindoles: Isomeric synthetic cannabinoids. *Sci Justice* 55(5):291–8.

26. Smith, F.T., DeRuiter, J., Abdel-Hay, K., Clark, C.R. 2014. GC-MS and FTIR evaluation of the six benzoyl-substituted-1-pentylindoles: Isomeric synthetic cannabinoids. *Talanta* 29:171–82.

27. Abdel-Hay, K.M., De Ruiter, J., Smith, F., Alsegiani, A.S., Thaxton-Weissenfluh, A., Clark, C.R. 2016. GC-MS differentiation of the six regioisomeric dimethoxybenzoyl-1-pentylindoles: Isomeric cannabinoid substances. *J Pharm Biomed Anal* 125:360–8.

28. Asada, A., Doi, T., Tagami, T., Takeda, A., Sawabe, Y. 2017. Isomeric discrimination of synthetic cannabinoids by GC-EI-MS: 1-adamantyl and 2-adamantyl isomers of N-adamantyl carboxamides. *Drug Test Anal* 9(3):378–388.

29. Diao, X., Scheidweiler, K.B., Wohlfarth, A., Zhu, M., Pang, S., Huestis, M.A. 2016. Strategies to distinguish new synthetic cannabinoid FUBIMINA (BIM-2201) intake from its isomer THJ-2201: Metabolism of FUBIMINA in human hepatocytes. *Forensic Toxicol* 34:256–267.

30. Kusano, M., Zaitsu, K., Nakayama, H. et al. 2015. Positional isomer differentiation of synthetic cannabinoid JWH-081 by GC-MS/MS. *J Mass Spectrom* 50(3):586–91.

31. Harris, D.N., Hokanson, S., Miller, V., Jackson, G.P. 2014. Fragmentation differences in the EI spectra of three synthetic cannabinoid positional isomers: JWH-250, JWH-302, and JWH-201. *Int J Mass Spectrom* 368:23–29.

32. Murakami, T., Iwamuro, Y., Ishimaru, R., Chinaka, S., Sugimura, N., Takayama, N. 2016. Differentiation of AB-FUBINACA positional isomers by the abundance of product ions using electron ionization-triple quadrupole mass spectrometry. *J Mass Spectrom* 51(11):1016–22.

33. Mohr, S., Weiß, J.A., Spreitz, J., Schmid, M.G. 2012. Chiral separation of new cathinone- and amphetamine-related designer drugs by gas chromatography-mass spectrometry using trifluoroacetyl-l-prolyl chloride as chiral derivatization reagent. *J Chromatogr A* 1269:352–9.

34. Weiß, J.A., Mohr, S., Schmid, M.G. 2015. Indirect chiral separation of new recreational drugs by gas chromatography–mass spectrometry using trifluoroacetyl-L-prolyl chloride as chiral derivatization reagent. *Chirality* 27(3):211–5.

35. Schürenkamp, J., Beike, J., Pfeiffer, H., Köhler, H. 2011. Separation of positional CPP isomers by chiral HPLC-DAD of seized tablets. *Int J Legal Med* 125(1):95–9.

36. Li, L., Lurie, I.S. 2015. Regioisomeric and enantiomeric analyses of 24 designer cathinones and phenethylamines using ultra high performance liquid chromatography and capillary electrophoresis with added cyclodextrins. *Forensic Sci Int* 254:148–57.

37. Mohr, S., Taschwer, M., Schmid, M.G. 2012. Chiral separation of cathinone derivatives used as recreational drugs by HPLC-UV using a CHIRALPAK AS-H column as stationary phase. *Chirality* 24(6):486–92.

38. Taschwer, M., Seidl, Y., Mohr, S., Schmid, M.G. 2014. Chiral separation of cathinone and amphetamine derivatives by HPLC/UV using sulfated ß-cyclodextrin as chiral mobile phase additive. *Chirality* 26(8):411–8.

39. Wolrab, D., Frühauf, P., Moulisová, A. et al. 2016. Chiral separation of new designer drugs (cathinones) on chiral ion-exchange type stationary phases. *J Pharm Biomed Anal* 120:306–15.

40. Rácz, N., Veress, T., Nagy, J., Bobály, B., Fekete, J. 2016. Separation of isomers of JWH-122 on porous graphitic carbon stationary phase with non-aqueous mobile phase using intelligent software. *J Chromatogr Sci* 54(10):1735–1742.

41. Uchiyama, N., Kikura-Hanajiri, R., Shoda, T., Fukuhara, K., Goda, Y. 2011. Isomeric analysis of synthetic cannabinoids detected as designer drugs. *Yakugaku Zasshi* 131(7):1141–7.

42. Kenner, B. 2015. Differentiation of isomers of synthetic cannabinoids by U(H)PLC-DAD, GC-MS and LC-QTOFMS methods. Bachelor thesis, Hochschule Fresenius, University of Applied Sciences, Idstein.

43. United Nations Office on Drugs and Crime (UNODC). 2015. Recommended Methods for the Identification and Analysis of Synthetic Cathinones in Seized Materials. New York. https://www.unodc.org/documents/scientific/STNAR49_Synthetic_Cathinones_E.pdf.

44. Belal, T., Awad, T., DeRuiter, J., Clark, C.R. 2009. GC-IRD methods for the identification of isomeric ethoxyphenethylamines and methoxymethcathinones. *Forensic Sci Int* 184(1–3):54–63.

45. McDermott, S.D., Power, J.D., Kavanagh, P., O'Brien, J. 2011. The analysis of substituted cathinones. Part 2: An investigation into the phenylacetone based isomers of 4-methylmethcathinone and N-ethylcathinone. *Forensic Sci Int* 212:13–21.

46. Christie, R., Horan, E., Fox, J. et al. 2014. Discrimination of cathinone regioisomers, sold as 'legal highs', by Raman spectroscopy. *Drug Test Anal* 6(7–8):651–7.

47. Merola, G., Fu, H., Tagliaro, F., Macchia, T., McCord, B.R. 2014. Chiral separation of 12 cathinone analogs by cyclodextrin-assisted capillary electrophoresis with UV and mass spectrometry detection. *Electrophoresis* 35(21–22):3231–41.

48. Siroká, J., Polesel, D.N., Costa, J.L., Lanaro, R., Tavares, M.F., Polášek, M. 2013. Separation and determination of chlorophenylpiperazine isomers in confiscated pills by capillary electrophoresis. *J Pharm Biomed Anal* 84:140–7.

49. Burrai, L., Nieddu, M., Pirisi, M.A., Carta, A., Briguglio, I., Boatto, G. 2013. Enantiomeric separation of 13 new amphetamine-like designer drugs by capillary electrophoresis, using modified-B-cyclodextrins. *Chirality* 25(10):617–21.

50. Mohr, S., Pilaj, S., Schmid, M.G. 2012. Chiral separation of cathinone derivatives used as recreational drugs by cyclodextrin-modified capillary electrophoresis. *Electrophoresis* 33(11):1624–30.

51. Aturki, Z., Schmid, M.G., Chankvetadze, B., Fanali, S. 2014. Enantiomeric separation of new cathinone derivatives designer drugs by capillary electrochromatography using a chiral stationary phase, based on amylose tris(5-chloro-2-methylphenylcarbamate). *Electrophoresis* 35(21–22):3242–9.

Section II

13 Mass Spectrometric Analysis Applied for the Determination of Synthetic Cannabinoids

Akira Namera, Takeshi Saito, and Yuji Fujita

CONTENTS

13.1 INTRODUCTION

Nowadays, various drugs are abused and this phenomenon is a serious worldwide problem. Although illegal stimulants and narcotics, such as heroin, cocaine, and methamphetamine, are abused by limited persons and populations, herbal mixtures containing new designer drugs with cannabinoid receptor activity, which have appeared on underground markets since 2008, are abused regardless of age or gender [1]. These drugs have selective and high affinity for the cannabinoid receptors (CB1 and CB2) and exhibit higher activity than true cannabinoids such as delta-9-tetrahydrocannabinol. Called cannabimimetics or synthetic cannabinoids (SCs), these drugs are sold on the market as "Spice" or "K2." The prevalence of SCs in various countries has been summarized in many reports [2–6]. The most common clinical features reported are tachycardia, reduced level of consciousness, agitation, vomiting, dizziness, confusion, mydriasis, and hallucination. Other symptoms include epileptic seizures and acute psychosis [7,8].

While various SCs are distributed and sold on the market, the core structures of abused SCs can be classified as several types, such as N-alkylindole-3-carbonyl and N-alkylindazole-3-carbonyl derivatives, which are well known as the JWH and AM series (Figure 13.1), respectively. Although these drugs are controlled in most countries, almost all regulations are limited by the structure of each drug. As a result, new analogs appear on the market shortly after regulations are published because the regulations are not applicable to compounds in which the structure of a side

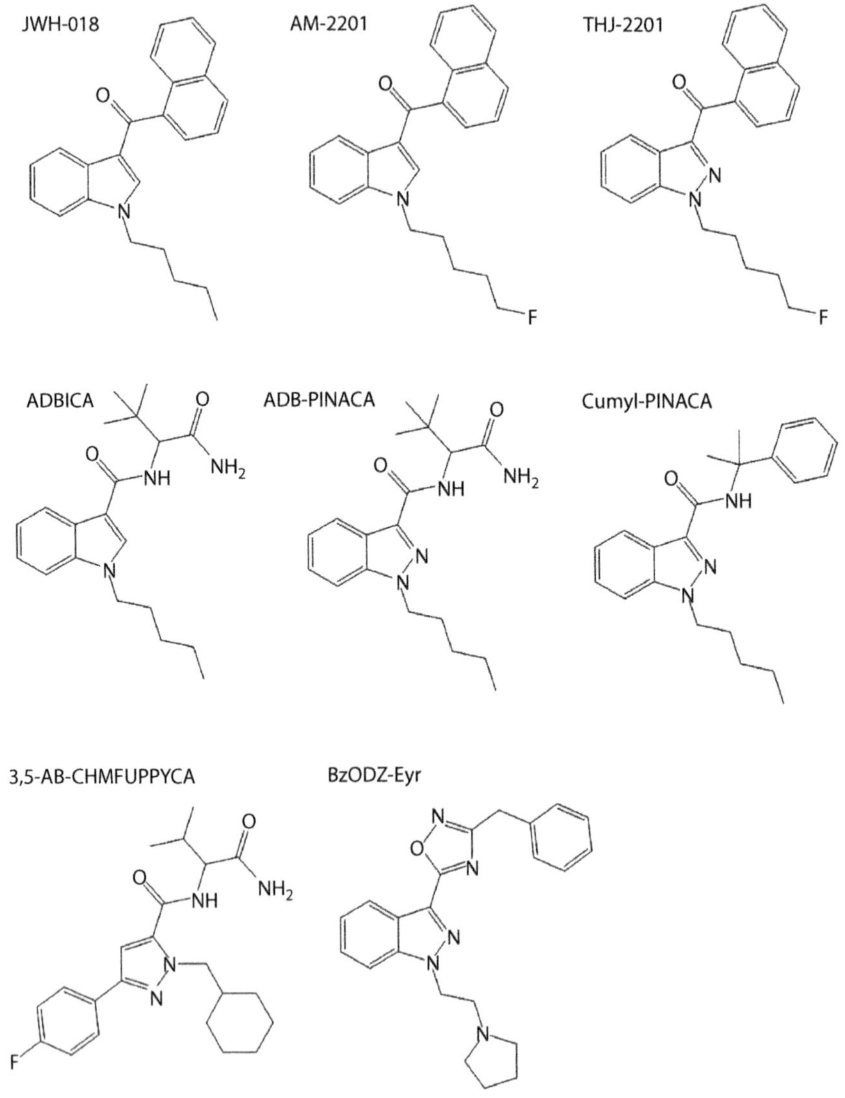

FIGURE 13.1   Structures of typical synthetic cannabinoids.

chain or substituent is slightly different from that in the regulated drugs (Figure 13.1); this cycle makes the regulation process ineffective. Owing to their structural similarities with controlled drugs, SCs always show psychoactivity and have stronger activity than conventional cannabinoids. However, the detailed pharmacological activity of each SC is not yet known. Moreover, as other ingredients with which SCs are packaged are also unknown, the use of these drugs is very dangerous for human health and some fatal intoxication cases of new drugs have been reported [9,10].

Many researchers have focused on developing methods for detecting SCs, and various detection methods have been published. Analytical reviews that summarize techniques for systematic identification and quantification of these drugs are helpful for quantitative and qualitative analysis of SCs in biological samples [11–14]. For many SCs, the gold standards for forensic and clinical analyses are gas chromatography–mass spectrometry (GC-MS) and liquid chromatography–mass spectrometry (LC-MS) techniques. In this chapter, we summarize the use of GC-MS and LC-MS for detecting SCs in seized and biological samples.

13.2   GAS CHROMATOGRAPHY–MASS SPECTROMETRY (GC-MS)

13.2.1   Identification of Synthetic Cannabinoids

GC-electron impact (EI)–MS is routinely used for drug identification because the inherent fragmentation patterns allow for comparison with drug spectral libraries. For naphthoylindoles, carbonyl group fragment ions, which are formed by α-cleavage of the alkylamino group of the indole, are typically observed. In addition, [M-17]⁺ ions are commonly observed for naphthoylindoles, as shown in Figure 13.2. For example, the fragment ions observed at m/z 284 and 214 for JWH-018 correspond to the indole moieties obtained by α-cleavage of the N-pentyl group of the indole and of the naphthoyl group, respectively. Further, fragment ions observed at m/z 127 and 155 for JWH-018 correspond to the naphthalene moieties obtained by α-cleavage of the carbonyl group. Moreover, [M-17]⁺ ions are observed at m/z 324. Similar to naphthoylindoles, fragment ions obtained following α-cleavage of the alkylamino group of the indole and carbonyl groups are observed for benzoylindoles, but [M-17]⁺ ions are not observed. For example, fragment ions are observed at m/z 264 and 214 for RCS-4, resulting from α-cleavage of the N-pentyl group of indole and of the 4-methoxybenzoyl group, respectively. Ions at m/z 127 and 155, which correspond to the naphthyl and naphthoyl moieties of naphthoylindoles, and ions at m/z 135, which correspond to 4-methoxybenzoyl moieties, are useful precursor ions for identifying these drugs by GC-MS-MS. On the other hand, when methylpiperidine and morpholine moieties are bound to the nitrogen of the indole, characteristic base peaks are observed at m/z 98 and 100. Unlike naphthoyl- and benzoylindoles, the base peak corresponding to the N-alkylindole 3-carbonyl moiety of phenylacetyl-, cyclopropyl-, and adamantylindoles is only observed in the full-scan spectra. Analogs in which the indole skeleton is changed to an indazole, such as THJ-018, have also appeared on the market. For these analogs, molecular and N-dealkylated ions are typically observed in the spectra. Recently, amide- or ester-type analogs

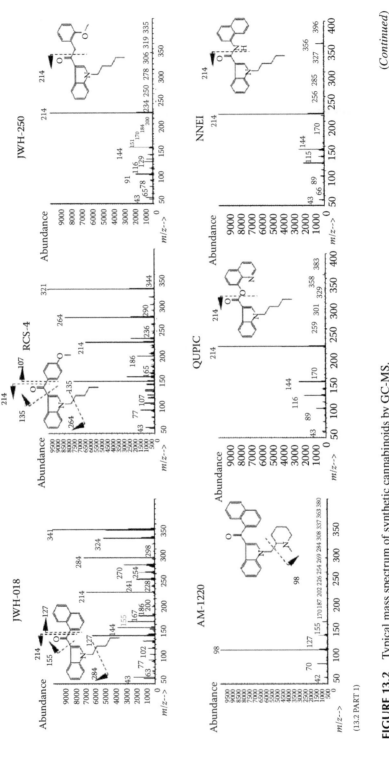

FIGURE 13.2 Typical mass spectrum of synthetic cannabinoids by GC-MS.

(13.2 PART 1)

(Continued)

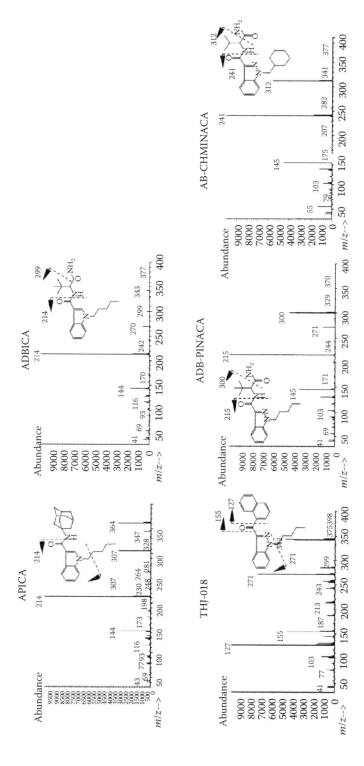

FIGURE 13.2 (CONTINUED) Typical mass spectrum of synthetic cannabinoids by GC-MS.

with *N*-alkylindole or *N*-alkylindazole-3-carbonyl moieties have appeared on the market. In these analogs, the abundance of molecular ions is low, and the fragment ion corresponding to the indoyl (or indazoyl) moiety is observed as the base peak. Although the fragment ion obtained by elimination of the terminal CO–NH$_2$ group is less intense than that obtained by cleavage of the amide moiety in indole analogs, such as ADBICA, these fragment ions have similar intensities in indazole analogs such as ADB-PINACA and AB-CHMINACA.

Commonly, herbal blend samples containing SCs are extracted with ethanol or methanol and analyzed using GC-EI-MS. The obtained EI spectra are evaluated against the Scientific Working Group for the Analysis of Seized Drugs (SWGDRUG) Mass Spectral Library and the Cayman Spectral Library. However, as replacement of SCs on the market occurs quickly, if the new SC is not already registered in these libraries, structure elucidation is required. Although structure elucidation is not possible without expensive instrumental analysis, such as nuclear magnetic resonance (NMR) spectrometry or high-resolution tandem mass spectrometry (HR-MS/MS), a few studies have been carried out using only GC-MS. Angerer et al. identified ME-CHMIMO and 5F-ADB using GC-MS, GC-solid-state infrared (GC-sIR), and NMR [15]. However, Shevyrin et al. failed to identify compounds using GC-MS alone because the spectra were not found in commercially available EI-MS libraries, and low-resolution spectra were not sufficient for reliable interpretation [16].

A common disadvantage of EI ionization is the small abundance of molecular ions usually encountered for SCs with naphthoyl-, benzoyl-, or cyclopropyl moieties. Moreover, conventional EI ionization does not provide an adequate molecular ion for amide- or ester-type analogs of SCs. Overcoming this disadvantage requires complementing GC-EI-MS with techniques such as GC–chemical ionization (CI)–MS [17]. GC-CI-MS is widely suitable for identification and structure elucidation of SCs. In positive CI and negative CI modes, many compounds produce [M+H]$^+$ and/or [M-H]$^-$ fragment ions, which can be used for molecular weight estimation. Moreover, the terminal structure of an unknown compound can be estimated using positive CI, allowing for efficient structure elucidation.

Over 90% of hydroxylated metabolites in urine are found as conjugated forms [18], and enzymatic or acid decomposition is required. The metabolites detected after derivatization can be structurally classified as four types: monohydroxylation, dihydroxylation, carboxylation, and *N*-dealkylation.

13.2.2 Sample Preparation for GC-MS

For GC-MS analysis, herbal mixtures are extracted using methanol or ethanol. After vortexing and centrifugation, the supernatant is evaporated to dryness under a gentle stream nitrogen. The residues are then reconstituted in ethyl acetate or methanol.

Urinary metabolites are extracted after enzymolysis using β-glucuronidase or hydrolysis using concentrated hydrochloric acid. Liquid-liquid extraction (LLE) or solid-phase extraction (SPE) is used to prepare urine samples. Extracted samples are then derivatized by trimethylsilylation, acetylation, or trifluoroacetylation.

13.2.3　Applications of GC-MS

Choi et al. used GC-MS to analyze 60 seized materials containing SCs, including 40 dried leaves, 6 bulk powders, and 14 tablets [19]. The dried leaves contained 6.8–46.9 mg/g of JWH-018 and 0.4–41.8 mg/g of JWH-073. The bulk powders were confirmed as crude JWH-019, JWH-073, or JWH-250 powders.

Herbal blends sold as containing SCs were extracted with methanol and steeped as teas, and two SCs (JWH-018 and JWH-073) were detected [20]. A small amount of the powdered material from the bags was diluted in methanol and injected directly into the GC-MS.

Sobolevsky et al. reported urinary metabolites of JWH-018 using LC- and GC-MS/MS after enzymolysis with β-glucuronidase [18]. The fragment ions observed in the MS/MS data clearly indicated that the metabolites have different hydroxylation patterns. Similarly, Emerson et al. synthesized a 5-hydroxypentyl JWH-018 metabolite (M1) and confirmed the presence of this metabolite in a urine sample from an individual suspected of consuming "K2" [21].

JWH-018 and its five regioisomeric 1-naphthoyl-substituted 1-*n*-pentylindoles were completely separated using a capillary column containing a trifluoropropyl methyl polysiloxane (Rtx-200) stationary phase [22].

13.2.4　Pitfalls

While an unknown compound can be estimated from the mass fragment, it is necessary to perform a comparison with standard reagents. Synthesis of a candidate compound or confirmation of the structure by NMR is necessary if an authentic compound is unavailable.

Some SCs undergo thermal degradation in the injector port (e.g., PB-22) or in the GC column (e.g., UR-144, QUPIC) [23,24]. Therefore, UR or XLR series should be modified using a suitable derivatization procedure.

13.3　LIQUID CHROMATOGRAPHY–MASS SPECTROMETRY (LC-MS)

13.3.1　Identification of Synthetic Cannabinoids by Tandem MS

To identify an analyte by LC-MS, both the retention time and mass spectra are usually compared with those of the pure standard. Although a pure standard of each SC is needed for proper identification, tandem MS has recently been used for the estimation of parent SCs and their metabolites in biological samples. The following two steps are used when identifying SCs using tandem MS. Firstly, data is accumulated using the multiple-reaction monitoring (MRM) or selected-reaction monitoring (SRM) mode to improve selectivity. The sensitivity of the MRM mode depends on the dwell time (cycle time) of the acquisition. Therefore, data can be acquired during the expected retention time of the target, using a method called scheduled MRM (sMRM) or dynamic MRM (dMRM). In the MRM mode, two transitions are usually considered (one quantifier ion and one qualifier ion). Secondly, the fragmentation of the analyte detected by MRM is acquired using the product ion scan mode. The observed fragmentation reflects the structure of the target, and this data is important for target identification. Some researchers have studied the fragmentation of SCs using LC-MS [25,26]. As only the protonated molecular ion is observed by LC-MS, this method provides much less information than GC-MS, and LC-MS/MS must be used to obtain further data for chemical structure determination. Fragment ions can be observed by product ion scanning when the protonated molecular ion is used as the precursor ion. For example, for *N*-alkylindole-3-carboxyl derivatives, a fragment ion is observed at *m/z* 144 corresponding to the indole-3-carbonyl moiety. Moreover, for naphthoylindole, fragment ions at *m/z* 127 and 155 correspond to naphthyl and naphthoyl moieties, respectively. It is possible to estimate the structure of SCs by considering such fragmentation information. However, as similar analogs that produce the same LC-MS/MS fragmentation patterns exist on the market, retention time information is needed for comparison with authentic drugs.

A unique approach has been reported for estimating SCs using precursor ion scanning (PIS). This approach is a powerful tool to search for the structure-modified SCs. The PIS methodology is applied to search for compounds that have similar core structures and is used for research on metabolites of drugs administrated in complex biological samples. In PIS mode, Q1 is set to scan from low to high masses, whereas Q3 is set to transmit only the selected product ion. Therefore, only precursor ions that generate the common product ion will be registered in the total ion chromatogram. Mazzarino et al. reported to estimate 15 SCs in serum at concentrations of 5 ng/mL by setting precursor ions of *m/z* 127, 144, and 155 [27]. The results confirm the authenticity of the samples, as the 15 SCs were clearly detected utilizing the PIS method (7 using the ions at *m/z* 127 and 155, which are specific to naphthyl and naphthoyl moieties, and 9 using the ions at *m/z* 144, which are specific to the indole-3-carbonyl moiety). For another example, the extracted ion chromatogram and extracted spectra obtained from the nine SCs standards (AM-2201, JWH-016, JWH-081, JWH-201, JWH-210, JWH-251, JWH-370, JWH-398, and RSC-4) mixture solution by PIC method are shown in Figure 13.3. Seven peaks (AM-2201, JWH-081, JWH-201, JWH-210, JWH-251, JWH-398, and RSC-4) were obtained by setting precursor ions of *m/z* 144, three peaks (AM-2201, JWH-016, and JWH-370) were obtained by setting precursor ions of *m/z* 127 and 155. As AM-2201 has both indole-3-carbonyl and naphthoyl moieties, all of the setting precursor ions (*m/z* 127, 144, 155) were detected. Since the spectra extracted from each peak reveal the protonated molecular ion, all the SCs can be estimated by the information of the molecular weight. As well as searching for SCs in biological samples, this technique can be used to estimate the related metabolites of SCs in urine, as SCs are metabolized in the human body and excreted in urine [28]. This methodology has been applied to search for metabolites in urine using LC-MS/MS.

The MRM method requires updated information about the targets. However, as structurally modified SCs appear one after another on the market, up-to-date information is unavailable and targets can be overlooked. The identification of an unknown drug without any relevant information is very difficult using conventional LC-MS methods.

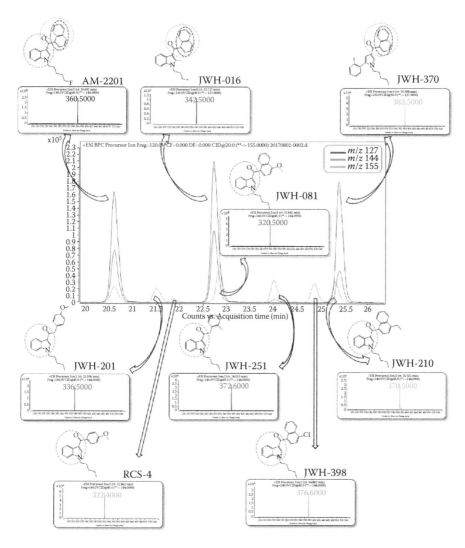

FIGURE 13.3 Extracted ion chromatogram and extracted spectra of a standard mixture of the nine synthetic cannabinoids (AM-2201, JWH-016, JWH-081, JWH-201, JWH-210, JWH-251, JWH-370, JWH-398, and RSC-4). Acquisition mode precursor ion scan of m/z 127, 144 and 155.

13.3.2 IDENTIFICATION OF SYNTHETIC CANNABINOIDS BY HIGH-RESOLUTION MS

Time-of-flight (TOF)/MS is a powerful tool for the identification of analytes of interest because accurate mass can be measured and compositional formulas can be calculated from the obtained data. However, TOF/MS is very expensive and cannot be used for routine bioanalysis in all clinical and toxicological laboratories. Recently, nontarget screening methods have been reported that use quadrupole linear ion trap or high-resolution quadrupole/TOF(QTOF)/MS. These methods are capable of recording full-scan accurate-mass product-ion spectra in MS/MS mode, which is one of the most valuable tools for confirmatory analysis. Moreover, as these methods are nontargeted in nature, updated information is not required to detect next-generation SCs [29,30]. Although these methods are a helpful tool for the tentative identification of parent drugs and their metabolites of SCs, they are dependent on the database used.

The use of mass defects is an interesting technique for identifying drug metabolites by high-resolution MS [31]. The mass defect is the difference between the exact mass of a compound and its closest integer value, e.g., the theoretical mass defect of carbon (nominal mass: 12 Da; exact mass: 12.0107 Da) is 0.0107 Da, and the mass defect shift between a parent and metabolite ion would be 0.0107 Da or 10.7 mDa. Generally, mass defect changes owing to biotransformation are less than 50 mDa and have a maximum value of 89 mDa [31]. This approach has been used to monitor metabolites that have molecular masses that are similar to that of the drug owing to minor transformations such as oxidation, reduction, and demethylation. Grabenauer et al. have applied this technique to detect analogs related to JWH-018 [27].

MRM and mass defect techniques are dependent on the information acquired by MS. Recently SWATH (sequential windowed acquisition of all theoretical fragment ions) has been applied to the nontargeted and simultaneous identification of the most current 47 SC metabolites from 21 SCs [32].

13.3.3   Sample Preparation for LC-MS

Sample preparation is one of the most important processes to obtain an accurate analytical result. In seized samples, methanol or acetonitrile is added to the sample, which is then sonicated for several minutes to achieve extraction owing to the high hydrophobicity of SCs. After extraction, the extractants are directly injected into a LC. However, SCs are usually metabolized in the human body via oxidative defluorination, hydroxylation, and carboxylation. Moreover, those compounds can be further metabolized by glucuronidation and excreted in urine. Direct analysis of the glucuronides of SCs is difficult because the pure standards are not available, and thus the glucuronides need to be hydrolyzed using β-glucuronidase. Various sample preparation techniques, including protein precipitation [33,34], LLE [35–38], and SPE [39], have been reported for the extraction and purification of SCs in biological samples. LLE is a popular extraction procedure. Hexane and ethyl acetate are usually used to extract the parent SCs, whereas ethyl acetate and chlorobutane, which has a lower hydrophobicity than hexane, are usually used to extract the metabolites of SCs. As extraction recoveries vary, the recovery of the selected technique should be confirmed before it is applied to real samples.

13.3.4   Applications of LC-MS

Often, commercial packages include a mixture of various SCs, and the ingredients of the package are unknown to both sellers and buyers. Thus, screening methods that can identify many kinds of SCs simultaneously are helpful for ingredient identification.

Following LLE from 1-mL serum samples, Kneisel and Auwärter demonstrated the simultaneous detection of 30 SCs in serum by MRM mode with limits of detection (LODs) of 0.01–2.0 ng/mL and limits of quantification (LOQs) of 0.1–2.0 ng/mL [35]. Subsequently, 833 serum samples were routinely applied for identification and quantification of SCs. Scheidweiler and Huestis reported the simultaneous detection of 20 SCs and 21 metabolites in urine by MRM mode with LODs of 0.05–1.0 ng/mL and LOQs of 0.1–1.0 ng/ml [36]. The samples were prepared by enzyme hydrolysis and SLE+ extraction (similar to LLE) of 0.2 mL of urine. Knittel et al. simultaneously detected 15 SCs and 17 metabolites in blood and urine following enzyme hydrolysis, and LLE of 1.0 mL of blood and 2.0 mL of urine by MRM mode with LODs of 0.01–0.5 ng/mL and lower limits of quantitation (LLOQs) of 0.025–0.1 ng/mL [37]. Following enzyme hydrolysis and protein precipitation from 1.0 mL of blood and 0.1 mL of urine, Wohlfarth et al. achieved simultaneous detection of 9 SCs and 20 metabolites in urine by MRM mode with LODs of 0.5–10 ng/mL [33]. Kneisel et al. demonstrated the simultaneous detection of 30 SCs in oral fluid, obtained by LLE of 0.3 mL of an oral fluid sample by MRM mode with LODs of 0.015–0.9 ng/mL and LOQs of 0.15–3.0 ng/mL [40,41]. Hutter et al. realized simultaneous detection of 22 SCs in human hair by MRM mode with LLOQs of 0.5 pg/mg for samples obtained by ethanol extraction from 50 mg of hair [42].

Like conventional cannabis, the most common modality of SC consumption is smoking. Inhalation, sublingual and rectal administration, and intramuscular and intravenous injection have also been reported. The detection of SCs in blood is very difficult because the concentration of parent SCs in blood reaches a maximum several minutes after administration and then dramatically decreases [43]. Although the concentration is influenced by the sampling time after drug intake and the intake amount, the concentration of a drug in serum has been found to be 0.1–190 ng/mL in intoxication cases [44].

In fatal cases, the concentrations of drugs in the blood were 0.1–199 ng/mL for JWH-018 and 0.1–68.3 ng/mL for JWH-073 [45], 12 ng/mL for AM-2201 [46], 1.1–1.5 ng/mL for 5F-PB-22 [47], 12.4 ng/mL for MAM-2201 [48], 0.64–0.99 ng/mL for NNEI [49], 5.6 ng/mL for MDMB-CHMICA [50], and 7.3 ng/mL for ADB-FUBINACA [51].

13.3.5 PITFALLS

Parent SCs only remain in blood for a short time and are smoothly metabolized to hydroxyl, carboxyl, and other metabolites. Thus, the concentration of SCs in blood is very low if any parent SCs remain in the blood at all. The metabolites of SCs are usually excreted in urine [13]. Thus, urine is used to detect SCs owing to the longer detection window and ease of collection. Metabolites with an N-alkylhydroxyl and carboxyl side chain are well known as major metabolites of many SCs, and sometimes the same metabolites are produced from different parent SCs. Determination of the parent SC is very important in forensic investigations. Unfortunately, even if the types and ratios of the metabolites in urine are determined, the ingested SC can be misidentified.

The hydrolysis activity of β-glucuronidase is different from the origin of enzyme used and the hydrolysis conditions. Recently, human liver microsomes (HLMs) have been used as a new approach to determine urinary metabolites. HLM models do not always extrapolate well to authentic human urine samples, owing to several intrinsic limitations. For example, cytochrome P-450 (CYP-450) enzymes and flavin-containing monooxygenase are rich in HLMs. However, non-CYP-450 enzymes located in cytosol are lacking in HLMs. As a result, the metabolic pathway during HLM incubation may be misleading when the non-CYP-450 metabolic pathway is dominant, such as during oxidative defluorination of AM-2210 and 5F-AKB48 [52].

13.4 FUTURE PROSPECTS

Various kinds of SCs are distributed worldwide and the number of SC abusers has increased remarkably. The chemical structures of SCs are skillfully changed so that the drugs exploit the limitations of the law and avoid detection by GC-MS and LC-MS screening. Simple screening methods are required for detection of these drugs in seized and biological materials. However, there are currently no commercial kits or devices for the routine screening of these drugs. Colorimetric, immunochemical, and chromatographic methods have been developed by many researchers, but a suitable method must be chosen for each laboratory. Of the various human sample matrices available for testing, urine and blood are often the first choices. However, many of these drugs, especially the parent SCs, only exist in urine and blood for a short period. Therefore, other matrices that can prove the consumption of these drugs, such as hair and saliva, are likely to receive more attention in the future.

ABBREVIATIONS

3,5-AB-CHMFUPPYCA: N-(1-amino-3-methyl-1-oxobutan-2-yl)-1-(cyclohexylmethyl)-3-(4-fluorophenyl)-1H-pyrazole-5-carboxamide

AB-CHMINACA: N-[(1S)-1-(aminocarbonyl)-2-methylpropyl]-1-(cyclohexylmethyl)-1H-indazole-3-carboxamide

ADB-FUBINACA: N-(1-amino-3,3-dimethyl-1-oxobutan-2-yl)-1-(4-fluorobenzyl)-1H-indazole-3-carboxamide

ADBICA: N-(1-amino-3,3-dimethyl-1-oxobutan-2-yl)-1-pentyl-1H-indole-3-carboxamine

ADB-PINACA: N-(1-amino-3,3-dimethyl-1-oxobutan-2-yl)-1-pentyl-1H-indazole-3-carboxamide

AM-2201: [1-(5-fluoropentyl)-1H-indol-3-yl]-1-naphthalenylmethanone

APICA: N-(1-adamantyl)-1-pentyl-1H-indole-3-carboxamide

BzODZ-Eyr: 3-benzyl-5-[1-(2-pyrrolidin-1-ylethyl)-1H-indol-3-yl]-1,2,4-oxadiazole

Cumyl-PINACA: N-(1-methyl-1-phenylethyl)-1-pentyl-1H-indazole-3-carboxamide

5F-PB-22: 1-(5-fluoropentyl)-8-quinolinyl ester-1H-indole-3-carboxylic acid

JWH-016: 1-naphthalenyl(1-butyl-2-methyl-1H-indol-3-yl)methanone

JWH-018: 1-naphthalenyl(1-pentyl-1H-indol-3-yl)methanone

JWH-019: 1-naphthalenyl(1-hexyl-1H-indol-3-yl)methanone

JWH-073: 1-naphthalenyl(1-butyl-1H-indol-3-yl)methanone

JWH-081: 4-methoxy-1-naphthalenyl(1-pentyl-1H-indol-3-yl)methanone

JWH-201: 2-(4-methoxyphenyl)-1-(1-pentyl-1H-indol-3-yl)ethanone

JWH-210: 4-ethyl-1-naphthalenyl(1-pentyl-1H-indol-3-yl)methanone

JWH-250: 2-(2-methoxyphenyl)-1-(1-pentyl-1H-indol-3-yl)ethanone
JWH-251: 2-(2-methylphenyl)-1-(1-pentyl-1H-indol-3-yl)ethanone
JWH-370: 1-naphthalenyl [5-(2-methylphenyl)-1-pentyl-1H-pyrrol-3-yl]methanone
JWH-398: 4-chloronaphthalen-1-yl(1-pentylindolin-3-yl)-methanone
MAM-2201: [1-(5-fluoropentyl)-1H-indol-3-yl](4-methyl-1-naphthalenyl)methanone
MDMB-CHMICA: methyl 2-(1-(cyclohexylmethyl)-1H-indole-3-carboxamido)-3,3-dimethylbutanoate
NNEI: N-1-naphthalenyl-1-pentyl-1H-indole-3-carboxamide
PB-22: 1-pentyl-8-quinolinyl ester-1H-indole-3-carboxylic acid
QUPIC: quinolin-8-yl 1-pentyl-1H-indole-3-carboxylate
RCS-4: (4-methoxyphenyl)(1-pentyl-1H-indol-3-yl)methanone
THJ-018: 1-naphthalenyl(1-pentyl-1H-indazol-3-yl)methanone
THJ-2201: [1-(5-fluoropentyl)-1H-indazol-3-yl](naphthalen-1-yl)methanone
UR-144: (1-pentyl-1H-indol-3-yl)(2,2,3,3-tetramethylcyclopropyl)methanone

REFERENCES

1. Auwärter, V., Dresen, S., Weinmann, W., Müller, M., Pütz, M., and Ferreirós, N. 2009. 'Spice' and other herbal blends: Harmless incense or cannabinoid designer drugs? *J. Mass Spectrom.*, 44: 832–837.
2. Barratt, M.J., Cakic, V., and Lenton, S. 2013. Patterns of synthetic cannabinoid use in Australia, *Drug Alcohol Rev.*, 32: 141–146.
3. Caudevilla-Gálligo, F., Ventura, M., Ruiz, B.I.I., and Fornís, I. 2013. Presence and composition of cathinone derivatives in drug samples taken from a Drug Test Service in Spain (2010–2012), *Hum. Psychopharmacol.*, 28: 341–344.
4. Kikura-Hanajiri, R., Uchiyama, N., Kawamura, M., and Goda, Y. 2013. Changes in the prevalence of synthetic cannabinoids and cathinone derivatives in Japan until early 2012, *Forensic Toxicol.*, 31: 44–53.
5. Seely, K.A., Patton, A.L., Moran, C.L., Womack, M.L., Prather, P.L., Fantegrossi, W.E., Radominska-Pandya, A., Endres, G.W., Channell, K.B., Smith, N.H., McCain, K.R., James, L.P., and Moran, J.H. 2013. Forensic investigation of K2, Spice, and "bath salt" commercial preparations: A three-year study of new designer drug products containing synthetic cannabinoid, stimulant, and hallucinogenic compounds, *Forensic Sci. Int.*, 233: 416–422.
6. Maxwell, J.C. 2014. Psychoactive substances—Some new, some old: A scan of the situation in the U.S., *Drug Alcohol Depend.*, 134: 71–77.
7. Wells, D.L., and Ott, C.A. 2011. The "new" marijuana, *Ann. Pharmacother.*, 45: 414–417.
8. Waugh, J., Najafi, J., Hawkins, L., Hill, S.L., Eddleston, M., Vale, J.A., Thompson, J.P., and Thomas, S.H. 2016. Epidemiology and clinical features of toxicity following recreational use of synthetic cannabinoid receptor agonists: A report from the United Kingdom National Poisons Information Service, *Clin. Toxicol.*, 54: 512–518.
9. Adamowicz, P. and Wrzesień, W. 2016. Simple approach for evaluation of matrix effect in the mass spectrometry of synthetic cannabinoids, *J. Anal. Chem.*, 71: 794–802.
10. Barceló, B., Pichini, S., López-Corominas, V., Gomila, I., Yates, C., Busardò, F.P., and Pellegrini, M. 2017. Acute intoxication caused by synthetic cannabinoids 5F-ADB and MMB-2201: A case series, *Forensic Sci. Int.*, 273: e10–e14.

11. Favretto, D., Pascali, J.P., and Tagliaro, F. 2013. New challenges and innovation in forensic toxicology: Focus on the "New Psychoactive Substances," *J. Chromatogr. A*, 1287: 84–95.

12. Presley, B.C., Jansen-Varnum, S.A., and Logan, B.K. 2013. Analysis of synthetic cannabinoids in botanical materials: A review of analytical methods and findings, *Forensic Sci. Rev.*, 25: 27–46.

13. Elsohly, M.A., Gul, W., Wanas, A.S., and Radwan, M.M. 2014. Synthetic cannabinoids: Analysis and metabolites, *Life Sci.*, 97: 78–90.

14. Namera, A., Kawamura, M., Nakamoto, A., Saito, T., and Nagao, M. 2015. Comprehensive review of the detection methods for synthetic cannabinoids and cathinones, *Forensic Toxicol.*, 33: 175–194.

15. Angerer, V., Bisel, P., Moosmann, B., Westphal, F., and Auwärter, V. 2016. Separation and structural characterization of the new synthetic cannabinoid JWH-018 cyclohexyl methyl derivative "NE-CHMIMO" using flash chromatography, GC-MS, IR and NMR spectroscopy, *Forensic Sci. Int.*, 266: e93–e98.

16. Shevyrin, V., Melkozerov, V., Nevero, A., Eltsov, O., Morzherin, Y., and Shafran, Y. 2014. 3-Naphthoylindazoles and 2-naphthoylbenzoimidazoles as novel chemical groups of synthetic cannabinoids: Chemical structure elucidation, analytical characteristics and identification of the first representatives in smoke mixtures, *Forensic Sci. Int.*, 242: 72–80.

17. Umebachi, R., Saito, T., Aoki, H., Namera, A., Nakamoto, A., Kawamura, and M., Inokuchi, S. 2017. Detection of synthetic cannabinoids using GC-EI-MS, positive GC-CI-MS, and negative GC-CI-MS, *Int. J. Leg. Med.*, 131: 143–152.

18. Sobolevsky, T., Prasolov, I., and Rodchenkov, G. 2010. Detection of JWH-018 metabolites in smoking mixture post-administration urine, *Forensic Sci. Int.*, 200: 141–147.

19. Choi, H., Heo, S., Choe, S., Yang, W., Park, Y., Kim, E., Chung, H., and Lee, J. 2013. Simultaneous analysis of synthetic cannabinoids in the materials seized during drug trafficking using GC-MS, *Anal. Bioanal. Chem.*, 405: 3937–3944.

20. Penn, H.J., Langman, L.J., Unold, D., Shields, J., and Nichols, J.H. 2011. Detection of synthetic cannabinoids in herbal incense products, *Clin. Biochem.*, 44: 1163–1165.

21. Emerson, B., Durham, B., Gidden, J., and Lay, J.O., Jr. 2010. Gas chromatography–mass spectrometry of JWH-018 metabolites in urine samples with direct comparison to analytical standards, *Forensic Sci. Int.*, 229: 1–6.

22. Thaxton, A., Belal, T.S., Smith, F., DeRuiter, J., Abdel-Hay, K.M., and Clark, C.R. 2015. GC-MS studies on the six naphthoyl-substituted 1-n-pentyl-indoles: JWH-018 and five regioisomeric equivalents, *Forensic Sci. Int.*, 252: 107–113.

23. Grigoryev, A., Kavanagh, P., Melnik, A., Savchuk, S., and Simonov, A. 2013. Gas and liquid chromatography-mass spectrometry detection of the urinary metabolites of UR-144 and its major pyrolysis product, *J. Anal. Toxicol.*, 37: 265–276.

24. Tsujikawa, K., Yamamuro, T., Kuwayama, K., Kanamori, T., Iwata, Y.T., and Inoue, H. 2014. Thermal degradation of a new synthetic cannabinoid QUPIC during analysis by gas chromatography–mass spectrometry, *Forensic Toxicol.*, 32: 201–207.

25. Hudson, S. and Ramsey, J. 2011. The emergence and analysis of synthetic cannabinoids, *Drug Test. Anal.*, 3: 466–478.

26. Sekuła, K., Zuba, D., and Stanaszek, R. 2012. Identification of naphthoylindoles acting on cannabinoid receptors based on their fragmentation patterns under ESI-QTOFMS, *J. Mass Spectrom.*, 47: 632–643.

27. Mazzarino, M., de la Torre, X., and Botrè, F. 2014. A liquid chromatography-mass spectrometry method based on class characteristic fragmentation pathways to detect the class of indole-derivative synthetic cannabinoids in biological samples, *Anal. Chim. Acta*, 837: 70–82.

28. Grabenauer, M., Krol, W.L., Wiley, J.L., and Thomas, B.F. 2012. Analysis of synthetic cannabinoids using high-resolution mass spectrometry and mass defect filtering: Implications for nontargeted screening of designer drugs, *Anal. Chem.*, 84: 5574–5581.

29. Sundström, M., Pelander, A., Angerer, V., Hutter, M., Kneisel, S., and Ojanperä, I. 2013. A high-sensitivity ultra-high performance liquid chromatography/high-resolution time-of-flight mass spectrometry (UHPLC-HR-TOFMS) method for screening synthetic cannabinoids and other drugs of abuse in urine, *Anal. Bioanal. Chem.*, 405: 8463–8474.

30. Ibáñez, M., Sancho, J.V., Bijlsma, L., van Nuijs, A.L.N., Covaci, A., and Hernández, F. 2014. Comprehensive analytical strategies based on high-resolution time-of-flight mass spectrometry to identify new psychoactive substances, *Trends Anal. Chem.*, 57: 107–117.

31. Zhang, H., Zhang, D., Ray, K., and Zhu, M. 2009. Mass defect filter technique and its applications to drug metabolite identification by high-resolution mass spectrometry, *J. Mass Spectrom.*, 44: 999–1016.

32. Scheidweiler, K.B., Jarvis, M.J., and Huestis, M.A. 2015. Nontargeted SWATH acquisition for identifying 47 synthetic cannabinoid metabolites in human urine by liquid chromatography-high-resolution tandem mass spectrometry, *Anal. Bioanal. Chem.*, 407: 883–897.

33. Wohlfarth, A., Scheidweiler, K.B., Chen, X., Liu, H., and Huestis, M.A. 2014. Qualitative confirmation of 9 synthetic cannabinoids and 20 metabolites in human urine using LC–MS/MS and library search, *Anal. Chem.*, 85: 3730–3738.

34. Adamowicz, P. and Tokarczyk, B. 2016. Simple and rapid screening procedure for 143 new psychoactive substances by liquid chromatography-tandem mass spectrometry, *Drug Test. Anal.*, 8: 652–667.

35. Kneisel, S. and Auwärter, V. 2012. Analysis of 30 synthetic cannabinoids in serum by liquid chromatography–electrospray ionization tandem mass spectrometry after liquid–liquid extraction, *J. Mass Spectrom.*, 47: 825–835.

36. Scheidweiler, K.B. and Huestis, M.A. 2014. Simultaneous quantification of 20 synthetic cannabinoids and 21 metabolites, and semi-quantification of 12 alkyl hydroxy metabolites in human urine by liquid chromatography-tandem mass spectrometry, *J. Chromatogr. A*, 1327: 105–117.

37. Knittel, J.L., Holler, J.M., Chmiel, J.D., Vorce, S.P., Magluilo, J., Jr, Levine, B., Ramos, G., and Bosy, T.Z. 2016. Analysis of parent synthetic cannabinoids in blood and urinary metabolites by liquid chromatography tandem mass spectrometry, *J. Anal. Toxicol.*, 40: 173–186.

38. Borg, D., Tverdovsky, A., and Stripp, R. 2017. A fast and comprehensive analysis of 32 synthetic cannabinoids using agilent triple quadrupole LC-MS-MS, *J. Anal. Toxicol.*, 41: 6–16.

39. Simões, S.S., Silva, I., Ajenjo, A.C., and Dias, M.J. 2014. Validation and application of an UPLC-MS/MS method for the quantification of synthetic cannabinoids in urine samples and analysis of seized materials from the Portuguese market, *Forensic Sci. Int.*, 243: 117–125.

40. Kneisel, S., Auwärter, V., and Kempf, J. 2013. Analysis of 30 synthetic cannabinoids in oral fluid using liquid chromatography–electrospray ionization tandem mass spectrometry, *Drug Test. Anal.*, 5: 657–669.

41. Kneisel, S., Speck, M., Moosmann, B., Corneillie, T.M., Butlin, N.G., and Auwärter, V. 2013. LC/ESI-MS/MS method for quantification of 28 synthetic cannabinoids in neat oral fluid and its application to preliminary studies on their detection windows, *Anal. Bioanal. Chem.*, 405: 4691–4706.

42. Hutter, M., Kneisel, S., Auwärter, V., and Neukamm, M.A. 2012. Determination of 22 synthetic cannabinoids in human hair by liquid chromatography–tandem mass spectrometry, *J. Chromatogr. B*, 903: 95–101.

43. Teske, J., Weller, J.P., Fieguth, A., Rothämel, T., Schulz, Y., and Tröger, H.D. 2010. Sensitive and rapid quantification of the cannabinoid receptor agonist naphthalen-1-yl-(1-pentylindol-3-yl)methanone (JWH-018) in human serum by liquid chromatography–tandem mass spectrometry, *J. Chromatogr. B*, 878: 2659–2663.

44. Hermanns-Clausen, M., Kneisel, S., Szabo, B., and Auwärter, V. 2013. Acute toxicity due to the confirmed consumption of synthetic cannabinoids: Clinical and laboratory findings, *Addiction*, 108: 534–544.

45. Shanks, K.G., Dahn, T., and Terrell, A.R. 2012. Detection of JWH-018 and JWH-073 by UPLC-MS-MS in postmortem whole blood casework, *J. Anal. Toxicol.*, 36: 145–152.

46. Patton, A.L., Chimalakonda, K.C., Moran, C.L., McCain, K.R., Radominska-Pandya, A., James, L.P., Kokes, C., and Moran, J.H. 2013. K2 toxicity: Fatal case of psychiatric complications following AM2201 exposure, *J. Forensic Sci.*, 58: 1676–1680.

47. Behonick, G., Shanks, K.G., Firchau, D.J., Mathur, G., Lynch, C.F., Nashelsky, M., Jaskierny, D.J., and Meroueh, C. 2014. Four postmortem case reports with quantitative detection of the synthetic cannabinoid, 5F-PB-22, *J. Anal. Toxicol.*, 38: 559–562.

48. Saito, T., Namera, A., Miura, N., Ohta, S., Miyazaki, S., Osawa, M., and Inokuchi, S. 2013. A fatal case of MAM-2201 poisoning, *Forensic Toxicol.*, 31: 333–337.

49. Sasaki, C., Saito, T., Shinozuka, T., Irie, W., Murakami, C., Maeda, K., Nakamaru, N., Oishi, M., Nakamura, S., and Kurihara, K. 2015. A case of death caused by abuse of a synthetic cannabinoid N-1-naphthalenyl-1-pentyl-1H-indole-3-carboxamide, *Forensic Toxicol.*, 33: 165–169.

50. Adamowicz, P. 2016. Fatal intoxication with synthetic cannabinoid MDMB-CHMICA, *Forensic Sci. Int.*, 261: e5–e10.

51. Shanks, K.G., Clark, W., and Behonick, G. 2016. Death associated with the use of the synthetic cannabinoid ADB-FUBINACA, *J. Anal. Toxicol.*, 40: 236–239.

52. Diao, X. and Huestis, M.A. 2017. Approaches, challenges, and advances in metabolism of new synthetic cannabinoids and identification of optimal urinary marker metabolites. *Clin. Pharmacol. Ther.*, 101: 239–253.

14 Cathinone Derivatives and Their Analysis

Milena Majchrzak and Rafał Celiński

CONTENTS

14.1 INTRODUCTION

Starting from mid-2000, new psychoactive substances (NPSs) have emerged on drug markets worldwide, popularly referred to as designer drugs. Currently, the most numerous representatives of this group are derivatives of the biologically active alkaloid known as cathinone and obtained from the shrub of *Catha edulis*. Synthetic cathinones dominate the illegal designer drug industry and the spectrum of modifications of the precursor's original structure steadily expands. In many countries, a growing number of the cathinone derivatives become legally banned and considered as psychotropic substances equivalent to such hard drugs as, e.g., amphetamine. Possibilities to structurally modify these compounds that are already present on the market are so large that immediately after having banned a given substance its newly modified derivatives appear. In view of a growing problem with these new psychoactive substances, both refinement of the already existing analytical methods for their identification and quantification and elaboration of the brand-new methods become an urgent task. Expansion of the existing databases that contain a given group of the designer drugs by adding to them physicochemical and pharmacological properties of novel compounds allows analytical chemists and toxicologists to rapidly identify a given compound.

This chapter provides a short history and the chemical and pharmacological characteristics of cathinone derivatives, and, in the first instance, the most frequently applied analytical techniques that have proved most efficient in the toxicological analysis of material evidence (i.e., powders and drugs), and of biological samples containing substances belonging to this group. Finally, analytical and physicochemical data are going to be provided referring to the novel derivatives that belong to the cathinone group and have been seized and described between 2014 and 2016.

14.1.1 SHORT HISTORY OF THE APPEARANCE OF CATHINONE AND EVOLUTION OF ITS DERIVATIVES

Cathinone is one of the biologically active alkaloids derived from the shrub *Catha edulis*. This plant has been known and utilized for centuries by the inhabitants of East Africa and the northeastern parts of the Arabian Peninsula, due to its psychoactive properties. Chewing of the freshly collected khat leaves (in that way liberating cathinone, which affects the central nervous system) in many geographical regions is considered as a matter of cult and local tradition [1–5]. Due to their structural resemblance to amphetamine (Figure 14.1), cathinone and its analogs are often denoted as "natural amphetamines" and the only structural difference between amphetamine and cathinone is the presence of a carbonyl group in the alpha position of the cathinone's side chain. Similar to amphetamine, cathinone and its analogs characterize with stimulating, euphoric, and empathogenic properties [1–3,5,6].

The first synthetic cathinone derivatives were synthesized for medicinal purposes in the early twentieth century, as potential curative agents. Due to their stimulating effect on the central nervous system (CNS), they started being used on a wide scale for recreational purposes, something that has officially been observed as starting from the year 2000 onward. Synthetic cathinones were included in a broader

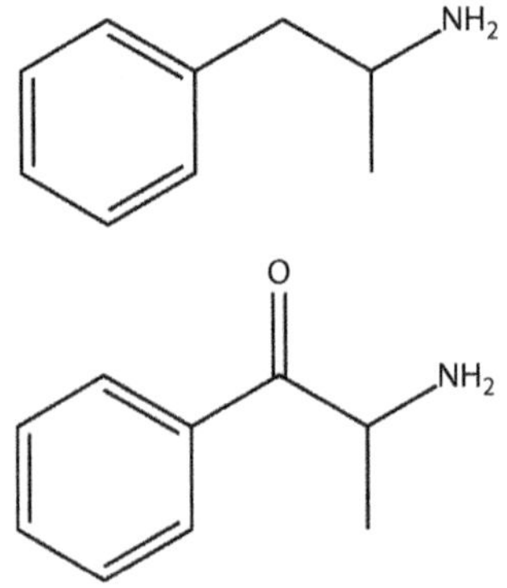

FIGURE 14.1   Structural similarity between amphetamine and cathinone.

group of psychoactive compounds denoted as so-called "legal drugs" or "designer drugs" [3,6–9]. In the course of the past 17 years, cathinone derivatives have gradually become available from the so-called "smart shops," online through the Internet, and now one can get them from selling spots advertising "Funny Items," "Aromas," "Magic Blow," or even "Xerox Spot" [10,11]. Most often the synthetic cathinones are sold as white or colored crystalline powders, and rather rarely as tablets or capsules, placed in zipped plastic foil bags, or in small Eppendorf tubes. In the past, products containing active ingredients from the cathinone group used to be advertised as "plant nutrients," "bath salts," "research chemicals," "coccolino," or "cherry." Nowadays, the same substances are frequently advertised as "conquerors of leeches," "driver's charms," "additives to sand," "bidet refreshers," etc. It can happen that among substances traded under these names not always the same psychoactive substances are found. Moreover, such preparations increasingly more often contain not only a single cathinone derivative, but a combination of two or more, plus caffeine, lidocaine, or benzocaine [12]. Appearance in one and the same analyzed sample of several compounds belonging to the same group and often very closely structurally related (e.g., isomers) needs application of highly sensitive instrumental techniques to be able to successfully identify all of them.

Although first synthesized for medicinal purposes only (in order to treat patients with Parkinsonism, obesity, or depression), at the beginning of the twenty-first century, synthetic cathinones have become recreational consumables as substitutes of proper drugs. After the year 2000, two pioneer representatives from this group emerged on an illegal drug market, namely CAT (methcathinone) and 4-MMC (mephedrone, 4-methylmetcathinone), followed by methylone (3,4-methylenedioxy-N-methylcathinone) and MDPV (3,4-methylenedioxypyrovalerone) [6,12,13]. Once full chemical and medicinal characteristics were obtained for these compounds (and, as a consequence, in many countries became illegal), synthetic chemists started modifying their structures to obtain new analogs. In that way, novel cathinones as substitute drugs were synthesized, such as butylone, ethylone, buphedrone, and an analog of the latter one, pentedrone, soon replaced by its constitutional isomer, 4-MEC (4-methyl-N-ethylcathinone). At about the same time, the chemical structure of mephedrone was modified by introducing new substituents to the aromatic ring, so that in 2009, 4-FMC (flephedrone, 4-fluoromethcathinone) and its positional isomer 3-FMC (3-fluoromethcathinone) were characterized. Currently, the most frequently encountered successor of 4-FMC is the compound with a chlorine atom replacing fluorine in the aromatic ring, i.e., 4-CMS (4-chloromethcathinone), and its constitutional isomers, 3-CMC and 2-CMC. Along with pentedrone, a third-generation synthetic cathinone, α-PVP (alpha-pyrrolidynopenthiophenone) appeared, belonging to the same group [3,6,12]. Subsequent and more recent inventiveness in the modification of cathinone derivatives is presented in Section 14.2.1.

14.1.2 Chemistry of Cathinone Derivatives

Structures of the first synthetic cathinones have been continuously modified to this day, so that each year several new derivatives emerge on an illegal designer drug market. Due to these circumstances, identification of these compounds and implementing the drug library with the new structures and their physicochemical and pharmacological characteristics becomes an analytical challenge, which is equally important for forensic chemists and toxicologists.

Structures of all synthetic cathinones are derived from that of natural cathinone and they can be considered as phenylalkylamino derivatives, structurally resembling the molecule of amphetamine with a carbonyl group in the alpha position of the side amino-alkyl chain substituted to the aromatic ring. From the chemical point of view, cathinone derivatives are divided into four groups. Group 1 includes N-alkyl compounds, or those with alkyl and halogen substituents in any possible position of the aromatic ring (Table 14.1). The first synthetic cathinones fall in this group and they are ethcathinone, ephedrone, mephedrone, flephedrone, buphedrone, and pentedrone, as well as their later successors, e.g., 4-chloroethcathinone (4-CEC) and 4-methyl-N,N-dimethylcathinone (4-N,N-DMC). Group 2 includes methylenedioxy-substituted compounds with substituents in any given position of aromatic ring, such as, e.g., methylone, penthylone, butylone, and one of the most recently introduced derivatives, N-ethylpentylone. Structurally and in terms of physiological effect, all these compounds are quite similar to 3,4-methylenedioxyamphetamine (Table 14.2). Cathinones from group 3 are analogs of natural cathinone with an N-pyrrolidinyl substituent and these compounds currently are the most frequently encountered group on the designer drug market. Representatives from this group are shown in Table 14.3 [6]. Finally, group 4, characterized based on the chemical structure of its representatives, includes in the molecules of its representatives both the methylenedioxyl and N-pyrrolidinyl substituent, and one of its representatives is 3-desoxy-3,4-methylenodioxypyrovalerone (3-desoxy-3,4-MDPV) (Table 14.4).

14.1.3 Metabolism and Mechanism of Action of Cathinones

In the *in vitro* experiments, it was shown that synthetic cathinones easily penetrate the blood-brain barrier (BBB) [13]. Cathinone and its derivatives (denoted as β-keto amphetamines) exert a stimulating and sympathomimetic effect on the central nervous system due to an increased concentration of catecholamines in the intersynapse spaces, and their effect is much stronger than that of amphetamine itself [2,14–21]. Similar to phenylethylamine, cathinones appear in two stereoisomeric forms and each characterizes with different potency [6]. The mechanism of action of synthetic cathinones consists in the inhibition of such monoamine transporters as dopamine transporter (DAT), noradrenaline transporter (NAT), and serotonin transporter (SERT). Depending on a given derivative and, more precisely, on its chemical structure, their affinity to the aforementioned transporters can be different. Different selectivity toward individual monoamines differentiates the synthetic cathinones in terms of their effect on neurotransmission [14,17,21]. Based on two properties of synthetic cathinones, i.e., on the one hand, on their potency of the reverse capture of dopamine, noradrenaline, and serotonin, and on the other, on their ability to liberate neurotransmitters, Simmler et al. [13] classified them into three groups. The first group includes cathinones that act in a similar way to cocaine and MDMA (methylenedioxymethamphetamine), and it is denoted as the "cocaine–MDMA–mixed cathinones" group. The action mechanism of the cathinones belonging to

TABLE 14.1

Common Names and Chemical Structures of Selected Cathinone Derivatives: Group 1

Common Name	Chemical Structure
Flephedrone (4-FMC)	
Buphedrone	
4-Chloroethcathinone (4-CEC)	
4-Methyl-N,N-dimethylcathinone	
Mephedrone	
Pentedrone	

Note: Group 1 consists of cathinone derivatives that share the common structural characteristics of N-alkyl compounds, or those with the alkyl and halogen substituents in any possible position of the aromatic ring.

TABLE 14.2

Common Names and Chemical Structures of Selected Cathinone Derivatives: Group 2

Common Name	Chemical Structure
Butylone	
Methylone	
Pentylone	
N-Ethylpentylone	

Note: Group 2 consists of cathinone derivatives that share the common structural characteristics of the methylenedioxy substituent in any given position of the aromatic ring.

TABLE 14.3

Common Names and Chemical Structures of Selected Cathinone Derivatives: Group 3

Common Name	Chemical Structure
MPHP	
α-PVP	
4-Fluoro-PV9	
4-Methoxy-α-PHPP	

Note: Group 3 consists of cathinone derivatives that share the common structural characteristics of the N-pyrrolidinyl substituent.

TABLE 14.4

Common Names and Chemical Structures of Selected Cathinone Derivatives: Group 4

Common Name	Chemical Structure
MDPBP	
MDPPP	
3,4-MDPV	
3-Desoxy-3,4-MDPV	

Note: Group 4 consists of cathinone derivatives that share the common structural characteristics of the methylenedioxy and N-pyrrolidinyl substituent.

this group consists in a rather non-selective inhibition of monoamine capture (in that way resembling cocaine, which shows greater selectivity toward the dopamine transporter than the serotonin one) and on inducing serotonin liberation (similar to MDMA) [13]. Substances falling into this group and showing action resemblance to cocaine are, e.g., mephedrone, methylone, ethylone, and butylone, whereas naphyrone acts in a way similar to MDMA. The second group embraces cathinones that act in a similar way to methamphetamine, and its representatives are denoted as "methamphetamine-like cathinones." Their action mechanism consists in the preferential capture of catecholamines and the liberation of dopamine, and the representatives of this group are methcathinone, flephedrone, and clephedrone [13,14]. The third type of pharmacological effect on neurotransmission is represented by synthetic cathinones with their structure based on that of pyrovalerone and therefore denoted as "pyrovalerone-cathinones." The representatives of this third group are MDPV and MDPBP, recognized as very potent and selective inhibitors of the catecholamine capture, which do not demonstrate the neurotransmitter liberating effect [13,15].

Reaction symptoms of the human organism after an intake of the discussed cathinone derivatives remain in conformity with the aforementioned action mechanisms for each individual cathinone group, as revealed in *in vitro* experiments and defined on the neurotransmission level basis [17]. The extent and strength of cathinone action on the central nervous system can be very broad and it depends on such factors as age, sex, degree of addiction, general health condition, medication, intake of other hallucinogenic or psychotropic agents, and the use of alcohol. However, subjective feelings of the cathinone takers are rather similar and described as strong excitation, euphoria, an increased empathy, an increased self-assurance and interpersonal openness, and an increased libido [2,3,6,12]. It has to be clearly stated, though, that both chronic exposure of the human organism to the action of synthetic cathinones and a unique or sporadic intake thereof can equally jeopardize human health and life. Among discomforts experienced by the consumers of "bath salts" and similar products are most often vomiting, sweating, short memory troubles, migraines and a giddy feeling, excessive heart beats, and muscle trembling. Neurologically, cathinone overdosing can result in memory disturbances and memory loss, fits of panic and aggression, hallucinations and depression, and even fits of psychosis with suicidal thoughts [13]. From the standpoint of cardiology, synthetic cathinones can evoke an elevated blood pressure, heart arrhythmia, tachycardia, and heart stoppage as a result. Among some more frequent effects of cathinone usage, one can mention hyponatremia, hyperthermia, anemia, and rhabdomyolysis [16].

The metabolism is relatively well known for these compounds, which are viewed as the precursors of this designer drugs group. The metabolism of mephedrone, presented by Meyer et al. [18], is understood as its N-demethylation to basic amines as the main decomposition step, followed by the reduction of the ketone functionality to 4-methylnorephedrine and hydroxylation of the methyl substituent of the aromatic ring, leading to its final oxidation to the respective carboxylic acid. Uralets et al. [22] worked on determination of the metabolites of 16 synthetic cathinones in human urine, upon dividing them into three cathinone derivative groups. The first group included mephedrone, buphedrone, 4-methylbuphedrone, pentedrone, 4-methyl-N-ethylcathinone, N-ethylbuphedrone, flephedrone, and ethcathinone, which metabolized following the pattern of synthetic cathinone precursors (i.e., methcathinone and cathinone). In urine of people treated with these compounds, the metabolites resulting from the reduction of the β-ketone group and N-dealkylation were detected, which was equivalent to the appearance of the two diastereoisomers, norephedrine and ephedrine, as the main metabolites. The second group included the 3,4-methylenedioxy-substituted cathinones (i.e., methylone, butylone, and ethylone), which characterize with the less effective β-keto reduction than the compounds from the first group, which might be due to the presence of the 3,4-methylenedioxyl substituent in the aromatic ring. Thus, in the analyzed urine, the presence of the parental molecules was detected. The third group included such pyrrolidinephenones as PVP (α-pyrrolidinovalerophenone) and PBP (α-pyrrolidinobutiophenone), which did not metabolize following the reduction of the ketone group pattern, but remained present and structurally unchanged in urine. Traceable metabolites thereof were the decomposition products of pyrrolidine to the primary amines [18,22,23].

14.2 APPLICATIONS

14.2.1 TRENDS IN THE MODIFICATION OF CATHINONE STRUCTURES AND CHARACTERIZATION THEREOF

Because of a wide possibility to introduce various different modifications to cathinone structures, novel compounds continuously appear on the designer drugs market. Identification and physicochemical characterization thereof are a real challenge for chemical analysts. Owing to information exchange on new derivatives (i.e., to the inter-laboratory transfer of complete characteristics of these compounds, the identification methods depending on different matrices, and the casualties recorded), toxicological analysis is a dynamically developing field of investigation. Starting from the year 2000, when the first synthetic cathinones emerged [3,4,6], in the world literature continuous announcements have appeared on new derivatives [8,9,11]. In the last three years, more than a dozen new cathinone derivatives have appeared on the market [11]. Early in 2014, the methoxy derivative of the already-known designer drug α-PVP appeared, i.e., 4-methoxy-α-PVP. This compound was detected in a product that additionally contained 4-methylbuphedrone. Then Uchiyama et al. described as many as seven new synthetic cathinones in their studies released in March 2014 [24]. Under the seized materials, different multi-color liquids were found, known as "aroma liquids," and also multi-color powders denoted as "fragrance powders." All these samples were prepared for analysis by means of liquid extraction. To this effect, the 2-mg aliquots of powders and the 20-μL aliquots of liquids were extracted with 1 mL methanol, with ultrasonication assistance. After centrifugation and filtration, the obtained supernatants were diluted (if necessary) and then analyzed by means of ultra-performance liquid chromatography coupled with the electrospray-type mass detector (UPLC-ESI-MS) and gas chromatography coupled with the electron impact mass detector (GC-EI-MS). The exact mass-to-charge ratios (m/z) of the analyzed compounds were determined with use of liquid chromatography coupled with quadrupole time-of-flight mass spectrometry (LC-Q-TOF-MS). Structures of all identified compounds were additionally confirmed with the use of hydrogen and carbon nuclear magnetic resonance spectroscopy. These seven new cathinone derivatives were identified as MPHP (4-methyl-α-pyrrolidinohexylphenone), α-PHPP (α-pyrrolidinoheptanophenone, PV8), α-POP (α-pyrrolidinooctanophenone, PV9), 3,4-dimethoxy-α-PVP (3,4-dimethoxy-α-pyrrolidinopentiophenone), 4-fluoro-α-PVP (4-fluoro-α-pyrrolidinopentiophenone), α-EAPP (α-ethylaminopenthiophenone), and N-ethyl-4-methylpentedrone (4-methyl-α-ethylaminopenthiophenone). Less than half-a-year later and using the same extraction and analytical approach, Uchiyama et al. [25] added four new compounds to the database of synthetic cathinones, i.e., α-PHP (α-pyrrolidinohexaphenone), 4-methoxy-α-POP (4-methoxy-α-pyrrolidino-octanophenone), 4-methoxy-α-PHPP (4-methoxy-α-pyrrolidino-heptanophenone), and 4-fluoro-α-PHPP (4-fluoro-α-pyrrolidinoheptanophenone).

Moreover, a growing tendency was observed and then described in the literature of delivering mixtures of the designer drugs in the commercialized products. These usually were the binary and ternary mixtures not necessarily of the compounds belonging to one group only, but of the drugs also belonging to different groups (e.g., as combinations of synthetic cathinones and synthetic cannabinoids) [24–26]. Due to an unknown action mechanism and toxicity of the new psychoactive compounds, combinations thereof can result in an unexpected synergism and, consequently, in jeopardizing the health and lives of potential drug takers.

In the second half of 2015, Doi et al. [27] for the first time reported a discovery of thienyl cathinone derivatives in commercialized designer drugs such as α-PBT (α-pyrrolidinobutiothiophenone), and, moreover, the bromothienyl analogs of α-PVT (α-pyrrolidinopentiothiophenone) and α-PBT. Approximately at the same time, Gambaro et al. [28] reported on a new cathinone derivative, thiothinone [(2-methylamino)-1-(2-thienyl)-1-propanone]. Complementary information and an expansion of the physicochemical database regarding α-PHP, and the first report on a new cathinone derivative, 4-fluoro-PV9 (4-fluoro-α-pyrrolidinooctanophenone), was provided in a paper [29] by the end of 2015. Apart from standard applications of HPLC-MS, GC-MS, and NMR, the authors characterized these two compounds with the use of tandem mass spectrometry with electrospray ionization (ESI-MS^n), Fourier-transform infrared spectroscopy, differential scanning calorimetry, and thermogravimetric analysis. Moreover, the analyzed materials (multicolor powders) were not extracted with the use of methanol, but they were treated with in-home elaborated solvent systems. At the first stage, 10 mg of a given powder was dissolved in 1 mL acetonitrile/methanol (50:50, v/v), then ultrasonicated, centrifugated, and the obtained supernatant was dissolved with methanol/water (80:20, v/v) [29]. In Table 14.5, chemical structures and selected analytical data are presented for the new compounds belonging to the group of cathinone derivatives and identified in the course of the past few years.

14.2.2 ANALYTICAL TECHNIQUES OF IDENTIFICATION OF CATHINONES

The steadily growing market for designer drugs poses a permanent analytical challenge for those who focus on physicochemical characteristics thereof and their identification in biological material. Apart from the key stage of sample preparation, a crucial role is played by analytical techniques used to assess the chemical composition of these samples. Considerable progress with chromatographic and spectroscopic techniques (resulting in sophisticated instruments able to identify hundreds of compounds in nanomole concentrations) allows expansion of the scope of toxicological investigations from the currently searched—both known and unknown—designer drug molecules towards the metabolites thereof. Each of the employed analytical techniques has its praiseworthy advantages and, at the same time, certain drawbacks and limitations, yet their rational combination can provide a powerful tool for the identification and quantification of material evidence and biological samples.

TABLE 14.5

Chemical Names, Common Names, and Chemical Structures of the Most Recently Reported Cathinone Derivatives

Chemical Name	Common Name	Chemical Structure
1-[2-(pyrrolidin-1-yl)-pentan-1-onyl]-4-methoxybenzene	4-methoxy-α-PVP, 4-methoxy-α-pyrrolidinopentiophenone	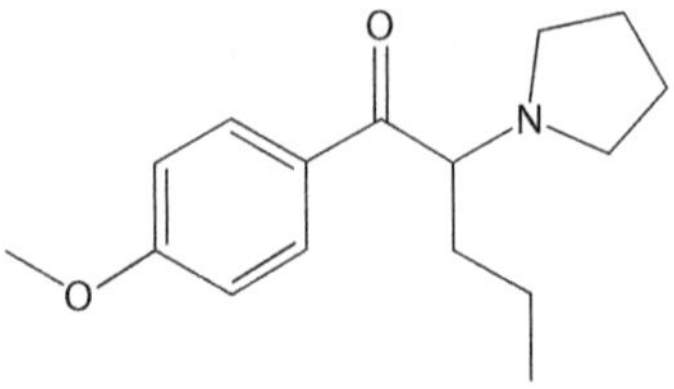
1-[2-(pyrrolidin-1-yl)-heptan-1-onyl]-benzene	α-PHPP, PV8, α-pyrrolidinoheptanophenone	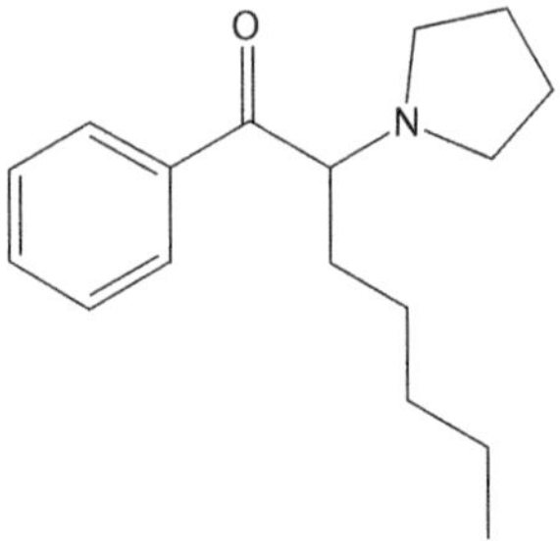
1-[2-(pyrrolidin-1-yl)-octan-1-onyl]-benzene	α-POP, PV9, α-pyrrolidinooctanophenone	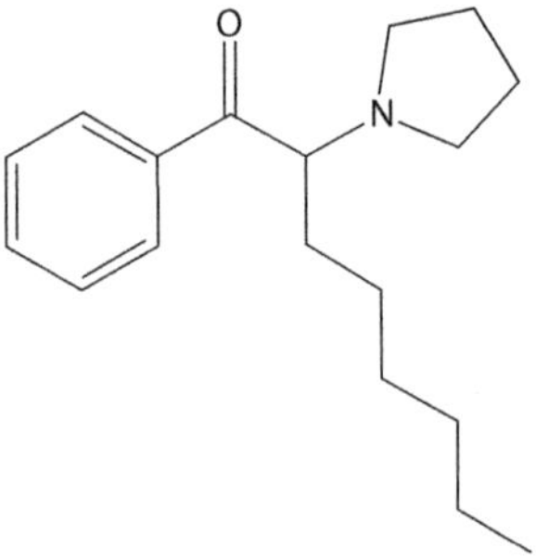
1-[2-(pyrrolidin-1-yl)-pentan-1-onyl]-3,4-dimethoxybenzene	3,4-dimethoxy-α-PVP, 3,4-dimethoxy-α-pyrrolidinopentiophenone	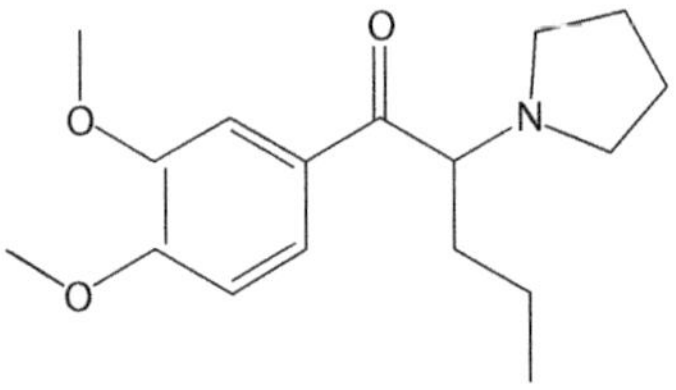
1-[2-(pyrrolidin-1-yl)-pentan-1-onyl]-4-fluorobenzene	4-fluoro-α-PVP, 4-fluoro-α-pyrrolidinopentiophenone	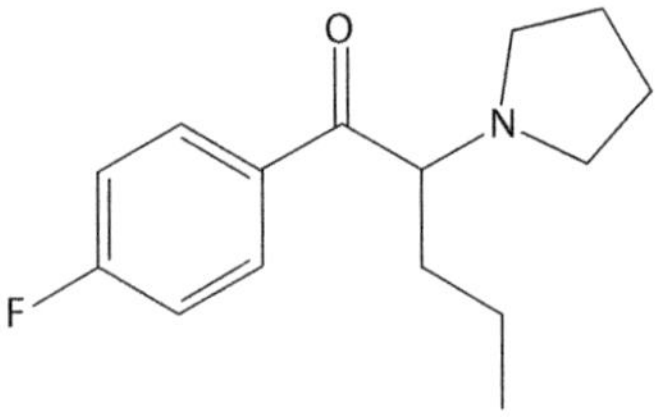

(Continued)

TABLE 14.5 (CONTINUED)

Chemical Names, Common Names, and Chemical Structures of the Most Recently Reported Cathinone Derivatives

Chemical Name	Common Name	Chemical Structure
1-[2-(N-ethylamino)-pentan-1-onyl]-benzene	α-EAPP, α-ethylaminopentiophenone	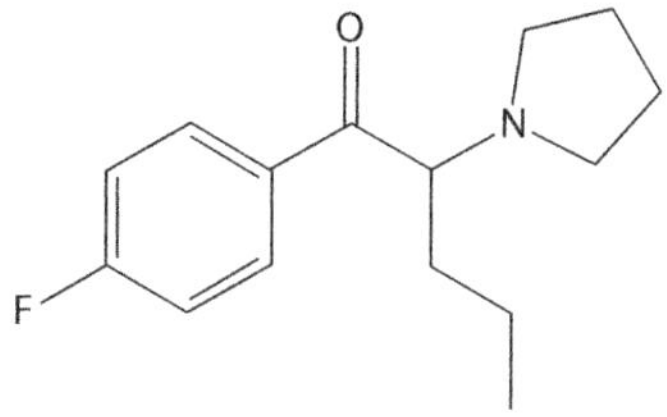
1-[2-(N-ethylamino)-pentan-1-onyl]-4-methylbenzene	N-ethyl-4-methylpentedron, 4-methyl-α-ethylaminopenthiophenone	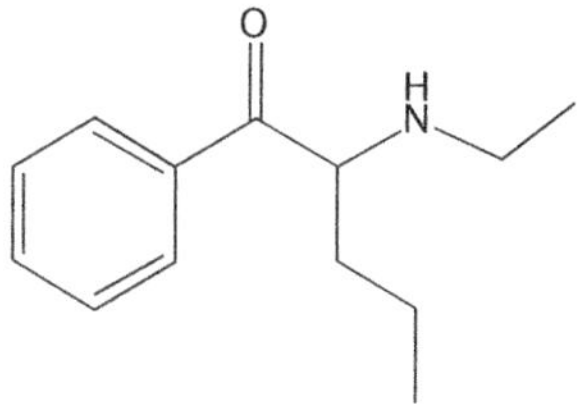
1-[2-(pyrrolidin-1-yl)-hexan-1-onyl]-benzene	α-PHP, α-pyrrolidinohexaphenone	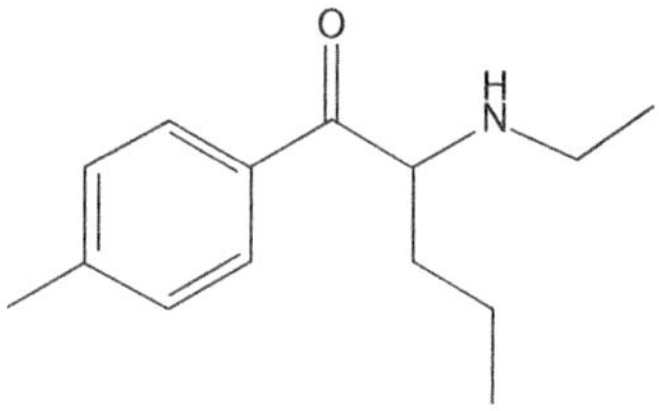
1-[2-(pyrrolidin-1-yl)-octan-1-onyl]-4-methoxybenzene	4-methoxy-α-POP, 4-methoxy-α-pyrrolidino-octanophenone	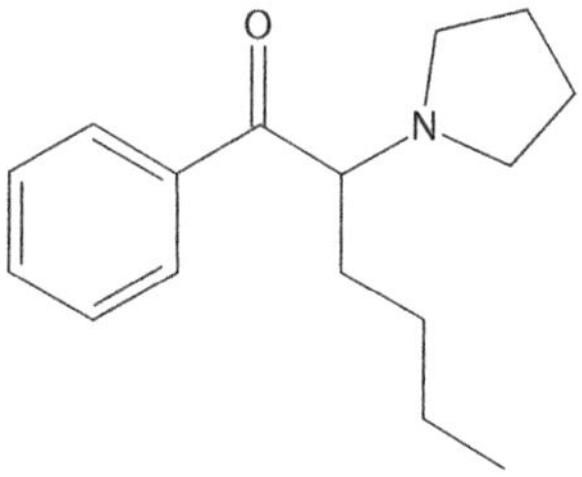
1-[2-(pyrrolidin-1-yl)-heptan-1-onyl]-4-methoxybenzene	4-methoxy-α-PHPP, 4-methoxy-α-pyrrolidinoheptanophenone	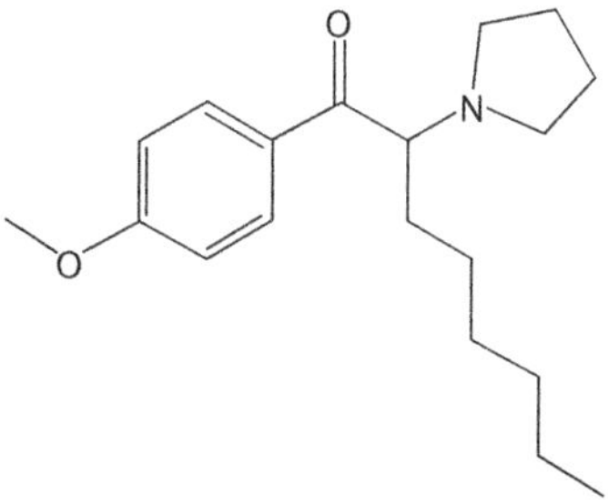

(Continued)

TABLE 14.5 (CONTINUED)
Chemical Names, Common Names, and Chemical Structures of the Most Recently Reported Cathinone Derivatives

Chemical Name	Common Name	Chemical Structure
1-[2-(pyrrolidin-1-yl)-heptan-1-onyl]-4-fluorobenzene	4-fluoro-α-PHPP, 4-fluoro-α-pyrrolidinoheptanophenone	
2-(pyrrolidin-1-yl)-1-(thiophen-2-yl)butan-1-one	α-PBT	
1-(5-bromothiophen-2-yl)-2-(pyrrolidin-1-yl)butan-1-one 1-(4-bromothiophen-2-yl)-2-(pyrrolidin-1-yl)butan-1-one 1-(3-bromothiophen-2-yl)-2-(pyrrolidin-1-yl)butan-1-one	5-Br-α-PBT x=Br y=z=H 4-Br-α-PBT y=Br x=z=H 3-Br-α-PBT z=Br x=y=H	
1-(5-bromothiophen-2-yl)-2-(pyrrolidin-1-yl) pentan-1-one 1-(4-bromothiophen-2-yl)-2-(pyrrolidin-1-yl) pentan-1-one 1-(3-bromothiophen-2-yl)-2-(pyrrolidin-1-yl) pentan-1-one 1-(4,5-bromothiophen-2-yl)-2-(pyrrolidin-1-yl) pentan-1-one	5-Br-α-PVT x=Br y=z=H 4-Br-α-PVT y=Br x=z=H 3-Br-α-PVT z=Br x=y=H 4,5-Br-α-PVT	
2-(M=methylamino)-1-(2-thienyl)-1-propanone	thiothinone	

(Continued)

TABLE 14.5 (CONTINUED)
Chemical Names, Common Names, and Chemical Structures of the Most Recently Reported Cathinone Derivatives

Chemical Name	Common Name	Chemical Structure
1-[2-(pyrrolidin-1-yl)-octan-1- onyl]-4-fluorobenzene	4-fluoro-α-PV9, 4-fluoro-α-POP, 4-fluoro-α-pyrrolidinooctanophenone	

14.2.2.1 Screening Techniques in the Analysis of Biological and Non-Biological Materials

Practically all attempts at identification of psychoactive compounds (cathinone derivatives included) start from the application of non-specific screening methods. In the case of material evidence (e.g., powders, tablets, and the contents of capsules), the standard colorimetric methods are used, playing the routine service in most analytical laboratories, police forensic laboratories included [30,31]. The most common test for the compounds that contain a nitrogen atom in their structure (widely used for the identification of amphetamine) makes use of the Marquis reagent (sulfuric acid and formaldehyde). It does not provide color reaction with synthetic cathinones derived from mephedrone, but it gives positive effect with the compounds containing the methylenedioxy substituent, like, e.g., MDPV. For this latter cathinone, another test, which includes the Chen reagent (acetic acid, copper monosulfide, and sodium hydroxide), is also applied, and is also suitable for ephedrine derivatives [31].

Colorimetric techniques are advantageous in the sense that they are rapid and easy to apply. However, they usually allow identification of a single structural fragment of a given molecule only, which is not a sufficient classification criterion permitting to ascribe the compound to a given group of derivatives. Due to this shortcoming, identification of synthetic cathinones cannot be carried out with use of colorimetric methods, which are generally not used for the preliminary screening of designer drugs.

Screening of biological material is most often carried out with use of immunoenzymatic assays. The most common one is ELISA (enzyme-linked immunosorbent assay), devised to detect certain psychoactive substances in an investigated material with use of mono- or polyclonal antibodies coupled with an enzyme. This technique is very popular in biomedical analyses, e.g., in virusology (the HIV test), bacteriology (the mycobacterium test), in food analysis targeting the contents of possible allergens, etc. [32]. In most commercial laboratories, an immunoenzymatic assay for toxicological purposes is also applied. It can be used as a screening technique for the detection of synthetic cathinones in biological material [33,34], yet it is considered as non-specific (due to possible cross reactions, like between MDPV and phenylcyclidine) [34].

Screening analyses can be used as a preliminary step in an assessment of material evidence and/or biological samples for the contents of psychoactive compounds. Their results often mark the direction for further investigations with a narrower group of suspect compounds, yet a decisive role is played by specific analytical techniques. With synthetic cathinones, the analytical techniques of first choice are gas chromatography (GC) and liquid chromatography (LC) coupled with different spectroscopic techniques.

14.2.2.2   Gas Chromatography Coupled with Mass Spectroscopy (GC-MS)

Gas chromatography coupled with mass spectrometry (GC-MS) is the most frequently used instrumental technique in toxicological analysis. It is applicable to many psychoactive compounds (cathinones included) that characterize with volatility [30,35–40]. Moreover, the time of a single analytical run is relatively short and in the period of *ca.* 40 minutes a screening can be done with a vast number of compounds [37]. In GC-MS, chemical ionization (CI) is occasionally applied, yet in most cases, the electron-impact (EI) ionization mode predominates [30,35–40]. The cathinone mass spectra originating from GC-MS and run in the positive ionization mode are very simple and characterize with molecular signals originating from the imine ions. Due to that, identification of different cathinone derivatives becomes rather complicated. For this reason, different modifications of the detection are most welcome, one of them being tandem mass spectrometry, which provides more information about the molecular structure and proves very helpful with identification. In 2012, Zuba [37] proposed a novel analytical procedure for the determination of synthetic cathinones by means of GC-EI-MS. According to his approach, if the aforementioned imine ion appears as a molecular ion ($m/z = 16 + 14n$, where $n = 1, 2,$ 3, etc.), it can be assumed that the cathinone present in the analyzed sample characterizes with a straight aliphatic chain. If in the mass spectrum a signal appears that corresponds with the pyrrolidine ion ($m/z = 70 + 14n$, where $n = 1, 2, 3$, etc.), then the identified cathinone should contain the pyrrolidine structural element. Due to a possible encounter of many different regioisomers with different cathinone derivatives, an essential task consists in an assessment of the aliphatic chain length and its possible substituents, or in defining the substituents of the aromatic ring. With unsubstituted rings, the presence of the fragmentation ions at $m/z = 70$ and 105 is characteristic. Signals at $m/z = 91$ and 119 prove characteristic of the methyl phenyl ring, and those at $m/z = 121$ and 149 bear witness to the presence of the methylenedioxy ring substituent. The GC-EI-MS technique is rapid, yet its main drawbacks are a possibility of identical fragmentation patterns with certain isomers and low intensity of molecular ions, when applying the EI ionization mode. These are the main reasons why application of the alternative mass spectrometric techniques often becomes inevitable [37].

In Table 14.6, the m/z number values for the base and fragmentation peaks obtained with use of the GC-EI/MS technique are provided for the selected new psychoactive substance from the group of cathinone derivatives.

TABLE 14.6

Common Names, Absorption Maxima, Molecular Weights, LC-ESI/MS Ions, GC-EI/MS Peaks, and References of the Most Recently Reported Cathinone Derivatives

Common Name	Absorption Maxima [nm]	Molecular Weight [Da]	Precursor Ion and Fragmentation Ions [M+H+] of LC-ESI/MS Spectra [m/z]	Base Peak and Other Peaks of GC-EI/MS Spectra [m/z]	References
4-methoxy-α-PVP	292	261.36	262	**126**, 135, 107	[26]
α-PHPP, PV8	253	259.39	260	**154**, 105, 77	[24]
α-POP, PV9	253	273.41	274	**168**, 105, 77	[24]
3,4-dimethoxy-α-PVP	286, 316	291.39	292	**126**, 137, 165	[24]
4-fluoro-α-PVP	256	249.32	250	**126**, 95	[24]
α-EAPP	251	205.30	206	**100**, 77	[24]
N-ethyl-4-methylpentedrone	264	219.32	220	**100**, 91	[24]
α-PHP	252, 251	245.36	246, 228, 175	**140**, 141, 105, 96, 77	[25,29]
4-methoxy-α-POP	292	303.44	304	**168**, 135	[25]
4-methoxy-α-PHPP	292	289.41	290	**154**, 135	[25]
4-fluoro-α-PHPP	255	277.38	278	**154**, 123, 95	[25]
α-PBT	No data	223.33	224		[27]
5-Br-α-PBT	No data	302.23	302/304	112	[27]
4-Br-α-PBT			302/304	112	
3-Br-α-PBT			302/304	**112**	
5-Br-α-PVT	No data	316.26	316/318	**126**, 189, 191	[27]
4-Br-α-PVT		395.15	316/318	**126**, 189, 191	
3-Br-α-PVT			316/318	**126**, 189, 191	
4,5-Br-α-PVT			395/318	**126**, 267, 269, 271	
thiothinone	No data	169.24	170	**58**, 83, 111	[28]
4-fluoro-α-PV9, 4-fluoro-α-POP	254, 253	291.40	292, 274, 221, 203, 189	**168**, 169, 123, 110, 95, 84, 55	[29]

14.2.2.3 Liquid Chromatography Coupled with Mass Spectroscopy (LC-MS)

Liquid chromatography coupled with mass spectrometry (LC-MS) is employed in toxicological analysis laboratories with practically equal frequency as GC-MS and its popularity is due to high sensitivity and selectivity of this particular technique. Most LC-MS analyses are carried out in the multiple reaction monitoring (MRM) mode and the selected reaction monitoring (SRM) mode, and the most frequently encountered ionization technique is by means of electrospray (ESI). With synthetic cathinones, a vast number of the analyses have been carried out in the ESI-MS^n mode, which enables observation of characteristic fragmentation patterns with the respective pseudomolecular ions. Characteristic features of the molecular ion fragmentation are the loss of the water molecule and the split-off of the pyrrolidine ring [24,28,29]. Figures 14.2 through 14.5 present the mass spectra of four cathinone derivatives obtained in the ESI-MS^n mode, along with the suggested fragmentation patterns. Each compound represents one out of four cathinone derivative groups, with classification based on structural characteristics of the members belonging to individual derivative groups (as presented in Section 14.1.2).

Lesiak et al. [41] presented the results of the analysis of a mixture of the cathinone derivatives in a commercial product labeled as "bath salt." These authors admit that the most popular and most frequently applied analytical techniques are GC-MS and LC-MS supported with the mass spectral libraries, yet the efficiency of these approaches tends to diminish in a clash with an avalanche of brand new cathinone derivatives that appear with a growing speed on the designer drugs market. As an alternative, these authors propose application of the DART (Direct Analysis in Real Time) ionization source coupled with mass spectrometry. The results obtained with this approach point to its increased usefulness and possibility to better differentiate among the compounds that are structurally closely related or even isomeric, both as individual species and in the mixtures.

A different technique that can be used in the analysis of active components of designer drugs is ultra-high-performance liquid chromatography (UHPLC) coupled with mass spectrometry with the time-of-flight mass analyzer (TOF-MS) and its quadrupole (or multi-quadrupole) TOF modification, UHPLC-QTOF-MS. Using these latter techniques, Ibáñez et al. [42] successfully identified compounds (including certain cathinone derivatives) present in numerous designer drugs commercialized as tablets, capsules, powders, and the dried herbs. The results of these authors demonstrate the high potential of the discussed techniques applied both in the target analysis and the non-target analysis of psychoactive compounds, where each consecutive compound is regarded as an unknown substance. An advantage of the QTOF-MS-based methods is that preliminary identification of the analyzed compounds can be performed without any reference standards, which are not needed at the initial stage of investigation. Standards are only acquired at the final stage of investigation, in order to ultimately confirm the presence of a given compound, once the solid instrumental evidence is already at hand.

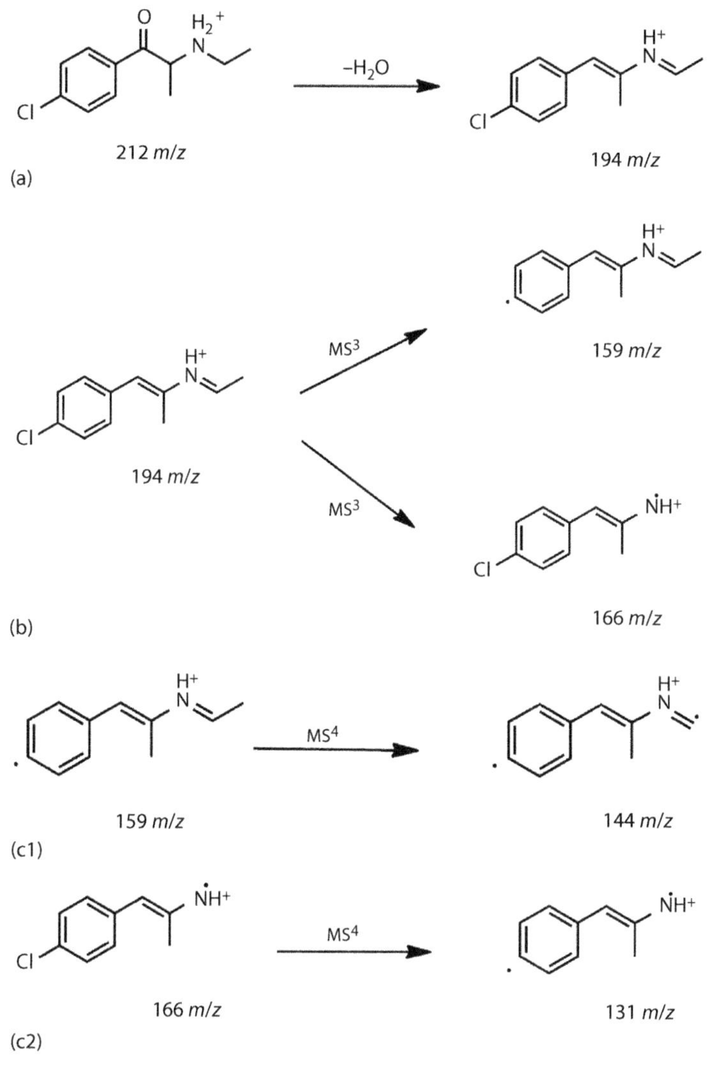

FIGURE 14.2 Mass spectra of 4-chloroethcathinone (4-CEC) obtained in the (a) ESI-MS^2, (b) ESI-MS^3, and (c) ESI-MS^4 modes, along with the suggested fragmentation patterns.

In Table 14.6, the *m/z* number values for the pseudomolecular and fragmentation peaks obtained with use of the LC-ESI/MS techniques are provided for selected new psychoactive substances from the group of the cathinone derivatives.

14.2.2.4 Liquid Chromatography Coupled with UV-Vis, NMR, and IR Spectroscopy

A less-frequently employed detection system, also applied in the investigations of the material evidence and biological samples, and coupled with liquid chromatography

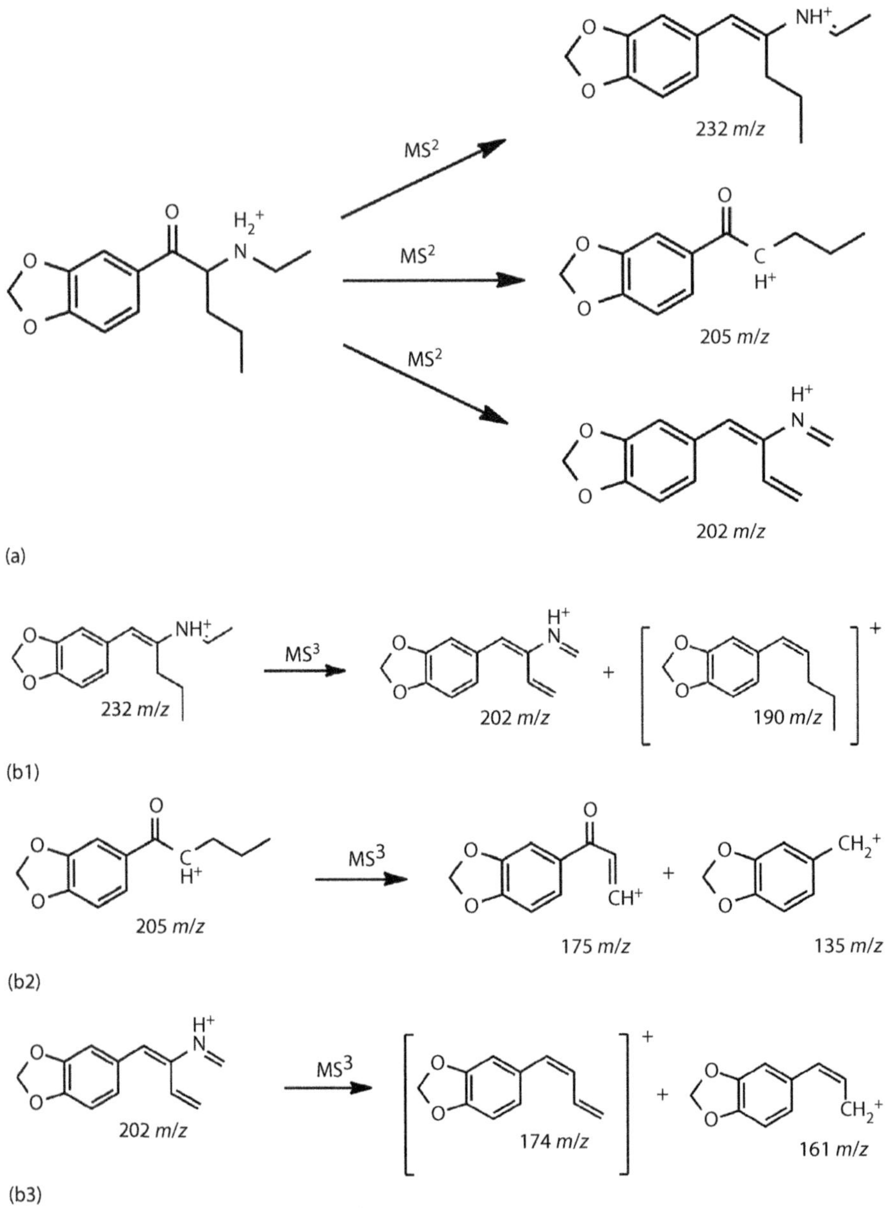

(a)

(b1)

(b2)

(b3)

FIGURE 14.3 Mass spectra of ethylpentylone obtained in the (a) ESI-MS^2 and (b) ESI-MS^3 modes, along with the suggested fragmentation patterns.

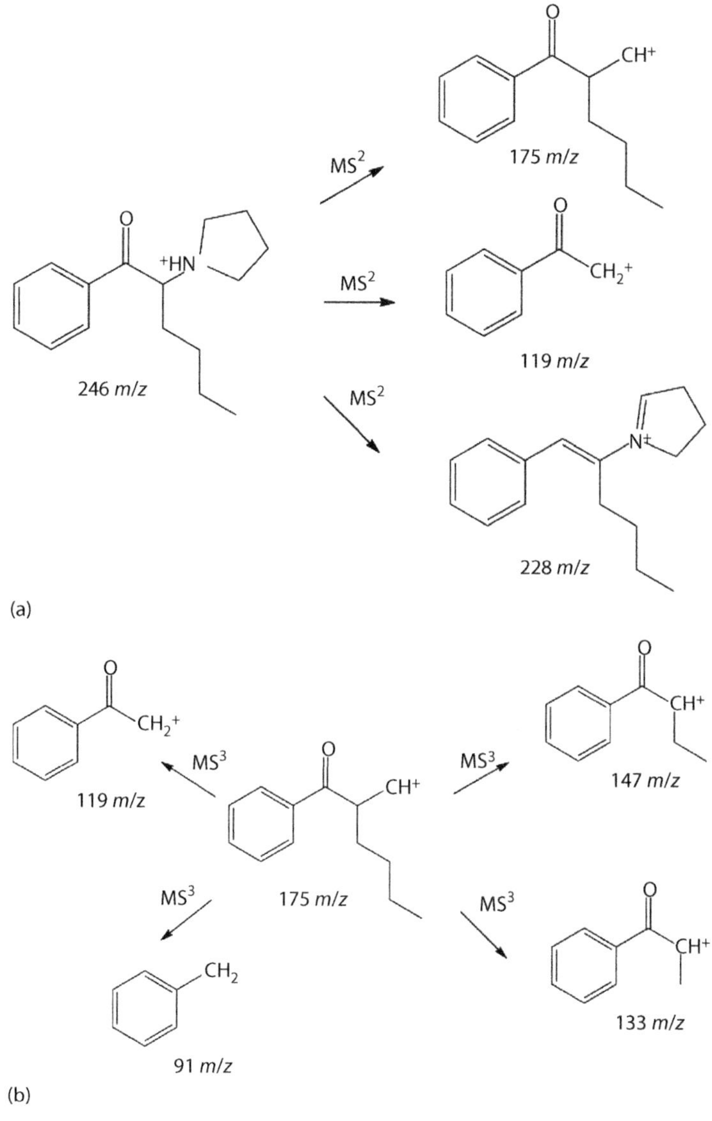

FIGURE 14.4 Mass spectra of α-pyrrolidinohexaphenone (α-PHP) obtained in the (a) ESI-MS² and (b) ESI-MS³ modes, along with the suggested fragmentation patterns.

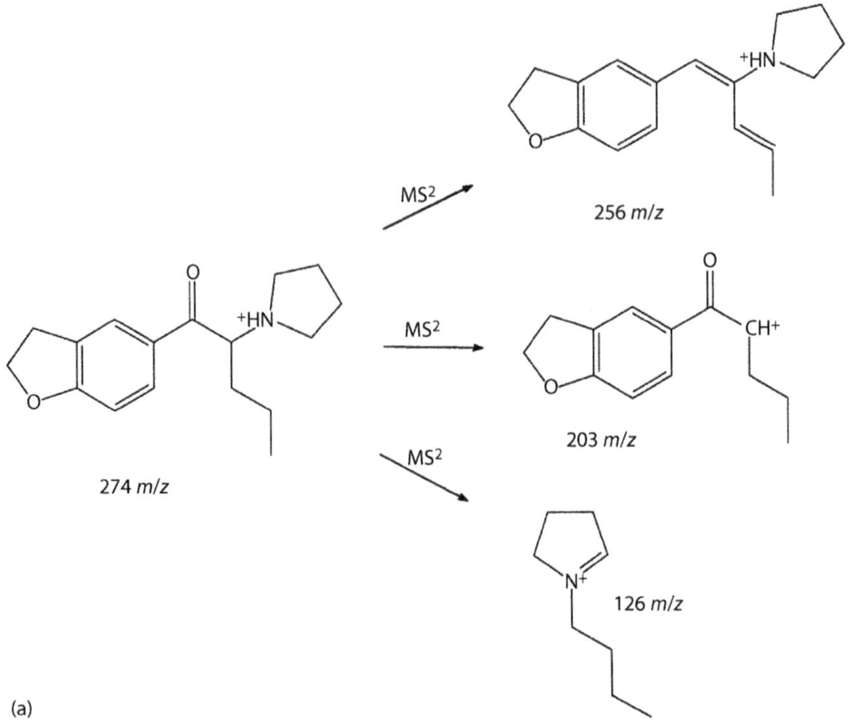

(a)

FIGURE 14.5 Mass spectra of 3-desoxy-3,4-methylenedioxypyrovalerone (3-desoxy-3,4-MDPV) obtained in the (a) ESI-MS^2, (b) ESI-MS^3, and (c) ESI-MS^4 modes, along with the suggested fragmentation patterns. *(Continued)*

(LC), is UV-Vis spectrometry in the form of diode array detection (DAD), or photodiode array detection (PDA) [24,29,43]. Owing to this particular detection system, one can record the UV-Vis spectra of the investigated cathinones and establish the absorption wavelength characteristics of the individual group representatives. These data can further be added to the library providing physicochemical characteristics of individual cathinone species. In Table 14.6, the maxima of the wavelengths (nm) are given for the selected new psychoactive substances from the group of cathinone derivatives.

Last but not least, in view of possible structural modifications of cathinone derivatives, one cannot forget such techniques used in the analysis of the evidence material as nuclear magnetic resonance (NMR) spectroscopy and absorption infrared (IR) spectroscopy [29,39,40]. Owing to NMR spectroscopy, the substitutional isomerism with a given molecule can be defined without using any standards for the sake of comparison [27,29,40,44]. Obviously, this technique cannot be employed to quantify the contents of psychoactive substances in biological material, yet ^{1}H and ^{13}C NMR spectroscopy are commonly used for a detailed assessment of the chemical structure of cathinone derivatives, including substitutional isomerism thereof.

FIGURE 14.5 (CONTINUED) Mass spectra of 3-desoxy-3,4-methylenedioxypyrovalerone (3-desoxy-3,4-MDPV) obtained in the (a) ESI-MS², (b) ESI-MS³, and (c) ESI-MS⁴ modes, along with the suggested fragmentation patterns.

REFERENCES

1. Brenneisen, R., Fisch, H.U., Koelbing, U., Geisshüsler, S., and P. Kalix. 1990. Amphetamine-like effects in humans of the khat alkaloid cathinone. *Br. J. Clin. Pharmacol.* 30:825–828.
2. Feyissa, A.M., and J.P. Kelly. 2008. A review of the neuropharmacological properties of khat. *Prog. Neuropsychopharmacol. Biol. Psych.* 32:1147–1166.
3. Katz, D.P., Bhattacharya, D., Bhattacharya, S., Deruiter, J., Clark, C.R., Suppiramaniam, V., and M. Dhanasekaran. 2014. Synthetic cathinones: "A khat and mouse game". *Toxicol. Lett.* 229:349–356.
4. Patel, N.B., 2015. "Natural amphetamine" khat: A cultural tradition or a drug of abuse? *Int. Rev. Neurobiol.* 120:235–255.
5. Szendrei, K., 1980. The chemistry of khat. *Bull. Narc.* 32:5–35.
6. Valente, M.J., Guedes de Pinho, P., de Lourdes Bastos, M., Carvalho, F., and M. Carvalho. 2014. Khat and synthetic cathinones: A review. *Arch. Toxicol.* 88:15–45.
7. Baumann, M.H., Solis, E., Watterson, L.R., Marusich, J.A., Fantegrossi, W.E., and J.L. Wiley. 2014. Baths salts, spice, and related designer drugs: The science behind the headlines. *J. Neurosci.* 34:15150–15158.
8. Favretto, D., Pascali, J., and F. Tagliaro. 2013. New challenges and innovation in forensic toxicology: Focus on the "new psychoactive substances". *J. Chromatogr. A.* 1287:84–95.
9. Weaver, M.F., Hopper, J.A., and E.W. Gunderson. 2015. Designer drugs 2015: Assessment and management. *Addict. Sci. Clin. Pract.* 10:8.
10. Vardakou, I., Pistos, C., and C. Spiliopoulou. 2011. Drugs for youth via Internet and the example of mephedrone. *Toxicol. Lett.* 201:191–195.
11. Debruyne, D., Loilier, M., Cesbron, A., Le Boisselier, R., and J. Bourgine. 2014. Emerging drugs of abuse: Current perspectives on substituted cathinones. *Subst. Abuse. Rehabil.* 5:37–52.
12. Zawilska, J.B., and J. Wojcieszak. 2013. Designer cathinones—An emerging class of novel recreational drugs. *Forensic. Sci. Int.* 231:42–53.
13. Simmler, L.D., Buser, T.A., Donzelli, M., Schramm, Y., Dieu, L.H., Huwyler, J., Chaboz, S., Hoener, M.C., and M.E. Liechti. 2013. Pharmacological characterization of designer cathinones in vitro. *Br. J. Pharmacol.* 168:458–470.
14. Baumann, M.H., Ayestas, Jr. M.A., Partilla, J.S., Sink, J.R., Shulgin, A.T., Daley, P.F., Brandt, S.D., Rothman, R.B., Ruoho, R.A.E., and N.V. Cozzi. 2012. The designer methcathinone analogs, mephedrone and methylone, are substrates for monoamine transporters in brain tissue. *Neuropsychopharmacology.* 37:1192–1203.
15. Baumann, M.H., Partilla, J.S., Lehner, K.R., Thorndike, E.B., Hoffman, A.F., Holy, M., Rothman, B., Goldberg, S.R., Lupica, C.R., Sitte, H.H., Brandt, S.D., Tella, S.R., Cozzi, N.V., and C.W. Schindler. 2013. Powerful cocaine-like actions of 3,4-methylenedioxypyrovalerone (MDPV), a principal constituent of psychoactive 'bath salts' products. *Neuropsychopharmacology.* 38:552–562.
16. Lopez-Arnau, R., Martinez-Clemente, J., Pubill, D., Escubedo, E., and J. Camarasa. 2012. Comparative neuropharmacology of three psychostimulant cathinone derivatives: Butylone, mephedrone and methylone. *Br. J. Pharmacol.* 167:407–420.
17. Martinez-Clemente, J., Escubedo, E., Pubill, D., and J. Camarasa. 2012. Interaction of mephedrone with dopamine and serotonin targets in rats. *Eur. Neuropsychopharmacology.* 22:231–236.
18. Meyer, M.R., Wilhelm, J., Peters, F.T., and H.H. Maurer. 2010. Beta-keto amphetamines: Studies on the metabolism of the designer drug mephedrone and toxicological detection of mephedrone, butylone, and methylone in urine using gas chromatography-mass spectrometry. *Anal. Bioanal. Chem.* 397:1225–1233.

19. Gibbons, S., and M. Zloh. 2010. An analysis of the "legal high" mephedrone. *Bioorg. Med. Chem.* 20:4135–4139.
20. Dargan, P.I., Sedefov, R., Gallegos, A., and D.M. Wood. 2011. The pharmacology and toxicology of the synthetic cathinone mephedrone (4-methylmethcathinone). *Drug. Test. Anal.* 3:454–463.
21. Liechti, M. 2015. Novel psychoactive substances (designer drugs): Overview and pharmacology of modulators of monoamine signaling. *Swiss. Med. Wkly.* 145:w14043.
22. Uralets, V., Rana, S., Morgan, S., and W. Ross. 2015. Testing for designer stimulants: Metabolic profiles of 16 synthetic cathinones excreted free in human urine. *J. Anal. Toxicol.* 38:233–241.
23. Lusthof, K.J., Oosting, R., Maes, A., Verschraagen, M., Dijkhuizen, A., and A.G.A Sprong. 2011. A case of extreme agitation and death after the use of mephedrone in The Netherlands. *Forensic. Sci. Int.* 206:93–95.
24. Uchiyama, N., Matsuda, S., Kawamura, M., Shimokawa, Y., Kikura-Hanajiri, R., Aritake, K., Urade, Y., and Y. Goda. 2014. Characterization of four new designer drugs, 5-chloro-NNEI, NNEI indazole analog, α-PHPP and α-POP, with 11 newly distributed designer drugs in illegal products. *Forensic Sci. Int.* 243:1–13.
25. Uchiyama, N., Shimokawa, Y., Kawamura, M., Kikura-Hanajiri, R., and T. Hakamatsuka. 2014. Chemical analysis of a benzofuran derivative, 2-(2-ethylamino-propyl)benzofuran (2-EAPB), eight synthetic cannabinoids, five cathinone derivatives, and five other designer drugs newly detected in illegal products. *Forensic Toxicol.* 32: 266–281.
26. Uchiyama, N., Matsuda, S., Kawamura, M., Kikura-Hanajiri, R., and Y. Goda. 2013. Identification of two new-type designer drugs, piperazine derivative MT-45 (I-C6) and synthetic peptide Noopept (GVS-111), with synthetic cannabinoid A-834735, cathinone derivative 4-methoxy-α-PVP, and phenethylamine derivative 4-methylbuphedrine from illegal products. *Forensic Toxicol.* 32:9–18.
27. Doi, T., Asada, A., Takeda, A., Tagami, T., Katagi, M., Matsuta, S., Kamata, H., Kawaguchi, M., Satsuki, Y., Sawabe, Y., and H. Obana H. 2015. Identification and characterization of α-PVT, α-PBT, and their bromothienyl analogs found in illicit drug products. *Forensic Toxicol.* 34:76–93.
28. Gambaro, V., Casagni, E., Dell'Acqua, L., Roda, G., Tamborini, L., Visconti, G.L., and F. Demartin. 2015. Identification and characterization of a new designer drug thiothinone in seized products. *Forensic. Toxicol.* 34:174–178.
29. Majchrzak, M., Rojkiewicz, M., Celiński, R., Kuś. P. and M. Sajewicz. 2015. Identification and characterization of new designer drug 4-fluoro-PV9 and α-PHP in the seized materials. *Forensic. Toxicol.* 34:115–124.
30. Namera, A., Kawamura, M., Nakamoto, A., Saito, T., and M. Nagao. 2015. Comprehensive review of the detection methods for synthetic cannabinoids and cathinones. *Forensic Toxicol.* 33:175–194.
31. Toole, K.E, Fu, S., Shimmon, R.G., and N. Kraymen. 2011. Color test for the preliminary identification of methcathinone and analogues of methcathinone. *Microgr. J.* 9:27–32.
32. Apollonio, L.G., Whittall. I.R., Pianca, D.J., Kyd, J.M., and W.A. Maher. 2007. Matrix effect and cross-reactivity of select amphetamine-type substances, designer analogues, and putrefactive amines using the Bio-Quant direct ELISA presumptive assays for amphetamine and methamphetamine. *J. Anal. Toxicol.* 31:208–213.
33. Ellefsen, K.N., Anizan, S., Castaneto, M.S., Desrosiers, N.A., Martin, L.T.M, Klette, C.K.L, and M.A. Huestis. 2014. Validation of the only commercially available immunoassay for synthetic cathinones in urine: Randox drugs of Abuse V Biochip Array Technology. *Drug. Test. Anal.* 6:728–738.

34. Swortwood, M.J., Lee Hearn, W., and A.P. DeCaprio. 2014. Cross-reactivity of designer drugs, including cathinone derivatives, in commercial enzyme-linked immunosorbent assays. *Drug. Test. Anal.* 6:716–727.

35. Saito, T., Namera, A., Osawa, M., Aoki, H., and Inokuchi, S. 2013. SPME–GC–MS analysis of α-pyrrolidinovaleorophenone in blood in a fatal poisoning case. *Forensic Toxicol.* 31:328–332.

36. Kudo, K., Usumoto, Y., Usui, K., Hayashida, M., Kurisaki, E., Saka, K., Tsuji, A., andN. Ikeda. 2013. Rapid and simultaneous extraction of acidic and basic drugs from human whole blood for reliable semi-quantitative NAGINATA drug screening by GC–MS. *Forensic Toxicol.* 32:97–104.

37. Zuba, D. 2012. Identification of cathinones and other active components of "legal highs" by mass spectrometric methods. *TRAC—Trend. Anal. Chem.* 32:15–30.

38. Zweipfenning, P.G., Wilderink, A.H., Horsthuis, P., Franke, J.P., and R.A. de Zeeuw. 1994. Toxicological analysis of whole blood samples by means of Bond-Elut Certify columns and gas chromatography with nitrogen–phosphorus detection. *J. Chromatogr. A.* 674:87–95.

39. Westphal, F., Junge, T., Klein, B., Fritschi, G., and U. Girreser. 2011. Spectroscopic characterization of 3,4-methylenedioxypyrrolidinobutyrophenone: A new designer drug with α-pyrrolidinophenone structure. *Forensic Sci. Int.* 209:126–132.

40. Westphal, F., Junge, T., Girreser, U., Greibl, W., and C. Doering. 2012. Mass, NMR and IR spectroscopic characterization of pentedrone and pentylone and identification of their isocathinone by-products. *Forensic Sci. Int.* 217:157–167.

41. Lesiak, A.D., Musah, R.A., Cody, R.B., Domin, M.A., Dane, A.J., and J.R.E. Shepard. 2013. Direct analysis in real time mass spectrometry (DART-MS) of "bath salt" cathinone drug mixtures. *Analyst.* 138:3424–3432.

42. Ibáñez, M., Sancho, J.V., Bijlsma, L., van Nuijs, A.L.N., Covaci, A., and F. Hernández. 2014. Comprehensive analytical strategies based on high-resolution time-of-flight mass spectrometry to identify new psychoactive substances. *TRAC—Trend. Anal. Chem.* 57:107–117.

43. Uchiyama, N., Shimokawa, Y., Kikura-Hanajiri, R., Demizu, Y., Goda, Y., and T. Hakamatsuka. 2015. A synthetic cannabinoid FDU-NNEI, two 2H-indazole isomers of synthetic cannabinoids AB-CHMINACA and NNEI indazole analog (MN-18), a phenethylamine derivative N–OH-EDMA, and a cathinone derivative dimethoxy-α-PHP, newly identified in illegal products. *Forensic Toxicol.* 33:244–259.

44. Westphal, F., Junge, T., Rosner, P., Fritschi, G., Klein, B., and U. Girresee. 2007. Mass spectral and NMR spectral data of two new designer drugs with an α-aminophenone structure: 4′-Methyl-α-pyrrolidinohexanophenone and 4′-methyl-α-pyrrolidinobutyrophenone. *For. Sci. Int.* 169:32–42.

15 2C Derivatives of Phenylethylamines and Their Analysis

Beril Anilanmert, Fatma Çavuş Yonar, and Ali Acar Özdemir

CONTENTS

15.1 INTRODUCTION

The use of new designer drugs has exploded over the last decade, creating an additional burden for health care systems [1]. There are multiple factors that have contributed to the rapid expansion of this epidemic use, but the most prominent one seems to be the ease with which these agents may be purchased on the Internet, at raves and night clubs, and in head shops. New synthetic or "designer" drugs are created by manipulating the chemical structures of other psychoactive drugs so that the resulting product is structurally similar but not identical to illegal psychoactive drugs. There is an overwhelming variety of new drugs, most of which until recent years were not covered by existing laws. The scale of synthetic drugs often seems limited only by the creativity of enterprising chemists [2]. One of the most important groups of new synthetic drugs is the phenethylamine group. Phenethylamines refer to a class of substances with documented psychoactive and stimulant effects, which includes stimulants (e.g., amphetamine itself) and entactogens (e.g., MDMA), all of which are controlled under the 1971 Convention [3,4]. Phenethylamines share a common phenylethan-2-amine structure and also include hallucinogenic ring-substituted substances such as the "2C series" (e.g., 2-CB, 2-CE and 2-CI) and NBOMe (25I-, 25B- and 25-C NBOMe), ring-substituted amphetamines such as the "D series" (e.g., DOI, DOC), benzodifurans (e.g., bromo-dragonfly, 2C-B-Fly), and others (e.g., p-methoxymethamphetamine [PMMA]) [4,5].

The 2C-X class has a basic structure of 2,5 dimethoxyphenethylamine and is an analog of mescaline [6]. They are similar in structure to 3,4-methylenedioxy-N-methylamphetamine (MDMA, ecstasy) [7]. The terminology "2Cs" was introduced by Alexander Shulgin to describe the two carbons between the amino group and the benzene ring in the chemical structure shown in Figure 15.1. The 2C drugs series is a large group of chemicals that have a phenethylamine base with methoxy substitutions on the 2- and 5-positions of the aromatic ring, as well as an alkyl, halo, or alkylthio substitution on the 4-position of the aromatic ring [3,6,8]. The hydrophobic substituent at the 4-position may be iodine in 2C-I or bromine in 2C-B [9]. Frequently, the 4-carbon position on the benzene ring is substituted to create a different compound [10]. Furthermore, by making alterations to the substituents at the 2-, 3-, 5-, and 6-positions of the aromatic ring, even more new compounds can be synthesized (e.g., "Fly" compounds).

Synthetic mescaline analogs, known as 2C designer drugs, became widespread in Europe in the 1970s and 1980s [6] and were integrated into American drug culture in the 1990s. In 1991, Alexander Shulgin published a book entitled *PIHKAL: A Chemical Love Story* (PIHKAL is an acronym for "Phenethylamines I Have Known and Loved") [1]. The book contains information on over 200 psychedelic compounds and synthesis instructions, determination methods, dosages, and commentaries describing their effects. This publication led to a rise in popularity of the 2C agents and the subsequent classification as Schedule I substances of 2C-B, 2C-T-7, 2C-E,

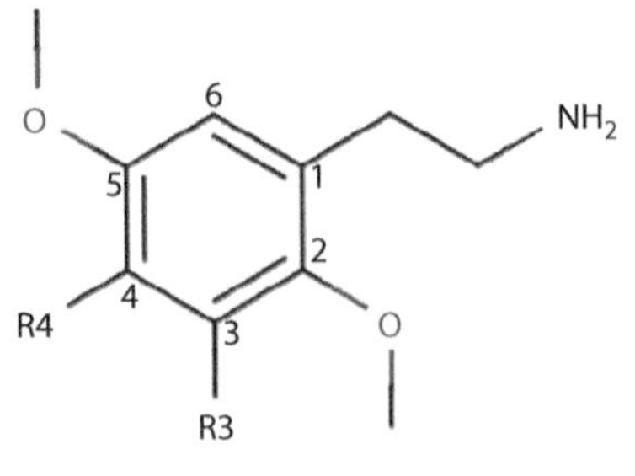

FIGURE 15.1   General structure of 2C with labeled positions 3 and 4. (Reprinted from Dean, B. V. et al., 2013, *J Med Toxicol.* 9:172–178.)

2C-D, 2C-C, 2C-I, 2C-T2, 2C-T4, 2C-H, 2C-N, and 2C-P. These agents also fall under the Analog Statute of the Controlled Substances Act, making it illegal to traffic any of these compounds that are structurally analogous to scheduled phenethylamines. 2C drugs such as 2C-T-7, 2C-T-2, and 2-CB have reportedly been detected in Europe and the United States over the last decade [11–14]. Individuals who look for ecstasy at "rave parties" and music festivals may be exposed to 2C drugs as contaminants or MDMA substitutes [10]. These substances, which generally come in tablet or powder form, can be purchased on the Internet and can be listed as "research chemicals." 2C users are more likely to be younger males and may have a history of polydrug use. True prevalence of 2C and other phenethylamine use is unknown. A survey of U.K. dance club frequenters reported that 17.6% had used 2C-B and 11.2% had used 2C-I. 2C-B was present in roughly 3% of drug materials analyzed in Spain between 2006 and 2009. Many 2C substances are listed as Schedule I substances. However, newer phenethylamine compounds are persistently being designed and introduced to divert existing legislation and regulatory oversight.

The 2C series with any hydrophobic substitution at the 4-position of the phenethylamine confers hallucinogenic activity [3,6,8]. These substances affect the 5-HT_2 receptors and reportedly create an experience similar to that of LSD [6]. Substitutions at the 4-position display varying potency, where halogen substitutions are the most potent, then alkyl substitutions, followed by alkylthio substitutions, and finally hydrogen substitutions. They differ from the D-series of substituted amphetamines only by the absence of a methyl group on the alpha carbon of the side chain [3,8]. The most active compounds identified to date possess an alkyl, alkylthio, or halogen group, and potency increases in the sequence: H < OR < SR < R < halogen [15].

Many different 2C agents have emerged over the years [1]. The most recent popular 2C agents are 2,5-dimethoxy-4-iodophenethylamine (2C-I) and 2-(4-chloro-2,5 dimethoxy-phenyl)-N-[(2-methoxyphenyl)methyl]ethanamine (2C-I-NBOMe). They are available in powder form and can be purchased online in the United States, Asia, and Western Europe and may be found at head shops, dance clubs, gas stations, truck stops, and raves. The range of effective dose of 2C-I per milligram is approximately ten times lower than for mescaline. 2C-I-NBOMe is structurally similar to 2C-I but has higher affinity for the serotonin $5\text{-HT}_2\text{A}$ receptor. 2-C analogs on the market are given in Table 15.1 [6,7,16], with their corresponding IUPAC names. Reports of 2C-phenethylamines increased 295% from 2011 to 2015. During 2011, 2C-E, 2C-I, and 2C-B accounted for 90% of 2C-phenethylamine reports, whereas 25I-NBOMe, 25C-NBOMe, and 25B-NBOMe accounted for 91% in 2015 [16].

15.1.1   2C-I Analog

The synthesis of 2C-I was published in 1991 and the drug became popular in tablet form as a club drug in the United Kingdom around 2003 [9]. 2C-I is known to be present in conjugated and metabolized forms in urine. The dose–toxicity relationship of 2C-I is unknown. We have met a toxicity case reported in the literature: A 39-year-old African-American woman in the emergency department on New Year's Day after a night of partying exhibited rapidly diminishing mental status,

TABLE 15.1
Some 2-C Analogs on the Market, with Their Corresponding IUPAC Names

Compound	IUPAC Name
Bromo-dragonfly	8-bromo-α-methyl-benzo[1,2-b:4,5-b']difuran-4-ethanamine
2C-B FLY	4-bromo-2,5-dimethoxy-benzenethanamine
2C-B-BZP	1-[(4-bromo-2,5-dimethoxyphenyl)methyl]-piperazine
2C-C	4-chloro-2,5-dimethoxy-benzeneethanamine
2C-D	2,5-dimethoxy-4-methyl-benzeneethanamine
2C-E	2,5-dimethoxy-4-ethylphenethylamine
2C-G	2,5-dimethoxy-3,4-dimethyl-benzeneethanamine
2C-G-3	2,5-dimethoxy-3,4-trimethyl-benzeneethanamine
2C-G-5	2,5-dimethoxy-3,4-norbomyl-benzeneethanamine
2C-H	2,5-dimethoxy-benzeneethanamine
2C-I	4-iodo-2,5-dimethoxy-benzeneethanamine
2C-N	2,5-dimethoxy-4-nitro-benzeneethanamine
2C-P	2,5-dimethoxy-4-propyl-benzeneethanamine
2C-SE	2,5-dimethoxy-4 methylseleno-benzeneethanamine
2C-T	2,5-dimethoxy-4-(methylthio)-benzeneethanamine
2C-T-2	4-(ethylthio)-2,5-dimethoxy-benzeneethanamine
2C-T-4	2,5-dimethoxy-4-[(1-methylethyl)thio]-benzeneethanamine
2C-T-7	2,5-dimethoxy-4-(propylthio)-benzeneethanamine
2C-T-8	2,5-dimethoxy-4-(cyclopropylmethylthio)-benzeneethanamine
2C-T-9	2,5-dimethoxy-4-[(t) butylthio)]-benzeneethanamine
2C-T-13	2,5-dimethoxy-4-(2-methoxyethylthio)-benzeneethanamine
2C-T-15	2,5-dimethoxy-4-(cyclopropylthio)-benzeneethanamine
2C-T-17	2,5-dimethoxy-4-[(s) butylthio]-benzeneethanamine
2C-T-21	2,5-dimethoxy-4-(2-fluoroehtylthio)-benzeneethanamine
3C-P	1-(3,5-dimethoxy-4-propoxyphenyl)propan-2-amine
25B-NBOMe	4-bromo-2,5-dimethoxy-N-[(2-methoxyphenyl)methyl]-benzeneethanamine
25C-NBOMe	2-(4-chloro-2,5-dimethoxyphenyl)-N-(2-methoxybenzyl)ethanamine
25D-NBOMe	2-(2,5-dimethoxy-4-methylphenyl)-N-(2-methoxybenzyl)ethanamine
25E-NBOMe	2-(4-ethyl-2,5-dimethoxyphenyl)-N-(2-methoxybenzyl)ethan-1-amine
25G-NBOMe	2,5-dimethoxy-N-[(2-methoxyphenyl)methyl]-3,4-dimethyl-benzeneethanamine
25H-NBOMe	2-(2,5-dimethoxyphenyl)-N-(2-methoxybenzyl)ethanamine
25I-NBOMe	4-iodo-2,5-dimethoxy-N-[(2-methoxyphenyl)methyl]-benzeneethanamine
25I-NBF	N-(2-fluorobenzyl)-2-(4-iodo-2,5-imethoxyphenyl)ethanamine
25T2-NBOMe	2,5-dimethoxy-N-[(2-methoxyphenyl)methyl]-4-(methylthio)-benzeneethanamine

Sources: Reprinted from McGonigal, M. K. et al., 2017, *Forensic Sci Int.* 275: 83–89., 7, 17; Dean, B. V. et al., 2013, *J Med Toxicol.* 9:172–178; U.S. Drug Enforcement Administration, Diversion Control Division. 2017. 2C-Phenethylamines, Piperazines, and Tryptamines Reported in NFLIS, 2011–2015. Springfield, VA: U.S. Drug Enforcement Administration.

agitation, hypothermia, emesis, urinary incontinence, severe hypertension, vaso-constriction, and extensor posturing [9]. Her friends provided a history of alcohol, cocaine, MDMA, and 2C-I ingestion sometime between 23:00 the night before and 11:00 on the day of admission. The patient reportedly synthesized the 2C-I at home using a recipe from the Internet. The amphetamine screen was found to have no cross-reactivity with 2C-I (up to 100 µg/mL urine concentration). The screen was negative for all other drugs, including cocaine metabolite, which was inconsistent with the history provided by the patient's friends. This suggested that the patient or the friends were not completely aware of the drugs they were using. The HPLC method with the scanning UV mode used for amphetamine confirmation was nega-tive for amphetamine and methamphetamine and an interfering substance prevented the detection of MDMA and MDA. However, MDA (5.56 µg/mL) and 2C-I (0.311 µg/mL) were detected using a multitargeted LC-MS/MS method in her urine. It is known that use of (2C-I) analogs in combination with monoamine oxidase inhibitors (MAOIs) is strongly associated with serotonin syndrome or toxicity [4].

15.1.2   2C-B Analog

2C-B (4-bromo-2,5-dimethoxyphenethylamine or desmethyl-DOB) is a phenylethyl-amine derivative with psychedelic/entactogenic effects in humans and is one of the famous members of the group of so called "2Cs" [17]. 2C-B, a bromine-substituted phenethylamine, became a Schedule I material in the United States of America in 1995 [6]. Internationally, 2C-B is a Schedule II drug under the Convention on Psychotropic Substances [11]. Since 2C-B became a controlled substance, dozens of analogs have entered the drug market, leaving law enforcement struggling to pros-ecute and crime labs backlogged with unknown materials [6]. It is also known as Nexus or Afro and it is an analog of mescaline that is becoming increasingly popular as a rave and club drug [11].

Subjective effects involved perceptual modifications analogous to those observed after ayahuasca and salvia, but absent after amphetamine and MDMA [11]. Pleasure and sociability effects did not differ from those after MDMA and incapacitation was lower than for the psychedelics used as comparators. It was first synthesized in the mid-1970s and became popular as a legal substitute for MDMA after its prohibition in 1985. In some European countries 2C-B was legally sold as an aphrodisiac with the brands Nexus, Erox, and Performax in "smart shops." 2C-B was legal in most countries until the mid-1990s, when it became a controlled drug in most countries. 2C-B has been detected in the illegal drug market in several countries over the last two decades.

15.1.3   Other 2-C Analogs and the D-Series

There are many other derivatives of the 2-C series, besides 2C-B and 2C-I (Figure 15.2) [3,8,18,19]. The hallucinogenic properties of these drugs are further enhanced by a methyl group at the α-carbon (D-Series or ring-substituted amphetamines). Example drugs included in this group are 2,5-dimethoxy-4-methylamphetamine (DOM), 2,5-dimethoxy-4-bromoamphetamine (DOB), and 2,5-dimethoxy-4-io-doamphetamine (DOI) (Figure 15.2). Nausea and tachycardia but also long-lasting

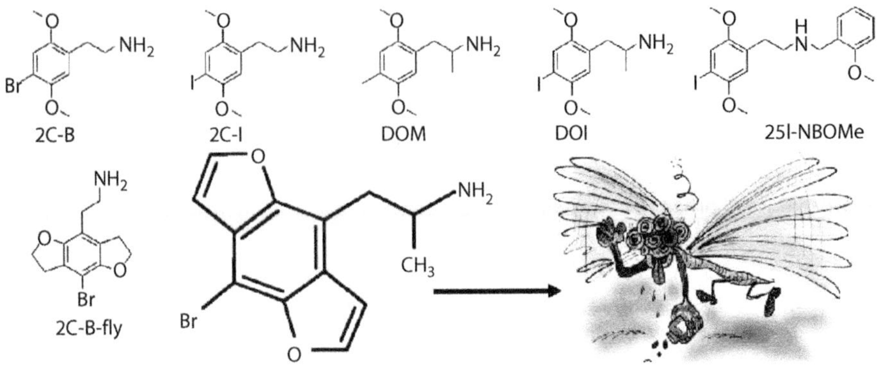

FIGURE 15.2   Structures of some phenethylamines of the 2-C series (2C-B, 2C-I, and 25I-NBOMe) and of the D-series (DOM and DOI), benzodifurans (2C-B-fly, bromo-dragonfly), 2C-B-fly, bromo-dragonfly, and similar molecules. These are investigated under the 2C group in some studies, because of the similarity with the 2C-B derivatives except for the difference of two ring formations with added ethyl groups. Pharmacologically, all presented substances are hallucinogens and potent serotonin 5-HT$_2$A receptor agonists. (Reprinted from Liechti, M. 2015. *Swiss Med Wkly.* 145: 14043; Hoffman, R. S. et al., Amphetamines, in *Goldfrank's Toxicologic Emergencies*, 10th Ed., McGraw-Hill Global Education Holdings, LLC., http://accesspharmacy.mhmedical.com/content.aspx?bookid=1163&sectionid=64554057, accessed May 2017; Graaf, Å. Vad är drogen bromodragonfly? 2017, I SAY NO DRUGS, http://www.isaynodrugs.org/vad-ar-drogen-bromodragonfly/, accessed May 2017.)

hallucinogenic effects, agitation, and vasospasms have been reported about these hallucinogenic amphetamines. The hallucinogenic activity of the 2-C and D-series drugs is mediated by an interaction with the serotonergic 5-HT$_2$A receptor. Further tinkering on the structure of 2C-x compounds produced analogs of the 2C-x-fly series, where x is the substituent at the 8-position and fly the presence of two dihydrofuran rings attached to both opposite sides of the benzene ring [20]. The compound with a bromine substituent found widespread use under the name 2C-B-fly. Aromatization of the dihydrofuran moieties of 2C-x-fly led to a class of ultrapotent analogs of mescaline, nicknamed "dragonflys" because their structures appear reminiscent of the homonymous insect (Figure 15.2). Bromo-dragonfly is prepared from the fly precursor by protection of the primary amino group, bromination, aromatization with DDQ (2,3-Dichloro-5,6-dicyano-1,4-benzoquinone), and deprotection.

Fly drugs (bromo-dragonfly, 2C-B-fly, etc.) are also hallucinogens and may cause paranoia, agitation, tachycardia, vasospasm, seizures, limb pain/ischemia, and hyperthermia and fatalities [8]. Bromo-dragonfly is a new hallucinogenic designer drug that is related to the phenylethylamines but with effects lasting up to three days [1]. It can be purchased from the Internet and is mainly sold as an acid "blotter," a powder, liquid, or tablet. The desired effects of bromo-dragonfly are similar to those of LSD, causing hallucinations, stimulation, and increase in mood. However, the psychoactive effects may differ among users, for example, some have described a feeling as being like "a ride to the moon" because it "lasts too long and leaves you drained." Published case reports have described effects like severe agitation, hallucinations, tonic–clonic seizures, renal failure, hepatic failure, profound vaso-constriction leading to multiple digital amputations, and death. Users often combine it with LSD, cannabis, 2C-B, ketamine, methylone, amphetamines, alprazolam, cocaine, and alcohol. It is not detected with commercially available immunoassay techniques, although GC-MS or LC-MS/MS may be used to confirm exposure.

15.1.4 Availability, Symptoms, Routes of Use, Pharmacology, and Pharmacokinetics

Various N-methoxybenzyl-substituted phenethylamines have emerged on the EU drug market since 2012 [3,8]. Routine screening methods do not detect 2C agents in urine or serum. Therefore, clinicians will need to evaluate the patient history, clinical signs, and symptoms for diagnosis [1].

25I-NBOMe (Figure 15.2), exhibits even higher potency on the 5-HT$_2$A receptor and possibly also on other receptors, when compared with the very potent classic hallucinogenic drugs [3,8]. Severe and fatal intoxications including agitation, hallucinations, seizures, and hyperthermia have been reported with 25I-NBOMe as a consequence of serotonergic but also sympathomimetic toxicity. These novel hallucinogens are extremely potent and psychoactive at microgram doses. This is likely to result in overdosing. Since they are a relatively newly emerged subgroup, NBOME derivatives are presented in Chapter 16, "NBOMe Derivatives of Phenylethylamines and Their Analysis." 2C-B, 2C-T-7, 2C-E, 2C-D, 2C-C, 2C-I, 2C-T2, 2C-T4, 2C-H, 2C-N, and 2C-P are in Schedule I [1]. Some 2C analogues have been linked to serious adverse reactions from delirium with agitation to violence, hyperactivity, hyperthermia; furthermore, even cardiopulmonary arrest (death) [1,3,11,21]. The formulations (capsule, powder, and liquid) depend on the specific 2C agent and how it is abused. 2C series are mainly ingested or insufflated. In Xenopus laevis oocytes, the 2C series chemicals had little or no efficacy on 5HT$_2$A receptors but were active on 5HT$_2$C receptors.

Typical doses of 12–24 mg are reported for 2C-B. The pharmacokinetic properties of the 2C-B are not well characterized but the onset of clinical effects occur after a few minutes when insufflated and after about one hour when taken orally [3]. Effects reach their peak at about two hours and last about five hours. Metabolism is hepatic via oxidative deamination and/or O-demethylation. There appears to be a wide variation in human hepatocyte susceptibility to 2C-B *in vitro*, suggesting that certain individuals may be at higher risk of toxicity than others. The dose response curve is steep and varies among individuals. A 10-mg dose is reported to be stimulant; doses of more than 10 mg tend to be psychoactive with hallucinogenic and entactogenic effects, and doses of 30 mg or more may cause such intense hallucinations or psychosis as to necessitate medical services for users.

15.1.5 Metabolism

Metabolism of 2C drugs occurs via o-demethylation with oxidative deamination to form a corresponding acid or reduction to a corresponding alcohol [10]. Deamination mainly occurs via the monoamineoxidase (MAO) enzyme, and as a result, 2C drugs may create drug interactions with MAO inhibitors. The 2C drugs tend to have a higher affinity for MAO-A than for MAO-B. Hepatic cytochrome P450 enzymes, in particular CYP2D6, also play a role in metabolism.

15.2   ANALYSIS OF 2C DERIVATIVES

15.2.1   Immunoassay Analysis

Traditionally, immunoassays are commonly used for drug screening purposes to identify presumptive positive samples [22]. The substituents on 2C drugs are constantly changed in order to avoid listing as controlled substances [6]. As these drugs are newly emerging substances of abuse, there are no accepted standard protocols for this analysis. Additionally, presumptive or screening tests have not been widely developed for these materials. For years, immunoassays were utilized for presumptive drug testing. In the immunoassay technique, cross reactivity, where a compound binds to an assay that is not the intended target compound, causes significant issues when dealing with designer drugs. The variety of substituents that change periodically can interfere with binding to the assay. Since new psychoactive substances are unique and rapidly changing, companies typically have difficulties in developing assays as fast as the speed of the changing drug market. Many of these substances are not detected and therefore labeled as "nondetect" because of the lack of specific assays. If the presumptive test is negative or inconclusive, often the case would become backlogged in the crime lab. However, the recent information on immunoassays show us that the results from immunoassays should be confirmed with additional analysis, especially for identification of new compounds or analogs. Using preexisting assays to try to determine patterns based on class of compounds is not robust and fails for the 2C drugs. Immunoassays do not illicit a response in the CEDIA DAU amphetamine/ecstasy assay for the NBOMe class. Larger substituents on the 4-position of the aromatic ring and the N-2-methoxybenzyl substituent on the amine are believed to interfere with binding to the assay. False negative possibility is more risky than false positive results, because a false negative result may cause the analyst to not confirm the identification, thus leading to a "nondetect."

15.2.2   GC-MS

The most used techniques in the analysis of 2C drugs in the literature are GC-MS and LC-MS/MS. Up to recent years, GC-MS was the gold standard; however, recently, especially high-resolution LC-MS/MS techniques take the place of GC-MS or are required to confirm the results of GC-MS because of the newly emerging derivatives. The application of GC-MS is limited with the volatiles, semi-volatiles, and the ones that can be derivatized; there may be false negatives, but for LC-MS/MS, this limitation does not exist. However, GC-MS is still an important component of the identification and structure elucidation studies for unknown compounds, their metabolites, and determination studies for isobaric compounds, especially because of the long bore columns where even the separation of some enantiomers can be succeeded.

In one GC-MS study in the literature, a screening and simultaneous quantification method was applied to some members of the 2C series such as 2C-D, 2C-E, 2C-P, 2C-B, 2C-I, 2C-T-2, and 2C-T-7, and their analog mescaline in human blood plasma [23]. Some fatal intoxications involving 2C-T-7 have been reported. Only scarce data have been published about analyses of these substances in human blood and/or plasma. The analytes were analyzed by GC-MS in the selected-ion monitoring mode, after mixed-mode solid-phase extraction (HCX) and derivatization with heptafluorobutyric anhydride. Validation data were acceptable except for 2C-T-2 and 2C-T-7. LOQ was 5 ng/mL and the ions can be detected at 1 ng/mL.

15.2.3 Capillary Electrophoresis (CE)

Analytical chemists have tried various techniques for the determination of designer drugs, while new designer drugs continuously emerge. In the last 15 years, application of capillary electrophoresis in the analysis of the 2C-series of phenethylamine designer drugs started to be encountered in the literature [24,25]. CE-MS, like LC-MS/MS, does not require sample derivatization and has been successfully applied to a wide variety of small molecules in biological matrices, including amphetamine-like compounds. An example for CE-ESI-MS method was developed in 2007, for the detection and quantification of 2C-T, 2C-T-2, 2CT-5, and 2C-T-7 in human plasma [24]. Prior to CE-ESI-MS analysis, a simple liquid extraction was used for sample cleanup. The method was validated. Using an aqueous pH 2.5 phosphate buffer, the CE-MS analysis provided data for unambiguous confirmation of these drugs in human plasma. Separations in capillary electrophoresis were performed using uncoated fused-silica capillary (120 cm × 50 μm i.d.). A separation voltage of 25 kV was applied. Samples were injected hydrodynamically with a pressure of 50 mbar for 10 seconds. The following mass spectrometry conditions were used: A capillary temperature of 200°C, a source-voltage of −3.5 kV, positive-ion mode, sheath gas of 20 arbitrary units nitrogen. Sheath liquid consisted of methanol–water–phosphate buffer 100 mM pH 2.5 (50:49.5:0.5). Sample preparation was performed through addition of 20 μL of diluted phosphoric acid (1/5) to 1 mL plasma in a test tube, ultrasonicating for 30 minutes and then extraction via the addition of 2 mL acetonitrile. The extract was evaporated to dryness and the residue was reconstituted in 1 mL separation buffer. LOD and LOQ values ranged from 11.3 to 23.0 ng/mL and from 27.3 to 43.0 ng/mL, respectively. Recoveries were between 76–85%. The only concern in this study is the use of a phosphate buffer in CE and phosphoric acid use in the sample preparation step, because of the risk of clogging the MS inlet capillary in routine analyses. In the future studies, it may be better to exchange the phosphate buffer and phosphoric acid with formate and formic acid, which have close pK_a constants, if this method is aimed to be used in routine analysis.

Tsai et al. separated and detected some of the 2C-series of phenethylamine designer drugs, including 2C-T-2, 2C-T-7, 2C-C, 2C-B, 2C-I, in mixture, by means of CE separation with native fluorescence and light-emitting diode (LED)-induced fluorescence (LIF) detection, respectively, for comparison [25]. Then the method for separating these drugs in a human urine sample by means of micellar elektrokinetic capillary chromatography (MEKC)–LIF was optimized. LED-induced fluorescence detection was used after derivatizing with fluorescein isothiocyanate isomer I. The derivative was directly subjected to CE separation. A blue LED (~2 mW) was used as the fluorescence excitation source. The detection limits were improved to

$\sim 10^{-7}$ and $\sim 10^{-8}$ M, respectively, when the MEKC and stacking-MEKC modes were applied. A urine sample spiked with the five standards, and, after alkaline extraction of 1 mL urine with ethyl acetate, was examined by means of the MEKC–LIF mode. The fluorescence excitation and fluorescence emission wavelengths were 300 and 340 nm. Then the urine sample extract was examined by means of GC/MS; a capillary column (30 m × 0.25 μm i.d.) with an HP-5MS phase and 0.25 μm in thickness was used in GC-MS analysis. Ion source temperature was 230°C. GC-MS results were in agreement with the results of CE method. The methods used were based on native fluorescence and LED-induced fluorescence detection, respectively. The former method is simple but results in a poor LOD; whereas the latter provides a much better LOD. When the MEKC mode in conjunction with fluorescence derivatization was applied to a urine extract, the above five 2C-series drugs could be determined successfully. This method has been presented by the authors as a sensitive, accurate, simple, and economic complementary method to GC/MS for use in forensic and clinical analysis.

15.2.4   LC-MS AND LC-MS/MS

Undoubtedly, MS techniques (especially HRMS/high-resolution MS) combined to liquid chromatography are becoming the preferred and required techniques in the detection, identification, and quantification of NPSs, metabolism studies, and discrimination of isomers or close analogs with small modifications. If a powerful method is developed and the suitable HRMS technique is used, LC-MS/MS may be adequate in some cases, both for screening and confirmation together. However, the persistently changing profile of newly introduced NPS analogs causes a challenge and the need of support from other techniques, such as GC-MS and/or NMR.

Pichini et al. [26] used LC-APCI-MS for the determination method for 2C-D, 2C-B, 2C-B-Fly, 2C-T-2, 2C-I, and 2C-E in urine, and validated. Sample preparation involved a solid-phase extraction (SPE) procedure at pH 6.0 of both nonhydrolyzed and enzymatically hydrolyzed urine samples. Analyses were performed with a C_{18} column using a linear gradient of 10 mM ammonium bicarbonate, pH 7.3, and acetonitrile as a mobile phase. Separated analytes were determined in LC-MS single-ion monitoring mode using an atmospheric pressure ionization–electrospray ionization (ESI) interface. Drying gas (nitrogen) (350°C) at a flow rate of 12.0 L/min; nebulizer gas (nitrogen) at a pressure of 50 psi; capillary voltage at 4000 V were used. Three ions for each compound were used for identification. LOQ values were varied between 20 and 60 ng/mL for the different analytes under investigation. Mean recoveries ranged between 55.4 and 95.6% for the different analytes. Higher analyte concentrations in hydrolyzed samples showed the presence of conjugated compounds in urine. From the results obtained, it can be said that phenethylamine compounds (2C-B, 2C-E, etc.) can be found in urine mainly as sulfate and glucuronide conjugates. Interestingly, in the case of coadministration of MDMA and 2C-B, the portion of this latter drug measured as conjugate (from 0 to 31%) appears less than that measured when 2C-B is consumed alone (from 46 to 92%).

Recent studies on the detection of 2C derivatives have been performed generally using EI-GC-MS or/and LC-MS/MS. In a study, 2C-B was identified in the samples from the illicit market of recreational drugs in Spain, between January 2006 and December 2009, through a combination of validated analysis techniques [11]. To detect 2C-B and check for potentially toxic adulterants, the authors used TLC and GC/MS. With the TLC technique, analytes were identified by comparing retention factor and color in the Marquis test with those of a reference standard, and confirmation was performed using GC-MS in 14.5 minutes. Drug material was analyzed for 2C-B and information on pattern of use and subjective effects was obtained from recreational users. To confirm the TLC results, samples were analyzed using the GC/MS technique with a quadrupole mass spectrometer detector. Samples were injected in split mode into a 12 m × 0.2 mm i.d., 0.33-µm film thickness, 5% phenylmethyl-silicone column. The injector and the interface were operated at 280°C. Helium was used as carrier gas (flow rate: 0.48 mL/min). Qualifying ions selected for analytes under investigation were: m/z 215, 230, and 259. To confirm the mass spectra, a reference library was used. UV spectrophotometry also was used to determine the purity of the samples. After verifying the existence of 2-CB derivatives in the samples of the participants who were recruited from among 2C-B users, the researchers confirmed their results through evaluation of their pattern of use of 2C-B and the subjective effects of the drug. During the routine analyses on real samples, 97 of 3303 samples coming to the laboratory between January 2006 and December 2009 contained 2C-B. 2C-B was detected in 96 samples (99%). The other samples contained 2C-I rather than 2C-B. Of the 96 confirmed 2C-B samples, 52 (54%) were in tablet form and the remaining 44 (46%) were in powder form or encapsulated.

15.2.5 DISCRIMINATION AND IDENTIFICATION OF ISOBARIC COMPOUNDS

GC separation on nonpolar stationary phases can successfully resolve the regioisomeric and isobaric compounds. However, with accurate-mass LC-MS/MS detection, a retention time parameter and/or more MRM transitions should be added for absolute identification for isobaric and regioisomeric compounds. Only LC-HRMS gives an accurate mass, which can be used for predicting formulas and structural elucidation [27]. The ion trap (IT) presents the chance of observation of subsequent MS^n fragmentation and TOF provides high resolution and mass accuracy for the unknown compounds and impurity profiling. With the HRMS technologies the mass-resolving and mass-determining capabilities of MS technology is enhanced and this gives the chance of fast analysis in combination with chromatographic separation techniques UPLC or HPLC with monolithic column. Chiral columns, which are used effectively in enantioseparation in chemistry, will be realized to help significantly in the future analysis of the isomers linked to 2C derivatives.

Ambach et al. reported a validated method in dried blood spots (DBSs) using QTRAP LC-MS/MS for 64 new psychoactive substances including amphetamine derivatives; the 2C family: 2C-B; 2C-D; 2C-E; 2C-H; 2C-I; 2C-P; 2C-T-2; 2C-T-4; 2C-T-7; aminoindanes, tryptamines; desoxypipradol; ephedrine; pseudoephedrine; ketamine; norephedrine; norpseudoephedrine; and PCP [28]. The method was used successfully also in the identification of isobaric molecules. For the method development and validation, 10 venous blood samples were used for the application on the blood spot cards and dried for at least three hours at room temperature.

A 1-cm-diameter DBS was punched out and collected in an Eppendorf tube. The extraction was performed by adding 500 µL of methanol and 10 µL of the IS solution (10 ng/mL) followed by 15 minutes vortexing. The methanolic solution was acidified and dried and reconstituted with 100 µL of water/formic acid (99.9/0.1; v/v) solution and analyzed. A gradient elution was performed using water and acetonitrile both containing 0.1% formic acid through a Synergi Polar-RP (100 × 2.0 mm, 2.5 µm) at 50°C. LC-ESI(+)-MS/MS system was used at an ion spray voltage of 5 kV, along with the following: curtain gas 30, collision gas 6, gas 1 to 40 and gas 2 to 60 psi. The ion source temperature was set to 700°C. For the analytes with isobaric precursor m/z ratio and fragments m/z ratio, three transitions were monitored. The LODs ranged from 1 to 10 ng/mL and LOQs from 2.5 to 10 ng/mL and the extraction efficiency was higher than 60%, for 2C derivatives. All isobaric compounds could be separated except flephedrone and 3-FMC, and MDDMA and MDEA. Similar to this study, there are also other methods giving good results for some isobaric compounds; however, if chiral separation is considered in the future studies, more successful and specific results could be obtained with the LC-MS/MS technique.

15.2.6 Impurity Analysis

Impurity analysis is also important in forensic sciences, to define the source of the manufacturer. If there are the same impurities in the samples from different cases, a relationship could be figured out among them, and the source of the drug manufactured may be found [27]. In recent years, there have been several publications on the identification of impurities in illegal drugs using the LC-MS/MS technique. Again in impurity analysis, the success of LC-IT-TOF-MS has been proved in the literature [27]. It would also be a very good way to use LC-MS/MS with GC-MS or GC-MS/MS as a double control to confirm the results, especially in routine forensic analysis. In the case of impurity profiling, structure elucidation of unknown new molecules or metabolites, other techniques as purification with preparative chromatography, NMR, FTIR, etc. should also be applied to support the studies.

A very good example of impurity profiling was published in 2016 [27]. A simple and effective LC/MS-IT-TOF method was established to identify the impurities of the drug 2C-E and to deduce its synthetic route. The fragment pathway of 2C-E was identified. The structures and fragment pathways of the impurities were investigated. The details are mentioned as follows, in order to shine a light on the methods of researchers who plan to study impurity profiling: for preparative HPLC, 2.2 g of the illicit 2C-E sample was extracted twice with methanol using an ultrasonic bath for 10 minutes, and filtered. Then the extracts were combined and concentrated to 5 mg/mL and pH was adjusted to 3.0. The sample was filtered and analyzed in a preparative HPLC instrument (column: 250 mm × 20.0 mm, I.D., 5 µm). Mobile phase: methanol–water (30:70, v/v), flow rate: 8 mL/min. For the ESI-LC/MS-IT-TOF detection, the illicit 2C-E sample was dissolved in HPLC-grade methanol to give a concentration of 20 mg/mL and filtered before analysis. A C_{18} column (150 mm × 4.6 mm, I.D.,

5 mm) at 40°C and mobile phases of 0.5% formic acid in water and methanol was used for gradient elution (flow rate: 0.5 mL/min). Positive and negative electrospray voltages of 4.5 and 3.5 kV were used, respectively. The detector voltage, pressures of the nebulizer gas and the drying gas were 1.65 kV, 1.5 L/min and 128 kPa, in order. The pressures of TOF and IT regions were set at 1.7×10^4 and 1.8×10^2 Pa, respectively. The ion accumulation time was set at 60 ms. The CID collision energy was set at 60% for MS^n. In positive-ion mode, MS^n data of the 2C-E standard were acquired. M + H$^+$ ion of 2C-E was observed at m/z 210.1432. The product ion observed at m/z 193.1182 was formed by the loss of the NH_3 moiety from the precursor ion. From the MS^3 and MS^4, the results indicated losses of 15 Da, which represented sequential losses of CH_3 (Figure 15.3). In TIC, ten impurities were detected in the 2C-E tablets. Based on the analysis of impurities, a synthetic route was proposed. In order to confirm the structures of impurities, some of them were separated and prepared by preparative HPLC and then identified by MS and NMR. Ten impurities were identified according to their MS^n spectra and elemental composition analysis. Besides, the synthetic route of 2C-E tablets was surmised through analysis of impurities.

Zuba et al. confirmed 2,5-dimethoxy-4-nitro-b-phenethylamine (2C-N) in a seized powder sample and gave a synthetic route according to the analyzed impurities [15]. Identification of 2C-N was carried out by means of GC-MS, LC-QTOF/MS, FTIR, and NMR. In the GC-MS analysis, the powder (10 mg) was dissolved in 10 mL methanol and processed with and without trifluoroacetic anhydride (TFAA) derivatization. Chromatography was carried out using a capillary column (30 m, 0.25 mm i.d.) coated with 0.25 mm film of (5%-phenyl)-methylpolysiloxane (HP-5MS). ESI(+) LC–QTOF/MS: The stock solution was diluted in 0.1% (v/v) formic acid in water and introduced to ESI(+)LC–QTOF/MS. C_{18} (7.5 cm × 2.1 mm × 2.7 mm) column (35°C) was used. A gradient was applied using a mixture of 0.1% (v/v) formic acid in water (a) and in acetonitrile (b) as the mobile phase constituents (flow rate: 0.3 mL/min). FTIR: The solid sample was directly measured by the transmission technique. NMR: The solid sample was dissolved in deuterated dimethyl sulfoxide (DMSO-d_6), placed in the NMR tube, and ^1H and ^{13}C NMR spectra were recorded. The molecular mass of 2C-N (226.0954 amu) was confirmed in the LC/ESI-QTOFMS and GC-EI/MS spectrum. A characteristic set of ions for the parent substance was found using both chromatographic methods. Based on the obtained ion masses and additional fragmentation of ions at $m/z = 195.0527$ and $m/z = 151.0755$ (fragmentor voltage = 200 V, CE = 15 eV), they proposed the structures of the product ions of 2C-N under the ESI conditions. Two broad dominant bands at 1520 cm^{-1} and 1342/1322 cm^{-1}, observed in the FTIR spectrum of 2C-N, originated from the nitro group. The vibrations of other bands occurring in the 2C-N molecule were also identified in this spectrum. NMR spectroscopy helped unequivocal elucidation of the structure. The applied identification procedure proved to be a powerful tool to determine the structure of a new designer drug. Comparison of impurity profiles of the evidence material and the standard of 2C-N prepared in-house revealed that the investigated sample was synthesized by nitration of the 2C-H, according to the protocol proposed by Shulgin.

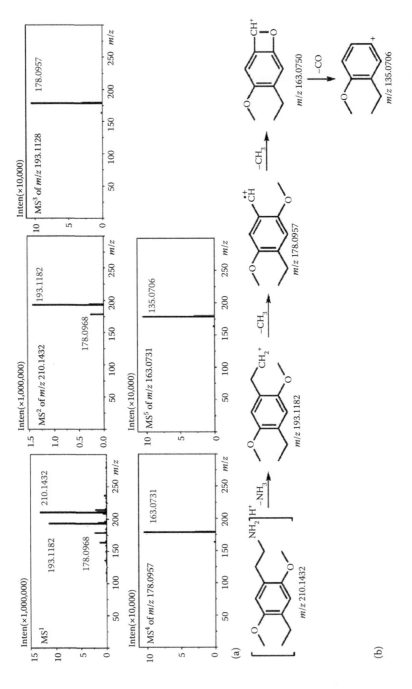

FIGURE 15.3 (a) MS^n of 2C-E and (b) the proposed fragmentation pathway of 2C-E. (Reprinted from Tsai, C. C. et al., 2006, *J Chromatogr A*. 1101:319–323.)

15.2.7 Collaboration of Clinical and Analytical Methods in Diagnosis of Intoxication

Due to the widespread use of the Internet, information regarding synthesis of and access to novel compounds is more accessible than ever [29]. This brings new challenges to the medical community in terms of treatment as well as identification of the abused substance, especially in patients unable to communicate. For diagnosis, an analytical method for detection and quantification is required for clinical chemistry and forensic toxicology, as well as the observation of the clinical symptoms. Vrancken et al. shared their results on a case: A 26-year-old white man with known polysubstance abuse and psychiatric problems was found unresponsive at a friend's house. According to the patient's friend, he had ingested 2C-E, but was not aware of any other drug or alcohol use. The patient had prescriptions for various psychiatric medications including sertraline, clonazepam, gabapentin, and zolpidem. Urine drug screens were positive for the compound known as 2C-E. 2C-E was confirmed with LC-MS/MS. In single MS, 2C-E has a molecular weight of approximately 209 and a charge of 1^+ following ionization. The mass spectrometer only allows ions with a specific m/z ratio to pass through and detect. However, other compounds may have very similar m/z ratios and detract from the specificity of a mass spectrometer (For example, morphine and hydromorphine have an identical m/z ratio of 286, thus making separation of the two compounds difficult by nontandem mass spectrometry.) Running mass spectrometry in tandem can help alleviate this issue.

15.2.8 Wide-Scope Screening and Confirmation

In 2015, a fast and simple ESI(+)LC-MS/MS screening procedure was developed by Adamowicz et al. for 143 new psychoactive substances from different groups, including 2C-B, 2C-C, 2C-D, 2C-E, 2C-H, 2C-I, 2C-N, 2C-T-2, and 2C-T-7, in blood [30]. The total preparation of one sample took about 30 minutes, followed by 14 minutes in LC-MS analysis. 20 µL of 1 µg/mL methanolic IS solutions was added to the 0.2 mL blood samples. 0.6 mL of acetonitrile was added in 50 µL portions for precipitation, and vortexed for 10 seconds after each addition. The samples were mixed for five minutes, and centrifuged for five minutes at 13,000 rpm. The isolated organic phase was evaporated to dryness under nitrogen at 37°C. The dry residues were dissolved in 100 µL of 0.1% formic acid in water (v/v) and analyzed. The gradient chromatographic elution was achieved with C_{18} (2.1 × 50 mm, 1.8 µm) column at 25°C and 0.1% formic acid in acetonitrile and 0.1% formic acid in water. Capillary voltage was 3.5 kV, gas flow (nitrogen) 10 L/min and gas temperature 325°C; sheath gas flow 10 L/min; sheath gas temperature 350°C; and nebulizer pressure was 40 psi. Three MRM transitions were monitored for all but three compounds. The total number of transitions monitored in dynamic mode was 432. LODs for 104 compounds were in the range 0.01–3.09 ng/mL. Although the recoveries determined for 32 compounds were between 1.8 and 133%, some of which were outside the analytical range, the method can be used for rapid screening of new drugs of abuse in forensic or as long as the analytical signals are sufficiently high, and for quantification of the ones with high recoveries. The procedure was successfully applied to the analysis of forensic blood samples in the routine casework; during the three-year period 2012–2014 (over 1000 blood samples were analyzed, 112 were positive for NPS), 2C-B and 2C-P were among the most detected compounds.

In 2017, Pasin et al. [31] analyzed 12 2C–X, 6 DOX, and 14 25X–NBOMe analogs by UPLC-QTOF-MS in order to evaluate their collision-induced dissociation (CID) pathways and assess the applicability of the generated product ions for nontargeted detection of novel hallucinogenic phenethylamine analogues. The panel of selected analytes also contained isotopically labelled 2C–B-d_6 and 25I–NBOMe-d_9, which were used to aid in the elucidation of product ions. CID experiments were performed using collision energies set at 10, 20, and 40 eV. Novel analogs of the selected classes could be detected by applying neutral loss filters (NLFs) and extracting the common product ions. Chromatographic separation was achieved on the Poroshell 120 C_{18} column (2.1 × 75 mm, 2.7 µm particle size) using a gradient elution with a flow rate of 0.4 mL/min with a total run time of 17 minutes. Mobile phases consisted of 20 mM ammonium formate aqueous solution and acetonitrile containing 0.1% (% v/v) formic acid. The ESI(+)-QTOF/MS was operated using Extended Dynamic Range (2 GHz) with capillary and fragmentor voltages set to 3500 V and 180 V, respectively. Hallucinogenic phenethylamines were successfully characterized by using CID at different CEs. The 2C–X and DOX derivatives had common losses of NH_3, CH_6N, and C_2H_9N and common product ions at m/z 164.0837, 149.0603, and 134.0732 for 2C–X derivatives and m/z 178.0994, 163.0754, 147.0804, and 135.0810 for DOX derivatives. The 25X–NBOMe derivatives had characteristic product ion spectra with abundant ions at m/z 121.0654 and 91.0548, together with minor neutral losses corresponding to 2-methylanisole and 2-methoxybenzylamine and •$C_9H_{14}NO$. Screening for these common neutral losses and product ions can be used in a nontargeted screening approach to detect and tentatively identify novel analogs.

In 2014, a wide scope of the method development of an LC-HRMS-QTOF analysis of new stimulant designer drugs (e.g., phenethylamine, amphetamine, cathinone, and piperazine derivatives), including 2-CB, 2C-I, 2C-T-2, 2C-T-4, 2C-T-7, 2-C-H, and common drugs of abuse (e.g., ketamine and ritalinic acid) in urine is reported [32]. Sample preparation was carried out by a fast and convenient salting-out liquid–liquid extraction (SALLE) procedure. 20 µL internal standard (concentration = 0.5 mg/L) and 250 µL of acetonitrile was added to 200 µL urine and vortexed. 50 µL 10 M ammonium acetate, followed by 50 µL 10 M KOH solution, was added successively, and vortexed and centrifuged for three minutes at 10000 (rpm). 100 µL organic phase was transferred to a LC vial. 100 µL of mobile phase A was added, vortexed, and analyzed. Target analytes were separated on a C_{18} (100 mm × 2.1 mm I.D., 1.8 µm) column at 40°C, using a mobile phase A consisting of a 5 mM ammonium formate buffer (pH 4.0) and mobile phase B consisting of 0.1% formic acid in acetonitrile (flow rate: 0.4 mL/min). The Q-TOF instrument with dual-spray ESI source was used. A quadrupole served for isolation of precursor ions (mass window 1.3 m/z) in data-dependent acquisition mode (auto-MS), and a linear hexapole collision cell (nitrogen as collision gas) was used for precursor fragmentation. Retention time data was also used. The application of the additional untargeted data dependent acquisition mode enabled the identification of metabolites of the preferred target list compounds ketamine and methylenedioxypyrovalerone (MDPV) without use of reference standards.

The approach presented here provides a very useful tool for the combined targeted and untargeted analysis of drugs of abuse in biological matrices such as urine. Untargeted data-dependent acquisition provided additional sample information (i.e., metabolites of target compounds that are not listed or drugs not stored in the database). The identification is realized by a fully automated data extraction algorithm, taking into account the accurate mass spectra, fragment masses, and retention times. Acceptable quantitative results were obtained for 35 of the 39 analytes. LOD and LOQ values for the 2C derivatives analyzed were in the ranges of 1–19 and 1–6 ng/mL, respectively. Since the recoveries were between 41.6–44.8% for 2C analogues (except for 2-C-H, which was 60.8%), the method can be used only for screening in case of 2C drugs.

An analytical method with high recoveries (71–100%) was developed and validated by Pasin et al. for the purpose of detecting and quantifying 37 new designer drugs, including cathinones, hallucinogenic phenethylamines (e.g., 2C-B, 2C-E, 2C-H, 2C-I), and piperazines [33]. The aim of their study was to develop and validate a method capable of detecting and quantifying 37 designer drugs in whole blood using LC-QTOF-MS. Using only 100 mL whole blood, a salting-out-assisted liquid-liquid extraction with acetonitrile was performed to isolate target compounds followed by chromatographic separation using a UPLC-QTOF/MS. I.S. was mephedrone-d_3. A C_{18} column (150 mm × 2.1 mm I.D., × 1.8 μm) at 50°C was used in a gradient elution mode (flow rate = 0.4 mL/min) with a total run time of 15 minutes. Mobile phase A consisted of 5 mM ammonium formate (pH 3.0) and mobile phase B consisted of acetonitrile containing 0.1% (v/v) formic acid. All compounds were identified using the ESI(+) mode with MSE (Mass spectrometry with ramping of collision energies in the same run) acquisition, based on accurate mass of the molecular ion, two product ions and retention time (Table 15.2). All analyte calibration curves were linear over the range of 0.05–2 mg/L and LOD and LOQ were between 0.007–0.07 mg/L and within 0.05–0.1 mg/L, respectively. Recoveries ranged from 71 to 100% and the matrix effects were assessed. The method was applied to several forensic

TABLE 15.2
Molecular Ions and 1st and 2nd Product Ions, and the Retention Times Used in the Discrimination of Each of the Analyzed 2C and NBOMe Derivatives

Drug Class	Drug	Retention Time (min)	Monoisotopic Mass [M+H]⁺	1st Product Ion (m/z)	2nd Product Ion (m/z)
2C	Mephedrone-d_3	2.94	181.1430 (IS)		
	2C-B	4.41	260.0286	227.9907	243.0270
	2C-E	5.35	210.1494	178.1042	163.0849
	2C-H	2.58	182.1181	150.0752	135.0489
	2C_I	5.04	308.0147	290.9965	275.9697
NBOMe	25B-NBOMe	8.21	380.0861	121.0688	91.0644
	25C-NBOMe	7.92	336.1366	121.0712	91.0967
	25H-NBOMe	6.71	302.1756	121.0706	91.0847
	25I-NBOMe	8.74	428.0722	121.0662	91.0648

Source: Reprinted from Pasin, D. et al., 2017. *Drug Test Anal* 9:1620–1629.

cases where the subject exhibited behavior characteristics of designer drug intoxication and where routine screening for a panel of drugs was negative and some drugs are detected in two cases, in one of which, 25C-NBOMe and 25I-NBOMe were detected in blood and urine.

Despite the increasing number of designer drugs available, there are few comprehensive screening techniques for their detection and quantification in biological specimens [34]. Swortwood et al. have used LC-QQQ-MS/MS (triple-quadrupole MS), and their method encompassed over 30 important compounds within the phenethylamine, tryptamine, and piperazine designer drug classes, including 2C-B, in their method that was developed in serum. 20 µL I.S. was added to 1 mL serum samples, which were diluted with 2 mL of sodium phosphate buffer (100 mM, pH 6.0). The samples were vortexed and extracted with SPE and eluted with basic solution mix where dichloromethane, IPA, and ammonium hydroxide (80:20:2 v/v/v) was used. Eluates were acidified with 100 µL of HCl–IPA (1:3, v/v) and reconstituted in 50 µL of mobile phase after evaporation. Analytes were determined by LC-QQQ-MS/MS using the MRM mode. Separation was performed using a C_{18} column (50 × 2.1 mm, 1.8 µm particle size) at 40°C, with gradient elution at a flow rate of 0.5 mL/min using 2 mM ammonium formate/0.1% formic acid in water as mobile phase A and acetonitrile/water (90:10, v/v) with 0.1% formic acid as mobile phase B. In MS/MS analysis; gas temperature was 320°C, gas flow was 8 L/min, nebulizer was 27 psi, sheath gas heater was 380°C, sheath gas flow was 12 L/min, capillary voltage was 3,750 V and charging voltage was 500 V. The LOQ values were in the range of 1–10 ng/mL and the LOD values were close to 10 pg/mL. The validated method was used to analyze postmortem specimens from two cases and it was able to identify and quantify seven of these compounds at concentrations as low as 11 ng/mL. Figure 15.4 is shown as the LC-MS chromatogram (intensity vs. retention time) of the primary MRM transitions for 32 targeted analytes.

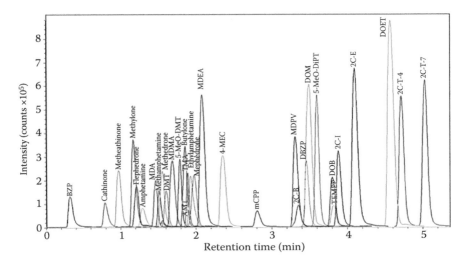

FIGURE 15.4 LC-MS chromatogram (intensity vs. retention time) of primary MRM transitions for 32 targeted analytes. (Reprinted from Paul, M. et al., 2014, *Anal Bioanal. Chem.* 406: 4425–4441.)

15.2.9 COMPUTATIONAL CHEMISTRY PROVIDING DATA FOR ANALYTICAL USE

The computational methods also started to gain importance in the prediction of physicochemical properties to make predictions on adsorption, distribution, and elimination of these drugs, where no pharmacological data is available. In 2016, synthetic phenethylamine drug analogs were synthesized using a shotgun method and rapidly characterized via ESI-MS for structural confirmation, determination of partition coefficients (LogP) by a chloroform/water system, and evaluation of blood–brain barrier permeability by parallel artificial membrane permeability assay (PAMPA-BBB) to determine the apparent permeability (Papp) [35]. Besides in reducing the experimental number in validation studies, computational methods are also used in calculating theoretical Log P values and molecular lipophilic potentials, to supplement experimental data. These data were combined with Lipinski's Rule of Five in a searchable ACD/Labs database for quick reference by forensic practitioners. To catch new trends in illicit drug manufacture, researchers have started to build large libraries of compounds that can later be compared to the data taken from samples in the field. Faster methods to equip these databases with relevant pharmacological information would enhance the ability of forensic researchers to keep up with rapidly changing new psychoactive drug trends. Partition coefficients based on classical experiments with n-octanol/water systems have been the standard measure of lipophilicity for decades, but these systems cannot be used for ESI-MS and have a tendency to form emulsions at the interface. As an alternative to Log P_{OW}, a lot of data have been collected on the partition coefficients of many compounds in a chloroform/water (Log P_{CW}) system. Although these solvents do not possess the difference in polarity that octanol and water do, their relative immiscibility and compatibility with ESI-MS make them ideal for rapid lipophilicity measurements. By utilizing this chloroform/water system with ESI-MS for Log P determination, time-consuming experimental procedures of a typical n-octanol/water system can be avoided. Ammonium bicarbonate buffer was chosen as the aqueous phase due to its large physiological pH range (pH 6.6–8.6) and compatibility with ESI. Two hours of extraction between 1 mL each of presaturated chloroform and 5 mM of ammonium bicarbonate buffer (pH 7.4) layers was performed before analysis in ESI-MS. Only those analogs that were detected in both phases were used for calculating partition coefficients.

While the analysis and clinical investigations continue, computational chemists also try to illustrate the 3D conformational structures of these new molecules using a series of programs. These more precise structures are important to serve as data to chemists. In a 2016 paper, Pratama et al. formulated more precise 3D Geometric Moment Invariants to represent the amphetamine-type stimulant drugs' molecular structure [36]. 2C-I was chosen as the model in Figure 15.5.

15.2.10 METABOLISM STUDIES/METABOLITE ANALYSIS

Metabolism studies have also begun to increase in the literature for 2C derivatives, as well as other new psychoactive drugs [37]. *In vitro* assays using microsomes, cells, and tissues, etc. can be conducted to obtain a primary idea on the metabolite formation. *In vivo* assays provide a more comprehensive evaluation of the overall

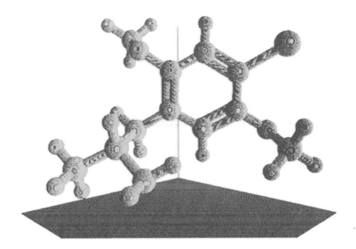

FIGURE 15.5   Voxelized molecular structure of 2C-I. (Reprinted from Swortwood, M. J. et al., 2013, *Anal Bioanal Chem.* 405:1383–1397.)

drug metabolism. Besides the experimental approaches, the *in silico* methods based on prediction rules are increasingly used for a rough approximation of drug metabolism. Once metabolites have been generated, they need to be identified and quantified using the possibilities of analytical techniques available in laboratories. Often, the analytes are purified and preconcentrated via sample preparation techniques based on protein removal, liquid and solid-phase (micro) extraction, etc. The resulting extracts should then better be analyzed using powerful techniques such as LC-MS/MS and NMR for performing separation, structural elucidation, and quantification of the components of interest. Structural elucidation of metabolites is essential to quickly identify and establish the metabolite profiles and the rate of drug biotransformation. Electrochemical methods may support metabolism studies through providing a straightforward and cheap picture of oxidative routes. NMR detection is especially successful for resolving the structures of metabolites and the isobaric molecules because of its ability to deduce the exact positions of functional groups in the molecules, which is, indeed, the recurrent problem of MS. In this regard, LC-NMR with ^{1}H and ^{13}C analysis may be used as a complementary technique to LC-MS.

The capacity of LC-ESI-MS^n and LC-NMR can be combined for the structural elucidation of metabolites. Though there are some drawbacks to be solved concerning sensitivity, rapid spectral acquisition throughout the chromatogram, and use of deuterated mobile phases, the time-consuming sample preparation step can be simplified greatly as the separation process removes most of the matrix interferences. Additionally, the online hyphenation of LC, SPE, and NMR has been also demonstrated to be a suitable alternative for the structural elucidation process [38]. The inclusion of an online SPE step is a good strategy in order to increase the amount and purity of a product entering the NMR cell, leading to an improved sensitivity.

As previous studies have shown, the 2Cs are mainly metabolized by O-demethylation, N-acetylation, or deamination. Meyer et al. [39] investigated the role of the recombinant human N-acetyltransferase (NAT) isoforms 1 and 2 in the phase II metabolism of 2Cs. For these studies, cDNA-expressed recombinant human NATs were used and formation of metabolites after incubation was measured using GC-MS. Concerning the isoenzymes of the P450-type, only CYP2D6 displayed an effect on the metabolism of 2C-D, 2C-E, 2C-T-2, and 2C-T-7, but with a low formation rate. However, the role of the N-acetyltransferase (NAT) isoforms in the metabolism of the 2Cs is still unclear. The N-acetylated derivatives of all 2Cs were synthesized in-house according to a previously published procedure with some modifications: Supernatant of incubation mixture including the synthesized acetylated products were diluted with 100 µL acetonitrile for GC-MS or 200 µL acetonitrile for LC-MS/MS. Identity and purity analyses were performed. Metabolites were formed through a 30 min incubation procedure of the substrates 2C-B, 2C-I, 2C-H,2C-D, 2C-E, 2C-P, 2C-T-2, and 2C-T-7 with recombinant human NAT1 or NAT2. Supernatant phase was analyzed in EI-GC-MS, then in UHPLC-APCI(+)-MS/MS. The following ions were used as quantifiers for product quantification in LC-MS/MS: m/z 242 (2C-B), m/z 290 (2C-I), m/z 164 (2C-H), m/z 178 (2C-D), m/z 192 (2C-E), m/z 206 (2C-P), m/z 224 (2C-T-2), and m/z 238 (2C-T-7). The results showed that human NAT2 was responsible for the N-acetylation of the "2C" family. Figure 15.6 shows the acetylation of 2C drugs as well as the molecular structure of some 2C derivatives. The enzyme kinetic studies showed that all metabolite formation reactions followed the classical Michaelis–Menten kinetics and the affinity to human NAT2 was increasing with the volume of the 4-substituent. In consequence, a slow acetylator phenotype or inhibition of NAT2 could lead to decreased N-acetylation and might lead to an increased risk of the side effects of 2Cs. Individual differences

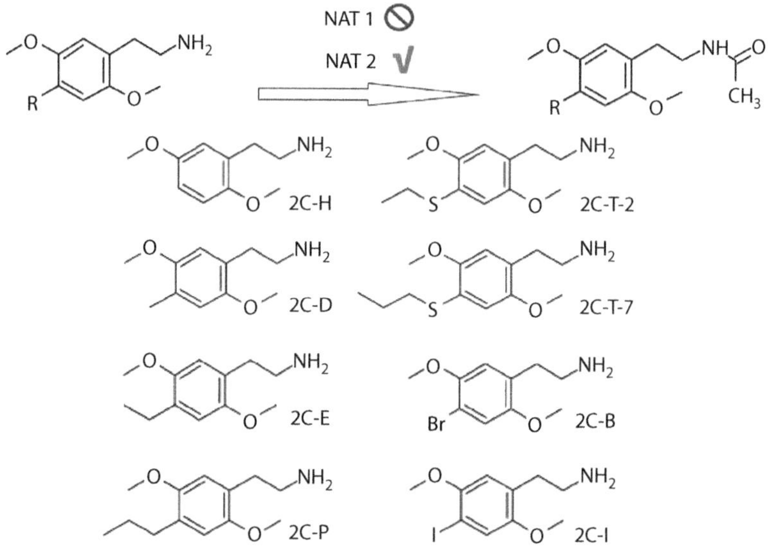

FIGURE 15.6 Chemical structures of some common 2C analogs. (Reprinted from Saurina, J. and Sentellas, S. 2017, *J Chromatogr B*. 1044–1045:103–111.)

in N-acetylation should be taken into consideration when evaluating the bioavailability of these phenethylamines.

15.2.11   Use of Multimethods

Multimethods have started to be commonly used for the identification or structural elucidation of unknown designer drugs that are newly introduced to the "legal highs" market. Chromatography helps to separate and purify the unknown compounds, and detection techniques such as full MS scan, MS/MS, UV-VIS, FTIR, elemental analysis, and NMR give information on the structure and the molecular formula. A study by Giroud et al. sought to identify 4-bromo-2,5-dimethoxyphenethylamine (2C-B), by means of several analytical methods (GC-MS, HPLC-DAD, CE-DAD, FTIR, and NMR), which was found in two sets of tablets obtained from the Swiss black market [40]. Reliable identification of 2C-B was only achieved by a combination of mass spectrometric and NMR analysis. Quantitation of 2C-B was performed by HPLC-DAD and CE-DAD. The amounts of 2C-B found in the tablets (3–8 mg) were in the range of the minimum quantity required to induce the effects characteristic of this drug.

A detection method was developed using GC-PCI-MS/MS (Positive Chemical Ionization), which has been rarely utilized to analyze new psychoactive drugs [41]. 0.5 g of right-heart whole blood from the postmortem autopsy was diluted with 0.5 mL of distilled water and 0.01 mL of an I.S. solution of caffeine d_3. Proteins were precipitated through adding acetonitrile (2.5 mL)+0.05 mL of formic acid and 0.01 mL of 5 M HCl vortexed and centrifuged. The filtrate was diluted with 2 mL of acetonitrile and the pH adjusted to 7 with ammonia water. The mixture was vortexed and centrifuged again. The upper fraction was extracted through a fast SPE and analyzed by GC-EI-MS and GC-PCI-MS/MS after an enrichment procedure. In the GC-PCI-MS/MS system, a tandem column was used consisting of a BPX5 (2 m, 0.25 mm i.d., 0.5-µm film thickness) coupled to a BPX5 (4 m, 0.15 mm i.d., 0.25-µm film thickness). The column utilized in LC-ESI(+)-MS/MS was a PFP column (50 mm × 2.1 mm i.d., 0.5 µm) at 40°C. The mobile phase consisted of a gradient between 0.1% formic acid in water (mobile phase A) and 0.2% formic acid in acetonitrile (mobile phase B) (flow rate: 0.2 mL/min). The ion transitions with highest intensities were selected. The three most prominent fragments (highest intensity, left to right) from the GC-EI-MS spectra are also included. GC-PCI-MS/MS and LC-ESI-MS/MS databases were formed for 104 psychotropic compounds, including 32 cannabinoid derivatives, 29 cathinone derivatives, 34 phenethylamine derivatives, and several other designer compounds. Using this database, 5 psychotropic compounds could be detected in an actual forensic autopsy case. The authors recommend the use of GC-PCI-MS/MS, together with the more established methods of GC-EI-MS and LC-ESI-MS/MS, for the forensic toxicology community to deal easier with the challenges of these ever-changing compounds.

The results of a study performed on samples of "legal highs" seized in head shops by law enforcement and health services in Poland between mid-2008 and mid-2011 are presented in another paper by Zuba et al. [42]. In total, 449 preparations that differed in labelling, net masses, forms of distribution, etc., were analyzed. GC-MS,

LC-QTOF-MS, high HPLC, and NMR were applied for the component identification and quantification of the designer drugs including 2C-B and 2C-E. Powders, tablets, and capsules were homogenized in a mortar. About 0.01 g samples were dissolved in 0.5 mL methanol (for GC-MS) and in 10 mL of methanol mixed with water (1:1, v/v) (for LC analyses) and centrifuged. When required, the solutions were dissolved with the appropriate solvent. The EI-GC-MS analyses of the extracts were carried out automatically in a splitless mode. Chromatographic separation was performed on a 30-m-long HP-5MS capillary column with 0.25 mm i.d., coated with a 0.25 µm film of (5%-phenyl)-methylpolysiloxane in a temperature gradient consisting of three segments. Positive ions were analyzed. HPLC-DAD quantitative analysis of detected substances was performed using a C_{18} endcapped monolithic column (5 mm × 100 mm), at 30°C. The LC-QTOF/MS separation was performed in 21 minutes on a C_{18} (7.5 cm × 2.1 mm × 2.7 mm) column, which was thermostated at 35°C. The mobile phase was composed of 0.1% (v/v) formic acid in acetonitrile and water (flow rate: 0.3 mL/min). Desolvation gas temperature (nitrogen) was 300°C, atomizing gas flow 10 L/min, atomizing gas pressure 45 psi, capillary voltage 3 kV, and fragmentor 100–300 V. Both scan and MS/MS mode were used. NMR: The samples (*ca.* 20 mg) were dissolved in deuterated dimethyl sulfoxide or deuterated chloroform. Complete assignments of ^1H and ^{13}C NMR chemical shifts were achieved by means of 1D and 2D NMR techniques, including H-H COSY, HSQC, and HMBC experiments at 300 K. GC-MS, LC-QTOF/MS, and, when required, NMR were applied for substance identification, whereas HPLC was used for quantification of selected active compounds of 449 analyzed preparations, 2C-B, 2C-E were found in a limited number of samples. The contents of the analyzed products were reported to change over time, and piperazines were often ousted by cathinones. In most, there were two or more ingredients. Cathinones and piperazines were mixed mainly within the chemical classes, caffeine both with piperazines (24 products) and cathinones (22 products), lidocaine only with the latter class (47 products). There was great inconsistency in the qualitative and quantitative composition of products with identical labelling and individual ingredients varied from several to hundreds of mgs.

Near-infrared spectroscopy (NIRs) is spreading as a tool of choice for fast and nondestructive analysis and detection of different compounds in complex matrices [43]. Risoluti et al. investigated the feasibility of using near-infrared (NIR) spectroscopy combined with chemometrics calibration to detect new psychoactive substances in street samples. The capabilities of this approach in forensic chemistry were assessed in the determination of newly emerging molecules. The study focused on synthetic cannabinoids and synthetic phenethylamines (including the NBOMe series, 25-I-NBOMe and 25-C-NBOMe, and 2-CP). The approach was validated comparing the results with the official methods and it has been successfully applied for "in-site" determination of illicit drugs in seized real samples. Drug samples in real matrices such as herbal spices and blotter papers are directly measured using NIR spectroscopy. For the parallel GC-MS analysis, standard and real samples were extracted from the matrices following the validated internal procedure and then analyzed. Simulated blotter paper or herbal spices were soaked in consecutive aliquots of 0.5 mL of methanol for six hours and the filtered supernatant was analyzed in EI-GC-MS in splitless mode. A HP-5 MS capillary column (30 m × 0.25 mm × 0.25 mm) was used for chromatographic separation. Helium at a flow rate of 1 mL/min was used as the carrier gas. For the blotter paper analysis, the initial column temperature (75°C) was maintained for one minute, then increased linearly at a rate of 25°C/min to 280°C, and finally maintained for two minutes. For the herbal sample analysis, the oven temperature was initially at 80°C for five minutes, increased to 290°C with a ramp of 1.0°C/min, and then held at 290°C for 15 minutes. The NIR spectra were collected in reflectance mode, through the use of an integrating sphere of an FT-NIR instrument equipped with a tungsten-halogen source and an InGaAs detector. The signals were recorded between 10,000 and 4000 cm^{-1}, collecting 82 scans at a resolution of 4 cm^{-1}. This in-site NIR analysis technique is time- and cost-saving, as it requires no or minimal sample preparation, nondestructive and easy to perform (no highly skilled personnel is necessary), allowing further investigation in forensic laboratories. Combining information from NIR spectroscopy and chemometrics provides a new approach as an analytical forensic tool for the screening of seized or unseized new psychoactive drugs.

The rapidly changing analogs make it difficult for laboratories to rely on traditional targeted screening methods [31]. In 2017, McGonigal et al. [6] analyzed 26 commercially available standards, including 2C-X and NBOMe, using a direct sample analysis technique coupled to the accurate-mass TOF MS (DSA-TOF-MS). The street sample consisted of ten blotter papers that were seized in an undercover buy operation. Two of the blotter papers were extracted with 500 μL of methanol, sonicated (30 minutes), vortexed (two minutes), filtered, and 5 μL of extract was then diluted into 95 μL of water. 2 μL was then applied to the mesh screen for direct sample analysis. The ionization parameters of AxION DSA-TOF-MS were as follows: +corona current: 7 mA, heater temperature: 325°C, drying gas temperature: 25°C, drying gas flow: 0 L/min, nebulizer gas pressure: 5.6 psi. Detector voltage was 3100 V. Under ambient conditions, DSA utilizes a direct charge transfer from nitrogen reagent gas thereby causing ionization followed by the in-source collision-induced dissociation. The fragmentation patterns, HRMS spectral data, and isotope ratios provide an adequate solution to qualitatively identify bromo-dragonfly, 2C-B Fly, 2C-B, 2C-B BZP, 2C-C, 2C-D, 2C-E, 2C-G, 2C-H, 2C-I, 2C-N, 2C-P, 2C-T, 2C-T-2, 2C-T-4, 2C-T-7, 3C-P, 25B, 25C, 25D, 25E, 25G, 25H, 25I, 25I NBF, 25-T-2 in unknown street samples in a very short time. Mass spectral data within the compound classes are similar, allowing for quick class identifications, but also unique and allow for individualization of compounds. The instrument uses the in-source collision-induced dissociation and ambient ionization to produce the [M+H]$^+$ ions and fragments. The authors recommend DSA-TOF-MS as a qualitative identification technique, with easier method development and screening. However, a disadvantage is that it is not available in many laboratories and LC-MS/MS has recently been a preferred method for the analysis of new drugs. Lately, especially LC-QTOF-MS, with its ability to both separate and identify the compounds, has been one of the more preferred techniques.

15.3 FUTURE PROSPECTS

Drugs of abuse are ever-changing and analytical chemists, forensic toxicologists, and physicians will be at the forefront of recognition of these substances. Given the

Internet age, many new psychoactive substances from around the world are available for individuals to use recreationally [10].

New analogs of ring-substituted phenethylamines, commonly known as 2Cs, are emerging as new drugs of abuse [7]. Currently, the amount of published literature concerning 2Cs is limited. There are a few 2Cs that have been studied, i.e., 2C-T-7, 2C-B, and 2C-E, so that information about the pharmacology, pharmacokinetics, and pharmacodynamics, although low in volume, is becoming available. A syndrome with excited delirium including severe agitation, aggression, violence, seizures, and hyperthermia is consistently depicted in lethal 2C cases. There are currently no antidotes for the 2C intoxication. Treatment is limited to supportive care, but it should include rapid sedation and aggressive treatment of severe hyperthermia in the 2C cases presenting with signs and symptoms consistent with a syndrome of an excited delirium.

Laboratory testing and imaging of these drugs should be directed toward clinical observations and symptoms, as needed [10]. The 2C phenethylamines are not detected with the standard commercial immunoassays. GC/MS or LC-MS/MS in combination with the other identification techniques as FTIR and NMR can be used to confirm the substance or exposure. LC-MS/MS has become more popular, especially as a support for the clinical practice.

The most used techniques also in the analysis of 2C drugs in the literature are GC-MS and LC-MS/MS. Up to recent years, GC-MS was the gold standard; however, recently, high-resolution LC-MS/MS techniques started to take the place of GC-MS or are required to confirm the results of GC-MS, because of the risky false negative probability of newly emerging derivatives.

Multimethods have started to be commonly used for the identification or structural elucidation of unknown designer drugs that are newly introduced to the market. Chromatography helps to separate and purify the unknown compounds, and the detection techniques such as full MS scan, MS/MS, UV-VIS, FTIR, elemental analysis, and NMR investigations help to understand the structure and to calculate the molecular formula. Structural information of metabolites is essential in drug discovery to quickly identify and establish the metabolite profiles and the rate of drug biotransformation. The NMR detection is especially successful for resolving the structures of unknown molecules, metabolites, and isobaric molecules because of its ability to deduce the exact positions of functional groups in the molecules—which is one of the problems of MS.

Impurity analysis is also important in forensic sciences, to define the source of the manufacturer. If there are the same impurities in the samples from different cases, a relationship could be figured out among them, and the source of the drug manufactured may be found. GC-MS still is an important component of the identification and structure elucidation studies for unknown compounds and their metabolites, and determination studies for isobaric compounds, if a better method has not been developed in LC-MS/MS with better enantiomeric separation. GC separation on nonpolar stationary phases in long bore columns can successfully resolve the regioisomeric and isobaric compounds. However, with an accurate-mass LC-MS/MS detection, a retention time parameter and/or more MRM transitions should be added for absolute identification for isobaric and regioisomeric compounds. Only LC-HRMS gives an accurate mass, which can be used for predicting the formulas and structural elucidation [25]. The ion trap (IT) brings the opportunity of MS^n and TOF, the chance of high-resolution and mass accuracy for the unknown compounds and impurity profiling. The mass resolving and mass determining power of MS technology is enhanced with HRMS techniques, so that fast analyses are possible via coupling with UPLC or HPLC with monolithic columns. Chiral columns, which are effectively used for the enantioseparation in chemistry, should gain importance also in the future analysis of the isomers linked to the 2C derivatives.

In the literature, HPLC is reported to be coupled simultaneously to NMR and MS detectors to form the LC-NMR-MS system [44]. In such a system, separation and purification of each analyte through chromatography makes the NMR analysis easier and faster; with an online system, if the library is adequate, this powerful technique can be applied to enantiomers, drug metabolism studies, and fast screening and identification of 2C analogs, as well as to the other NPSs.

REFERENCES

1. Musselman, M. E. and Hampton J. P. 2014. Not for human consumption: A review of emerging designer drugs, *Pharmacother*. 34:745–757.
2. Carpenter, T. G. 2015. Designer drugs, carpenter a new, futile front in the war on illegal drugs, *Policy Anal*. 774:1–13.
3. Hill, S. L. and Thomas, S. H. 2011. Clinical toxicology of newer recreational drugs, *Clin Toxicol*. 49:705–719.
4. Abdulrahim, D. and Bowden, J. O. 2015. On behalf of the NEPTUNE Expert Group. Chapter 7 Amphetamine-type substances (ATS): An overview Drug group: Stimulant, in: Guidance on the Management of Acute and Chronic Harms of Club Drugs and Novel Psychoactive Substances. Novel Psychoactive Treatment UK Network (NEPTUNE), London.
5. Smith, P. R. and Morley, S. R. 2017. New Psychoactive Substances, Ed. Guy N. Rutty, In: *Essentials of Autopsy Practice: Reviews, Updates, and Advances*, Springer, Switzerland, p. 69.
6. McGonigal, M. K., Wilhide, J. A., Smith, P. B., Elliott, N. M. and Dorman, F. L. 2017. Analysis of synthetic phenethylamine street drugs using direct sample analysis coupled to accurate mass time of flight mass spectrometry, *Forensic Sci Int*. 275:83–89.
7. Dean, B. V., Stellpflug, S. J., Burnett, A. M., and Engebretsen, K. M. 2013. 2C or Not 2C: Phenethylamine designer drug review, *J Med Toxicol*. 9:172–178.
8. Liechti, M. 2015. Novel psychoactive substances (designer drugs): Overview and pharmacology of modulators of monoamine signaling, *Swiss Med Wkly*. 145:14043.
9. Drees, J. C., Stone, J. A. and Wu, A. H. B. 2009. Morbidity involving the hallucinogenic designer amines MDA and 2C-I, *J Forensic Sci*. 54:6.
10. Nelson, M. E., Bryant, S. M., and Aks, S. E. 2014. Emerging drugs of abuse, *Disease-A-Month*. 60:110–132
11. Gálligo, F. C., Riba, J., Ventura, M., González, D., Farré, M., Barbanoj, M. J. and Bouso, J. C. 2012. 4-Bromo-2,5-dimethoxyphenethylamine (2C-B): Presence in the recreational drug market in Spain, pattern of use and subjective effects, *J Psychopharmacol*. 26:1026–1035.
12. De Boer, D. and Bosman, I. 2004. A new trend in drugs-of-abuse: The 2C-series of phenethylamine designer drugs, *Pharm World Sci*. 26:110–113.
13. Andreasen, M. F., Telving, R., Birkler, R. I., Schumacher, B., and Johannsen, M. 2009. A fatal poisoning involving Bromo-Dragonfly, *Forensic Sci Int*. 183:91–96.

14. Wood, D. M., Looker, J. J., Shaikh, L., Button, J., Puchnarewicz, M., and Davies, S. 2009. Delayed onset of seizures and toxicity associated with recreational use of Bromo-dragon fly, *J Med Toxicol.* 5:226–229.
15. Zuba, D., Sekula, K., and Buczek, A. 2012. Identification and characterization of 2,5-dimethoxy-4-nitro-betaphenethylamine (2C-N)-a new member of 2C-series of designer drug, *Forensic Sci Int.* 222:298–305.
16. U.S. Drug Enforcement Administration, Diversion Control Division. 2017. 2C-Phenethylamines, Piperazines, and Tryptamines Reported in NFLIS, 2011–2015. Springfield, VA: U.S. Drug Enforcement Administration.
17. Páleníček, T., Fujáková, M., Brunovský, M., Horáček, J., Gorman, I., Balíková, M., Rambousek, L., Syslová, K., Kačer, P., Zach, P., Valešová, V.B., Tylš, F., Kubešová, A., Puskarčíková, J., and Höschl, C. 2013. Behavioral, neurochemical and pharmaco-EEG profiles of the psychedelic drug 4-bromo-2,5-dimethoxyphenethylamine (2C-B) in rats, *Psychopharmacol.* 225:75–93.
18. Hoffman, R. S., Howland, M. A., Lewin, N. A., Nelson, L. S., Goldfrank, L. R. 2015. Amphetamines, *in Goldfrank's Toxicologic Emergencies*, 10th Ed., McGraw-Hill Global Education Holdings, LLC., http://accesspharmacy.mhmedical.com/content.aspx?bookid=1163&sectionid=64554057, accessed May 2017.
19. Graaf, Å. Vad är drogen bromodragonfly? 2017, I SAY NO DRUGS, http://www.isaynodrugs.org/vad-ar-drogen-bromodragonfly/, accessed May 2017.
20. Appendino, G., Minassi, A., and Taglialatela-Scafati, O. 2014. Recreational drug discovery: Natural products as lead structures for the synthesis of smart drugs, *Nat Prod Rep.* 31:880–904.
21. Weaver, M. F., Hopper, J. A. and Gunderson, E. W. 2015. Designer drugs 2015: Assessment and management, *Addict Sci Clin Pract.* 10:1–9.
22. Lin, D. 2016. Designer drugs—A brief overview, *Therapeut Toxins News.* 2:1–10.
23. Habrdova, V., Peters, F. T., Theobald, D. S. and Maurer, H. H. 2005. Screening for and validated quantification of phenethylamine type designer drugs and mescaline in human blood plasma by gas chromatography/mass spectrometry, *J Mass Spect.* 40: 785–795.
24. Boatto, G., Nieddu, M., Dessì, G., Manconi, P., and Cerri, R. 2007. Determination of four thiophenethylamine designer drugs (2C-T-series) in human plasma by capillary electrophoresis with mass spectrometry detection, *J Chromatogr A.* 1159:198–202.
25. Tsai, C. C., Liu, J. T., Shu, Y. R., Chan, P. H. and Lin, C. H. 2006. Optimization of the separation and on-line sample concentration of phenethylamine designer drugs with capillary electrophoresis–fluorescence detection, *J Chromatogr A.* 1101:319–323.
26. Pichini, S., Pujadas, M., Marchei, E., Pellegrini, M., Fiz, J., and Pacifici, R. 2008. Liquid chromatography-atmospheric pressure ionization electrospray mass spectrometry determination of "hallucinogenic designer drugs" in urine of consumers, *J Pharm Biomed Anal.* 47:335–342.
27. Li, Y., Wang, M., Li, A., Zheng, H., and Wei, Y. 2016. Identification of the impurities in 2, 5-dimethoxy-4-ethylphenethylamine tablets by high-performance liquid chromatography mass spectrometry-ion trap-time of flight, *Anal Methods.* 8:8179–887.
28. Ambach, L., Redondo, A. H., König, S. and Weinmann W. 2014. Rapid and simple LC-MS/MS screening of 64 novel psychoactive substances using dried blood spots, *Drug Test Anal.* 6:367–375.
29. Van Vrancken, M. J., Benavides, R. and Wians, F. H. 2013. Identification of designer drug 2C-E (4-ethyl-2, 5-dimethoxy-phenethylamine) in urine following a drug overdose, *Proceedings*, 26: 58, Baylor University Medical Center.
30. Adamowicz, P. and Tokarczyk, B. 2015. Simple and rapid screening procedure for 143 new psychoactive substances by liquid chromatography–tandem mass spectrometry, *Drug Test Anal.* 8:652–667.

31. Pasin, D., Cawley, A., Bidny, S., and Fu, S. 2017. Characterization of hallucinogenic phenethylamines using high-resolution mass spectrometry for non-targeted screening purposes. *Drug Test Anal.* 9:1620–1629.

32. Paul, M., Ippisch, J., Herrmann, C., Guber, S., and Schultis, W. 2014. Analysis of new designer drugs and common drugs of abuse in urine by a combined targeted and untargeted LC-HR-QTOFMS approach, *Anal Bioanal Chem.* 406:4425–4441.

33. Pasin, D., Bidny, S., and Fu, S. 2015. Analysis of new designer drugs in post-mortem blood using high-resolution mass spectrometry, *J Anal Toxicol.* 39:163–171.

34. Swortwood, M. J., Boland, D. M., and DeCaprio, A. P. 2013. Determination of 32 cathinone derivatives and other designer drugs in serum by comprehensive LC-QQQ-MS/MS analysis, *Anal Bioanal Chem.* 405:1383–1397.

35. McBride, E. M., Kretsch, A., Garibay, L. K., Brigance, K., Frey, B., Buss, B. and Verbeck, G. F. 2016. Rapid experimental and computational determination of phenethylamine drug analogue lipophilicity, *Forensic Chem.* 1:58–65.

36. Pratama, S. F., Muda, A. K., Choo, Y. H., and Abraham, A. 2016. 3D geometric moment invariants for ATS drugs identification: A more precise approximation. In: Abraham A., Haqiq A., Alimi A., Mezzour G., Rokbani N., Muda A. (eds.), Proceedings of the 16th International Conference on Hybrid Intelligent Systems (HIS 2016). HIS 2016. *Advances in Intelligent Systems and Computing*, vol. 552, pp. 124–133, Springer, Cham.

37. Saurina, J. and Sentellas, S. 2017. Strategies for metabolite profiling based on liquid chromatography, *J Chromatogr B.* 1044–1045:103–111.

38. Gillotin, F., Chiap, P., Frederich, M., Van Heugen, J. C., Francotte, P., Lebrun, P., Pirotte, B., and de Tullio, P. 2010. Coupling of liquid chromatography/tandem mass spectrometry and liquid chromatography/solid-phase extraction/NMR techniques for the structural identification of metabolites following in vitro biotransformation of SUR1-selective ATP-sensitive potassium channel openers, *Drug Metab Dispos.* 38:232–240.

39. Meyer, M. R., Robert, A., and Maurer, H. H. 2014. Toxicokinetics of novel psychoactive substances: Characterization of N-acetyltransferase (NAT) isoenzymes involved in the phase II metabolism of 2C designer drugs, *Toxicol Lett.* 227:124–128.

40. Giroud, C., Augsburger, M., Rivier, L., Mangin, P., Sadeghipour, F., Varesio, E., Veuthey, J. L., and Kamalaprija, P. 1998. 2C-B: A new psychoactive phenylethylamine recently discovered in ecstasy tablets sold on the Swiss black market, *J Anal Toxicol.* 22:345–354.

41. Waters, B., Ikematsu, N., Hara, K., Fujii, H., Tokuyasu, T., Takayama, M., and Kubo, S. I. 2016. GC-PCI-MS/MS and LC-ESI-MS/MS databases for the detection of 104 psychotropic compounds (synthetic cannabinoids, synthetic cathinones, phenethylamine derivatives), *Legal Med.* 20:1–7.

42. Zuba, D. and Byrska, B. 2013. Prevalence and co-existence of active components of 'legal highs,' *Drug Test Analysis.* 5:420–429.

43. Risoluti, R., Materazzi, S., Gregori, A., and Ripani, L. 2016. Early detection of emerging street drugs by near infrared spectroscopy and chemometrics, *Talanta.* 153:407–413.

44. Corcoran, O. and Spraul, M. 2003. LC–NMR–MS in drug discovery, *Drug Discov. Today* 8:624–631.

16 NBOMe Derivatives of Phenylethylamines and Their Analysis

Beril Anilanmert, Fatma Çavuş Yonar,
and Sena Çağlar Andaç

CONTENTS

16.1 INTRODUCTION

Recently, a new class of "2C" serotonin 5-HT_{2A} receptor-agonist designer drugs, dimethoxyphenyl-N-[(2-methoxyphenyl) methyl] ethanamine (NBOMe) derivatives, became popular and easily obtainable over the Internet, which has resulted in their abuse in the U.S., Europe and Asia [1]. These drugs have been the cause of severe intoxications and even deaths. The NBOMe class of compounds are N-2-methoxybenzyl derivatives of the 2C-X series [2]. NBOMe compounds include 25B-N(BOMe)2, 25B-NBOMe, 25C-NBOMe, 25D-NBOMe, 25E-NBOMe, 25G-NBOMe, 25H-NBOMe, 25I-BOMe, 25N-NBOMe and 25iP-NBOMe [3]. There are reports of their use as new psychoactive substances and associated acute toxicity from Europe, the United States and elsewhere over the last five years.

They act as 5-HT$_{2A}$ receptor agonists and have been reported to produce a serotonin-like syndrome, with bizarre behavior, severe agitation, and seizures persisting for as long as three days [1]. They are frequently abused due to the intense hallucinations that they induce [4]. The hallucinogenic effects experienced following use are due in part to the effect that they have on the 5-HT$_{2A}$ receptor. Hallucinogens, by definition, are drugs that can induce hallucinations or other psychedelic effects [5].

NBOMe, sometimes referred to as 25x-NBOMe, includes various phenethylamine derivatives of the 2C class of hallucinogens. The most commonly reported derivatives are 25I-NBOMe, 25B-NBOMe, and 25C-NBOMe, with chemical names: 2-(4-iodo-2,5-dimethoxyphenyl)-N-[(2-methoxyphenyl)methyl]ethanamine, N-(2-methoxybenzyl)-2,5-dimethoxy-4-bromophenethylamine, and N-(2-methoxybenzyl)-2,5-dimethoxy-4-chlorophenethylamine, in order. The other NBOMe derivatives are given in Chapter 15, Table 15.1. Like many low-dose hallucinogenic drugs, these compounds are often sold on blotter paper. Currently, the most widely abused NBOMe derivative, which is sold as a powder or on blotter paper, exists under the names 25I-NBOMe, "N-Bomb," and "Smiles" [1]. Figure 16.1 shows these blotter papers containing illicit substances.

Several of these derivatives, including 25I-NBOMe and 25B-NBOMe, were first synthesized by Ralf Heim at the Free University of Berlin as part of a series of pharmacological tools to study the 5-HT$_2$A receptor. NBOMe derivatives could be easily synthesized. 2C-I-NBOMe can be synthesized via a reductive alkylation of 2C-I with 2-methoxybenzaldehyde [6]. This synthesis is done stepwise, first by synthesizing the imine and then by reducing this imine with sodium borohydride or by direct reaction with sodium triacetoxyborohydride. Since this designer drug is also

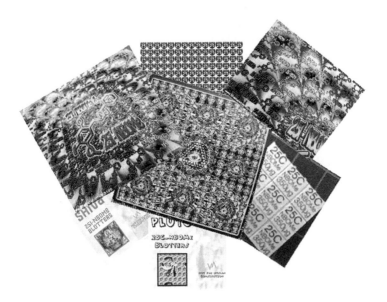

FIGURE 16.1 Blotter papers advertised in the Internet sites like shamanicharmonics.com, ivolabs.com, etsy.com, containing NBOMe derivatives. One dose of N-Bomb is normally between £2 and £4.

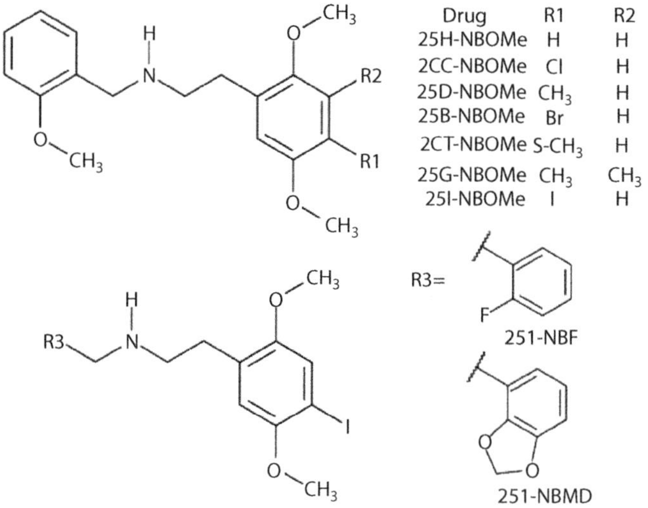

FIGURE 16.2 Structure of the NBOMe derivatives. (Reprinted from Poklis, J. L. et al., 2014, *J Anal Toxicol*, 38:113–121.)

produced clandestinely by amateur cookers, its purity and consequently its potency could vary widely. The Drug Enforcement Administration placed 25I-NBOMe, 25B-NBOMe, and 25C-NBOMe into Schedule I of the Controlled Substances Act on 10 October 2013.

General molecular structures of different types of NBOMe derivatives are demonstrated in Figure 16.2. The molecular names of most of the NBOMe derivatives are also given in Chapter 15, Table 15.1. Sharing a core phenethylamine structure, the NBOMe compounds differ from the 2C-substances (2C-I, 2C-C, and 2C-B) by the addition of a 2-methoxybenzyl group on the nitrogen atom [4].

16.1.1 PHARMACOLOGICAL AND TOXICOLOGICAL EFFECTS AND DOSES

Among the 2C derivatives, N-Benzyl-Oxy-Methyl (NBOMe) derivatives are concerning mainly due to several reports in the literature associated with acute toxicity over the last six years; the majority of these cases are related to 25I-NBOMe [7]. The reported effects of 25I-NBOMe are similar to those of prototypical serotonergic hallucinogens such as lysergic acid diethylamide or psilocybin [8]. It has even emerged as a common replacement for LSD [1]. Users report hallucinations with a varying degree of stimulating effects [8]. Depersonalization has been reported as well. In contrast to prototypical serotonergic hallucinogens, 442 users responding to an Internet survey reported that 25I-NBOMe had greater "negative effects while high," but with more "value for money." In addition to the anticipated visual and auditory hallucinations, many users experience psychiatric consequences, prompting them to access medical services. Some of these consequences include delirium, agitation, aggression, violence, paranoia, dysphoria, severe confusion, and self-harm. Some patients have presented with a serotonergic or sympathetic toxidrome consisting of an "excited delirium" with severe agitation, aggression, and violence. Tachycardia, hypertension, and mydriasis are frequently described in the few clinical reports of 25I-NBOMe users. Hyperreflexia and clonus have also been reported in several cases. Seizures occurred in many of the cases that eventually required medical attention. Severe toxicity has included hyperthermia, pulmonary edema, and death from trauma.

In one case, a reportedly hallucinogen-naïve 19-year-old man died fall after ingesting 25I-NBOMe and developing paranoid and bizarre behavior. In another fatal case, a 21-year-old male driver who ingested 25I-NBOMe developed sudden rage, pulled his car off the road, and began to destroy the inside of the vehicle before dying from an unknown cause. After fatal exposure, a 15-year-old girl became unresponsive after ingesting 25I-NBOMe outside a rave; on arrival at a local hospital she was in a systole with a rectal temperature of 39.9°C [3]. Long-term physiologic effects of 25I-NBOMe are not known [8]. Anecdotal reports indicate that the powder in doses of 50–250 mg may be administered sublingually, by insufflation or applied to the buccal cavity [9]. Blotter paper usually contains higher doses of 500–800 mg, apparently due to low bioavailability of the drug. Several published abstracts and a few papers have described signs and symptoms of 25I-NBOMe intoxication.

16.1.2 New NBOMe Analogs

As in other groups, newer NBOMe analogs are persistently produced by black science laboratories to escape from screening tests and the good guys try to catch the fast-changing trends in analogs, in their laboratory, using challenging methods and techniques.

In 2016, 1-(benzofuran-5-yl)-N-(2-methoxybenzyl)propan-2-amine analog, namely, 5-APB-NBOMe, was identified by liquid chromatography quadrupole time of flight mass spectrometry (LC-QTOF-MS), gas chromatography–mass spectrometry (GC-MS), and nuclear magnetic resonance spectrometry (NMR) [10]. 5-APB-NBOMe was previously reported by Westphal et al. [11] without high-resolution LC-MS data.

In 2017, Shevyrin et al. detected and identified N-(2-methoxybenzyl)-2-(2,4,6-trimethoxyphenyl)ethanamine (2,4,6-TMPEA-NBOMe), a new compound in the NBOMe series [12]. Identification was achieved by means of GC-MS, including high-resolution mass spectrometry with tandem experiments (GC-HRMS and GC-HRMS/MS), ultra-high-performance liquid chromatography/high-resolution mass spectrometry with tandem experiments (UHPLC-HRMS and UHPLC-HRMS/MS), and ^{1}H and ^{13}C NMR spectroscopy.

In 2012, Zuba and Sekula reported analytical properties of three potent serotonin 5-HT$_2$A receptor agonist hallucinogenic substances from a new class of N-methoxybenzyl derivatives of methoxyphenylethylamine (NBOMe) derivatives identified in blotter papers seized from the drug market in Poland: 2-(2,5-dimethoxy-4-methylphenyl)-N-(2-methoxybenzyl) ethanamine (25D-NBOMe), 2-(4-ethyl-2,5-dimethoxyphenyl)-N-(2-methoxybenzyl) ethanamine (25E-NBOMe) and 2-(2,5-dimethoxy-3,4-dimethylphenyl)-N-(2-methoxybenzyl)ethanamine (25G-NBOMe) [13]. In recent years, N-(2-methoxybenzyl)-2,5-dimethoxy-4-chlorophenethylamine (2CC-NBOMe) and 2-(4-iodo-2,5-dimethoxyphenyl)-N-[(2,3-methylenedioxyphenyl)methyl]ethanamine (25I-NBMD) have also been identified on blotter paper. GC-EI-MS (without derivatization and after derivatization with trifluoroacetic anhydride [TFAA]), LC-ESI-QTOF-MS, FTIR, and NMR techniques were used. The GC-MS spectra of the compounds were very similar. $m/z = 150$, 121, and 91 were observed as the dominant ions. The remaining ions were analogous to 2C-D, 2C-E, and 2C-G, but their intensities were low. Derivatization allowed determination of molecular masses of the investigated substances. The exact masses and chemical formulas were confirmed by LC-ESI-QTOF/MS. The fragmentation patterns were also determined. NMR and FTIR spectroscopy were used in final elucidation of the structures and to confirm the identifications. Zuba and Sekula used derivatization with trifluoroacetic anhydride (TFAA) in the identification of 25D-, 25E-, and 25G-NBOMe. However, derivatization did not allow separation of the isomers 25E- (containing H, C_2H_5) and 25G- (containing CH_3 and CH_3). Lum et al. [14] have demonstrated the use of derivatization by heptafluorobutyric anhydride (HFBA) in the differentiation of the isomers 25E- and 25G-NBOMe. It has been shown that the NBOMe series undergoes a characteristic and diagnostic McLafferty rearrangement during mass spectral analysis.

16.2   ANALYSIS OF NBOMᴇ DERIVATIVES

16.2.1   An Overview of Techniques Used for Screening and Identification

In forensic analyses, immunoassays are important screening techniques; however, difficulty arises in detecting the relatively newer 2C class compounds, such as 25I-NBOMe, particularly at lower drug concentrations [5].

LC-UV and HPTLC might be helpful in screening but still there is the possibility of false negatives, because of the low concentrations of NBOMe in blotters and of course in biofluids after intake. If LC-UV and HPTLC are preferred for use in screening, the results should be confirmed with a mass spectroscopy technique in order not to miss the low concentrations of new unknown derivatives, if they exist. Consumption of drug-impregnated blotters is very frequent to address this issue with readily accessible laboratory equipment, a new HPTLC method was developed and validated for the identification and quantitation of 25C-NBOMe in seized blotters and its confirmation by GC-MS [15]. Fifteen real samples of seized blotters were separately submerged in 25.0 mL of methanol and extracted for 15 minutes in an ultrasonic bath and analyzed in HPTLC and GC-MS. HPTLC was performed on 20 × 10 cm precoated silica gel F254 plates previously activated at 80°C for 30 minutes. The first application x-axis was 15 mm, the y-axis was 8.0 mm, and the distance between tracks was 5.8 mm. Plates were developed in an automatic developing chamber to a distance of 70 mm with cyclohexane/toluene/diethylamine (75/15/10 v/v/v) as the mobile phase (10 mL, without saturation of the chamber). After a drying time of 5.0 minutes, the bands were scanned with densitometer and the absorbances were measured at 298 nm. The GC-MS analysis was performed using EI mode at 70 eV after elution through a HP-5 capillary column (30 m × 0.32 mm × 0.25 μm). The column temperature was set at 75°C for one minute, then programmed to 280°C at a rate of 25°C/minute and kept constant at 280°C for 20 minutes. The injector temperature was 250°C, and flow rate of helium was 1 mL/minute with a splitless mode injection. Identification of the compound was performed by a comparison of its mass spectrum and retention time with the injected pure reference material. With the validated method, 15 real samples were analyzed, in all cases finding 25-C-NBOMe in a wide dosage range (701.0–1943.5 µg per blotter). LOD and LOQ were 7.1 µg per band and 21.63 µg per band, respectively. Mean recovery was 99.26%.

Again, MS/MS techniques, especially library-supported LC-HRMS/MS techniques or double/multi-instrument uses such as GC-MS+LC-MS/MS, GC-MS+LC-MS/MS+NMR, LC-MS/MS+NMR, and GC-MS+LC-MS/MS+FTIR+NMR (or, one of the best, LC-MS/MS+LC-NMR), should be considered according to the information that is expected to be obtained. As mentioned in Chapter 15, the advantages of different required techniques started to be brought together in multimethods and commonly used for identification or structural elucidation of unknown designer drugs that are newly introduced to the market, impurity profiling to predict the synthesis route and the source of the manufacturer, metabolite profiling or detection of the coexisting enantiomeric forms. As in other 2C derivatives, NMR detection is required if unknown metabolites and isobaric molecules should be recognized. FTIR and NMR predict the exact positions of functional groups in the molecules, which is one of the problems of MS. Use of LC-HRMS (high-resolution MS) methods will provide accurate masses of the newly introduced compounds, which can be used for predicting formulas and structural elucidation [16]. Time-of-flight (TOF) and Ion Trap (IT) instruments offer superior advantages to LC-MS/MS as the ability to conduct multiple fragmentation MS cycles (MS^n), high-resolution, mass accuracy, and fast analysis. The IT offers increased sensitivity and there is almost no surface charging effect because low voltage is needed for ion extraction, so nonconducting surfaces (e.g., a glass slide) or thick tissue samples can be used in combination with a MALDI instrument.

Several studies in recent years have demonstrated the utility of mass spectrometry–based methods in the screening and detection of NBOMe derivatives. Direct analysis techniques such as Direct Analysis in Real Time AccuTOF (DART-MS) gain importance in the field, because of the chance of no or minimum sample preparation and fast analysis. In 2015, it was used for screening of NBOMe class hallucinogens on blotter paper [1]. Rapid detection was performed on both the blotter paper directly and the blotter papers dissolved in methanol. NBOMe derivatives including: 25I-NBOMe imine, 25I-NBOMe, 25I-NBF, 25G-NBOMe, 25D-NBOMe, 25H-NBOMe, 25H-NBOMe imine, 25B-NBOMe, 25C-NBOMe, and 2CT-NBOMe were evaluated. DART-MS operated in (+) ion mode and the ion source had the helium gas flow rate at 2.0 L/minute, gas heater temperature of 300°C, discharge electrode needle at 4.000 V; electrode 1 was set at 150 V and electrode 2 at 250 V. Measurements were taken with the ion guide peak voltage of 800 V and reflectron voltage of 900 V. Orifice 1 was operated at 300°C in 20, 60, or 90 V using a switching mode that created a single file for all three voltages, Orifice 2 was set at 5 V and the ring lens was set at 3 V. The blotter papers were analyzed by DART-MS with and without sample preparation. The same samples prepared for DART-MS were analyzed in LC-MS/MS (Q trap). All of the analyzed NBOMe derivatives, with the exception of the NBOMe imines, produced the same or similar product ions. The 25H-NBOMe imine and 25I-NBOMe imine

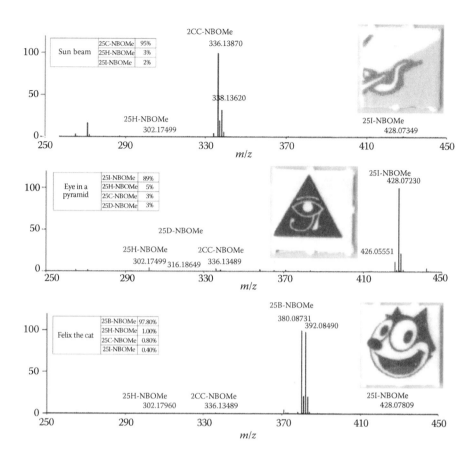

FIGURE 16.3 DART-MS spectra of each blotter paper. (Reprinted from Poklis, J. L. et al., 2015, *J Anal Toxicol*, 39:617–623.)

produced product ions similar to each other through the same fragmentation pattern. Therefore, $[M + H]^+$ ions (Orifice 1: 20 V) were used to characterize the derivatives in each blotter paper (Figure 16.3).

16.2.2 GC-MS ANALYSIS OF NBOME DERIVATIVES

Although laboratories tend to use the LC-HRMS technique to find out the structures of new unknown molecules in the samples as the new synthetic psychoactive analogs emerge, GC-MS is still the most widely and routinely utilized analytical technique for drug sample analyses. However, routine GC-MS methods remain inadequate in catching new psychoactive molecules. For example, they can misidentify 25I-NBOH (and indeed the other members of the NBOH series) as 2C-I because of its degradation into 2C-I (and corresponding 2C for the other members of the series) within the injector, unless a derivatization procedure is employed, which is often nonstandard [17]. Neto et al. have proposed an alternative procedure enabling an indirect identification of NBOHs that could discriminate it from 2C-I, using routine GC-MS methods, without derivatization or other analytical techniques. A HP-1MS fused-silica capillary column (30 m × 0.25 mm i.d., 0.25-µm film thickness) was used. 1 µL sample injection volume with a 25:1 split ratio and helium with a flow rate of 1 mL/minute were used. Injector temperature was 280°C. The oven program started at 150°C with a hold for 1.5 minutes, ramped up at 30°C/minute to reach 250°C with a hold for one minute, and then ramped up at 50°C/minute to 300°C with a hold for three minutes. The transfer line temperature was 300°C. Attempts to avoid thermal degradation of 25I-NBOH were performed by lowering injector temperature (280–150°C), changing the solvents used for extracting the samples (methanol, acetonitrile, ethyl acetate, and acetone) and testing different injection split ratios. Considering the degradation of 25I-NBOH into 2C-I during standard routine GC injection as a fact, it was noticed that other by-products formed by degradation would enter the GC column, possibly generating other peaks in a total ion current chromatogram. Consequently, the GC oven program was adjusted, methanol was used, and the solvent delay shortened to 1.5 minutes, in order to monitor potentially fast-eluting compounds. These enabled the detection of an additional peak due to 25I-NBOH degradation. The presence of this secondary early chromatographic peak allowed for the distinction between 25I-NBOH and 2C-I using routine GC-MS without resorting to derivatization (or other analytical processes), thus preventing misidentification of 25I-NBOH as 2C-I. Although it may be argued that the use of methanol or other alcohol as a solvent enhances thermal degradation, such enhancement likely benefits the identification of the original molecule, because the degradation now occurs through a specific reaction. Since the approach is interesting and extraordinary, the chromatogram of a blotter paper sample where the modified method was applied has been given in Figure 16.4.

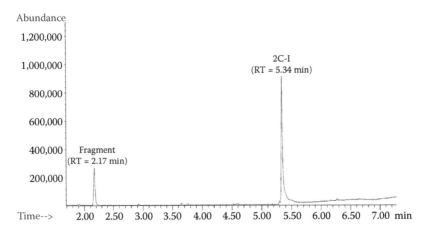

FIGURE 16.4   TIC obtained from the methanol extract of a street sample of blotter paper seized by local authorities, in GC-MS analysis. The primary peak (rt = 5.34 minutes) matched the retention time from 25I-NBOH and 2C-I certified reference standards injected under the same conditions. The secondary peak (rt = 2.17 minutes) was detected only for the 25I-NBOH certified standard and the seized blotter paper standard. (Reprinted from Neto, J. C. et al., 2017, *Forensic Toxicol*, 1–6.)

16.2.3 ANALYSIS OF NBOH COMPOUNDS

The NBOH compounds are proposed as a legal alternative for the NBOMe compounds, even if their physiological and toxicological properties are not fully reported. Since NBOH compounds are very similar with small differences from NBOMe molecules, we have also included the studies pertaining to the analysis of their derivatives. Scientific data regarding NBOH compounds are very scarce. Only a few pharmacological studies on mice or computational models have been conducted on these drugs. Only a single paper reported the recovery of 25C-NBOH, together with 25C-NBOMe and 2C-C from urine samples in three adult males, one of whom died experiencing respiratory difficulties [18]. Another approach for the identification of 25I-NBOH and discrimination from 2-CI in blotter papers is using LC-(ESI(+)–QTOF/MS, FTIR, and NMR analyses besides GC-MS. 25I-NBOH was identified in blotter papers by several state-level forensic laboratories in Brazil [19]. Blotter papers were extracted with methanol. For GC-MS analysis, a fused-silica capillary GC column (30 m × 0.25 mm i.d., 0.25-µm film thickness) was used. The injection volume was 0.5 µL at a split ratio of 50:1. Helium was used at 1 mL/minute, inlet line temperature at 280°C, and injector temperature at 200°C. The gradient started with an initial temperature of 100°C and initial hold time of 0 minute. It was then increased to 300°C at a rate of 6°C/minute and was held for 5.67 minutes. The known doses for 2C-I that have any hallucinogen effect in the human body are above 14 mg, which required further analysis. In the LC-(ESI(+)–QTOF/MS method, mobile phase consisted of 0.1% formic acid in water, and 0.1% formic acid in methanol. Separation was performed with a C_{18} (100 × 2.1 mm i.d., particle size 1.8 µm) column. Drying gas temperature was 350°C; drying gas flow, 8.0 L/minute; nebulizer, 35 psi; sheath gas temperature, 400°C; sheath gas flow, 11 L/minute; VCap, 3500 V. However, the ESI mass spectrum of the chromatogram peak showed two exact mass fragments: m/z 308.0142 [M + H]$^+$ (2C-I) and m/z 414.0561 [M + H]$^+$. The ion at m/z 414.0561 was compatible with the molecular formula of 25I-NBOH. Despite the higher degree of similarity with 25I-NBOMe than with 2C-I, a systematic decrease of the spectral band centered around 1250 cm^{-1}, characteristic of asymmetric C–O–C vibrations of NBOMe compounds, was observed. The decrease of this peak indicates the substitution of -OCH$_3$ by -OH in the NBOH structure, resulting in an NBOH compound. The complete ^{1}H and ^{13}C signals assignment and the ^{1}H-^{13}C correlations supported that the compound was 25I-NBOH. The GC-MS chromatogram and LC-QTOF-MS spectrum are given in Figure 16.5.

16.2.4 LC-MS/MS ANALYSIS OF NBOME DERIVATIVES

In a 2018 study, Casper et al. developed a highly sensitive method for the identification and determination of a group of designer drugs including 25B-NBOMe, 25C-NBOMe, 25E-NBOMe, 25I-NBOMe, 25H-NBOMe 2C-H, 2C-N, 2C-D, 2C-P, 2C-E, 2C-B, 2C-B-fly, 2C-I, 2C-T-2, and 2C-T-7, using Orbitrap-based LC-HRMS with alternating HR full scan (HRFS) MS and "All ions fragmentation" (AIF) MS [20]. Since they are consumed at very low doses resulting in low ng/mL blood concentration, analysis of these compounds in blood is a challenge for reliable

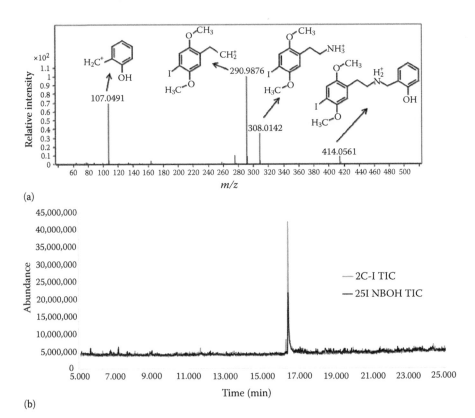

(a)

(b)

FIGURE 16.5   (a) Product ion spectrum of 25I-NBOH showing the corresponding chemical structure at 80 and 20 V of fragmentor voltage and collision energy, respectively, obtained using LC-QTOF-MS; (b) total ion current chromatogram for 25I-NBOH and 2C-I showing the same retention time in GC-MS analyses. (Reprinted from Arantes, L. C. et al., 2017, *Forensic Toxicol*, 1–7.)

detection and quantification. A gradient elution using a mobile phase consisted of 2 mM aqueous ammonium formate at pH 3.4 containing 0.1% formic acid (eluent A) and methanol:acetonitrile (50:50, v:v) containing 0.1% formic acid (eluent B), was performed. The heated electrospray ionization (HESI-II) source settings were as follows: sheath gas, nitrogen, at 53 arbitrary units; auxiliary gas, nitrogen, at 14 arbitrary units; temperature, 437°C; spray voltage, 4 kV; ion transfer capillary temperature, 269°C; and S-lens RF level, 60.0. Positive polarity mode using HRFS data and a subsequent AIF mode was carried out. In HRFS data acquisition: resolution was set to 35,000; microscans, 1; automatic gain control (AGC) target, 3×10^6; maximum injection time (IT), 200 ms; and scan range, m/z 100–600. For the AIF mode: resolution was 17,500; microscans, 1; AGC target, 3×10^6; maximum IT, 200 ms; HCD with stepped normalized collision energy, 17.5%, 35%, and 52.5%; scan range, m/z 50–600; and profile mode was used in spectral data collection. Two steps extraction of 1 mL plasma, first in neutral and second in alkaline pH, was performed using diethyl ether-ethyl acetate (1:1). Trimipramine-d3 was used as the internal standard (IS). Accurate mass and the signal intensity of the protonated molecule were acquired in HRFS mode (= qualifier 1 and quantifier) and additionally, at least two characteristic fragment ions with mass accuracies ≤5 ppm acquired in the AIF mode (= qualifier 2 and 3) were used for confirmation. LLOQ values were between 0.18 and 0.27 ng/mL. The method used is based on identification of selective fragment ions. The HRFS/AIF system, which was preferred instead of HRFS targeted MS-MS, allowed to add new compounds into the method without modification of the MS conditions. In this mode, the number of monitored compounds are not limited and also in HRFS/AIF gives the chance of searching previous MS data files to identify former unknown molecules utilizing their retention time and qualifier ions, after obtaining their standards. The most interesting benefit of this mode is the possibility of identification of the unexpected molecules, using the group-indicating fragment ions as in m/z 121.0653, which is a sign of the existence of NBOMes and this ion can also be characteristic for the main metabolites for 25B-NBOMe, 25C-NBOMe, and 25I-NBOMe.

NBOMe compounds are extremely potent and highly hallucinogenic at very low dosages [21]. Several independent reports suggest that doses of swallowed 25C-NBOMe range between 50 and 1200 μg. However, drug forums on the Internet show that some users may also choose fearsome doses; for example, for 25I-NBOMe, probably without adequate knowledge on the effective dose. Since in most instances the concentration of these NBOMe compounds reported in postmortem cases is exceedingly low, often as low as <0.50 ng/mL, sensitive and rapid methods are needed to solve the cases [4,22]. For this aim, a UPLC-ESI(+)-MS/MS method was developed in 2014 to be able to detect 25I-NBOMe, as well as other NBOMe classes of drugs (e.g., 25B-, 25C-, 25D-, 25H-, 25I-, and 25T2-NBOMe), in both ante- and postmortem specimens, including blood, urine, and other body fluids [22]. No sample preparation was needed for whole blood and vitreous humor. Bile, gastric contents, and urine were diluted 1:10 with water. SPE extraction with Clean Screen ZSDUA020 solid phase was used for serum. The analytes were eluted using 3 mL of 2% ammonia in an 80:20 mixture of methylene chloride/isopropanol, into disposable culture tubes containing 200 μL water and 100 μL 1% HCl in methanol. Eluents were concentrated to ~100 μL and analyzed in UPLC-MS-MS. A PFP (50 × 2.1 mm i.d., 1.7-μm particles) column and a gradient elution with 2.0 mM ammonium formate containing 0.2% formic acid and acetonitrile containing 0.1% formic acid were used. All compounds were eluted in three minutes. Q trap and MRM mode were used. Ion spray voltage was 5,000 V; entrance potential, 10 V; collisionally activated dissociation gas, medium; curtain gas, 10; ion source gas 1, 70; ion source gas 2, 30; and source temperature was 500°C. LOD's and LOQ's for these analogs ranged from 0.005 to 0.01 ng/mL and 0.01 to 0.02 ng/mL, in order. The typical linear dynamic ranges were 0.01–20 ng/mL. Recoveries were between 54% and 94%.

16.2.5 METABOLISM STUDIES/METABOLITE IDENTIFICATION

Metabolism studies on NBOMe derivatives are also increasing in the literature, especially LC-HRMS techniques are included in most of them. In some studies, results are confirmed using both GC-MS and LC-MS/MS.

In 2015, LC-HRMS/MS was used to study the phase I and II metabolism of 25I-NBOMe and its detectability in urine screening approaches [23]. After application of 25I-NBOMe to male Wistar rats, urine was collected over 24 hours. The phase I and II metabolites were identified by LC-HRMS/MS in urine after suitable workup. For the detectability studies, standard urine screening procedures by GC-MS, LC-MS^n, and LC-HRMS/MS were applied to rat and also to authentic human urine samples submitted for toxicological analysis. Finally, an initial CYP activity screening was performed to identify CYP isoenzymes involved in the major metabolic steps. Since 25I-NBOMe was extensively metabolized, it could be detected only by the LC-MS screening approaches. Since CYP2C9 and CYP3A4 are involved in initial metabolic steps, drug–drug interactions might be possible if used in combinations. Metabolization was mainly by O-demethylation, O,O-bis-demethylation, hydroxylation, and combinations of these reactions as well as by glucuronidation and sulfation of the main phase I metabolites. More than 60 metabolites for 25I-NBOMe were detected. However, instead of animal studies, metabolites and isoenzymes involved in the major metabolic steps could be investigated in NBOMe users, because there are a serious number of users brought or applying to hospitals while under drug effect or seized after use, some self-reporting their drug use. As long as there is chance to reach real samples or users, it is possible to detect metabolites in the samples obtained from these users.

In 2017, a case was reported on severe intoxication with a liquid mixture of NBOMes consumed nasally that has been retrospectively documented through 25I-NBOMe and metabolites determinations [7]. A 29-year-old man experienced unconsciousness within one hour after a drop of pink liquid through the nose. Hypertonia and tremors, then partial seizure with secondary generalization, bilateral and reactive mydriasis, tachycardia, hypertension, hyperthermia, and profuse sweating, suggestive of serotonin syndrome, were observed in initial examination. He presented persistent cognitive and psychiatric abnormalities. LC-HRMS revealed the presence of three NBOMes in the pink liquid with 25I-NBOMe as the main component, 25I-NBOMe in serum (LC-MS/MS quantitation: 0.9 ng/mL), and seven 25I-NBOMe metabolites in his urine samples: two desmethyl-25I-NBOMes, a desmethyl-hydroxy-25I-NBOMe, a hydroxy-25I-NBOMe, a di-desmethyl-25I-NBOMe, a desmethyl-25I-NBOMe glucuronide, and a hydroxy-25I-NBOMe glucuronide. These metabolites were retrospectively identified following *in vitro* investigations of new psychoactive drug (NPS) metabolism using analysis of samples collected after 25I-NBOMe incubation with a pool of human liver microsomes (HLMs) and cross-checking to *in silico* predicted biotransformations of the acquired data. Finally, the MS spectra of all the identified metabolites were in HRMS and MS-MS libraries in order to perform routinely efficient detection of new psychoactive drugs users.

Frequent confusion may occur between NBOMes and LSD, because of the renewed interest for LSD due to the popularity of NBOMes [24]. This was the case for a 25-year-old man crushed by a train while he was returning from a rave party. The study indicated that the consumption of 25I-NBOMe was rapidly evoked by people who had participated at the party. A broad screening was applied in the postmortem blood and urine samples of the corpse. A targeted screening in MRM mode was also performed on a smaller number of hallucinogens. Ethanol was measured at 0.71 and 1.59 g/L in blood and urine samples, respectively. The screening by LC-HRMS did not reveal the presence of NPS, including NBOMes. The targeted screening in MRM mode revealed the presence of LSD and its metabolite, the 2-oxo-3-hydroxy-LSD. LSD was quantified at 0.2 ng/mL in blood. This case therefore encourages the screening of all hallucinogenic substances, even when the context is evocative.

In another two fatality cases of acute intoxication associated with NBOMe derivatives reported by Tang et al. [25], a QTrap triple-quadrupole MS was used. Two male patients (17 and 31 years of age) had ingested drugs labelled as "NBOMe" or "Holland film" and developed confusion, agitation, hypertension, tachycardia, hyperthermia, sweating, and dilated pupils. The patients required benzodiazepines and other drugs for the control of symptoms. Urine samples from both patients were analyzed using LC-MS/MS, following glucuronidase digestion and solid-phase extraction. A QTrap triple-quadrupole MS equipped with turbo ion spray source coupled to a HPLC system was used along with a XDB-C_8 column (4.6 mm × 150 mm, 5 µm) and gradient elution (flow rate: 1 mL/minute) comprising 1 mM ammonium formate, 0.1% formic acid in water (A), and 1 mM ammonium formate, 0.1% formic acid in acetonitrile (B). The analysis time was 32 minutes. Mass parameters: source temperature was 650°C, curtain gas flow rate was 30 mL/minute, ion spray voltage was 5000 V, flow rates of ion source gas 1 and 2 were at 25 mL/minute each. Two MRM transitions were monitored for each compound: 25B-NBOMe: 380 > 121 and 91; 25C-NBOMe: 336 > 121 and 91; 25H-NBOMe: 302 > 121 and 91; and 25I-NBOMe: 428 > 121 and 91. The declustering potential was 20 V for 25H-NBOMe and 40 V for the remaining compounds. Further confirmation was performed using enhanced product ion scan at collision energy 20 and 50. The identity of the analytes was confirmed by comparison with reference standards. Besides specific analysis for NBOMe compounds, the urine samples were also subjected to general toxicology screening covering the common pharmaceuticals and drugs of abuse, using LC-MS/MS and GC-MS. Identification was based upon comparison of the retention time and enhanced product ion scan with reference standards. In both urine samples, 25B-NBOMe was detected. Additionally, 25C-NBOMe was identified in one of the urine samples. Clinicians as well as laboratory staff play an important role in facilitating the detection of this group of potentially dangerous emerging drugs.

16.2.6 Has the User Really Taken the Compound on the Label?

In 2015, LCMS/MS with Q trap technology was used for the analysis of NBOMe drugs in two cases [26]. These cases of NBOMe intoxication were reported in patients who believed they had used 25I-NBOMe, while lab confirmation proved otherwise in their serum and urine samples. Fast SPE extraction was used for sample preparation. The NBOMes were extracted using Clean Screen FASt® extraction columns. The following transition ions (m/z) were monitored in multiple reaction monitoring (MRM) mode: 25C-NBOMe: 336 > 121 and 336 > 91; 25B-NBOMe: 381 > 121 and 381 > 91; and 25I-NBOMe-d_3: 431 > 124 and 431 > 92. 25C-NBOMe was measured in urine using a previously published method. Patients believed they were taking

25I-NBOMe but it was confirmed via mass spectrometry that the drugs they have taken were the chlorine and bromine derivatives, 25C-NBOMe and 25B-NBOMe, respectively.

16.2.7   MULTIMETHODS FOR IDENTIFICATION: AN EFFECTIVE TOOL IN THE CHARACTERIZATION OF NEW ANALOGS

A combination of multitechniques such as GC-MS, LC-MS/MS, and NMR was also used for the identification of new designer drugs. Uchiyama et al. reported two new types of synthetic cannabinoids, an AM-2201 benzimidazole analog (FUBIMINA, 1) and (4 methylpiperazin-1-yl)(1-pentyl-1H-indol-3-yl)methanone (MEPIRAPIM, 2), and three newly emerged phenethylamine derivatives, 25B-NBOMe (3), 2C-N-NBOMe (4), and a 25H-NBOMe 3,4,5-trimethoxybenzyl analog (5), in illegal products distributed in Japan, in their paper in 2014 [27]. ten mg of each herbal-type product was crushed into powder, a 2 mg portion of the powder-type product and a 20 mL portion of the liquid-type product were taken, and each were mixed with 1 mL of methanol under ultrasonication for ten minutes. After centrifugation, the supernatant solution was filtered. Each sample solution was analyzed by UPLC-ESI-MS and GC-EI-MS according to our previous report. In the GC-MS analysis, the oven temperature program was 80°C (one-minute hold) with an increase at a rate of 5°C/minute to 190°C (15-minute hold), followed by an increase at 10°C/minute up to 310°C (20-minute hold). They also used their in-house EI-MS library of designer drugs generated from their continuous survey of illegal products and commercially available reagents for structural elucidation. Accurate mass numbers for the target compounds were determined by LC-ESI-QTOF-MS. For isolation of compounds 4 and 5, they used preparative gel permeation liquid chromatography (GPLC). Assignments were made via ^{1}H NMR, ^{13}C NMR, heteronuclear multiple quantum coherence (HMQC), heteronuclear multiple-bond correlation (HMBC), ^{15}N HMBC, double quantum filtered correlation spectroscopy (DQF-COSY), and rotating-frame nuclear Overhauser effect (ROE) spectra. In the GC-MS and LC-MS analyses, unknown peaks were investigated in the products. Accurate masses of protonated molecular ions [M + H]$^+$ were found in LC-MS/MS, QToF-MS was used for the accurate masses to predict the molecular formulas. [M + H]$^+$ and the observed fragment ions were compared with those of GC-MS results and the data of the purchased authentic compound. The ^{1}H, ^{13}C NMR, and 2D NMR spectra of compounds helped to elucidate the structures through giving the number of protons and carbons and gave an idea on the functional groups and their distance from each other. UV spectrum of compounds again helped to elucidate some other parts of the molecule. For example, the 2C-N-NBOMe (4) (λmax 244, 276, 371 nm), which was different from that of 25B-NBOMe (3) (λmax 296 nm) and similar to that of 2C-N (λmax 245, 279, 375 nm), supported the existence of a -NO$_2$ group. The connections of some moieties were suggested by the HMBC and the ^{15}N HMBC correlations. As a result of the above chromatographic and spectroscopic analyses, the structures of compounds were elucidated successfully. The molecular structures of 25B-NBOMe (3), 2C-N-NBOMe (4), and a 25H-NBOMe 3,4,5-trimethoxybenzyl analog (5) are given in Figure 16.6.

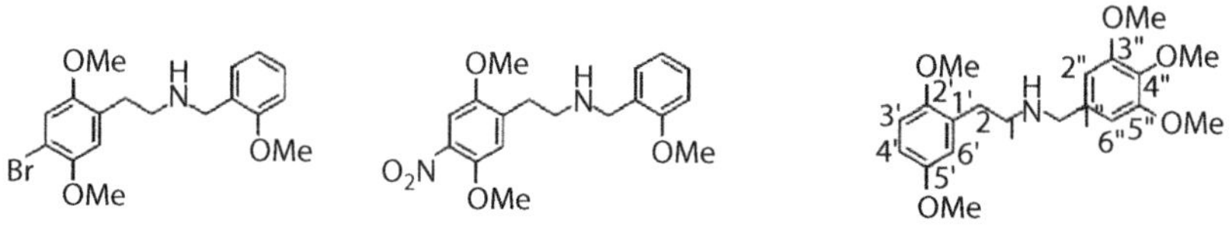

25B-NBOMe (3)
$C_{18}H_{22}BrNO_3$: 3380

2C-N-NBOMe (4)
$C_{18}H_{22}N_2O_5$: 3346

25H-NBOMe 3, 4, 5-trimethoxybenzyl analog (5)
2-(2, 5-dimethoxyphenyl)-N-(3,4,5-
trimethoxybenzyl) ethanamine
$C_{20}H_{27}NO_5$: 3361

FIGURE 16.6 The molecular structures of 25B-NBOMe (3), 2C-N-NBOMe (4), and a 25H-NBOMe 3,4,5-trimethoxybenzyl analog (5). (Reprinted from Uchiyama, N. et al., 2014, *Forensic Toxicol*, 32:105–115.)

Twenty-two samples of ingredients of recreational drugs before being sold on the drug market obtained from a drug dealer were analyzed by GC-MS, HRMS, and NMR [28]. This study demonstrated 15 novel designer drugs, which have not been described in scientific literature. The analyzed drugs included three NBOMe drugs: 25H-NBOMe, 25D-NBOMe, and 25E-NBOMe; three other phenethylamine-type drugs: 25I-NBMD, RH34, and escaline, and 25I-NBOMe, ADB-CHIMINACA, 5F-ADB, etc. In GC-EI-MS analysis, a 10-μL aliquot of each liquid sample or 10 mg of each powder sample was initially dissolved in 2 mL methanol, diluted and analyzed. A DB5MS capillary column (30 m × 0.25 mm i.d., 0.25-μm film thickness) with a helium flow at 1.56 mL/minute was used. The injection port was 260°C and the injection volume was 1 μL in the splitless mode. The initial oven temperature was set to 60°C, which was held for two minutes, and then increased by 10°C/minute up to 320°C and held for 10 minutes. Ion source was set to 200°C and operation was in full-scan (*m/z* 40–700) mode. For NMR analysis, a 10-mg sample of each powder (nine samples, which could not be identified by GC/MS) was dissolved in 1 mL methanol-d_4 (99.8%) or pyridine-d_5 (99.8%). NMR spectra were measured at 500 MHz for ^{1}H and 125 MHz for ^{13}C. The signals were assigned on the basis of 2D NMR experiments, which involved correlated spectroscopy (COSY), distortionless enhancement by polarization transfer (DEPT135), heteronuclear multiple quantum coherence (HMQC), and heteronuclear multiple-bond coherence (HMBC) spectral analyses. HRMS operated by fast atom bombardment (FAB) in the positive mode with xenon gas was used for the analyses of nine samples (which could not be identified by GC/MS). Glycerol or 3-nitrobenzyl alcohol was used as matrix. Every sample was composed of a single compound, showing a purity of more than 90%. Twelve samples (ten compounds) matched with the data of the spectral libraries and were identified as specific compounds. Among 25-NBOMe designer drugs, 25I-NBOMe and 25B-NBOMe were most commonly detected from seized materials.

16.2.8 DETERMINATION AND DISCRIMINATION OF CLOSE ANALOGS AND POSITIONAL ISOMERS

A major challenge in identifying newly emerging substances is the lack of analytical data and the ability to accurately identify positional isomers [29].

Six N-benzylphenethylamines based on the 2,5-dimethoxy-4-iodophenethylamine structure ("25I") and 12 substituted N-benzyl-5-methoxytryptamines ("5MT") have been prepared and extensively characterized. Techniques used for characterization were GC-IT-MS in electron and chemical ionization mode, HPLC-DAD, IR, electrospray high-mass-accuracy QToF-MS, and triple-quadrupole MS. The analytical characterizations were carried out using GC-IT-MS in electron impact (EI) and chemical ionization (CI) mode, LC-diode-array detection, IR spectroscopy, electrospray high-mass-accuracy QToF-MS, and triple-quadrupole MS. GC/MS data for all 18 compounds (0.5 mg/mL in methanol) were obtained in EI and CI mode. An injector (275°C) was used in split mode (1:50). Transfer line, manifold and ion trap temperatures were set at 310, 80 and 220°C, respectively. The carrier gas was helium at a flow rate of 1 mL/minute. A VF-5ms GC column (30 m × 0.25 mm, 0.25-µm film thickness) was employed for separation. The starting temperature was set at 130°C and held for one minute, then increased at 20°C/minute to 280°C and held constant for 11.50 minutes to give a total run time of 20 minutes. Electrospray triple-quadrupole tandem mass spectrometry experiments were carried out by direct infusion (10 µL/minute at 0.01 mg/mL) of compounds. MS optimizations were performed in MS scan and in product ion scan (+) modes. Capillary voltage was 3.12 kV; cone voltage 28 V; RF lens voltage 0.1 V; source temperature 100°C; desolvation temperature 200°C; and multiplier voltage 650 V. Nitrogen was used as the cone gas (50 L/h flow rate) and desolvation gas (200 L/h flow rate) while the collision gas was argon (0.3 mL/minute flow rate). The $[M + H]^+$ ions obtained for all 18 compounds were selected for MS/MS experiments. The collision energy used was 20 eV while a value of 10 eV was chosen for the tryptamine-based compounds. Compounds were characterized by UPLC/QTOF-MS/MS, as described previously. ESI-QTOF-MS data were acquired in positive mode scanning from m/z 100 to 1000 with and without auto MS/MS fragmentation. ESI-QTOF-MS parameters were: gas temperature 325°C; drying gas (N_2) flow rate 10 L/minute; and sheath gas (N_2) temperature 400°C. For HPLC/DAD analysis, a detection window was set at 200 to 595 nm. The characterization of 18 "NBOMe" compounds provided a comprehensive collection of chromatographic and spectral data. Four groups of three positional isomers, i.e., 25I-NB2OMe, 25I-NB3OMe, 25I-NB4OMe, 25I-NB2B, 25I-NB3B, 25I-NB4B, and their 5-methoxytryptamine counterparts, were included and assessed for ability to obtain differentiation. Six meta-substituted N-benzyl derivatives of 5-methoxytryptamine (-CF₃, -F, -CH₃, -Cl, -I, -SCH₃) were also studied. Structural representations of 25I-NBOMes and 5MT-NBOMes, and suggested key ions formed during MS analysis, are given in Figure 16.7.

In 2017, Pasin et al. investigated the comprehensive characterization of hallucinogenic phenethylamines and synthetic cannabinoids in a nontargeted screening approach for the 2C-X, DOX, and 25X-NBOMe derivatives [30]. The determination of common product ions and losses through collision-induced dissociation (CID) studies using HRMS for identification of product ion formulae was carried out for these molecules. The QTOF-MS was operated in ESI+ mode using Extended Dynamic Range (2 GHz) with capillary and fragmentor voltages set to 3500 and 180 V, respectively. An Auto-MS/MS (data-dependent) acquisition mode was used over a mass range of m/z 50–1000 for both MS and MS/MS experiments with scan

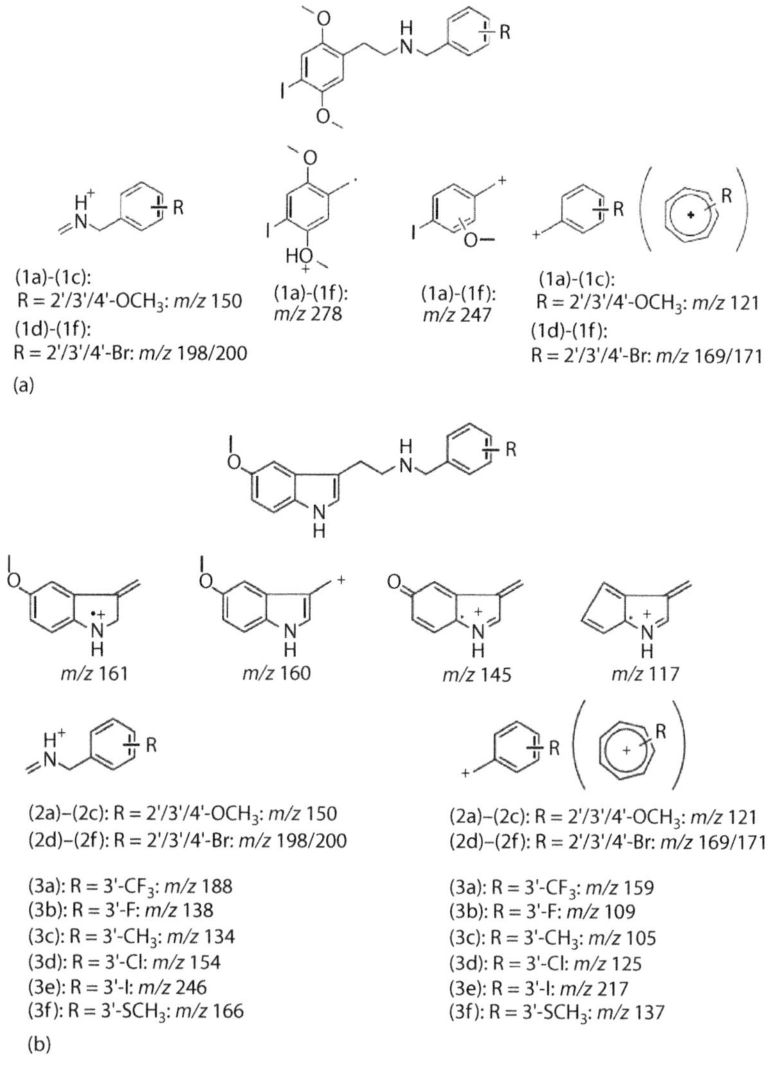

(1a)-(1c):
R = 2'/3'/4'-OCH$_3$: m/z 150
(1d)-(1f):
R = 2'/3'/4'-Br: m/z 198/200

(a)

(1a)-(1f):
m/z 278

(1a)-(1f):
m/z 247

(1a)-(1c):
R = 2'/3'/4'-OCH$_3$: m/z 121
(1d)-(1f):
R = 2'/3'/4'-Br: m/z 169/171

m/z 161   m/z 160   m/z 145   m/z 117

(2a)-(2c): R = 2'/3'/4'-OCH$_3$: m/z 150
(2d)-(2f): R = 2'/3'/4'-Br: m/z 198/200

(3a): R = 3'-CF$_3$: m/z 188
(3b): R = 3'-F: m/z 138
(3c): R = 3'-CH$_3$: m/z 134
(3d): R = 3'-Cl: m/z 154
(3e): R = 3'-I: m/z 246
(3f): R = 3'-SCH$_3$: m/z 166

(b)

(2a)-(2c): R = 2'/3'/4'-OCH$_3$: m/z 121
(2d)-(2f): R = 2'/3'/4'-Br: m/z 169/171

(3a): R = 3'-CF$_3$: m/z 159
(3b): R = 3'-F: m/z 109
(3c): R = 3'-CH$_3$: m/z 105
(3d): R = 3'-Cl: m/z 125
(3e): R = 3'-I: m/z 217
(3f): R = 3'-SCH$_3$: m/z 137

FIGURE 16.7 Structural representations of suggested key ions formed during analysis by electron ionization ion trap mass spectrometry: (a) 25I-NBOMes (1a)–(1f) and (b) 5MT-NBOMes (2a)–(3f). (Reprinted from Brandt, S. D. et al., 2015, *Rapid Commun Mass Spectrom*, 29:573–584.)

rates of 1 and 3 spectra/second, respectively. A maximum of three precursors from the MS scan were selected for CID per cycle with a cycle time of 2.1 seconds and an abundance threshold of 200 counts. CID experiments were performed at collision energies (CE) of 10, 20, and 40 eV in separate analyses with nitrogen as the collision gas. A Poroshell 120 C$_{18}$ column (2.1 × 75 mm, 2.7 μm) was used for separation, using a gradient elution with a flow rate of 0.4 mL/minute with a total run time of 17 minutes. Mobile phase A consisted of 20 mM ammonium formate and mobile phase B consisted of acetonitrile containing 0.1% (% v/v) formic acid. In this study, twelve 2C–X, six DOX, and fourteen 25X–NBOMe analogs were analyzed by UPLC-QTOF-MS in order to evaluate their CID pathways and assess the applicability of the generated product ions for nontargeted detection of novel hallucinogenic phenethylamine analogs. Isotopically labelled 2C-B-d_6 and 25I-NBOMe-d_9 were utilized in the elucidation of product ions. Hallucinogenic phenethylamines were successfully characterized by LC-QTOF-MS using CID at different CEs. The 2C-X and DOX derivatives had common losses of NH_3, CH_6N, and C_2H_9N, and common product ions at m/z 164.0837, 149.0603, and 134.0732 for 2C-X derivatives and m/z 178.0994, 163.0754, 147.0804, and 135.0810 for DOX derivatives. The 25X-NBOMe derivatives had characteristic product ion spectra with abundant ions at m/z 121.0654 and 91.0548, together with minor neutral losses corresponding to 2-methylanisole and 2-methoxybenzylamine and $\cdot C_9H_{14}NO$. Screening for these common neutral losses and product ions can be used in a nontargeted screening approach to detect and tentatively identify novel analogs.

In June 2015, a clandestine laboratory was dismantled in Hubei Province, China, and about 20 kg of new psychoactive powder samples were seized. About 200 unknown samples were submitted to the national narcotics laboratory of the Ministry of Public Security for analysis [31]. Nineteen species of substituted phenethylamine derivatives were identified, including: 5-(2-methylaminopropyl)-2,3-dihydrobenzofuran (5-MAPDB), 5-(2-aminoethyl)-2,3-dihydrobenzofuran (5-AEDB), N,2-dimethyl-3-(3,4-methylenedioxyphenyl)propan-1-amine (MDMA methylene homolog), 6-bromo-3,4-methylenedioxymethamphetamine (6-Br-MDMA), and 1-(benzofuran-5-yl)-N-(2-methoxybenzyl)propan-2-amine (5-APB-NBOMe). Analytical properties of five substituted phenethylamine derivatives were identified by LC-QTOF-MS, GC-MS, and NMR. For GC-MS, LC-QTOF-MS, and NMR analyses, the samples were prepared in methanol, 0.1% formic acid (v/v) in water, and in deuterated methanol, respectively. For LC-QTOF-MS analysis, separation was performed at 40°C with an Acquity UPLC CSHTM C_{18} column (10 cm × 2.1 mm i.d., 1.7-μm particle diameters). For gradient elution, the mobile phases 0.1% formic acid in water and acetonitrile were mixed, with a flow rate of 0.4 mL/minute. The QTOF instrument was operated by ESI (+) mode with these parameters: ion spray voltage, 5.5 kV; turbo spray temperature, 600°C; nebulizer gas (gas 1), 50 psi; heater gas (gas 2), 50 psi; and curtain gas, 30 psi. Nitrogen was used as the nebulizer and auxiliary gas. The optimized declustering potential and collision energy were 80 and 5 V, in order. In the second experiment, a sweeping collision energy setting at 25 ± 15 V was applied for CID to obtain the fragment ions from the ions in the preceding scan. Under these conditions the retention times of the investigated compounds were: 2.0 minutes for 5-AEDB; 2.9 minutes for 5-MAPDB; 3.1 minutes for MDMA methylene homolog; 3.3 minutes for 6-Br-MDMA; and 4 minutes for 5-APB-NBOMe. Gas chromatographic separation was carried out on a DB-5 MS capillary column (30 m × 0.25 mm i.d., 0.25-μm film thickness) with a constant flow rate of helium at 1.0 mL/minute. The filtered solutions were injected in split mode (20:1). The initial column temperature (60°C) was increased to 280°C at a rate of 20°C/minute, and held at this temperature for 20 minutes, then ramped up to 300°C at a rate of 10°C/minute, and finally held at 300°C for 20 minutes. The GC injector and transfer line were maintained at 280°C and 250°C, respectively. Ionization energy was set at 70 eV. Acquisition was carried out in a scan mode range of m/z 35–500. Injection volume was 1 μL. Under these conditions the retention times were 8.0 minutes for 5-MAPDB; 7.9 minutes for 5-AEDB; 8.1 minutes for MDMA methylene homolog; 9.2 minutes for 6-Br-MDMA; and 11.9 minutes for 5-APB-NBOMe. NMR spectra were obtained on an Avance III 400 spectrometer at 300 K with 400 MHz for ^{1}H and 100 MHz for ^{13}C. Assignments were made via ^{1}H-NMR, ^{13}C-NMR, ^{13}C-distortionless enhancement by polarization transfer (^{13}C-DEPT), ^{1}H/^{1}H correlation spectroscopy (^{1}H/^{1}H-COSY), ^{1}H/^{13}C-heteronuclear single-quantum correlation spectroscopy (^{1}H/^{13}C-HSQC), and ^{1}H/^{13}C-hetero nuclear multiple-bond correlation spectroscopy (^{1}H/^{13}C-HMBC) spectra. The chemical shifts for ^{1}H and ^{13}C NMR spectra were referenced to the residual solvent peak of CD$_3$OD at 3.31 ppm for ^{1}H NMR spectra and 49.0 ppm for ^{13}C NMR spectra.

Two new NPSs, a phenethylamine derivative 2-(4-iodo-2,5-dimethoxyphenyl)-N-[(3,4-methylenedioxyphenyl)methyl]ethanamine (25I-NB34MD) and a piperazine derivative 1-(3,4-difluoromethylenedioxybenzyl)piperazine (DF-MDBP), were identified in illicit products distributed from January to March 2015 in Japan [32]. 25I-NB34MD has a 3,4-methylenedioxybenzyl moiety that is an analog of N-benzylmethoxy derivatives of 25I-NBOMe. DF-MDBP is a difluoromethylenedioxy analog of the known designer drug 1-(3,4-methylenedioxybenzyl) piperazine (MDBP). The analyzed samples were obtained between January and March 2015 as 15 liquid-type products called "liquid aroma." UPLC-ESI-MS analysis was carried out with a binary mobile phase consisting of solvent A (0.1% formic acid in water) and solvent B (0.1% formic acid in acetonitrile). Two elution programs were used at a flow rate of 0.3 mL/minute in the LC-MS analysis. ESI mode was performed to obtain accurate mass numbers, according to the authors' previous report [33]. GC-EI-MS was performed on a HP-1MS capillary column (30 m × 0.25 mm i.d., 0.25-μm film thickness) with helium gas as a carrier at 0.7 mL/minute. As the GC-MS conditions, the electron energy was 70 eV; injector temperature, 220°C; injection, duration of splitless mode, 1.0 minute; transfer line temperature, 280°C; scan range, m/z 40–550. The oven temperature was held at 80°C for one minute and increased at a rate of 5°C/minute to 190°C, held for 15 minutes followed by an increase at 10°C/minute up to 310°C and held for 15 minutes. The obtained GC mass spectra were compared to those of an EI-MS library. The authors also used the in-house EI-MS library of designer drugs obtained by their continuous survey of illegal products and commercially available reagents for the structural elucidation. In the NMR study, assignments were made via ^{1}H NMR, ^{13}C NMR, heteronuclear multiple quantum coherence (HMQC), heteronuclear multiple-bond correlation (HMBC), ^{15}N HMBC, double quantum filtered correlation spectroscopy (HH-COSY) and nuclear Overhauser effect (NOE) spectra. Considering obtained results, when analyzing NPSs in illicit products, it is possible to misidentify them as isomers of other NPSs. Unknown substances should be compared with references standards by using multiple instruments, such as GC-MS and LC-MS.

16.3 CONCLUSIONS

NBOMe derivatives, especially 25I-NBOMe, are relatively new designer drugs that continue to remain popular drugs of abuse, especially among young adults. New synthetic analogs of these are continuously launched to the market, to avoid legal issues and detection in drug tests [8]. They are also similar in their common adverse reactions, especially clinically significant psychotic reactions. Hallucinogenic properties resemble LSD. Even though users had greater "negative effects while high," they prefer these drugs because they can purchase them at a lower price. Detection of these drugs with urine tests is challenging, so clinicians should consider designer drug use in young adults with agitation and psychosis. This class of drugs imitates LSD and raises significant health concerns, since a recreational LSD user may consume NBOMe unintentionally [34]. In fact, both in severe intoxications and fatal cases where the ingestion of NBOMe compounds was analytically confirmed, the individuals involved incorrectly thought they had consumed "acid." The effects produced after the use of "traditional" hallucinogens, including LSD, are treated with the use of benzodiazepines. Moreover, no fatal cases attributed to LSD overdose have been known. In contrast, despite the fact that NBOMes have only recently entered the drug market scene, a number of severe intoxications and deaths have been linked to this group. Results obtained by Coelho [35] suggest great inhomogeneity in drug dosages even within blotters presenting the same artwork, taken from the same blotter sheet. This could be attributed to the fact that these drugs are produced in clandestine laboratories by amateur cooks. Taking into account the potency of this class, the variation of purity and inhomogeneity in drug dosages, users can easily overdose. Moreover, the high potency and small dose ingested makes analytical detection of NBOMes exceedingly difficult. Even for laboratories with highly sensitive methods, the detection of these hallucinogens is challenging, since the signals from these drugs are very low and are easily lost in the background noise of the sample. Therefore, it is primarily important for clinicians, forensic pathologists, and investigators in intoxication and fatality cases to provide relevant information and the circumstances of their cases to the toxicology laboratory.

As well as 25I-NBOMe, which seems the most dangerous NBOMe type, intoxications and fatalities related to 25C-NBOMe use are also increasing worldwide, despite the fact that its use and trafficking are banned in many countries [36]. 25B-NBOMe has also previously been implicated in clinical intoxications and fatalities [25]. Clinicians as well as laboratory staff play an important role in facilitating the detection of this group of potentially dangerous emerging drugs. They should train themselves on these new drugs and should recognize the symptoms in order to properly direct the analysis to a certain drug group. The public should be informed as various new substituted analogs continuously appear. Clinicians and medical doctors should strongly discourage the purchase and use of psychedelic drugs and inform the public of all the severe consequences associated with their use. The development of sensitive, rapid, routine analytical methods with high accuracy and resolution for the determination of NBOMe derivatives and their metabolites in biological fluids is necessary for proper toxicological investigation of forensic and clinical cases.

Until recently, clinical and forensic toxicology laboratories used immunoassay for screening and MS techniques for confirmation for a limited number of classes of compounds. However, the specificity of immunoassay is effected from the affinity and the cross-reactivity of the antibodies used for the parent drug, its metabolites, and analogs [37]. Confirmatory analysis only applied to presumably positive samples and is not applied to ones that screened presumptively negative, which may cause practitioners to miss some new psychoactive compounds that may be included in the sample. If widescope screening methods are developed and used, they could help to save time in solving a case, but the method must provide accurate recognition and in some cases true quantification of the target compounds, as well. Today, HRMS offers the possibility to overcome the limitations of multitarget screening by enabling accurate-mass determination of ionic species obtained from drugs and their metabolites with an accuracy of at least four decimal numbers. Analyte identification is based on searching against a database of elemental formula of pertinent compounds. Formula of new drugs and their possible metabolites taken from the literature may be easily added to the database. The most applicable types of HRMS instruments for toxicological analysis are time-of-flight (TOF) MS and electrostatic traps. At the present time, high-end TOF instruments afford a resolution of up to 40,000–50,000 and accuracy <3 ppm, while standard MS devices have a resolution up to 10,000–30,000 and an accuracy <5 ppm. Resolution can be up to 100,000 in electrostatic traps with fourier transform MS detection, according to different instrumental arrangements, and 240,000 (with a <3 ppm accuracy) with a more specialized equipment.

Although laboratory testing is expanding, widespread standardized designer drug testing is not yet available in most clinical settings and laboratories; furthermore, even routine tests for the identification of NBOMe derivatives in biological fluids or tissues are not available at most hospitals or outside labs [6,8]. The analytical challenge is compounded by heterogeneity in designer drug product contents, concentration, and chemical constituents, all of which may vary between and within products. Therefore, new, fast, and adequately sensitive methods should be developed and properly validated for this purpose. Moreover, forensic toxicologists should be aware of the recent trends in drug abuse, including the extensive recreational use of substituted NBOMEs, as well as the recent trends in analytical technologies. Also, when they investigate a drug-related case, they should continuously enlarge the scope of their tests, including the search for newly emerging substances. Because of their high potencies and ease of synthesis, NBOMe use is seemingly becoming more widespread and more new derivatives will probably appear in the future. The publication of new data, case reports, and evaluation of the NBOMe metabolites at least in human hepatocytes and urine matrices is necessary in order to improve knowledge and awareness within the clinical and forensic community.

REFERENCES

1. Poklis, J. L., Raso, S. A., Alford, K. N., Poklis, A., and Peace, M. R. 2015. Analysis of 25I-NBOMe, 25B-NBOMe, 25C-NBOMe and other dimethoxyphenyl-N-[(2-methoxyphenyl) methyl] ethanamine derivatives on blotter paper. *J Anal Toxicol*, 39:617–623.
2. McGonigal, M. K., Wilhide, J. A., Smith, P. B., Elliott, N. M., and Dorman, F. L. 2017. Analysis of synthetic phenethylamine street drugs using direct sample analysis coupled to accurate mass time of flight mass spectrometry. *Forensic Sci Int*, 275:83–89.

3. Wood, D. M., Sedefov, R., Cunningham, A., and Dargan, P. I. 2015. Prevalence of use and acute toxicity associated with the use of NBOMe drugs. *Clin Toxicol*, 53: 85–92.

4. Johnson, R. D., Botch-Jones S. R., Flowers, T., and Lewis, C. A. 2014. An evaluation of 25B-, 25C-, 25D-, 25H-, 25I- and 25T2-NBOMe via LC-MS-MS: Method validation and analyte stability. *J Anal Toxicol*, 38:479–84.

5. Lin, D. 2016. Designer drugs—A brief overview. *Therap Toxins News, Newsl TDM Toxicol Div AACC*, 2:1–10.

6. Nikolaou, P., Papoutsis, I., Stefanidou, M., Spiliopoulou, C., and Athanaselis, S. 2015. 2C-I-NBOMe, an "N-bomb" that kills with "Smiles." Toxicological and legislative aspects. *Drug Chem Toxicol*, 38:113–119.

7. Richeval, C., Boucher, A., Humbert, L., Phanithavong, M., Wiart, J. F., Moulsma, M., Citterio-Quentin, A., Coulon, T., Hernu, R., and Vial, T. 2017. Retrospective identification of 25I-NBOMe metabolites in an intoxication case. *Toxicol Anal Clin*, 29:71–81.

8. Weaver, M. F., Hopper, J. A., and Gunderson, E. W. 2015. Designer drugs 2015: Assessment and management. *Addict Sci Clin Pract*, 10:1–9.

9. Poklis, J. L., Clay, D. J., and Poklis, A. 2014. High-performance liquid chromatography with tandem mass spectrometry for the determination of nine hallucinogenic 25-NBOMe designer drugs in urine specimens. *J Anal Toxicol*, 38:113–121.

10. Liu, C., Jia, W., Qian, Z., Li, T., and Hua, Z. 2017. Identification of five substituted phenethylamine derivatives 5-MAPDB, 5-AEDB, MDMA methylene homolog, 6-Br-MDMA, and 5-APB-NBOMe, *Drug Test Anal*, 9:199–207.

11. Westphal, F., Girreser, U., and Waldmüller, D. 2015. Analytical characterization of four new ortho-methoxybenzylated amphetamine-type designer drugs. *Drug Test Anal*, 8:910–919.

12. Shevyrin, V., Kupriyanova, O., Lebedev, A. T., Melkozerov, V., Eltsov, O., Shafran, Y., Morzherin, Y. and Sadykova, R. 2016. Mass spectrometric properties of N-(2-methoxybenzyl)- 2- (2, 4, 6-trimethoxyphenyl) ethanamine (2, 4, 6-TMPEA-NBOMe), a new representative of designer drugs of NBOMe series and derivatives thereof. *J Mass Spectrom*, 51:779–789.

13. Zuba, D. and Sekuła, K. 2013. Analytical characterization of three hallucinogenic N-(2-methoxy)benzyl derivatives of the 2C-series of phenethylamine drugs. *Drug Test Anal*, 8:634–645.

14. Lum, B. J., Brophy, J. J. and Hibbert, D. B. 2016. Identification of 4-substituted 2-(4-x-2, 5-dimethoxyphenyl)-N-[(2-methoxyphenyl) methyl] ethanamine (25X-NBOMe) and analogues by gas chromatography–mass spectrometry analysis of heptafluorobutyric anhydride (HFBA) derivatives. *Aust J Forensic Sci*, 48:59–73.

15. Duffau, B., Camargo, C., Kogan, M., Fuentes, E., and Cassels, B. K. 2016. Analysis of 25C NBOMe in seized blotters by HPTLC and GC–MS. *J Chrom Sci*, 54:1153–1158.

16. Li, Y., Wang, M., Li, A., Zheng, H., and Wei, Y. 2016. Identification of the impurities in 2, 5-dimethoxy-4-ethylphenethylamine tablets by high performance liquid chromatography mass spectrometry–ion trap–time of flight. *Anal Methods*, 8:8179–8187.

17. Neto, J. C., Andrade, A. F. B., Lordeiro, R. A., Machado, Y., Elie, M., and Arantes, F. E. J. L. C. 2017. Preventing misidentification of 25I-NBOH as 2C-I on routine GC–MS analyses. *Forensic Toxicol*, 35:415–420.

18. Kristofic, J. J., Chmiel, J. D., Jackson, G. F., Vorce, S. P., Holler, J. M., Robinson, S. L., and Bosy, T. Z. 2016. Detection of 25C-NBOMe in three related cases. *J Anal Toxicol*, 40:466–472.

19. Arantes, L. C., Ju'nior, E. F., Souza, L. F., Cardoso, A. C., Alcantara, T. L. F., Liao, L. M., and Machado, Y. 2017. 25I-NBOH: A new potent serotonin 5-HT$_2$A receptor agonist identified in blotter paper seizures in Brazil. *Forensic Toxicol*, 35:408–414.

20. Caspar, A. T., Kollas, A. B., Maurer, H. H., and Meyer, M. R. 2018. Development of a quantitative approach in blood plasma for low-dosed hallucinogens and opioids using LC-high resolution mass spectrometry. *Talanta*, 176:635–645.
21. Bersani, F. S., Corazza, O., and Albano, G. 2014. 25C-NBOMe: Preliminary data on pharmacology, psychoactive effects, and toxicity of a new potent and dangerous hallucinogenic drug. *BioMed Res Int*, 2014:734–749.
22. Poklis, J. L., Devers, K. G., Arbefeville, E. F., Pearson, J. M., Houston, E., and Poklis, A. 2014. Postmortem detection of 25I-NBOMe [2-(4-iodo-2,5-dimethoxyphenyl)-N-[(2- methoxyphenyl)methyl]ethanamine in fluids and tissues determined by UPLC-ESI(+)-MS/MS from a traumatic death. *Forensic Sci Int*, 234:14–20.
23. Caspar, A. T., Helfer, A. G., Michely, J. A., Auwärter, V., Brandt, S. D., and Meyer, M. R. 2015. Studies on the metabolism and toxicological detection of the new psychoactive designer drug 2-(4- iodo-2,5-dimethoxyphenyl)-N-[(2-methoxyphenyl)methyl]ethanamine (25I-NBOMe) in human and rat urine using GC-MS, LC-MS[(n)], and LC-HR-MS/MS. *Anal Bioanal Chem*, 407:6697–6719.
24. Bodeau, S., Bennis, Y., Régnaut, O., Fabresse, N., Richeval, C., Humbert, L., Alvarez, J. C., and Allorge, D. 2017. LSD instead of 25I-NBOMe: The revival of LSD? A case report. *Toxicol Anal Clin*, 29:139–143.
25. Tang, M. H. Y., Ching, C. K., Tsui, M. S. H., Chu, F. K. C., and Mak, T. W. L. 2014. Two cases of severe intoxication associated with analytically confirmed use of the novel psychoactive substances 25B-NBOMe and 25C-NBOMe. *Clin Toxicol*, 52:561–565.
26. Laskowski, L. K., Elbakoush, F., Calvo, J., Bernard G. E., Fong J., Poklis, J. L., Poklis, A., and Nelson, L. S. 2015. Evolution of the NBOMes: 25C- and 25B- Sold as 25I-NBOMe. *J Med Toxicol*, 11:237–241.
27. Uchiyama, N., Shimokawa, Y., Matsuda, S., Kawamura, M., Hanajiri R. K., and Goda, Y. 2014. Two new synthetic cannabinoids, AM-2201 benzimidazole analog (FUBIMINA) and (4-methylpiperazin-1-yl)(1-pentyl-1H-indol-3- yl)methanone (MEPIRAPIM), and three phenethylamine derivatives, 25H-NBOMe 3,4,5-trimethoxybenzyl analog, 25B-NBOMe, and 2C-N-NBOMe, identified in illegal products. *Forensic Toxicol*, 32:105–115.
28. Kaizaki, A., Noguchi, M. N., Yamaguchi, S., Odanaka, Y., Matsubayashi, S., Kumamoto, H., Fukuhara, K., Funada, M., Wada, K., and Numazaw, S. 2016. Three 25-NBOMe-type drugs, three other phenethylamine-type drugs (25I-NBMD, RH34, and escaline), eight cathinone derivatives, and a phencyclidine analog MMXE, newly identified in ingredients of drug products before they were sold on the drug market. *Forensic Toxicol*, 34:108–114.
29. Brandt, S. D., Elliott, S. P., Kavanagh, P. V., Dempster, N. M., Meyer, M. R., Maurer, H. H., and Nichols, D. E. 2015. Analytical characterization of bioactive N-benzyl-substituted phenethylamines and 5-methoxytryptamines. *Rapid Commun Mass Spectrom*, 29:573–584.
30. Pasin, D., Cawley, A., Bidny, S., and Fu, S. 2017. Characterization of hallucinogenic phenethylamines using high-resolution mass spectrometry for non-targeted screening purposes. *Drug Test Anal*, 9:1620–1629.
31. Liu, C., Jia, W., Qian, Z., Li, T., and Hua, Z. 2016. Identification of five substituted phenethylamine derivatives 5-MAPDB, 5-AEDB, MDMA methylene homolog, 6-Br-MDMA, and 5-APB-NBOMe. *Drug Test Anal*, 9:199–207.
32. Uchiyama, N., Hanajiri, R. K., and Hakamatsuka, T. 2016. A phenethylamine derivative 2-(4-iodo-2,5-dimethoxyphenyl)-N-[(3,4-methylenedioxyphenyl)methyl]ethanamine (25I-NB34MD) and a piperazine derivative 1-(3,4-difluoromethylenedioxybenzyl) piperazine (DF-MDBP), newly detected in illicit products. *Forensic Toxicol*, 34:166–173.

33. Uchiyama, N., Matsuda, S., Kawamura, M., Hanajiri R. K., and Goda, Y. 2013. Two new-type cannabimimetic quinolinyl carboxylates, QUPIC and QUCHIC, two new cannabimimetic carboxamide derivatives, ADB-FUBINACA and ADBICA, and five synthetic cannabinoids detected with a thiophene derivative a-PVT and an opioid receptor agonist AH-7921 identified in illegal products. *Forensic Toxicol*, 31:223–240.
34. Kyriakou, C., Marinelli, E., Frati, P., Santurro, A., Afxentiou, M., Zaami, S., and Busardo, F. P. 2015. NBOMe: New potent hallucinogens—Pharmacology, analytical methods, toxicities, fatalities: A review. *Eur Rev Med Pharmacol Sci*, 19:3270–3281.
35. Coelho, N. J. 2015. Rapid detection of NBOME's and other NPS on blotter papers by direct ATR-FTIR spectrometry. *Forensic Sci Int*, 252:87–92.
36. Nikolaou, P., Papoutsis, I., Dona, A., Spiliopoulou, C., and Athanaselis, S. 2014. Beware of 25C-NBOMe: An N-benzyl substituted phenethylamine. *J Forensic Toxicol Pharmacol*, 3:3.
37. Smith, P. R., and Morley, S. R. 2017. New psychoactive substances, in *Essentials of Autopsy Practice: Reviews, Updates, and Advances*, Ed. Rutty, Guy N. Springer International Publishing, Cham, Switzerland, DOI: 10.1007/978-3-319-46997-3_4.

17 Forensic Analysis of Piperazines

Chipo Kuleya and Michael D. Cole

CONTENTS

17.1 INTRODUCTION TO PIPERAZINES

Benzyl and phenylpiperazines are classes of compounds that offer the user similar effects to those sought from amphetamines and ring-substituted amphetamines, reportedly without some of the side effects. Chemically derived from piperazine, there are now a number of these drugs on the clandestine drug market, as shown in Table 17.1. These drugs were originally investigated as antihelminthics in the 1950s and as antidepressants in the 1970s [1]. Documentation of benzylpiperazine (BZP) as a drug of abuse began in 1996 [2]. The amphetamine-like activity of this drug class has been known since the 1970s with the drugs mimicking the effect of 3,4-methylenedioxymethamphetamine (MDMA) [3]. Whilst piperazines have acquired a reputation for being "safe" and legal, both assumptions are incorrect. Today their licit use includes serving as precursors or intermediates in the synthesis of ciproflaxin, quinolone antibiotics, phenothiazines, and the antidepressants trazodone, nefazodone, and etoperidon [4].

Street samples of piperazines are often found in combination with each other in varying proportions. For example, a common mixture is one of BZP and trifluoromethylpiperazine (TFMPP). They are then mixed with a number of adulterants and diluants,

TABLE 17.1

Structures and Examples of Commonly Encountered Benzylpiperazines and Phenylpiperazines

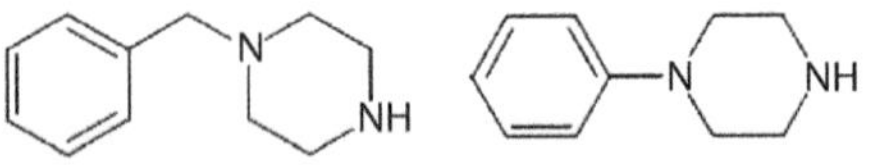

General structure of a   General structure of a
benzylpiperazine   phenylpiperazine

Benzylpiperazines

1-Benzylpiperazine (BZP)

1-Benzyl-4-methylpiperazine (MBZP)

1-(4-Bromo-2,5-dimethoxybenzyl)piperazine (2C-B BZP)

Phenylpiperazines

1-Phenylpiperazine

1-(3-Chlorophenyl)piperazine (3-CPP)

1-(4-Chlorophenyl)piperazine (4-CPP)

1-(3-Chlorophenyl)-4-(3-chloropropyl)piperazine (3-CPCPP)

1-(4-Fluorophenyl)piperazine (4-FPP)

4-Methylphenylpiperazine (4-MePP)

1-(4-Methoxyphenyl)piperazine (4-MeOPP)

1-(3-Trifluoromethylphenyl)piperazine (3-TFMPP)

1-(4-Trifluoromethylphenyl)piperazine (4-TFMPP)

which can include, for example, caffeine, cocaine, dapoxitine, dextromethorphan, diazepam, ephedrine, MDMA, methamphetamine, and nicotinamide. As a consequence, both monodrug and polydrug toxicity associated with piperazine abuse has been reported [1,3,5]. Street samples are often found in tableted form, examples of which are shown in Figure 17.1, although they can also be found in capsule, powder, and liquid forms. When used in combination, these drugs can have an effect for up to eight hours [6].

17.2 MANUFACTURE OF PIPERAZINES

Piperazines are relatively simple compounds to manufacture from readily available starting materials. For example, the manufacture of BZP is achieved by dissolving piperazine hexahydrate in absolute ethanol and adding benzyl chloride. After vigorously shaking for 25 minutes at 65°C white crystals are formed. This is the reaction product BZP, which is further "cleaned up" by cooling in an ice bath and washing with ice-cold ethanol. The hydrochloride salt is formed by the subsequent reaction of the free base with hydrochloric acid. A common by-product of this route often seen in samples manufactured this way is dibenzylpiperazine.

The manufacture of, for example, fluorinated piperazines such as 1-(4-fluorophenyl)piperazine (4-FPP), can be manufactured via the reaction of a fluorinated aniline with bis(2-chloroethyl)amine in the presence of a weak base at elevated

Mass of tablet (mg)	Image	Drugs found in the tablet	Mass of drug (free base) in tablet (mg) and percentage of tablet mass
333	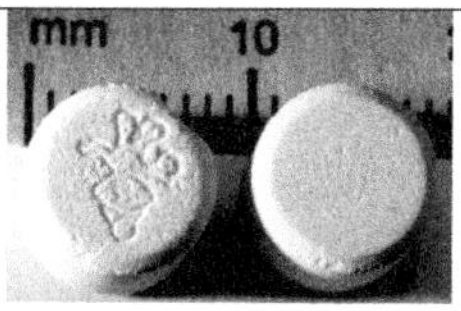	BZP	56 (17%)
		3-TFMPP	106 (32%)
		DBZP	8 (2%)
224		BZP	17 (8%)
		3-TFMPP	21 (9%)
		DBZP	1 (0.5%)
		Caffeine	Trace
355		Ephedrine	36 (10%)
		3-TFMPP	6 (2%)
		Caffeine	101 (28%)

FIGURE 17.1 Examples of tableted dose forms of piperazines.

temperatures. Other ring-substituted piperazines can be manufactured in an analogous fashion using the appropriately substituted aniline starting material.

17.3 LEGAL STATUS AND CONTROL OF PIPERAZINES

Piperazines are not currently under international control. No piperazines are listed in the United Nations 1971 Convention on Psychotropic Substances but several were pre-reviewed by the WHO Expert Committee on Drug Dependence in 2012, which may lead to stricter international controls in the future.

Piperazine-based drugs are not currently controlled by individual countries on a global basis. Risk assessments have been conducted, for example, in the EU by the European Monitoring Centre for Drugs and Drugs Addiction (EMCDDA) in 2008, after which they came under European-wide control. Outside of Europe, controls are also in place in Australia, Canada, Japan, New Zealand, and the U.S.

As a consequence of these control measures it is necessary, as a minimum, to identify any controlled substances in materials thought to be piperazines. It is for this reason that a logical sequence of analytical methods must be followed, including the use of chromatography.

17.4   ANALYTICAL SCHEMES FOR PIPERAZINES

In order to establish that a controlled substance is present it is necessary to carry out a definitive identification of any drugs that are present in the sample. Piperazine-based drugs are, generally, bulk samples of drugs. This means that a standard scheme for analysis can be applied as follows:

- Physical description of the samples
- Presumptive tests to identify the class(es) of drugs present
- Thin-layer chromatography to identify which piperazines are present
- Confirmation of the identity of the piperazines present using gas chromatography/mass spectrometry (GC-MS), high-performance liquid chromatography with diode array detection (HPLC-DAD), or liquid chromatography/mass spectrometry (LC-MS)
- Quantification of the drugs present (GC-MS, HPLC-DAD, or LC-MS)

Chemical impurity profiling of piperazine-based drugs is rarely, if ever, undertaken and there is a paucity of published literature in this area.

It is possible that a very large number of tableted dose forms of the drug could need to be examined. In this case the drugs should be separated into visually identical groups and then the method of sampling described by the United Nations Office of Drug Control (UNODC) could be applied to each group. An alternative approach has been suggested in that, regardless of size of the population, if three samples are taken the drugs to be identified will be found [7]. Whichever the protocol for sampling the materials to be analyzed, the analysts should satisfy themselves that the method itself satisfies the legislative requirements of the jurisdiction where the work is being undertaken.

17.5   PHYSICAL DESCRIPTION OF PIPERAZINES

Prior to carrying out any chemical examinations of the drugs it is good practice to describe the sample in full. The starting point for this is a description of the packaging in which the sample is received and the state of the packaging. This is to ensure the continuity of evidence and demonstrate that if controlled substances are found they came from within the sample and no other source. Once the packaging has been described it should be correctly opened. How this is achieved will depend upon the packaging in question but each entry to the packaging should be independent of any previous access. The samples should be removed and divided into visually identical groups that are then described. The description should identify key physical characteristics including, for tableted samples, the details of the "ballistics" of the tablets, which might be used to relate samples to each other [8]. When taking material for chemical analysis, these "ballistic" marks should be avoided since they can be used to relate samples to each other and they may be required for second and subsequent examinations of the items. Where tablets are seized the number of tablets in each group of visually identical tablets should be recorded, as should their size (dimensions), weight, shape, color, breaklines, logo, and any other markings.

17.6 PRESUMPTIVE TESTS

Color tests are often used in the first stages of the identification of drugs prior to instrumental analyses. They allow the determination of the class or classes of drugs found in a sample. There is currently no specific colorimetric (presumptive) test for piperazines and certainly none that individualizes them. This would reduce the need for expensive instrumental methods or facilitate identification where modern analytical equipment is not available or its use is not practical—where, for example, power supplies are not constant.

An early study of some less frequently encountered piperazines indicated that they reacted with the Lieberman–Burchard reagent. However, the color obtained was the same as that obtained from a number of different drug classes [9]. Using a polythetic approach and a number of tests in combination, namely the Marquis, the Scott, and the Simons tests, it is possible to identify piperazines as a class of drug. More recently, the use of sodium 1,2-napthoquinone-4-sulfonate (NQS) has been considered for the identification of piperazines [10]. In that study, a test was developed that gave an orange-red color when NQS was reacted with the piperazines. No other group of compounds tested gave the same color reaction. The test, with a detection limit of 40 μg, was sensitive enough to be applied to street samples but it did not individually identify the drugs present in the samples.

However, it is because there are no specific tests available that sophisticated instrumental analyses are required to achieve definitive and quantitative identification of members of this group of drugs and their congeners. It should also be remembered that other drug classes may also be present in the samples. For this reason, tests other than those for piperazines should be applied. This includes, for example, the use of the cobalt isothiocyanate test for cocaine and the Zimmerman reagent for benzodiazepines (for example, diazepam).

17.7 INSTRUMENTAL ANALYSIS OF PIPERAZINES

17.7.1 EXTRACTION OF PIPERAZINES FROM STREET SAMPLES

The extraction of controlled substances from street samples presents particular challenges because of the mixtures of drugs found in such samples. It is important that any solvent used to extract the piperazines and other drugs with which they have been mixed in street samples satisfy a number of criteria. The solvent must dissolve the drugs in question quantitatively over a wide concentration range. They must not react with the drugs either during the extraction phase or during any analytical process. They must not cause the drug to break down. In terms of health and safety they must be easy to handle and not present significant health and safety risks. From an operational point of view solvents must be readily available and their use must be cost-effective—in essence, they must be inexpensive. Finally, they should be compatible with any analytical technique used.

The UNODC-recommended methods for the analysis of piperazines [4] include a discussion of the use of water, acetone, chloroform, diethylether, and hexane. However, none of these solvents satisfy the criteria described above. In each case, at least one of the piperazines was not soluble in these solvents. Methanol was identified as a solvent in which all of the samples dissolved but there are problems with this solvent since it may dissolve and then the water that methanol absorbs may react with congeners in the samples—for example, cocaine resulting in the hydrolysis of this compound to benzoyl ecgonine and then ecgonine.

The low-molecular-weight ether methyl *tert*-butyl ether (MTBE) has been used to extract BZP from an aqueous solution prepared from street samples sold as A2 [11]. MTBE, being a low-molecular-weight ether, is sufficiently polar to dissolve the piperazines but at the same time will not react with the drug to form artefacts or cause breakdown of the BZP.

Other authors have used hydrochloric acid in methanol to dissolve the piperazines in street samples prior to analysis [12]. This is intended to convert the drug into hydrochloride salt, which is more readily soluble in polar solvents and mobile phases used for HPLC. However, the challenge with this method is that some of the piperazines are acid-labile (although not the ones encountered in the street samples in that study). Additionally, some of the congeners found today in street samples of piperazines (for example, cocaine) would hydrolyze in the methanol solution. Any analytical data where methanol is used may contain results from artefacts as well as the original drugs under consideration.

Recent studies of the stability of piperazines and congeners have shown that a number of solvents are not suitable for the extraction of piperazines prior to analysis [13]. The solvents examined were methanol, ethyl acetate, dichloromethane, and 2-methyl-propan-2-ol. Methanol was found not to be suitable as a solvent. Cocaine, which can be mixed with piperazines, is known to hydrolyze in methanol. The use of ethyl acetate carries with it health and safety considerations and it can potentially react with primary amines forming imines and Schiff's bases. On paper this is not a suitable solvent since the samples may contain primary amines. Dichloromethane (DCM) is not suitable as a solvent because the fluorinated piperazines are known to break down to their synthetic precursors. For example, 4-fluoroaniline is found in 4-FPP samples dissolved in DCM. In this study, the use of 2-methyl-propan-2-ol was recommended because it satisfied all of the requisite criteria for a good analytical solvent.

The conclusions that can be drawn from these studies are that low-molecular-weight ethers, or tertiary alcohols, are suitable for the extraction of piperazines prior to analysis. Other more commonly used solvents, for example primary alcohols, esters, and halogenated solvents, are not suitable for this group of compounds.

17.7.2 Thin-Layer Chromatography

It is possible to analyze piperazines by thin-layer chromatography. Authors have used a silica gel stationary phase and mobile phases used typically for alkaloid analysis, including methanol/ammonia (100:1.5 v/v), cyclohexane/toluene/diethylamine (75:15:10 v/v/v), and chloroform/acetone (4:1 v/v). The compounds were visualized using reagents typical of alkaloid analysis, namely potassium iodoplatinate reagent and Dragendorff's reagent [9]. However, there will be many compounds that have the same chromatographic properties as the piperazines in these systems and as such TLC does not provide a definitive means of drug identification. It is for this reason that more sophisticated instrumental techniques are required.

17.7.3 LIQUID CHROMATOGRAPHIC METHODS

High-performance liquid chromatography with diode array detection has been applied to the analysis of piperazines on a number of occasions. An early study [14] demonstrated that LC could be applied to piperazines. However, mixtures of isomers of the piperazines could not, in the system described, be resolved to baseline and the mass spectra of the isomers of TFMPP, methoxyphenylpiperazine (MeOPP), chlorophenylpiperazine (CPP), and FPP could not be discriminated. Indeed, this is a common difficulty encountered when using liquid and gas chromatographic techniques hyphenated to mass spectrometry.

A subsequent study [15] demonstrated that BZP, 2-CPP, 3-CPP, 4-CPP, and 3-TFMPP could be differentiated on the basis of their retention time and DAD spectra. This method was applied to toxicological samples but the methodology could be equally applied to street samples. This is particularly useful since, in the same study, using LC-MS, there were only minor differences in the mass spectra of the isomeric compounds, whereas these UV spectra between 200 nm and 595 nm were quite distinct.

One study has used octadecyl silica (ODS) columns with phosphoric acid buffers and with, separately, heptafluorobutyric acid buffers and diode array detection [13]. Whilst it was possible to separate some of the piperazines, the isomers of some were difficult to differentiate. When used in conjunction with GC/MS it proved possible to differentiate the compounds using complex data sets including GC retention times, MS data, HPLC capacity factors, and ultraviolet (UV) spectra from diode array detection. However, whilst they did not completely coelute in the LC systems, the retention times of the isomers of CPP were very similar and so if the sample contained more than one isomer its definitive identification would be extremely difficult.

In summary, whilst liquid chromatographic methods have been applied to the piperazines, there is a general problem of resolution of different isomers of the drugs. This may be required by some legislative systems and is certainly required if drug profiling is to be undertaken. Additionally, whilst diode array detection can be used to differentiate some of the piperazines, the preferred technique of mass spectrometry presents problems because under liquid chromatographic conditions the mass spectra of piperazine isomers cannot be readily and reliably differentiated.

17.7.4 GAS CHROMATOGRAPHIC METHODS

One of the difficulties in early studies of the analysis of piperazines by GC was that the retention times of different isomers was the same under a given set of analytical GC conditions [11]. Additionally, the mass spectra of underivatized piperazines under electron impact conditions are very similar, as shown in Figure 17.2. These factors precluded the definitive identification of these drugs using the system

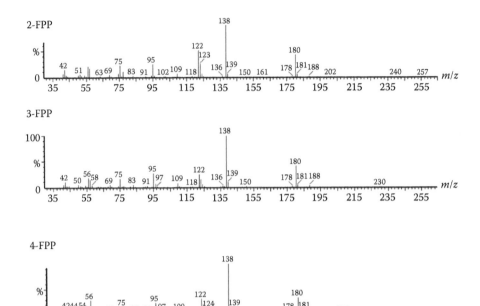

FIGURE 17.2 Electron impact mass spectra of 2-FPP, 3-FPP, and 4-FPP demonstrating that the EI spectra of underivatized piperazines can be very similar.

described. Improvement in the degree of chromatographic resolution or greater differences in the mass spectra was required.

Both acetylation and trifluoroacetylation of piperazines has been attempted prior to the analysis of piperazines by gas chromatography and GC/MS in order to address this issue and to improve the chromatographic behavior of the drugs. Acetylation led to the formation of stable derivatives and improved resolution by GC but the MS data for different isomers was not sufficiently differentiated. Trifluoroacetylation resulted in derivatives for which the MS data was much more distinct [11]. However, in that study, stable derivatives of *m*MeOPP could not be achieved. As a consequence, the use of N-trifluoroacetylation is not a viable possibility for precolumn derivatization of piperazines prior to GC or GC/MS analysis.

A more recent study has considered both the separation of the piperazine isomers and the congeners of the drugs [13]. In this study, it was decided not to derivatize the drugs for a number of reasons. These include the fact that it is difficult to prove that all of the drugs have completely derivatized, derivatization increases the time and cost of each analysis, and that there is always the risk of contamination of the sample during the derivatization process. After preparing samples in 2-methyl-propan-2-ol it was possible to separate all of the isomers of the drugs in addition to a large number of the reported congeners, as shown in Figure 17.3. It was also possible to separate the drugs in street samples of piperazines, as shown in Figure 17.4. It was also possible to identify the impurities in the mixture demonstrating that the drugs could also be chemical impurity profiled.

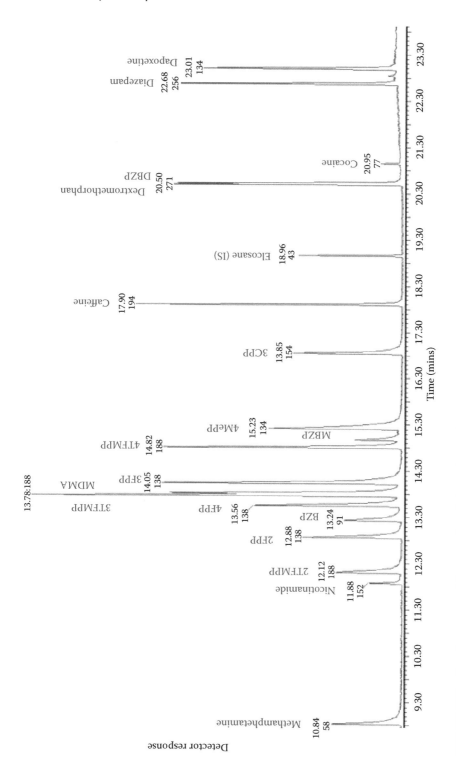

FIGURE 17.3 Gas chromatographic separation of piperazines and congeners.

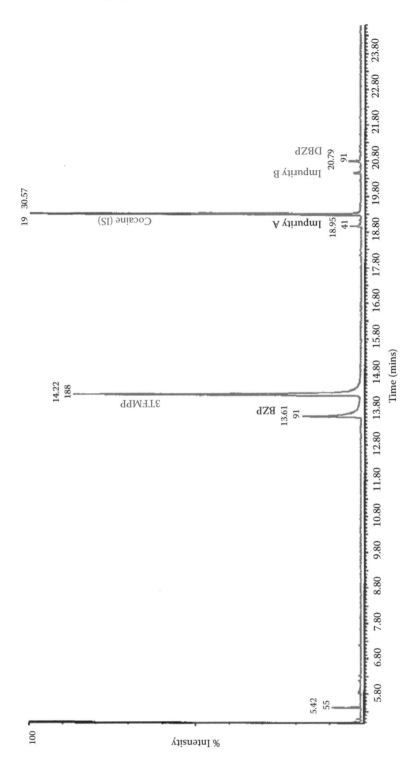

FIGURE 17.4   Gas chromatographic separation of a street sample of BZP.

17.7.5 CAPILLARY ZONE ELECTROPHORESIS

A very limited study has been made of the analysis of piperazines by capillary zone electrophoresis (CZE) [4]. In this study, samples of TFMPP (unstated isomer), 2-MeOPP, 3-MeOPP, and 4-MeOPP were analyzed by CZE with UV detection at 210 nm. Whilst this study demonstrated that the piperazines could be analyzed by this technique, it lacked: (i) definitive identification of the drug, (ii) data for the range of piperazines found in current drug samples, (iii) data on the adulterants and diluents likely to be found in street samples, and (iv) application to street samples themselves.

17.8 SPECTROSCOPIC METHODS

Infrared spectroscopy has been applied to piperazines [14] using either the KBr disc technique or by obtaining the spectrum from a thin film coated onto a NaCl disc. It was clearly possible to differentiate between pure samples of the isomers of TFMPP, MeOPP, CPP, and FPP. However, the method was not applied, in the paper, to street samples when the spectrum of the matrix would have been added to that of the drug(s) in the materials analyzed. What this does suggest, however, is that in addition to GC-MS the technique of GC-FTIR could be applied to the definitive and quantitative analysis of piperazines.

17.9 CONCLUSIONS

It is clear that piperazines present challenging samples in terms of their analysis in support of the legal process. It is necessary to identify whether or not there are controlled substances present in a seizure—in this case focusing on piperazines. In some jurisdictions, it is also necessary to identify which isomer of a particular piperazine is present. Up until recently this has been difficult to achieve. Now, however, by following a straightforward analytical sequence of physical description, presumptive testing and chromatographic methods to identify and quantify the drugs it is possible to achieve the complete identification and quantification of piperazines.

REFERENCES

1. Arbo, M. D., Bastos, M. L., and Carmo, H. F. 2012. Piperazine compounds as drugs of abuse, *Drug Alcohol Depend.*, 122: 174–185.
2. Austin, H. and Monasterio, E. 2004. Acute psychosis following ingestion of 'Rapture', *Australas. Psychiatr.*, 12: 406–408.
3. Staack, R. F. 2007. Piperazine designer drugs of abuse, *Lancet*, 369: 1411–1413.
4. Recommended Methods for the Identification and Analysis of Piperazines in Seized Materials. Manual for use by national drug analysis laboratories (2013). UNODC, New York.
5. Maurer, H. H. 2004. Mass spectra of select benzyl- and phenyl-piperazine designer drugs. Technical note. *Microgr. J.* 2: 22–26.
6. Nikolova, I. and Danchev, N. 2008. Piperazine based substances of abuse: A new party pills on Bulgarian drug market, *Biotechnol. Biotechnol. Equip.*, 22: 652–655.

7. Aitken, C. G. G. and Lucy, D. 2002. Estimation of the quantity of a drug in a consignment from measurements on a sample, *J. Forensic Sci.*, 47: 1–8.

8. Zingg, C. 2005. The analysis of ecstasy tablets in a forensic drug intelligence perspective. Institut de Police Scientifique, University of Lausanne.

9. Uchiyama, N., Kawamura, M., Kamakura, H., Kikura-Hanajiri, R., and Goda, Y. 2008. Analytical data of designated substances (Shitei-Yakubutsu) controlled by the Pharmaceutical Affairs Law in Japan, part II: Color test and TLC, *Yakugaku zasshi J Pharm Soc Jpn.*, 128: 981–987.

10. Morgan, P., Shimmon, R., Stojanovska, N., Tahtouh, M., and Fu, S. 2013. Development and validation of a presumptive colour spot test method for the detection of piperazine analogues in seized illicit materials, *Anal. Methods*, 5: 5402–5410.

11. de Boer, D., Bosman, I. J., Hidvégi, E., Manzoni, C., Benkö, A. A., dos Reys, L. J. A. L., and Maes, R. A. A. 2001. Piperazine-like compounds: A new group of designer drugs-of-abuse on the European market, *Forensic Sci. Int.*, 121: 47–56.

12. Misako, T., Nagashima, M., Suzuki, J., Seto, T., Yasuda, I., and Yoshida, T. 2009. Creation and application of psychoactive designer drugs data library using liquid chromatography with photodiode array spectrophotometry detector and gas chromatography–mass spectrometry, *Talanta*, 77: 1245–1272.

13. Kuleya, C., Hall, S., Gautam, L., and Cole, M. D. 2014. An optimised gas chromatographic-mass spectrometric method for the chemical characterisation of benzylpiperazine and 1-arylpiperazine based drugs, *Anal. Methods*, 6: 156–163.

14. Inoue, H., Iwata, Y. T., Kanamori, T., Miyaguchi, H., Tsujikawa, K., Kuwayama, K., Tsutsumi, H., Katagi, M., Tsuchihashi, H., and Kishi, T. 2004. Analysis of benzylpiperazine-like compounds, *Japn. J. Sci. Tech. Identif.*, 9: 165–184.

15. Elliott, S. and Smith, C. 2008. Investigation of the first deaths in the United Kingdom involving the detection and quantitation of the piperazines BZP and 3-TFMPP, *J. Anal. Toxicol.*, 32: 172–177.

18 Chromatographic Analysis of Fentanyl and Its Analogs

Thomas A. Brettell and Matthew R. Wood

CONTENTS

18.1 INTRODUCTION

Fentanyl was first synthesized in 1960 by Dr. Paul A. Janssen as a more potent and faster-acting analgesic alternative to both morphine and meperidine [1]. Dr. Janssen and his research group recognized that morphine was most likely more powerful than meperidine, due to its higher fat solubility [2]. At the time, it was known that increased lipophilicity enabled drug compounds to more effectively cross the blood–brain barrier. Dr. Janssen modified the structure of meperidine (Figure 18.1a) by adding a phenyl ring in order to increase the lipid solubility of the drug, and then optimized the distance between the ring and the parent compound by adding saturated carbons. The addition of a hydroxyl group to the carbon adjacent to the phenyl ring led to the successful development of phenoperidine (Figure 18.1b), a potent analgesic [3]. Fentanyl (Figure 18.1c) was synthesized by displacing the phenyl ring from the piperidine, replacing the ester moiety with an amide with the phenyl attached, instead, to the N atom.

Throughout the 1970s, Janssen Pharmaceutical continued to develop fentanyl analogs for a variety of different purposes. Their group designed sufentanil (Figure 18.2a) [4] and carfentanil (Figure 18.2b), both substantially more potent than fentanyl. These compounds were intended to lower the amount of analgesia required, be more reliable with less variation in patient response, and cause less stress and

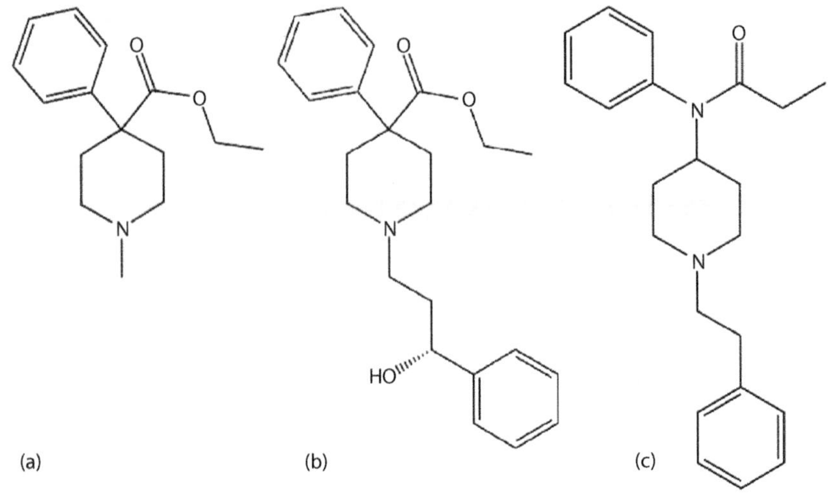

FIGURE 18.1 (a) Meperidine, (b) phenoperidine, and (c) fentanyl.

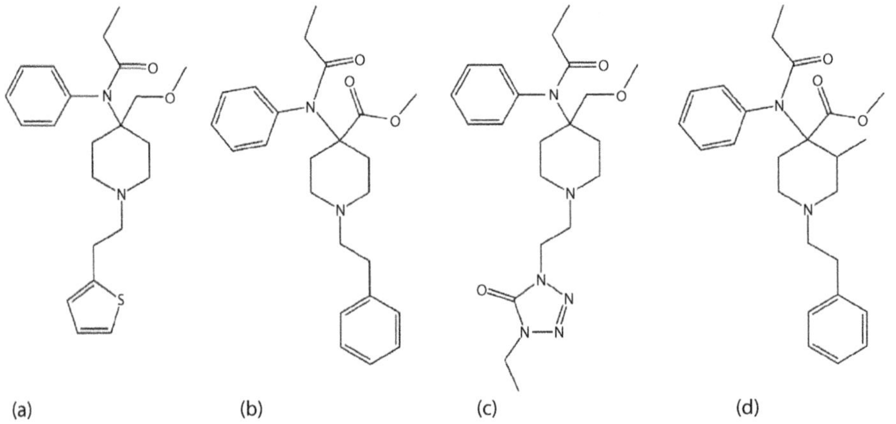

FIGURE 18.2 (a) Sufentanil, (b) carfentanil, (c) alfentanil, and (d) lofentanil.

hypertension. Carfentanil is so powerful, though, that it has only been designated for the immobilization of large animals such as elephants, buffalo, and rhinoceros. Lofentanil (Figure 18.2d), also developed by this group, has found only limited practical uses. Unfortunately, these compounds, fentanyl, sulfentanil, carfentanil and lofentanil, all exhibit long-duration respiratory effects [5]. Alfentanil (Figure 18.2c) was created to satisfy the need for a fast-acting, short-duration analgesic [6]. Alfentanil does not have the strength of fentanyl, but the onset of action is approximately 2 minutes and, due to its short duration, can be effectively reversed by naloxone. The previous compounds require postoperative intubation or maintained use of reversal agents.

In the patent application (3,164,600), Janssen describes the α-methyl substitution of fentanyl and separately, the preparation of substitutions at the 3-position (e.g.,

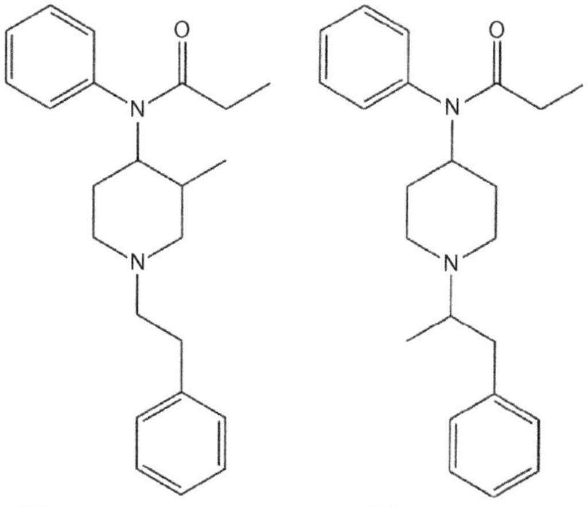

FIGURE 18.3   (a) 3-Methylfentanyl and (b) α-methylfentanyl.

3-methylfentanyl) of the piperidine ring [7] and 4-position [8]. This work demonstrated the enhanced analgesia of these analogs. The μ-opioid receptors continue to be the target of research in analgesia and pain management and analogs of fentanyl have proven to be very efficacious μ-receptor agonists. Many other fentanyl analogs have recently emerged, some from legitimate sources seeking to improve upon the analgesic properties, while others have come from clandestine labs seeking to subvert law enforcement, regulation, and control, and to enhance the drug's euphoric side effects [9]. Fentanyl and many of the fentanyl analogs are easily synthesized from the base 4-anilinopiperidine skeleton structure. The high potency, combined with the ease of synthesis, makes this class of compounds very attractive to illicit manufacturers. Through the early 1980s, the United States experienced its first epidemic of "designer" fentanyls [10]. It appears very credible that 3-methylfentanyl (Figure 18.3a) and α-methylfentanyl (Figure 18.3b) were chosen by clandestine chemists as designer drugs based on the early studies of structure activity relationships done by Janssen Pharmaceutical and the ease of synthesis [11]. Since the mid-1990s, the European Monitoring Centre for Drugs and Drug Addiction (EMCDDA) began to see sporadic incidences of the manufacture of fentanyl, para-fluorofentanyl, and 3-methylfentanyl [12], with the exception of Estonia, where the reported use has been consistent.

18.2   FENTANYL AND ANALOGS

Since 1979, a number of these illegal laboratories have been producing and selling fentanyl and its analogs to consumers involved in the illicit sale of the drug. An increasing number and percentage of fentanyl overdose deaths in the United States in the last few years have been attributed to illicit versions of fentanyl produced by these clandestine laboratories [13]. In 2015, the United States reported over 52,000 drug overdose deaths; 63% involved opioids. Death by synthetic opioids other than methadone increased by 72% from 2014 to 2015 [14].

There has been an enormous variety and amount of new psychoactive substances (NPSs) that have emerged over recent years that have posed a public health threat. The Special Testing and Research Laboratory's Emerging Trends Program [15] compiles data and reports the data through a query of archived seizure and analysis information from drug evidence analyzed by the Drug Enforcement Administration's laboratory system. This data is representative of drug evidence seized and analyzed in the date range of the report. According to the Emerging Threat Report for 2016 there were 1299 identifications of fentanyl, fentanyl-related substances, and other new opioids [15]. Fentanyl accounted for approximately 68% of the identifications. The next most prominent opioid, furanylfentanyl, accounted for 11% of the identifications. Of the 877 fentanyl identifications, fentanyl was found as the only controlled substance in 46.5% of the identifications and was found in combination with heroin in approximately 42% of the identifications. Of the 15 substances identified, nine of these substances were reported for the first time in 2016 (Table 18.1). A more comprehensive list of fentanyl-related compounds is given in Table 18.2.

There are a number of synthetic routes available for fentanyl, published in academic journals and also on the Internet. The most common route proceeds through an N-phenethyl-piperidone (NPP) intermediate reacted with aniline and reduced to the 4-anilino-N-phenethyl-piperidine (4-ANPP) precursor. The fentanyl synthesis is completed by reaction with propionyl chloride. The majority of fentanyl analogs can be made by substitutions to starting materials in the synthesis of NPP, replacement of aniline with substituted aniline (e.g., *para*-fluoroaniline), and by substitution

TABLE 18.1
Fentanyl-Related Compounds Reported by DEA in 2016

Compound Name	Number of Identifications
Fentanyl	877
Furanylfentanyl*	142
Acetylfentanyl	112
U-47700*	50
4-ANPP	32
4-Fluoroisobutyrylfentanyl*	20
Carfentanil	17
Acrylfentanyl*	13
Butryrlfentanyl	13
Valerylfentanyl	10
o-Fluorofentanyl*	4
Benzylfentanyl*	3
p-Fluorobutyrylfentanyl*	3
3-Methylfentanyl*	2
Acetylnorfentanyl*	1

Source: Emerging Threat Report Annual. Drug Enforcement Administration, 2016.

* Compounds reported by DEA for first time in 2016.

TABLE 18.2
Fentanyl-Related Compounds

	Compound Name	Molecular Weight (Da)	Chemical Formula
1	3-Allylfentanyl	376.53	$C_{25}H_{32}N_2O$
2	3-Methylbutyrfentanyl (3-MBF)	364.533	$C_{24}H_{32}N_2O$
3	3-Methylfentanyl(3-MF, mefentanyl)	350.236	$C_{23}H_{30}N_2O$
4	3-Methyl-thiofentanyl	356.526	$C_{21}H_{28}N_2OS$
5	4-Methoxy-butyrylfentanyl (MeO-BF)	380.246	$C_{24}H_{32}N_2O_2$
6	4-Methoxyfentanyl	366.231	$C_{23}H_{30}N_2O_2$
7	N-Phenylethyl-4-piperidinone (4-NPP)	203.28	$C_{13}H_{17}NO$
8	4-Phenylfentanyl	412.577	$C_{28}H_{32}N_2O$
9	Acetylfentanyl-4-methyphenethyl analog	336.4	$C_{22}H_{28}N_2O$
10	Acetylnorfentanyl	218.29	$C_{13}H_{18}N_2O$
11	Acetylfentanyl (desmethylfentanyl)	322.204	$C_{21}H_{26}N_2O$
12	Acrylfentanyl (acryloyl fentanyl)	334.204	$C_{22}H_{26}N_2O$
13	Alfentanyl (alfentanil, R-39209)	416.517	$C_{21}H_{32}N_6O_3$
14	4-Aminophenyl-1-phenethylpiperidine (4-ANPP)	280.4	$C_{19}H_{24}N_2$
15	Benzylfentanyl (R-4129)	322.204	$C_{21}H_{26}N_2O$
16	Brifentanil (A-3331)	420.481	$C_{20}H_{29}FN_6O_3$
17	Butyrylfentanyl	350.236	$C_{23}H_{30}N_2O$
18	Carfentanyl (carfentanilm, wildnil)	394.226	$C_{24}H_{30}N_2O_3$
19	Benzylcarfentanyl	380.4	$C_{23}H_{28}N_2O_3$
20	Cyclopentylfentanyl	376.251	$C_{25}H_{32}N_2O$
21	Despropionylfentanyl	280.41	$C_{19}H_{24}N_2$
22	Despropionyl-3-methylfentanyl	294.442	$C_{20}H_{26}N_2$
23	Despropionyl-p-fluorofentanyl	298.184	$C_{19}H_{23}FN_2$
24	Fentanyl	336.22	$C_{22}H_{28}N_2O$
25	Furanylfentanyl (FU-F)	374.199	$C_{24}H_{26}N_2O_2$
26	Isobutyrylfentanyl	350.236	$C_{23}H_{30}N_2O_2$
27	Lofentanil	408.533	$C_{25}H_{32}N_2O_3$
28	Mirefentanil	376.452	$C_{22}H_{24}N_4O_2$
29	N-Methylcarfentanyl (R-32395)	304.384	$C_{17}H_{24}N_2O_3$
30	Norfentanyl	232.327	$C_{14}H_{20}N_2O$
31	Normethylfentanyl		$C_{15}H_{22}N_2O$
32	Norsufentanil	276.4	$C_{16}H_{24}N_2O_2$
33	Ocfentanil (A-3217)	370.468	$C_{22}H_{27}FN_2O_2$
34	Ohmefentanil (OMF, RTI-4614-4)	366.505	$C_{23}H_{30}N_2O_2$
35	p-Fluoroacrylfentanyl	352.5	$C_{22}H_{25}FN_2O$
36	o-Flurobutyrylfentanyl (2-FBF, o-FBF)	368.226	$C_{23}H_{29}FN_2O$
37	p-Flurobutyrylfentanyl (4-FBF, p-FBF)	368.226	$C_{23}H_{29}FN_2O$
38	o-Fluorofentanyl(2-FF, o-FF)	354.469	$C_{22}H_{27}FN_2O$
39	p-Fluorofentanyl (4-FF, p-FF)	354.469	$C_{22}H_{27}FN_2O$
40	Phenaridine (2,5-dimethylfentanyl)	364.533	$C_{24}H_{32}N_2O$
41	4-Methoxymethylfentanyl (R-30490)	380.522	$C_{24}H_{32}N_2O_2$

(Continued)

TABLE 18.2 (CONTINUED)
Fentanyl-Related Compounds

Compound Name	Molecular Weight (Da)	Chemical Formula
42 Remifentanyl (remifentanil)	376.447	$C_{20}H_{28}N_2O_5$
43 Sufentanyl (sufentanil, R-30730)	386.554	$C_{22}H_{30}N_2O_2S$
44 Thiofentanyl	342.501	$C_{20}H_{26}N_2OS$
45 Trefentanyl (A-3665)	456.551	$C_{25}H_{31}FN_6O_2$
46 Valerylfentanyl	364.251	$C_{24}H_{32}N_2O$
47 Acetyl-α-methylfentanyl	336.479	$C_{22}H_{28}N_2O$
48 α-Methylfentanyl (china white)	350.497	$C_{23}H_{30}N_2O$
49 α-Methylthiofentanyl	356.528	$C_{21}H_{28}N_2OS$
50 β-Methylthiofentanyl	356.528	$C_{21}H_{28}N_2OS$
51 β-Hydroxyfentanyl	352.47	$C_{22}H_{28}N_2O_2$
52 ω-1-Hydroxyfentanyl	352.47	$C_{22}H_{28}N_2O_2$
53 β-Hydroxythiofentanyl	358.171	$C_{20}H_{26}N_2O_2S$
54 β-Methylfentanyl	350.506	$C_{23}H_{30}N_2O$
55 ω-Hydroxynorfentanyl	248.32	$C_{14}H_{20}N_2O_2$

of propionyl chloride with acetyl chloride, butyryl chloride, or other reactants to achieve modifications at the amide. Two interesting articles, one by Lurie et al. [16] and a more recent article by Mayer et al. [17], looked at characterizing the provenance of fentanyl samples by examining the reaction by-products and residual precursors. Both studies used liquid chromatography–tandem mass spectrometry (LC-MS/MS) in the analytical scheme; the second article also included both gas chromatography–mass spectrometry (GC-MS) and inductively coupled mass spectrometry (ICPMS).

18.3   GAS CHROMATOGRAPHY

Due to the emergence and widespread occurrence of fentanyl compounds there has been a lot of effort put into the development of analytical procedures for their analysis including a variety of separation and chromatographic methods. Due to the continued appearance of new fentanyl analogs, crime laboratories have been challenged to develop methods that are sensitive and specific for this class of compounds. Gas chromatography–mass spectrometry (GC-MS) has been the gold standard for the analysis of drugs of abuse for the last few decades so it is not surprising that many of the methods that have been developed for the analysis of fentanyl and analogs has been done using GC-MS.

One of the first gas chromatographic assays for fentanyl compounds was developed by Gillespie et al. [18]. They reported the gas chromatographic determination of fentanyl and its analogs in human plasma using a packed column and a nitrogen–phosphorus detector (NPD). Around the same time, GC-MS played an important role in the identification of α-methylfentanyl ("China White"). The combination of GC-MS, GC–chemical ionization mass spectrometry (CIMS), and spectroscopy confirmed the identification of what proved to be the first fentanyl analog [19]. Of course, these early methods using packed columns have been replaced with more sensitive and more efficient separation methods using capillary columns. Ohta et al. [20] were able to discriminate 25 fentanyl compounds by the combination of GC–flame ionization detection (FID) and direct inlet MS. Twenty-three of the 25 compounds were able to be discriminated by GC alone using a 50-m nonpolar capillary column. The N-methylfentanyls showed early elution using these conditions. Manral et al. [21] reported retention indices of fentanyl and 18 analogs relative to the homologous n-alkane series on both moderately polar BP-5 and nonpolar BP-1 capillary gas chromatographic columns using GC-FID. The effects of chromatographic conditions like temperature programming rate, carrier gas flow rate, and oven temperature were studied. The analogs differ in the substituent attached to the piperidine ring nitrogen, and retention indices are found to vary according to the nature of the substituent (Table 18.3).

GC-MS was used with reversed-phase liquid chromatography (RPLC) with diode array detection (DAD) to identify acetaminophen, caffeine, and ocfentanil in the powder of the first reported death involving ocfentanil [22]. Quantification of ocfentanil in biological samples was performed using a target analysis based on liquid–liquid extraction (LLE) and ultra-performance LC-MS/MS. Simultaneous analytical methods for 18 compounds of fentanyl and its analogs by thin-layer chromatography (TLC), GC/MS, and liquid chromatography/mass spectrometry (LC-MS) were developed. The fentanyl analogs were well separated using TLC with a developing solvent of toluene–acetone–28% aqueous ammonia (20:10:0.3, by vol.). Using GC-MS, fentanyl analogs, except for fentanyl and acetyl-α-methylfentanyl, could be separated on the extracted ion chromatograms (EICs) using the characteristic fragment ions for each compound. All 18 compounds could be separated and identified using the conditions reported for TLC, GC-MS, and LC-MS [23].

Dipyrone is an analgesic and antipyretic drug that is often used as an adulterant in illicit fentanyl samples. It undergoes thermal decomposition to aminopyrine and 4-methylaminoantipyrine during analysis by gas chromatography–flame ionization detection (GC-FID) and GC-MS and can complicate the analysis. Good chromatography and spectroscopic analysis is essential for these samples [24].

Drug seizures have shown to contain various drug(s) including illicit fentanyl only, illicit fentanyl and heroin, illicit fentanyl and cocaine and illicit fentanyl, heroin, and cocaine [25]. Fentanyl analogs have also been found in these drug combinations and sometimes the concentrations are very small compared to some of the other drugs. Figure 18.4 shows a chromatogram of a methanol extract of a seized sample containing a relatively small amount of carfentanil compared to the adulterants, procaine and quinine. This is a good example of the challenges that drug analysts face in the analysis of powders and contraband for these controlled substances. If the chromatography is not good, small peaks, as in this example, can be easily overlooked and the drug not identified. Figure 18.5 shows the mass spectrum for the peak at $t_R = 4.68$ minutes (top) and the spectral search match results for carfentanil. Figures 18.6 and 18.7 show the results from another case positive for heroin and furanyl fentanyl. Figure 18.6 shows a chromatogram of a methanol extract of a seized sample with peaks indicating procaine, 6-monoacetylmorphine (6-MAM),

TABLE 18.3

GC Conditions for the Identification of Fentanyl Compounds

Stationary Phase	Column Dimensions	Carrier Gas	Matrix	Detector	Analyte(s)	Reference
3% OV-17	2 m × 2 mm i.d.	Helium	Blood	NPD	3 fentanyls	18
Hi-Cap CBP-1	(50 m × 0.2 mm, 0.25 μm)	Nitrogen	Standards	FID	25 fentanyls	20
BP-5	(25 m × 0.22 mm, 0.25 μm)	Nitrogen	Standards	FID	19 fentanyls	21
BP-1	(25 m × 0.22 mm, 0.25 μm)	Nitrogen	Standards	FID	19 fentanyls	21
				MS	Ocfentanil	22
DB-5MS	(30 m × 0.25 mm, 0.25 μm)			MS	18 fentanyls	23
Rxi-5SilMS	(30 m × 0.25 mm, 0.25 μm)	Helium	Urine	MS	Acetylfentanyl	33
			Body fluids	MS	Acetylfentanyl	34
Zebron ZB-5MS	(15 m × 0.25 mm, 0.25 μm)	Helium	Body fluids	MS	Acetylfentanyl	35
Zebron ZB-5MS	(15 m × 0.25 mm, 0.25 μm)	Helium	Body fluids	MS	Butyr-fentanyl	36
DB-5MS	(15 m × 0.25 mm, 0.25 μm)	Helium	Powder	MS	Ocfentanil	37
DB-1MS	(25 m × 0.25 mm, 0.4 μm)	Helium	Urine	MS	Fentanyl	38
HP-5	(15 m × 0.25 mm, 0.25 μm)	Helium	Urine	MS	Fentanyl	38
DB-5	(12 m × 0.20 mm, 0.33 μm)	Nitrogen	Powder	FID	Dipyrone	24
DB-5MS	(30 m × 0.25 mm, 0.25 μm)	Helium	Powder	MS	Dipyrone	24
DB-5		Helium	Urine	MS	Acetylfentanyl metabolites	39
DB-5	30 m		Blood	MS	Fentanyl	42
Rtx-5	(30 m × 0.25 mm, 0.25 μm)	Helium	Blood	MS	Fentanyl	43
			Powder	MS	Fentanyl	28

(Continued)

TABLE 18.3 (CONTINUED)
GC Conditions for the Identification of Fentanyl Compounds

Stationary Phase	Column Dimensions	Carrier Gas	Matrix	Detector	Analyte(s)	Reference
Equity-5	(30 m × 0.25 mm, 0.25 μm)	Helium	Patches	MS	Fentanyl	29
						30
						31
Rtx-5	(30 m × 0.25 mm, 0.25 μm)	Helium	Blood	MS	Butyryl fentanyl Acetyl fentanyl	44
Ultra-1	(50 m × 0.2 mm, 0.25 μm)	Helium	Standards	FTIR	5 fentanyls	26
Elite-5	(25 m × 0.32 mm, 0.52 μm)	Helium	Blood	MS	Fentanyl	46
DB-5	(10 m × 0.18 mm, 0.18 μm)	Helium	Urine	MS	Fentanyl	47
1% OV-17	(2 m × 3 mm, 100-120 mesh)	Nitrogen	Urine	CIMS	α-methylfentanyl Metabolites	48
DB-35MS	(30 m × 0.25 mm, 0.15 μm)	Helium	Urine	MS	3 fentanyls	49
DB-5MS	(30 m × 0.25 mm, 0.1 μm)	Helium	Urine	MS	Fentanyl metabolites	49
DB-5MS	(30 m × 0.25 mm, 0.1 μm)	Helium	Patches	MS	Fentanyl	51
DB-5	(15 m × 0.25 mm, 0.25 μm)	Helium	Syringes	MS	Fentanyl	50

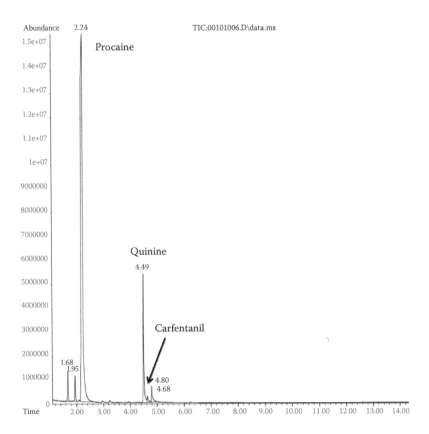

FIGURE 18.4   TIC of a methanolic extract of a seized sample containing carfentanil. (From Broward Sheriff's Office Crime Laboratory, Ft. Lauderdale, FL. With permission.)

heroin, and furanyl fentanyl. Figure 18.7 shows the mass spectrum for the peak at t_R = 8.50 minutes (top) and the spectral search match results for furanyl fentanyl.

The identification and discrimination of five mono-methylated fentanyl-related compounds were studied using GC-vapor-phase Fourier-transform infrared spectroscopy (FTIR) [26]. The compounds studied were α-methylfentanyl, 3-methylfentanyl, n-propylfentanyl, i-propylfentanyl, p-tolylfentanyl, and fentanyl. The GC-FTIR analysis for vapor phase spectra was acquired with a Hewlett Packard 5890 A gas chromatograph equipped with an Ultra-1 fused silica capillary column (50 m × 0.2 mm × 0.25 μm). The chromatographic conditions were operated with a column oven temperature programmed from 100°C to 320°C (10°C/minute) with a helium carrier gas flow of 20 mL/minute and a split ratio of 1:15. The interface and light pipe length, diameter, and temperature were 150 mm, 1 mm i.d., and 280°C, respectively. The FTIR spectrometer utilized a Nicolet FTIR 5SXC system with a mercury–cadmium telluride (MCT) detector. Infrared spectra were measured from 700 to 4000 cm^{-1} (26).

Bell [27] has discussed the pyrolysis of fentanyl in a review on pyrolysis of drug substances. Fentanyl has been shown to be stable up to ~500°C, with thermal

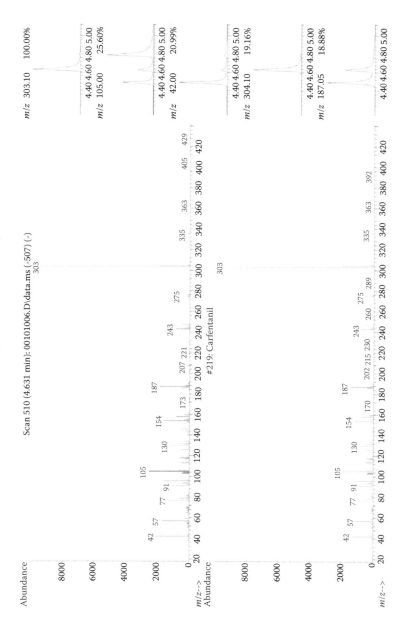

FIGURE 18.5 Mass spectral peak (t_R = 4.68 minutes) in TIC (top) from Figure 18.4 and digital library search result for carfentanil (bottom). (From Broward Sheriff's Office Crime Laboratory, Ft. Lauderdale, FL. With permission.)

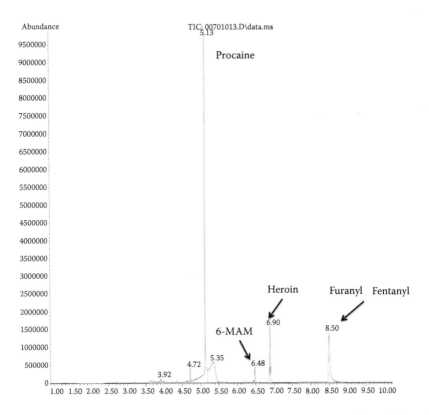

FIGURE 18.6   TIC of a methanolic extract of a seized sample containing furanyl fentanyl. (From Broward Sheriff's Office Crime Laboratory, Ft. Lauderdale, FL. With permission.)

degradation products appearing above this temperature [28]. n-Phenethyl-1,2,5,6-tetrahydropyridine and phenylpropanamide were tentatively identified in fentanyl samples by GC-MS and library comparisons. A method using an analytical pyro-probe operated at 750°C was developed to identify possible biomarkers associated with smoked fentanyl and fentanyl transdermal patches by carrying out pyrolysis experiments under anaerobic and aerobic conditions using helium and air coupled to GC-MS [29]. Styrene, aniline, pyridine, phenylacetaldehyde, propionanilide, benzaldehyde, and two metabolites, despropionyl fentanyl and norfentanyl, were identified. The patch matrix did not interfere with the analytical findings. Another compound, 1-phenylpyridinium (1-PEP), was reported in three separate studies using pyrolysis GC-MS [28,30,31].

With the influx of fentanyl-laced counterfeit pills and powders and toxic fentanyl-related compounds the CDC advises analyzing overdose victim samples by a two-tiered testing scheme to identify specific fentanyl compounds. Current guidelines recommend an enzyme-linked immunoassay (ELISA) screen for fentanyl followed by GC-MS [32]. Evidence from a fatal poisoning case was submitted to a police laboratory, which identified acetylfentanyl and PV8 (4-methoxy PHPP) in both a powder and a liquid. Scan analysis by GC-MS and LC-MS was used to identify acetylfentanyl

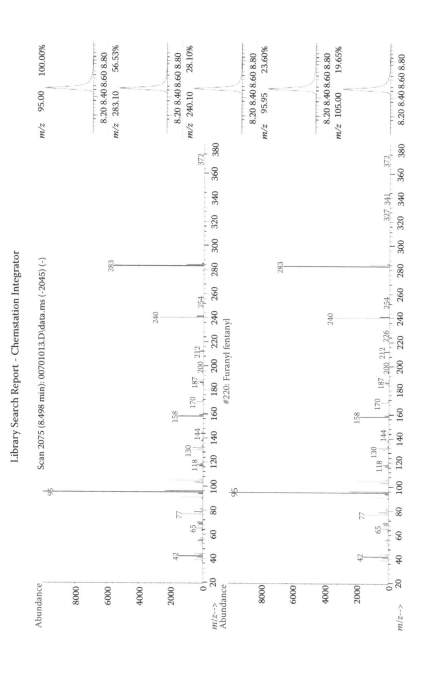

FIGURE 18.7 Mass spectral peak (t_R = 8.50 minutes) in TIC (top) from Figure 18.6 and digital library search result for furanyl fentanyl (bottom). (From Broward Sheriff's Office Crime Laboratory, Ft. Lauderdale, FL. With permission.)

and PV8 in the urine of an overdose victim. LC-MS/MS in the selected reaction ion mode was used to quantify both drugs simultaneously [33]. As mentioned previously, fentanyl can be confirmed by GC-MS in the presence of heroin and fentanyl in drug submissions after presumptive testing. Fentanyl and norfentanyl have been identified and quantified in toxicology specimens by GC-MS using solid-phase extraction (SPE) with the eluent derivatized with acetic anhydride and pyridine. The method has a limit of quantitation (LOQ) of 1 ng/mL [25]. An n-butyl chloride extraction has been used with GC-MS to confirm acetyl fentanyl in postmortem blood after screening with ELISA [34].

The distribution of postmortem concentrations of acetylfentanyl [35] and butyr-fentanyl [36] in peripheral blood, central blood, liver, and vitreous humor were confirmed after an ELISA screen using LLE and GC-MS with specific-ion monitoring (SIM). Ocfentanil was identified in powder by GC-MS after dissolving in methanol and filtering the sample. A second analysis of the powder was performed after acetylation [37]. The stability and analysis of fentanyl in urine was demonstrated over 12 weeks upon analysis by GC-MS with a deuterated internal standard (fentanyl-D_5) utilizing three different extraction techniques: LLE, SPE, and dispersed liquid–liquid microextraction (DLLME). The method was validated, showing good intra- and interday precision as well as excellent accuracy and reasonable LOQs [38]. Acetylation of urine extracts for the detection of acetylfentanyl metabolites by GC-MS has been recommended. The major metabolites of acetylfentanyl were identified in urine by this procedure using GC-MS and HPLC. Structures of the metabolites were proposed and confirmed by their mass spectral fragmentation of electron impact and atmospheric pressure chemical ionization (APCI). The mass spectral and chromatographic properties of some derivatives of acetylfentanyl metabolites were determined. The main path of the biotransformation of acetyl fentanyl is the hydroxylation of the phenyl moiety in the molecule [39]. Postmortem samples were analyzed by GC-MS [40,41]. Postmortem fentanyl concentrations may be affected by antemortem factors, postmortem redistribution, and laboratory variability. Forensic pathologists must use caution when interpreting fentanyl levels as part of death investigations [41]. Fentanyl was quantified after SPE from postmortem whole blood by GC-MS in 23 postmortem cases [42]. Fentanyl and acetylfentanyl immunoassay results were confirmed by full-scan GC-MS analysis of alkaline postmortem blood extracts from 23 overdose cases [43].

Acetylfentanyl and butyrylfentanyl were isolated from postmortem specimens of two fatal intoxication victims by LLE. Two milliliters of specimen was extracted with saturated borate buffer and a mixture of toluene, hexane, and isoamyl alcohol (78:20:2), back-extracted with sulfuric acid, neutralized, and concentrated in ethyl acetate for analysis. The alkaline extracts were separated on an Rtx-5 column. The initial oven temperature of 100°C was held for one minute, followed by a 15°C ramp to 230°C, then a 12°C ramp to 300°C followed by a 10-minute hold. Identification was by electron impact full-scan GC-MS analysis [44]. Acetylfentanyl, acetylnorfentanyl, and 4-anilino-N-phenethyl-4-piperidine (ANPP) were detected in postmortem urine from a fatal intoxication victim. The method involved extracting the urine with an alkaline extraction using Toxi-tubes A® and analysis by GC-MS. Analysis of a methanolic rinse of a syringe also resulted in detection of acetylfentanyl by GC-MS [45].

Fentanyl was identified in postmortem samples from a fatal intoxication by the application of a transdermal fentanyl patch upon a superficial bleeding abrasion of a two-year-old girl. The samples were analyzed in the TIC mode with splitless injection. Fentanyl was identified according to its retention time and three characteristic ions (m/z 245, 146, 189). The most abundant ion was used for quantification, and the second and third ions were used for the confirmation [46].

A fast GC-MS method has been developed and validated for the simultaneous screening of different classes of drugs of abuse in urine, including fentanyl [47]. The chromatographic separation was performed using a short GC column (DB-5, 10 m × 0.18 mm × 0.18 μm). The oven temperature was held at 140°C for 0.4 minute, then at 35°C/minute to 280°C, then at 118°C/minute to 320°C and held 1.5 minutes (total time, 6.2 minutes). Helium was used as carrier gas at a constant flow of 0.7 mL/minute; injection port was set at 280°C in pulsed splitless mode (pulse pressure, 40 mL/minute for 0.5 minute; purge time, 0.5 minute). The mass detector operated in electron ionization at 70 eV in SIM/SCAN mode. Full-scan acquisition range was m/z 51–550 and the SIM characteristic ions for fentanyl were m/z 245, 146, 189, and 202. The method was successfully applied for screening forensic toxicology samples [47].

Structures of four novel and minor metabolites that reflect α-methylfentanyl were elucidated with the assistance of GC–chemical ionization (CI) MS. Trifluoroacetyl derivatives were prepared of rat urine extracts and analyzed by packed-column GC-CIMS with isobutane as the reaction gas [48]. GC-MS procedures were developed and validated for the determination of fentanyl, sufentanil, and alfentanil and their major nor-metabolites in urine of potentially exposed opioid production workers. A simple, one-step extraction protocol was developed using SPE to recover all analytes from urine. The secondary amine functionalities of the nor-metabolites were derivatized to form stable pentafluorobenzamide (PFBA) derivatives with good chromatographic properties. Using the penta-deuterated analogs as internal standards, a limit-of-detection (LOD) of 2.5 pg fentanyl/mL, 2.5 pg sufentanil/mL, and 7.5 pg alfentanil/mL urine was achieved. For the opioid metabolites, the LODs were found to be <50 pg/mL urine. Upon storage at −30°C urine samples were found to be stable for at least two months [49].

Multiple specimen types (blood, urine, vitreous humor, and other tissues and fluids) from autopsies were screened for fentanyl and other drugs by enzyme multiplied immunoassay technique (EMIT® II Plus), followed by confirmation with alkaline extract and GC-MS [50].

A GC-MS procedure was developed to determine residual fentanyl in used Durogesic® reservoir patches and Durogesic® D-Trans® matrix technology-based systems to estimate the actual rate of transdermal fentanyl delivered in individual patients. The mass selective detection system was operated in the SIM mode. Base ion fragments occurring at m/z 245 for fentanyl, and m/z 250 for $^2H_5$-fentanyl, were monitored and used for subsequent quantification. Individual ion dwell times were set at 50 milliseconds for both ion fragments [51].

A simple quantitative method for the analysis of fentanyl citrate (Sublimaze®) in syringes returned to the pharmacy following surgery has been described that used Toxi-A® tubes and GC-MS [52]. The method was useful in verifying that any unused fentanyl was discarded according to narcotic regulations, thereby avoiding the possibility of diversion for illicit consumption.

18.4   LIQUID CHROMATOGRAPHY

There are many approaches to the separation of compounds by liquid chromatography available to the analyst. This section will discuss two strategies: the first being from the standpoint of the forensic toxicologist or forensic drug analyst confronted with a sample of unknown seized drug material or a biological sample with little to no contextual clues for guidance, i.e., a general unknown. In this approach, a more comprehensive assay would be preferred, where the analyst achieves high resolution at the cost of a rapid analysis time. The second approach will be taken as a targeted screening and confirmation where the analyst is concerned with rapidly identifying the presence of a fentanyl-type compound in the sample, perhaps in the setting of a high-throughput laboratory or in cases where the potential intoxicating substance belongs to a narrow pool of candidates. For example, in some postmortem toxicology cases, context clues can assist the analyst in many incidences; preliminary immunoassay drug screen, unconsumed drug substances found at a crime scene, drug residue in the nares, paraphernalia found near the deceased, past history of drug use, to name a few. In this instance, the analyst may choose a rapid assay to confirm the presence or identity of a suspected fentanyl-type compound.

A number of sample preparation techniques have appeared in literature for the analysis of fentanyl, metabolites, and analogs. The choice of sample prep will depend largely upon the matrix. For powder samples, dilute the sample down to the concentration appropriate to the column and system with the appropriate mobile phase. For a 2.1-mm i.d. narrow-bore column, a sample of 50 to 120 µg per injection volume dissolved in a 50:50 mixture of H_2O and acetonitrile or H_2O and methanol may be most effective in dissolving most of the sample components and avoiding column overload. Several published studies have examined a number of different matrices for postmortem toxicology sampling and a review of this material is beyond the scope of this chapter. The two most common samples, peripheral blood and urine specimens, are addressed here. Sample preparation methods vary widely. For example, urine analysis procedures have been successfully validated with sample prep as basic as dilute-and-shoot methods. More elaborate extraction and concentration techniques for whole-blood or postmortem blood samples may use fully automated SPE. Due to the high sensitivity required for detecting fentanyl compounds and metabolites, sample preparation and cleanup is preferred, if not required.

A number of different LLE methods have been suggested, ranging from a simple shake and separate method using n-chlorobutane and preserving the organic layer [37] to more elaborate methods. For example, alkalization of urine can be performed by addition of 1.0 mL potassium carbonate solution followed by agitation in a vortex mixer. Extraction can be performed with 5 mL of a mixture of n-hexane: ethyl acetate (7:3, v/v). After vortex mixing for two minutes and centrifugation at 3000 rpm for five minutes, the upper organic layer is evaporated. The sample can then be reconstituted in the mobile phase or a suitable solvent [22]. The method chosen should preserve the recovery of the compound(s) of interest while reducing the amount of unnecessary background material. In cases where the potentially intoxicating substance is a complete unknown, a general acid/base extraction method may be preferred, as to limit the deleterious effect on any of the possible drugs.

18.4.1 HPLC

The most commonly used liquid chromatographic column for the analysis of drugs of abuse is the very reliable, reversed-phase C-18 column. This column is a very good all-purpose column, with a broad pH operating range, and compatible with a number of mobile phase formulations. A laboratory concerned with developing an assay that covers a broad range of compounds would likely choose the C-18 column for their work. Other bonded phase columns, phenyl (biphenyl or hexyl phenyl) and pentafluorophenyl (PFP), provide differences in selectivity that can be taken advantage of to improve the resolution of structurally similar compounds with aromatic functions with the former or basic, ionizable compounds with the latter. When the analyst is concerned with distinguishing between fentanyl analogs that are positional isomers or analogs with slight modifications, these columns should be considered due to their advantages in selectivity.

For the chromatography of fentanyl and related compounds, it is recommended to start method development with a 2.1-mm i.d. narrow-bore column. The high potency of fentanyl requires increased sensitivity and this diameter column works well with electrospray ionization (ESI) mass spectrometry. If the increased system pressure is too great, such as in a traditional HPLC system, the analyst may be able to move up to a 3.0-mm i.d. column depending on the system and detector requirements.

Column length is directly proportional to column efficiency and analysis time. If the goal of the analysis is to quickly separate a few analytes of interest with significantly different structural features, a shorter 50-mm column would be the appropriate choice. Studies have shown that in both fatal and nonfatal overdoses, the victim ingested multiple drug compounds in approximately 44% of these cases, with alcohol being the most common drug. However, it is rare that the victim ingested more than five drug compounds. The differentiation of such a small number of compounds is well within the capabilities of shorter columns for routine, high-throughput analysis. Alternatively, if the goal is to separate a complex mixture of closely related fentanyl analogs then the appropriate choice would be longer HPLC columns of 150 mm or greater. A comprehensive assay can be developed with long columns with a small i.d. and small particle sizes, but the analyst must be willing to accept the lengthened analysis time. Clinical toxicology of pain management compounds has led many of the major chromatography column manufacturers to market their columns with methods for the analysis of 100+ compounds. The toxicologist must keep in mind that column length will also affect system pressure, but to a lesser degree than column diameter and particle size.

When choosing a particle size for chromatography of drugs of abuse, it is recommended to start with the smallest particle size that the system can tolerate. Smaller packing material particles will improve resolution and efficiency due to their smaller theoretical plate size and will allow for a broad range of flow rates. The use of smaller particle sizes is beneficial to both the high-throughput strategy, by reducing analysis time, and the comprehensive panel strategy, by increasing resolution of complex mixtures. Choice of particle size is generally only limited by the pressure tolerance of the liquid chromatography instrument. The smaller the particle size, the greater the efficiency in separation but this results in an increase in system backpressure. Particle sizes of approximately 2.0 μm or smaller are reserved for ultra-high-pressure chromatography systems that can tolerate pressures up to and above 10000 PSI or 700 bar. Traditional liquid chromatographic systems (<6000 psi or 400 bar) will be limited to particle sizes of 3.5 μm or larger, but can still achieve high efficiency by compensating with longer column lengths and lower flow rates.

Particles have two other characteristics that contribute to their separation efficiency: pore size and particle size distribution. The forensic chemist or toxicologist concerned with squeezing the most resolution out of the chromatography system can optimize their column selection by carefully evaluating all available column parameters. This may be necessary to separate closely related fentanyl positional isomers. Pore diameter is inversely related to the surface area of the column particle. The smaller the pore size, the larger the surface area of the particle, therefore the column will have a greater retention capability. For compounds of interest to the forensic chemist or toxicologist, pore diameters 60 to 120 Å are ideal for small molecules, molecular weights of 1000 amu or less. The range of molecular weights for fentanyl metabolites and analogues is within 200 to 500 amu.

The (advertised) particle size is actually the average diameter of the range of silica particle sizes used for the column packing. Generally speaking, the broader the range of particle sizes, the broader the peak width of the analyte. Therefore, a narrow range of particle size distribution will result in a high number of theoretical plates and greater column efficiency. The increased efficiency for small-molecule, liquid chromatography may be negligible; however, it is always a good idea to utilize well-manufactured columns from a provider with good quality control.

The choice of mobile phase is important and should be optimized for the particular separation. A common method that will separate a wide variety of drug compounds, including fentanyl-related drugs, is to use a two-phase system: (A) water with 0.1% formic acid and (B) either methanol or acetonitrile, also with 0.1% formic acid. Acetonitrile is more expensive to use than methanol, but has greater elution strength, less UV absorbance, and lower viscosity (therefore lower system pressures). In general, a desired mobile phase has a pH value 1 or 2 lower than the pK_a value of a basic analyte such as fentanyl (pK_a ~8.6) or other analytes. Mobile phase flow rates can vary widely from 0.100 μL/minute to 10 mL/minute depending upon the LC system, column selection, and pump capabilities. Most published assays for LC/MS of fentanyl compounds recommend average flow rates of 0.3 to 0.4 mL/minute. Higher flow rates will reduce peak broadening, but raise system pressures. The choice of solvent gradient will be based on the goals of the method, faster gradient ramping for high-throughput analysis, long slower gradient for more compounds resolution. A typical starting point for method development would be 95% mobile phase A, 5% mobile phase B, held for 1.5 to 2 minutes; then slowly ramp up the proportion of mobile phase B to 100% over the next 16 to 20 minutes; hold at 100% mobile phase B for approximately 2 to 4 minutes, before returning back to 95% mobile phase A and 5% mobile phase B to equilibrate the column for 4 to 8 minutes. This is just a simple method gradient, meant for broad general assay for method development.

Each application will need to have a method developed specifically based on the particular system, column, and analytical goals and requirements.

Rojkiewicz et al. [53] described an assay for the quantitation of the fentanyl analog 4-fluorobutyrylfentanyl by HPLC-MS. Initially the samples were subjected to LLE using pH 9 TRIS buffer, acetonitrile, and ethyl acetate. The extraction was performed on both a hydrolyzed urine sample and postmortem blood. The extracted sample was tested by both GC-MS and HPLC-MS (ion trap). The liquid chromatography was performed using a 150 mm × 4.6 mm-i.d. C-18 column. The mobile phases used in the group's method were (mobile phase A) 50 mM ammonium formate and 20 mM formic acid (0.084%) in deionized H_2O, and (mobile phase B) a 10% solution of mobile phase A and 90% acetonitrile. The authors used a solvent gradient that began at 95%A:5%B for two minutes, ramped to 30%A:70%B until 30 minutes, held for two minutes, before returning to the initial values from 32 to 40 minutes. The mobile phase flow rate was 100 µL/minute. In this case report, the analyte of interest eluted at just under 35 minutes of a 40-minute chromatographic analysis. This assay is an example of a more comprehensive method designed to resolve a large number of compounds from biological matrices.

Poklis et al. [54] developed and validated a method for the quantitation of acetyl fentanyl, fentanyl, acetyl norfentanyl, and norfentanyl, in various body fluids and tissue samples. Initial toxicological screening was done on the postmortem blood, vitreous fluid, and urine samples by immunoassay with positive presumptive results for fentanyl and confirmation by full-scan GC-MS. It is important to consider cross-reactivity of immunoassay screening when selecting a method for confirmation. Various fentanyl analogs may or may not react with fentanyl antibodies in the immunoassay. Samples for LC-MS/MS analysis were extracted from pH 6.0 phosphate buffered fluids and homogenized tissues using SPE columns. The samples were run on an UHPLC/MS/MS system utilizing a 100 mm × 2.5 mm i.d., 5 µm column. The mobile phases used in their method were (mobile phase A) 10 mM ammonium formate and 0.1% formic acid in deionized H_2O, and (mobile phase B) methanol. The column particle size and ID allows for higher mobile phase flow rates and shorter analysis time. In this case, the authors used a gradient starting at 95%A:5%B ramped to 60%A:40%B after 1.5 minutes, to 100%B after three minutes, before returning to 95%A:5%B. The complete chromatography took place in four minutes. The authors subsequently applied this method to the quantitation of butyryl fentanyl when this analog was encountered in casework [44]. This is an excellent example of a rapid, high-throughput, targeted method for quantitation of fentanyl analogs when their presence has been either presumptively identified, or confirmed by prior methods.

18.4.2   LC-MS

The integration of the mass spectrometer with a liquid chromatograph has vastly improved the analytical utility of LC for identification of drugs of abuse. The LC-MS has several advantages over GC-MS and is quickly gaining widespread use in forensic and toxicology labs. In general, the LC-MS is more sensitive, requires less sample prep, and easily handles thermally labile compounds. These advantages are particularly helpful for the analysis of fentanyl and fentanyl analogs. Minimizing sample preparation steps reduces opportunity for loss of target analyte and the high-potency low dosages of fentanyl compounds require highly sensitive analytical techniques. In LC, electrospray ionization (ESI), atmospheric pressure chemical ionization (APCI), or photo ionization (APPI) can be used to apply a charge to the analytes. Electrospray (or heated electrospray) ionization is the most popular ionization technique for drugs of abuse, although APCI has been successfully used to identify fentanyl in whole blood and urine [55]. Fentanyl and the fentanyl analogs are basic drugs with amine functions that can be protonated, therefore positive ion mode is the appropriate choice of ionization for methods concerned with fentanyl detection.

A liquid chromatograph connected to a single quadrupole mass spectrometer can be used to either scan a range of mass-to-charge (*m/z*) ratios in full-scan mode or scan for selected ions in selected-ion monitoring (SIM) mode for increased sensitivity. While not as selective or sensitive as an LC-MS/MS or LC-HRAMS instrument, a single quad LC-MS can still be used for routine analysis. If a single quad LC-MS is to be used for the analysis of structurally related fentanyl analogues, it is critical that the developed method has sufficiently high chromatographic separation in order to distinguish between the fentanyl analogs with identical nominal masses, such as the positional isomers 3-methyl fentanyl and α-methyl fentanyl, MW 351. Venisse et al. [56] developed a general unknown screening procedure for HPLC-MS in full-scan mode. Their method relied upon robust sample preparation in order to reduce extraneous background noise while achieving high analyte recovery. The chromatographic system used a reversed-phase C-18 column (150 mm × 1 mm i.d. with 5-μm particles), a long 50-minute mobile phase gradient, and a low 50 μL/minute flow rate in order to achieve the chromatographic resolution necessary for the method. In their design, product ions were generated by alternating the in-source collision-induced dissociation (CID) voltages. Fragmentation occurs in the electrospray interface when ions are accelerated into gas molecules within the source. This research team developed their own mass spectral libraries based on reconstructing spectra collected at both low and high CID voltages and optimized data processing. Compound identification was made by comparing the unknown fragmentation pattern and retention times to the custom-built library generated under identical conditions.

A scanning method results in a loss of sensitivity, but may identify other compounds present in the matrix. Selective ion monitoring (SIM) results in higher sensitivity, but requires more method development. In using SIM mode, only a small number of ions can be monitored at a given time. A SIM method needs to be developed for each fentanyl compound to determine the expected retention times and the appropriate cone voltage(s) to preserve the precursor ion while generating sufficient product ions. A successful identification requires that the presence of specific ions, in a specific ratio, at a specific retention time. The acceptance criteria are set by the applicable regulatory body. As an example, the College of American Pathologists (CAP) recommends ion ratios must be within 20% of a known calibrator using at least two separate ion ratios. One ratio with two characteristic ions is acceptable for analytes with a low number of characteristic ions. Care must be taken when selecting ions to monitor. Returning to the previous example, 3-methyl fentanyl and α-methyl fentanyl share a common product ion of 202 *m/z*. The analyst would need to rely on different ions to monitor and have adequate resolution of these analytes.

18.4.3 LC-MS/MS

The most common type of detector for liquid chromatography is now the triple-quadrupole mass spectrometer. The first quadrupole isolates specific precursor masses, the second quadrupole (or collision cell) induces fragmentation at various collision energies, and the third quadrupole filters the product ions. This is sometimes referred to as tandem mass spectrometry, due to the presence of the two mass filters, one before, and one after the collision cell. Fragmentation is reliable and reproducible when the same collision energies and other parameters are applied consistently from sample to sample. This allows the analyst to search for the intact mass of the precursor or original compound with the first quadrupole, Q1, induce fragmentation of the all ions in a mass range (all-ion fragmentation—AIF) or only a set of targeted ions, for example, 337 m/z for fentanyl in Q2, and then generate a spectrum of fragments in Q3. This method of analysis, known as multiple reaction monitoring (MRM), is the most common use of the triple-quadrupole mass spectrometer. MRM analysis allows the practitioner the ability to search for a precursor ion, apply a known collision energy to generate one or two ions for confirmation purposes, and a second or third ion for quantitation.

The prevalence of fentanyl, fentanyl analogs, and other opioids has led to the development of several robust methods of analysis using LC-MS/MS. Gergov et al. [57] developed a comprehensive method of screening and confirmation of opioids that included fentanyl, norfentanyl, and seven other fentanyl derivatives using a quadrupole-linear ion trap mass spectrometer. In their method, the authors obtained sufficient chromatographic resolution to enable them to distinguish between three compounds with identical nominal masses: α-methylfentanyl, and the stereoisomers *cis*- and *trans*-3-methylfentanyl, a critical component of identifying compounds with similar precursor ions and MRM transitions to similar product ions. High-resolution chromatography was achieved using a narrow-bore C-18 column with a moderately small particle size (100 mm × 2.0 mm i.d., 3-μm particle size) and a low 150 μL/minute flow rate. The authors used a gradient of mobile phase A, 0.1% formic acid in 10 mM ammonium acetate (pH 3.2), and mobile phase B, 0.1% formic acid in acetonitrile, starting at 85%A:15%B for nine minutes to equilibrate the column. A 10-μL sample was injected into the LC and the percent of mobile phase B was gradually increased to 30% over 13 minutes. The gradient was then ramped to 80% mobile phase B in ten minutes, before the final 95% mobile phase B for the last minute. Collision energies and other parameters were previously determined and optimized individually for each of the 25 compounds in the panel. For identification, two MRM transitions were monitored for each compound within select retention time windows. For example, the transitions for remifentanil 377→345 and 377→317 (CE = 20) were monitored during the time period from 5.5 minute to 12 minutes. The transitions of 337→188 and 337→105 (CE = 30) for fentanyl were monitored in the retention time window from 12 minutes until the end of the chromatographic run. The partitioning of the MRM transition monitoring into separate time windows results in faster scan times, which increases resolution and generates more data points per peak. A list of some example MRM transitions for fentanyl-related compounds can be found in Table 18.4.

TABLE 18.4
MRM Transitions of Selected Fentanyl Compounds

	[M+H]	Product Ion	Product Ion	Collision Energy	Declustering Potential (or Cone Voltage)	Reference
Norfentanyl	233	84	177	25	40	57
Remifentanil	377	345	317	20	30	57
Alfentanil	417	268	385	25	40	57
Fentanyl	337	188	105	30	50	57
p-Fluorofentanyl	355	206	216	30	50	57
α-Methyl fentanyl	351	202	119*	30 & 35*	50 & 60*	57
Trans-3-Methyl fentanyl	351	202	230	30	60	57
Cis-3-Methyl fentanyl	351	202	230	30	60	57
Sufentanil	387	238	355	25	40	57
Acetyl norfentanyl	219	85	56*	17 & 25*	32	54
Acetyl fentanyl	323	105	188*	35 & 23*	48	54
Butyryl fentanyl	351	105	188*	36 & 24*	50	54
4-fluorobutyryl fentanyl	369	188	299	Unspecified	Unspecified	53 / 53
Furanyl fentanyl	375.1	188	105*	25 & 40*	125 (V) Fragmentor	58
Ocfentanil	371	188	105*	24 & 32*	28	22
Norcarfentanil[†]	291.1	113	231.2*	33 & 17*	121	59
Norsufentanil	277.2	184.2	128.2	17	60	59
Carfentanil	395	113	335.2*	37 & 23*	176	59
Lofentanil	409.1	105.1	200.2*	59 & 38*	188	59

* Indicates that the second product ion was generated by the second collision energy also indicated by the
  * symbol.
† Norcarfentanil is also a major metabolite of remifentanil.

Tandem mass spectrometry (MS/MS) has proven to be a valuable tool for fentanyl detection because it achieves high selectivity between fentanyl, its analogs, and the metabolites while maintaining low detection limits (estimated range from 0.003 to 0.027 ng/mL).

18.4.4 LC-HRAMS

While tandem mass spectrometry is the most widely used detection system for liquid chromatography in forensic toxicology, high-resolution accurate mass spectrometry (HRAMS) is becoming more popular. LC-MS/MS requires either prior knowledge of potential analytes to establish MRM transitions and compound dependent parameters, or reduced sensitivity by operation in full-scan mode. Additionally, LC-MS/MS

TABLE 18.5

Selected Fentanyl Compounds by HRAMS

Compound Name	Experiment Type	Chemical Formula	Extracted Mass	Adduct	Fragment 1	Fragment 2	Fragment 3	Fragment 4
Alfentanil	XIC	$C_{21}H_{32}N_6O_3$	417.26087	M+H	314.18534	268.176	197.12799	170.10319
α-methylfentanyl	XIC	$C_{23}H_{30}N_2O$	351.24309	M+H	216.13776	202.15855	119.08548	91.05454
Benzylfentanyl	XIC	$C_{21}H_{26}N_2O$	323.21179	M+H	216.13825	174.12774	91.05475	82.06573
3-methylfentanyl	XIC	$C_{23}H_{30}N_2O$	351.24309	M+H	230.15381	202.15896	150.09135	105.07025
Fentanyl	XIC	$C_{22}H_{28}N_2O$	337.22744	M+H	216.13774	188.14291	134.09616	105.07001
Hydroxyfentanyl	XIC	$C_{22}H_{28}N_2O_2$	353.22235	M+H	216.1383	204.13832	202.12274	186.12776
Norfentanyl	XIC	$C_{14}H_{20}N_2O$	233.16484	M+H	177.13826	150.09104	84.08117	
Normethylfentanyl	XIC	$C_{15}H_{22}N_2O$	247.18049	M+H	191.1542	150.09134	98.09684	69.07057
Ohmefentanyl	XIC	$C_{23}H_{30}N_2O_2$	367.238	M+H	349.22723	218.15381	216.13822	200.14331
ω-hydroxynorfentanyl	XIC	$C_{14}H_{20}N_2O_2$	249.15975	M+H	84.08137	56.05029		
Remifentanil	XIC	$C_{20}H_{28}N_2O_5$	377.2071	M+H	317.18583	285.15961	261.15959	228.12297

is limited to the nominal mass of the peaks detected. High-resolution accurate mass spectrometry (HRAMS), such as time-of-flight (TOF) or Orbitrap® instruments, provides the analyst the exact mass of the detected compound with a mass resolution down to 2 ppm. Consider the pain medications oxymorphone ($C_{17}H_{19}NO_4$) and dihydrocodeine ($C_{18}H_{23}NO_3$): both have a nominal mass of 302.2 m/z, but their exact masses allow for discrimination by HRAMS. Oxymorphone and dihydrocodeine have exact masses of 301.13141 and 301.16779, respectively, for a difference of 120 ppm. Due to the increased resolution, these mass spectrometers are also capable of resolving isotope patterns of the identified compounds. LC-HRAMS allows for compound identification by accurate mass, chromatographic retention time, fragmentation pattern, isotope patterns, and by comparison to a published or user-defined library or database. A list of selected fentanyl compounds with HRAMS properties is given in Table 18.5.

18.5 FUTURE PERSPECTIVES

Since its synthesis in 1960, fentanyl has been a useful and important drug for the treatment of severe to moderate pain. It is a more potent and faster-acting analgesic to both morphine and meperidine. Over the past 50 years, and especially more recently, many novel analogs and derivatives have emerged. What has resulted has been a pandemic of dangerous drugs causing an overdose rate never seen before by humankind. Although there has already been a plethora of fentanyl-related compounds, there are still many possibilities for future development with different and, in some cases, more potent biological activity. It is a frightening thought that more potent fentanyl, or other drugs for that matter, could be synthesized, which are inevitably going to be abused and cause more overdoses. Hopefully, future developments will bring about new fentanyl-related compounds that can be used in a positive way to relieve pain for mankind.

It is ironic that chromatography has been around just a little longer than fentanyl. It has matured and continues to develop as a powerful separation technique for the analysis of controlled substances, including fentanyl-related compounds. The ability to interface chromatographic techniques to spectroscopy has revolutionized the usefulness of the separation technique and has proved invaluable to the analytical chemist and toxicologist who must analyze unknown and dangerous substances. The ability to automate the technique is a great advantage to the laboratories that are overwhelmed with a large workload. High-throughput sensitive analysis is very important for this type of application. The future looks bright for these techniques as tandem mass spectrometric techniques continue to develop and microsampling and microextraction procedures improve.

REFERENCES

1. Stanley, T.H. 1992. The history and development of the fentanyl series. *J. Pain Symptom Mgmt.* 7(3 Suppl):S3–S7.
2. Stanley, T.H., Egan, T.D., and Van Aken, H. 2008. A tribute to Dr. Paul A. J. Janssen: Entrepreneur extraordinaire, innovative scientist, and significant contributor to anesthesiology. *Anesth. Analg.* 106(2): 451–462.
3. Janssen, P.A.J. 1962. A review of the chemical features associated with strong morphine-like activity. *Br. J. Anaesth.* 34: 260–268.

4. Niemegeers, C.J.E., Schellekens, W.F.M, Van Bever, W.F.M., and Janssen, P.A.J. 1976. Sulfentanil, a very potent and extremely safe intravenous morphine-like compound in mice, rats and dogs. *Arzneimittelforsching.* 26: 1551–1556.

5. Janssen, P.A.J. 1982. Potent, new analgesics, tailor-made for different purposes. *Acta Anaesth. Scand.* 26: 262–268.

6. Niemegeers, C.J.E. and Janssen, P.A.J. 1981. Alfentanil (R 39 209)—A particularly short-acting intravenous narcotic analgesic in rats. *Drug Develop. Res.* 1: 83–133.

7. Van Bever, W.F., Niemegeers, C.J.E., and Janssen, P.A.J. 1974. Synthetic analgesics. Synthesis and pharmacology of the diastereoisomers of N-[3-methyl-1-(2-phenylethyl)-4-piperidyl]-N-phenylpropanimide and N-[3-Methyl-1-(1-methyl-2-phenylethyl)-4-piperidyl]-N-phenylpropanimde. *J. Med. Chem.* 17:1047–1051.

8. Van Bever, W.F., Niemegeers, C.J.E., Schellekens, K.H.L., and Janssen, P.A.J. 1976. N-(4-substituted-1-(2-arylethyl)-4-piperidinyl)-N-phenylpropanamides, a novel series of extremely potent analgesics with unusually high safety margin. *Arzneimittelforschung.* 26: 1548–1551.

9. Al-Hasani, R. and Bruchas, M.R. 2011. Molecular mechanisms of opioid receptor-dependent signaling and behavior. *Anesthesiol.* 115: 1363–1381.

10. Henderson, G.L. 1988. Designer drugs: Past history and future prospects. *J. Forensic Sci.* 33: 569–575.

11. Carroll, F.I., Lewin, A.H., Mascarella, S.W., Seltzman, H.H., and Reddy, P.A. 2012. Designer drugs: A medicinal chemistry perspective. *Ann. N. Y. Acad. Sci.* 1248: 8–38.

12. EMCDDA. 2012. *Fentanyl in Europe: Trendspotter Study.* The European Monitoring Centre for Drugs and Drug Addiction.

13. Stanley, T. 2014. The fentanyl story. *J. Pain* 15:1215–1226.

14. CDC. 2016. *Increases in Drug and Opioid-Involved Deaths—US 2010–2015.* US Department of Health and Human Services/Centers for Disease Control and Prevention, 65: 1445–1452.

15. 2016. *Emerging Threat Report Annual.* Drug Enforcement Administration.

16. Lurie, I.S., Berrier, A.L., Casale, J.F., Iio, R., and Bozenko, J.S., Jr. 2012. Profiling of illicit fentanyl using UHPLC-MS/MS. *Forensic Sci. Int.* 220: 191–196.

17. Mayer, B.P., DeHope, A.J., Mew, D.A., Spackman, P.E., and Williams, A.M. 2016. Chemical attribution of fentanyl using multivariate statistical analysis of orthogonal mass spectral data. *Anal. Chem.* 88: 4303–4310.

18. Gillespie, T.J., Gandolfi, A.J., Maiorino, R.M., and Vaughan, R.W. 1981. Gas chromatographic determination of fentanyl and its analogues in human plasma. *J. Anal. Toxicol.* 5: 133–137.

19. Kram, T.C., Cooper, D.A., and Allen, A.C. 1981. Behind the identification of china white. *Anal. Chem.* 53(12): 1379A–1386A.

20. Ohta, H., Suzuki, S., and Ogasawara, K. 1999. Studies on Fentanyl and related compounds IV. Chromatographic and Spectrometric Discrimination of fentanyl and its derivatives. *J. Anal. Toxicol.* 23: 280–285.

21. Manral, L., Gupta, P.K., Ganesan, K., and Malhotra, R.C. 2008. Gas chromatographic retention indices of fentanyl and analogues. *J. Chromatogr. Sci.* 46: 551–555.

22. Coopman, V., Codonnier, J., De Leeuw, M., and Cirimele, V. 2016. Ocfentanil overdose fatality in the recreational drug scene. *Forensic Sci. Int.* 266: 469–473.

23. Kanamori, T., Iwata, Y.T., Tsujikawa, K., Kuwayama, K., Yamamuro, T., Segawa, H., and Inoue, H. 2016. Simultaneous analysis of 18 compounds of fentanyl and its analogues by TLC, GC/MS and LC/MS. *Japn. J. Forensic Sci. Technol.* 21(1): 139–147.

24. Isaacs, R.C.A., Harper, M.M., and Miller, E.C. 2017. Analytical challenges in the confirmation identification of dipyrone as an adulterant in illicit drug samples. *Forensic Sci. Int.* 270: 185–192.

25. Marinetti, L.J. and Ehlers, B.J. 2014. A series of forensic toxicology and drug seizure cases involving fentanyl alone and in combination with heroin, cocaine or heroin and cocaine. *J. Anal. Toxicol.* 38: 592–598.

26. Suzuki, S. 1989. Studies on fentanyl and related compounds: II. Spectrometric discrimination of five monomethylated fentanyl isomers by gas chromatography/Fourier transform-infrared spectrometry. *Forensic Sci. Int.* 43: 15–19.

27. Bell, S.C. and Nida, C. 2015. Pyrolysis of drugs of abuse: A comprehensive review. *Drug Test. Anal.* 7: 445–456.

28. Manral, L., Gupta, P.K., Suryanarayana, M.V.S., Ganesan, K., and Malhotra, R.C. 2009. Thermal behavior of fentanyl and its analogues during flash pyrolysis. *J. Therm. Anal. Calorim.* 96: 531–534.

29. Nishikawa, R.K., Bell, S.C., Kraner, J.C., and Callery, P.S. 2009. Potential biomarkers of smoked fentanyl utilizing pyrolysis gas chromatograph–mass spectrometry. *J. Anal. Toxicol.* 33: 418–422.

30. Rabinowitz, J.D., Wensley, M., Lloyd, P., Myers, D., Shen, W., Lu, A., Hodges, C., Hale, R., Mufson, D., and Zaffaroni, A. 2004. Fast onset medications through thermally generated aerosols. *J. Pharm. Exp. Ther.* 309: 769–775.

31. Garg, A. Solas, D.W., Takahashi, L.H., and Cassella, J.V. 2010. Forced degradation of fentanyl: Identification and analysis of impurities and degradants. *J. Pharmaceut. Biomed.* 53: 325–334.

32. Influx of Fentanyl-Laced Counterfeit Pills and Toxic Fentanyl-Related Compounds Further Increases Risk of Fentanyl-related Overdose and Fatalities, Centers for Disease Control and Prevention, 2016. https://emergency.cdc.gov/han/han00395.asp, accessed 27 May 2017.

33. Yonemitsu, K., Sasao, A., Mishima, S., Ohtsu, Y., and Nishitani, Y. 2016. A fatal poisoning case by intravenous injection of "bath salts" containing acetyl fentanyl and 4-methoxy PV8. *Forensic Sci. Int.* 267: e2–e6.

34. Lozier, M.J., Boyd, M., Stanley, C., Ogilvie, L., King, E., Martin, C., and Lewis, L. 2015. Acetyl fentanyl, a novel fentanyl analog, causes 14 overdose deaths in Rhode Island, March–May 2015. *J. Med. Toxicol.* 11: 208–217.

35. McIntyre, I.M., Trochta, A., Gary, R.D., Malamatos, M., and Lucas, J.R. 2015. An acute acetyl fentanyl fatality: A case report with postmortem concentrations. *J. Anal. Toxicol.* 39: 490–494.

36. McIntyre, I.M., Trochta, A., Gary, R.D., Wright, J., and Mena, O. 2016. An acute butyrfentanyl fatality: A case report with postmortem concentrations. *J. Anal. Toxicol.* 40: 162–166.

37. Dussy, F.E., Hangartner, S., Hamberg, C., Berchtold, C., Scherer, U., Schlotterbeck, G., Wyler, D., and Breillman, T.A. 2016. An acute ocfentanil fatality: A case report with postmortem concentrations. *J. Anal. Toxicol.* 40: 761–766.

38. Gardner, M.A., Sampsel, S., Jenkins, W.W., and Owens, J.E. 2015. Analysis of fentanyl in urine by DLLME-GC-MS. *J. Anal. Toxicol.* 39: 118–125.

39. Melent'ev, A.B., Kataev, S.S., and Dvorskaya, O.N. 2015. Identification and analytical properties of acetyl fentanyl metabolites. *J. Anal. Chem.* 70(2): 240–248.

40. Krinsky, C.S., Lathrop, S.L., and Zumwalt, R. 2014. An examination of the postmortem redistribution of fentanyl and interlaboratory variability. *J. Forensic Sci.* 59 (5): 1275–1279.

41. Algren, D.A., Monteilh, C.P., Punja, M., Schier, J.G., Belson, M. Hepler, B.R., Schmidt, C.J., Miller, C.E., Patel, M., Paulozzi, L.J., Straetemans, M., and Rubin, C. 2013. Fentanyl-associated fatalities among illicit drug users in Wayne County, Michigan (July 2005–May 2006). *J. Med. Toxicol.* 9: 106–115.

42. Thompson, J.G., Baker, A.M., Bracey, A.H., Seningen, J., Kloss, J.S., Strobl, A.Q., and Apple, F.S. 2007. Fentanyl concentrations in 23 postmortem cases from the Hennepin County Medical Examiner's Office. *J. Forensic Sci.* 52 (4): 978–981.
43. Pearson, J., Poklis, J., Poklis, A., Wolf, C., Mainland, M., Hair, L., Devers, K., Chrostowski, L, Arbefeville, E., and Merves, M. 2015. Postmortem toxicology findings of acetyl fentanyl, fentanyl and morphine in heroin fatalities in Tampa, Florida. *Acad. Forensic Pathol.* 5(4): 676–689.
44. Poklis, J., Poklis, A., Wolf, C., Hathaway, E.A., Chrostowski, L., Devers, K., Hair, L., Mainland, M., Merves, M., and Pearson, J. 2016. Two fatal intoxications involving butyryl fentanyl. *J. Anal. Toxicol.* 40: 703–708.
45. Cunningham, S.M, Haikal, N.A., and Kraner, J.C. 2016. Fatal intoxication with acetyl fentanyl. *J. Forensic Sci.* 61 (S1): S276–S280.
46. Bakovic, M., Nestic, M., and Mayer, D. 2015. Death by band-aid: Fatal misuse of trans-dermal fentanyl patch. *Int. J. Legal Med.* 129: 1247–1252.
47. Srano-Rossi, S., Bermejo, A.M., de la Torre, X., and Botrê, F. 2011. Fast GC-MS method for the simultaneous screening of THC-COOH, cocaine, opiates and analogues including buprenorphine and fentanyl, and their metabolites in urine. *Anal. Biochem. Chem.* 399: 1623–1630.
48. Sato, S., Suzuki, S., Lee, X.-P., and Sato, K. (2010). Studies on 1-(2-phenethyl)-4-(N-propionylanilino)piperidine (fentanyl) related compounds. VII. Quantification of α-methylfentanyl metabolites excreted in rat urine. *Forensic Sci. Int.* 195: 68–72.
49. Van Nimmen, N.F.J., Poels, K.L.C., and Veulemans, H.A.F. 2004. Highly sensitive gas chromatographic–mass spectrometric screening method for the determination of picto-gram levels of fentanyl, sufentanil and alfentanil and their major metabolites in urine of opioid exposed workers. *J. Chromatogr. B.* 804: 375–387.
50. Hull, M.J., Juhascik, M., Mazur, F., Flomenbaum, M.A., and Behonick, G.S. 2007. Fatalities associated with fentanyl and co-administered cocaine and opiates. *J. Forensic Sci.* 52 (6): 1383–1388.
51. Van Nimmen, N.F.J. and Veulemans, H.A.F. 2007. Validated GC-MS analysis for the determination of residual fentanyl applied Durogesic® reservoir and Durogesic® D-Trans® matrix transdermal fentanyl patches. *J. Chromatogr. B* 846: 264–272.
52. Kingsbury, D.P., Makowski, G.S., and Stone, J.A. 1995. Quantitative analysis of fen-tanyl in pharmaceutical preparations by gas chromatography–mass spectrometry. *J. Anal. Toxicol.* 19: 27–30.
53. Rojkieewicz, M., Majchrzak, M., Celinski, R., Kus, P., and Sajewicz, M. 2016. "Identification and physiochemical characterization of 4-fluorobutyrfentanyl (1-((4-fluorophenyl)(1-phenethylpiperidin-4-yl)amino)butan-1-one, 4-FBF) in seized materi-als and postmortem biological samples. *Drug Test. Anal.* Online. 8(8).
54. Poklis, J. et al. 2015. Postmortem tissue distribution of acetyl fentanyl, fentanyl and their respective nor-metabolites analyzed by ultrahigh performance liquid chromatog-raphy with tandem mass spectrometry. *Forensic Sci. Int.* 257: 435–441.
55. Skulska, A., Kala, M., Adamowicz, P., Chudzikiewicz, E., and Lechowicz, W. 2007. Determination of fentanyl, atropine, and scopolamine in biological materials using LC-MS/APCI methods. *Przegl Lek.* 64(4–5): 263–267.
56. Venisse, N., Marquet, P. Duchoslav, E., Dupuy, J.L., and Lachâtre, G. 2003. A general unknown screening procedure for drugs and toxic compounds in serum using liquid chro-matography–electrospray single quadrupole mass spectrometry. *J. Anal. Toxicol.* 27(1): 7–14.
57. Gergov, M. Nokua, P., Vuori, E., and Ojanperä, I. 2009. Simultaneous screening and quantification of 25 opioid drugs in postmortem blood and urine by liquid chromatography-tandem mass spectrometry. *Forensic Sci. Int.* 186(1–3): 36–43.

58. Mohr, A.L.A., Friscia, M., Papsun, D., Kacinko, S.L., Buzby, D., and Logan, B.K. 2016. Analysis of novel synthetic opioids U-47700, U-50488, and furanyl fentanyl by LC-MS/MS in postmortem casework. *J. Anal. Toxicol.* 40: 709–717.
59. Shaner, R.L., Kaplan, P., Hamelin, E.I., Bragg, W.A., and Johnson, R.C. 2014. Comparison of two automated solid phase extractions for the detection of ten fentanyl analogs and metabolites in human urine using liquid chromatography tandem mass spectrometry. *J. Chromatogr. B.* 962: 52–58.

19 Modern Techniques for the Identification of Tryptamines

Dariusz Zuba

CONTENTS

19.1 CHARACTERIZATION OF TRYPTAMINE-TYPE DESIGNER DRUGS

Tryptamine is the parent compound for one of the most popular families of the designer drugs. It is a monoamine alkaloid, and it is structurally similar to the amino acid tryptophan. Tryptamine contains an indole ring structure, which is a bicyclic structure, consisting of a six-membered benzene ring fused to a five-membered nitrogen-containing pyrrole ring (Figure 19.1). It can be found in trace amounts in the brains of mammals, where it plays a role as a neuromodulator or neurotransmitter [1].

Many derivatives of tryptamine are found in nature, but many, if not all, can be produced synthetically. Substituted tryptamines, or simply tryptamines, include many biologically active compounds, including neurotransmitters and psychedelic drugs. Well-known representatives are psilocybin (O-phosphoryl-4-hydroxy-N, N-dimethyltryptamine) and psilocin (4-hydroxy-N,N-dimethyltryptamine), which are obtained from certain mushrooms indigenous to tropical and subtropical regions of South America, Mexico, and the United States. As pure chemicals, these

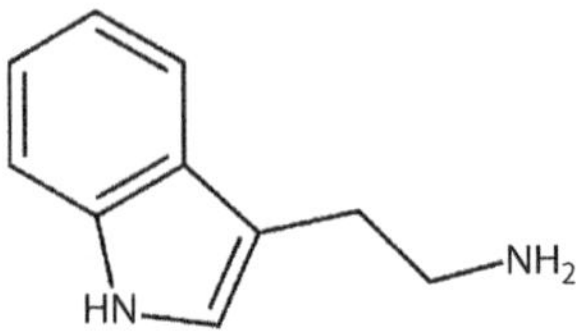

FIGURE 19.1   Chemical structure of tryptamine.

hallucinogens produce muscle relaxation, dilation of pupils, vivid visual and audi-tory distortions, and emotional disturbances. On the other hand, there are many spe-cies of "magic" mushrooms that contain varying amounts of tryptamines, as well as different amounts of other chemicals. Another representative of this class is dimeth-yltryptamine (DMT), which is also found in a variety of plants and seeds, and it can also be produced synthetically. A number of other hallucinogens have very similar structures and properties to those of DMT. Diethyltryptamine (DET) produces the same pharmacological effects, but it is somewhat less potent than DMT. Bufotenine (5-hydroxy-N-N-dimethyltryptamine) is a substance found in certain mushrooms, seeds, and skin glands of Bufo toads. N,N-Diisopropyl-5-methoxytryptamine (referred to as Foxy-Methoxy) is an orally active tryptamine, recently encountered in the United States [2]. On the other hand, the tryptamine nucleus serves as a template for many drugs used therapeutically.

Tryptamines became popular as designer drugs after publication in 1997 of the book *TIHKAL: The Continuation*, written by Alexander Shulgin and Ann Shulgin [3]. TIHKAL is an acronym that stands for "Tryptamines I Have Known and Loved". This book is a sequel to *PIHKAL: A Chemical Love Story*, which was about a fam-ily of phenethylamine derivatives. *TIHKAL* contains the details for synthesis for 55 psychedelic compounds (many discovered by Alexander Shulgin), including their chemical structures, dosage recommendations, and qualitative comments. Although the first designer tryptamines entered the drug market in the 1990s, many others appeared in the 2008–2012 period. To the end of 2016, a total of 33 tryptamines were notified to the European Monitoring Centre for Drugs and Drug Addiction (EMCDDA) via the Early Warning System (EWS) and their full list is presented in Table 19.1.

When analyzing the structures of designer tryptamines sold on the drug market, it is clearly seen that there are five different locations in the structure of tryptamines where substituents are usually attached. The substituents in the side chain are located on the nitrogen atom and it is common that both groups are identical: dimethyl, diethyl, diallyl, diisopropyl or dipropyl, and only some designer tryptamines contain mixed alkyl groups. There are two preferred positions in the indole moiety. Hydroxyl and acetoxyl groups are usually attached at the 4-position, while the methoxyl group is attached at the 5-position. 4-Hydroxy-DMT is psylocin, and its 5-isomer is called bufotenine. The remaining substances are also strictly related to the aforementioned classes. AMT means that the methyl group is attached to the alpha carbon atom. The less common compound is the first one, 2-Me-DMT, which was detected in 2014 in two small packages, one in Germany and one in Finland.

TABLE 19.1

List of Designer Drugs Being Derivatives of Tryptamine, Which Were Detected on the European Market in the Last Ten Years (2005–2016)

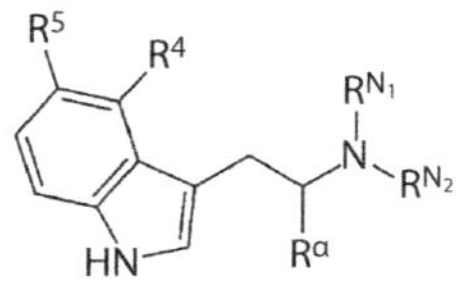

Name	R⁴	R⁵	Rᵅ	R^N₁	R^N₂	Other	Systematic Chemical Name
2-Me-DMT	–	–	–	Methyl	Methyl	2-methyl	N,N-dimethyl-2-(2-methyl-1H-indol-3-yl)ethanamine
4-AcO-DALT	Acetoxy	–	–	Allyl	Allyl	–	4-acetoxy-N,N-diallyltryptamine
4-AcO-DET	Acetoxy	–	–	Ethyl	Ethyl	–	4-acetoxy-N,N-diethyltryptamine
4-AcO-DIPT	Acetoxy	–	–	Isopropyl	Isopropyl	–	4-acetoxy-N,N-diisopropyltryptamine
4-AcO-DMT	Acetoxy	–	–	Methyl	Methyl	–	4-acetoxy-N,N-dimethyltryptamine
4-AcO-DPT	Acetoxy	–	–	Propyl	Propyl	–	4-acetoxy-N,N-dipropyltryptamine
4-AcO-MET	Acetoxy	–	–	Methyl	Ethyl	–	4-acetoxy-N-methyl-N-ethyltryptamine
4-AcO-MIPT	Acetoxy	–	–	Methyl	Isopropyl	–	4-acetoxy-N,N-methylisopropyltryptamine
4-HO-DET	Hydroxy	–	–	Ethyl	Ethyl	–	4-hydroxy-N,N-diethyltryptamine
4-HO-DIPT	Hydroxy	–	–	Isopropyl	Isopropyl	–	4-hydroxy-N,N-diisopropyltryptamine
4-HO-DPT	Hydroxy	–	–	Propyl	Propyl	–	4-hydroxy-N,N-dipropyltryptamine
4-HO-MET	Hydroxy	–	–	Methyl	Ethyl	–	4-hydroxy-N-methyl-N-ethyltryptamine
4-HO-MIPT	Hydroxy	–	–	Methyl	Isopropyl	–	4-hydroxy-N,N-methylisopropyltryptamine
5-HO-DMT	–	Hydroxy	–	Methyl	Methyl	–	5-hydroxy-dimethyltryptamine
5-MeO-AMT	–	Methoxy	Methyl	–	–	–	5-methoxy-alphamethyltryptamine
5-MeO-DET	–	Methoxy	–	Ethyl	Ethyl	–	5-methoxy-N,N-diethyltryptamine
5-MeO-DIPT	–	Methoxy	–	Isopropyl	Isopropyl	–	5-methoxy-di-isopropyl-tryptamine

(Continued)

TABLE 19.1 (CONTINUED)
List of Designer Drugs Being Derivatives of Tryptamine, Which Were Detected on the European Market in the Last Ten Years (2005–2016)

Name	R^4	R^5	R^α	R^{N1}	R^{N2}	Other	Systematic Chemical Name
5-MeO-DMT	–	Methoxy	–	Methyl	Methyl	–	5-methoxy-N,N-dimethyltryptamine
5-MeO-DPT	–	Methoxy	–	Propyl	Propyl	–	5-methoxy-N,N-dipropyltryptamine
5-MeO-EIPT	–	Methoxy	–	Ethyl	Isopropyl	–	N-ethyl-N-[2-(5-methoxy-1H-indol-3-yl)ethyl]propan-2-amine
5-MeO-MALT	–	Methoxy	–	Methyl	Allyl	–	N-[2-(5-methoxy-1H-indol-3-yl)ethyl]-N-methyl-prop-2-en-1-amine
5-MeO-MET	–	Methoxy	–	Methyl	Ethyl	–	5-methoxy-N-ethyl-N-methyl-tryptamine
5-MeO-MiPT	–	Methoxy	–	Methyl	Isopropyl	–	5-methoxy-N-methyl-N-isopropyltryptamine
5-MeO-NiPT	–	Methoxy	–	–	Isopropyl	–	N-[2-(5-methoxy-1H-indol-3-yl)ethyl]-propan-2-amine
5-MeO-tryptamine	–	Methoxy	–	–	–	–	5-methoxytryptamine
5-MeO-DALT	–	Methoxy	–	Allyl	Allyl	–	N,N-diallyl-5-methoxytryptamine
AMT	–	–	Methyl	–	–	–	alpha-methyltryptamine
DALT	–	–	–	Allyl	Allyl	–	N-allyl-N-[2-(1H-indol-3-yl)ethyl]prop-2-en-1-amine
DIPT	–	–	–	Isopropyl	Isopropyl	–	Diisopropyltryptamine
DMT	–	–	–	Methyl	Methyl	–	N,N-dimethyltryptamine
DPT	–	–	–	Propyl	Propyl	–	N,N-dipropyltryptamine
MET	–	–	–	Methyl	Ethyl	–	N-methyl-N-ethyltryptamine
MIPT	–	–	–	Methyl	Isopropyl	–	N-methyl-N-isopropyltryptamine
McPT	–	–	–	Methyl	Cyclopropyl	–	N-(2-(1H-indol-3-yl)ethyl-N-methylcyclopropanamine
6α-TMT	–	–	Methyl	Methyl	Methyl	–	(2-(1H-indol-3-yl)-1-methyl-ethyl)dimethylamine

Source: EMCDDA.

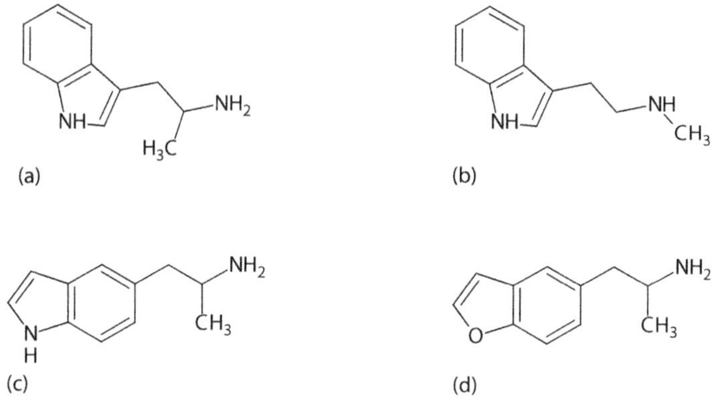

FIGURE 19.2 Chemical structures of (a) AMT, (b) NMT, (c) 5-IT, and (d) 5-APB.

When considering the identification of tryptamines seized from the drug market or found in biological material as a result of intoxication, it is important to take into account their isomers originating from other chemical classes. Of the substances used as "legal highs", the most important is 5-(2-aminopropyl)indole, often called 5-IT. It is a positional isomer of the tryptamine drugs AMT and NMT, but the compound is not itself a tryptamine, as the indole ring is substituted at the 5-position rather than at the 3-position. The compound is chemically closer to phenethylamine derivatives such as 5-APB (5-(2-aminopropyl)benzofuran). This is reflected in the compound's effects when used as a drug, which are reportedly stimulating rather than psychedelic. 5-IT is a designer drug that has been openly sold as a recreational drug by online vendors from 2011 and since its discovery it has been attributed to 14 deaths in Sweden [4]. Chemical structures of AMT, NMT, 5-IT, and 5-APB are compared in Figure 19.2.

A number of analytical methods can be applied for the detection and identification of tryptamines. Simple chemical tests, so-called spot tests, or other drug/narcotic detection kits, aerosol sprays/cans, or collection paper dispensers are commercially available for the detection of tryptamines, but due to their limited selectivity and sensitivity the application for clinical or forensic purposes is limited. Other commercially used assays that have been developed for identification of tryptamines include a fluorescence polarization immunoassay [5] and an immunochromatographic assay [6], but due to the fact that "legal highs" are often the mixtures of several substances of similar structure, coupled techniques are most commonly used in forensic laboratories. Usually, gas or liquid chromatography is used for the separation of components, while mass spectrometry is applied for their identification. The range of applications of different techniques is discussed below.

19.2 IDENTIFICATION OF TRYPTAMINES BY GC-MS

19.2.1 Chromatographic Data

Gas chromatography–mass spectrometry (GC-MS) is a very popular assay used in forensic laboratories investigating samples seized from the drug market. GC-MS combines two powerful techniques to provide the identification of compounds with low detection limits and the potential for quantitative analysis. Gas chromatography is used for the highly efficient gas-phase separation of components in complex mixtures, and mass spectrometry for the confirmation of identity of these components as well as for the identification of unknowns.

As the active doses of tryptamines are relatively high in comparison to other designer drugs and the substances are usually sold in a pure form as powders, the use of a standard analytical protocol enables their detection in seized samples. Complex extraction procedures are normally not required for simple products and the most common approaches involve a straightforward acid–base extraction or dissolution into a suitable solvent. Sample preparation is usually limited to dissolution of the powder in an appropriate solvent, e.g., methanol, or to liquid–liquid extraction (LLE) with hexane or the mixture of hexane and dichloromethane from the basic solution [7].

The retention time of a substance from the chromatographic column is a very important parameter used for verification of the presence of a compound in a sample. For many years, even in forensic laboratories an agreement in the retention times of a substance and the standard on two chromatographic columns was accepted as a sufficient proof for this purpose. Nowadays, however, this feature is more useful in exclusion than in the confirmation of presence, because it could happen that other substances have similar retention times, which often happens for isomers, especially stereoisomers. Therefore, as the chemical structures of many tryptamines are similar, it is very important to control separation conditions in order to assure as high as possible repeatability of the retention times. One of the most practical and useful approaches is the application of retention time locking (RTL), which can be done using modern instruments manufactured by Agilent Technologies. The details of RTL application are presented in Chapter 9.

In the author's laboratory, the chromatographic separation of tryptamines (and other designer drugs) has been carried out in the following conditions:

- Initial temperature of the column was 75°C and kept for one minute.
- Ramped increase in temperature by 25°C per minute to 280°C.
- Temperature was maintained for 21.8 minutes.

The total analysis time was 31 minutes. The flow rate of helium used as a carrier gas was 1 mL/minute. Diphenylamine was used as a RTL agent, whose retention time was set at 6.90 minutes. A Hewlett Packard HP 6890N Series GC system Plus equipped with autosampler Agilent 7638, coupled to Agilent 5973 Network mass spectrometer was used. Separation was achieved on the Agilent HP-5MS capillary column (length, 30 m; inner diameter, 0.250 mm; film thickness, 0.25 μm). The mass spectrometer was operated in the electron impact (EI) ionization mode with energy of 70 eV. The ions were collected in a broad range, from m/z 29 to 600. The collection of ions with low values of m/z is important, because—as it will be presented later—the ions characteristic of many tryptamines are recorded at m/z 30 and 44, while it is often recommended to record the GC-MS spectra starting from m/z 50 (the explanation for this recommendation is possible interferences, as, e.g., the molecular mass of carbon dioxide is 44, but the use of proper protocol prevents or at least significantly reduces this phenomenon). The developed method covered 19 tryptamines and 5-IT. Differences in the retention times of some compounds, including isomers, were very slight. The smallest differences in the retention times were at the 0.03-minute level, which corresponds to two seconds (0.4%). Such differences pose a risk of false identification, when identification is based exclusively on the retention times. Therefore, it is very important to control instrument parameters and validate the method properly, including assessment of the method repeatability. After a series of experiments supported by data analysis using the Valistat 1.0 software, it was proved that the developed method was highly reproducible. Intraday precision expressed by RSD was 0.06%, and interday precision was 0.1%. Even for different instruments, the precision was better than 0.5%. This confirms that the retention time windows can be set very tightly. Such an approach is very useful in distinguishing among constitutional isomers, e.g., 5-MeO-DMT and 4-MeO-DMT, DIPT and DPT, as well as NMT, AMT, and 5-IT, because the differences in their mass spectra are very slight, which could lead to improper identification.

19.2.2 ANALYSIS OF FRAGMENTATION UNDER EI

In the GC-MS assay, the most commonly used method of ionization is electron impact (EI). A great number of organic compounds are amenable to EI. To give an EI spectrum, the compound must be volatile. The sample may be solid, liquid, or gaseous, and it can be heated to achieve the necessary pressure. Ions are formed when a 70 eV beam of electrons hits the sample molecules in the gas phase. This gives the sample molecules a great deal of an excess energy and many fragment ions are formed. The mass spectra are very reproducible and are widely used for spectral libraries. Fragmentation is useful, because it provides structural information for interpreting unknown spectra.

In the mass spectrum, the heaviest ion (the one with the greatest m/z value) is likely to be the molecular ion. It is formed when the vaporized organic sample passes into the ionization chamber of a mass spectrometer and it is bombarded by a stream of electrons. These electrons have a high enough energy to knock an electron off an organic molecule to form a positive ion. Detection of the molecular ion is very important in the identification of tryptamines. A characteristic feature of tryptamines is that they typically contain two nitrogen atoms, in contrast with most representatives of the other classes of designer drugs sold in the form of powders, e.g., the derivatives of phenethylamine or cathinone. According to the nitrogen rule, an even number of nitrogen atoms in its structure causes that the molecular mass of the compound is even. Therefore, the molecular ion for tryptamines is observed to possess the even m/z values. Unfortunately, EI in standard conditions causes extensive fragmentation, so that the molecular ion is not observed for many compounds, or its intensity is very low. Chemical ionization (CI) is a lower energy process than EI. The lower energy yields less or sometimes no fragmentation. A typical CI spectrum has an easily identifiable protonated molecule peak $[M+1]^+$, which allows for determination of molecular mass. Another way to determine this mass is derivatization of the analytes, which makes the compounds more stable.

Ionization of molecules of the tryptamine-type designer drugs by a beam of electrons leads to characteristic breakdown of the bonds between the carbon atoms of the side chain. At the first stage, the bond cleavage occurs between the α and β carbon atoms in relation to the nitrogen atom in the side chain. The process can be presented as given below:

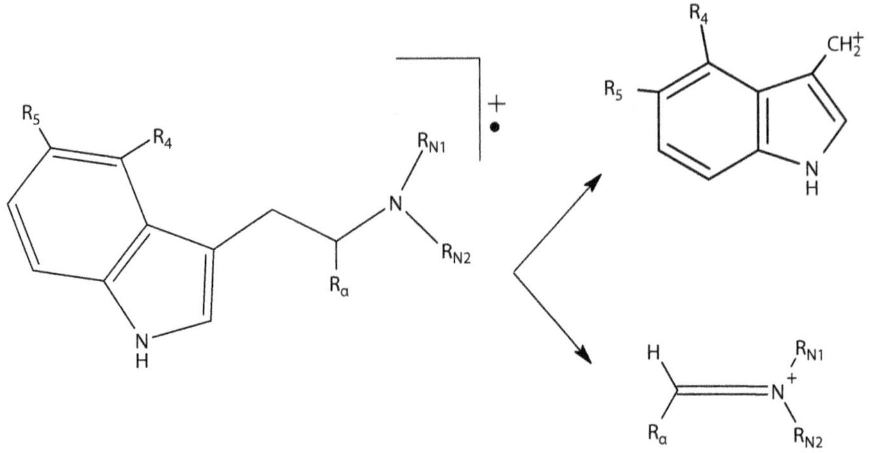

Usually, the formed immonium ions are the most intense in the GC-EI/MS spectra of tryptamines. Their general formula is $C_nH_{2n+2}N^+$ (n = 1, 2, etc.) and these ions are observed at $m/z = 16 + 14n$, that is, 30, 44, 58, 72, 86, etc. Of the designer drugs, the iminium ions are also formed for phenethylamines and the straight-chained cathinones, but not for piperazines (which also have an even molecular mass) [8]. The kind of immonium ion does not depend on substituents in the core structure of tryptamine. The formed iminium ions are even-electron ions and can show secondary and tertiary fragmentations in alignment with the ion series characteristically found with aliphatic amines (also 16 + 14n), which can help to distinguish among the isomers.

It has to be noted that for the substances with the same number of carbon atoms in the side chain of tryptamine, the mass of the primarily formed immonium ion is identical. For example, common tryptamines containing both N,N-dipropyl and N,N-diisopropyl substituent groups have the most intense peak at $m/z = 114$, but further fragmentation of the immonium ion allows for differentiation among these isomers. In the case of the N,N-dipropyl substituent, the ion with $m/z = 86$ is obtained with a relatively high efficiency, whereas for N,N-diisopropyl substituent, an ion is observed at $m/z = 72$. This process occurs following the McLafferty rearrangement [9]:

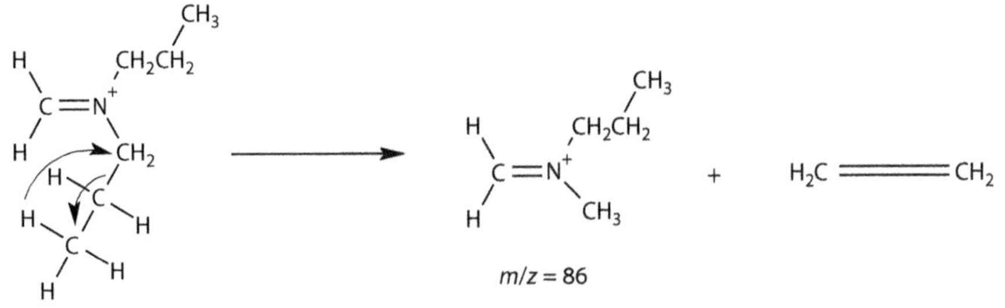

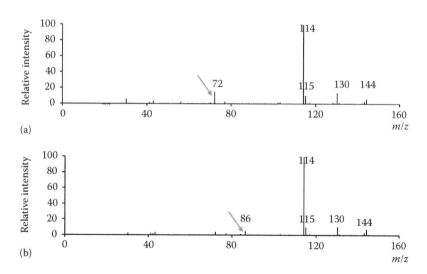

DIPT

The mass spectra of DIPT and DPT are presented in Figure 19.3. The characteristic ions were marked.

The situation is similar for tryptamines containing two ethyl groups substituted to the nitrogen atom of the side chain (*N,N*-diethyl tryptamines, e.g., DET) and those containing one methyl and one propyl or isopropyl group located at this position. In all cases, we obtain an intense peak with $m/z = 86$. Further fragmentation following the McLafferte rearrangement allows identification of the substituents; an *N*-methyl-*N*-isopropyl isomer fragments to form an ion with $m/z = 86$, an *N*-methyl-*N*-propyl isomer to form an ion with $m/z = 72$, and an *N,N*-diethyl one to form an ion with $m/z = 58$.

As mentioned earlier, there are three main substituent groups which can be attached to the indole moiety in tryptamines. The characteristic ions for the derivatives containing an acetoxy-, methoxy-, and hydroxy- group substituted to the indole ring are presented in Table 19.2. It is worth mentioning that the acetoxy group degrades easily, therefore the intensity of the ions at m/z 188 and 202 usually is very low.

FIGURE 19.3 Mass spectra of (a) DIPT and (b) DPT. Characteristic secondary immonium ions were marked with arrows.

TABLE 19.2

List of Characteristic Ions Depending on the Substituent in the Indole Ring

R^4/R^5	Characteristic Ions [m/nz]
-(H,H)	**130**, 144
-OH	**146**, 160
-OCH$_3$	145, **160**, 174
-OOCH$_3$	**145**, 160, (188), (202)

Note: The most intensive ion is marked in bold.

In the spectra of most tryptamines, a peak with $m/z = 130$ is observed, which is connected to the indole moiety and probably originates from the following structure:

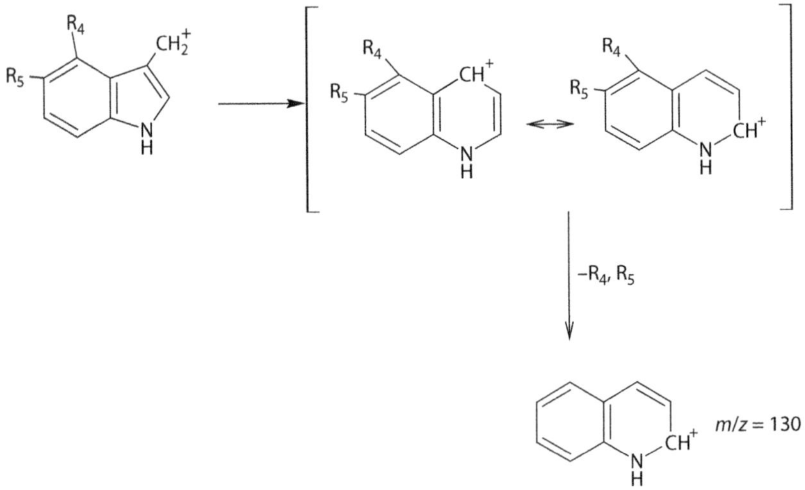

Another characteristic ion in the mass spectra of tryptamines is recorded at $m/z = 117$. However, its structure has not been unequivocally established so far, because two possible structures of the ion can be considered:

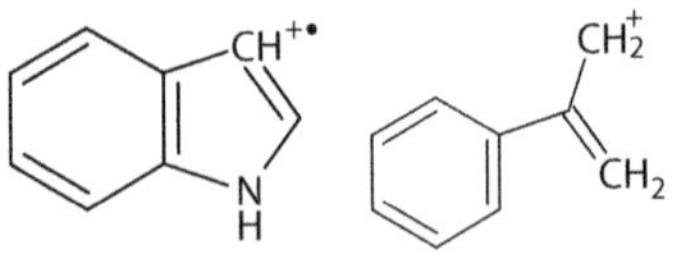

Final determination of the structure of this ion could be possible, based on theoretical calculations of the ions' stability. Further degradation of the indole ring results in the formation of ions with m/z values of 115, 103, 91, and 77.

Knowledge of the fragmentation routes can be used for the prediction of the spectra of new possible tryptamines. This statement can be illustrated with an example of 5-methoxy-ethyl-isopropyltryptamine (5-MeO-EIPT), whose EI spectrum was not available for the author. Taking into account the earlier considerations, the following ions were predicted for this substance:

- 100 (dominant), 72, 58, 30—immonium ions,
- 174, 160, 145—characteristic ions for methoxy-substituted indole ring,
- 130, 117, 115, 103, 91, 77—ions formed from degradation of the indole ring.

Several months later, 5-MeO-EIPT was notified to the EMCDDA and the Swedish forensic laboratory kindly provided us with the mass spectrum, which is presented in Figure 19.4. It turned out that the theoretical considerations were fully confirmed.

19.3   LIQUID CHROMATOGRAPHY WITH DIFFERENT DETECTORS USED FOR IDENTIFICATION OF TRYPTAMINES

19.3.1   HIGH-PERFORMANCE LIQUID CHROMATOGRAPHY WITH DIODE-ARRAY DETECTION

High-performance liquid chromatography with diode-array detection (HPLC-DAD) is the other technique that is commonly used in forensic and clinical laboratories. It turned out that detailed analysis of the UV/VIS spectra is very useful in distinguishing among the isomers, especially in determination of the location of the substituent group in the indole moiety.

As it was presented in Section 19.1, tryptamines detected on the European drug market contain a substituent group located at the 4-position or 5-position of benzene ring. The UV/VIS spectra of the two pairs of positional isomers, 4-MeO-DMT and 5-MeO-DMT, are presented in Figure 19.5.

It was noted that tryptamines with a methoxyl substituent located at the 5-position have two maxima located in the range of 200–225 nm. It was proved for all the tested compounds, i.e., for 5-MeO-AMT, 5-MeO-DALT, 5-MeO-DIPT, 5-MeO-DMT, and 5-MeO-MIPT, for which two bands with very similar intensities were recorded at 203 nm and 219 nm. The tested tryptamines with the hydroxyl or acetoxyl group at the 4-position have only one maximum in this region. This feature is very important in the identification of tryptamines, because such isomers cannot be distinguished using the routine mass spectrometric techniques.

19.3.2   IDENTIFICATION OF TRYPTAMINES BY LC-QTOFMS

As was presented in Chapter 7, LC-QTOFMS is a very useful tool in the identification of unknown compounds seized from the drug market, or detected in biological materials, because although some information can be obtained from the chromatographic data, an in-depth analysis of fragmentation is much more useful in the identification process for this assay.

In LC-QTOFMS, electrospray ionization (ESI) is the most commonly used technique. ESI belongs to the group of soft ionization techniques, which means that pseudomolecular ions can be recorded. There are many advantages of LC-QTOFMS, including high resolution and thus accuracy of ion mass determination. It is also

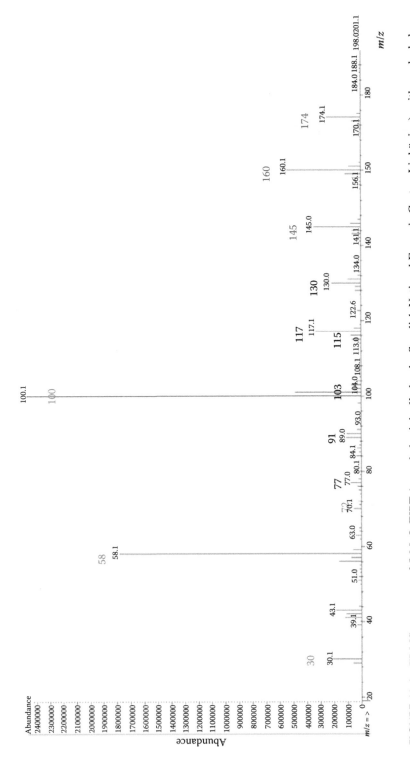

FIGURE 19.4 EI-MS spectrum of 5-MeO-EIPT (recorded originally by the Swedish National Forensic Centre, Linköping), with marked characteristic ions theoretically anticipated.

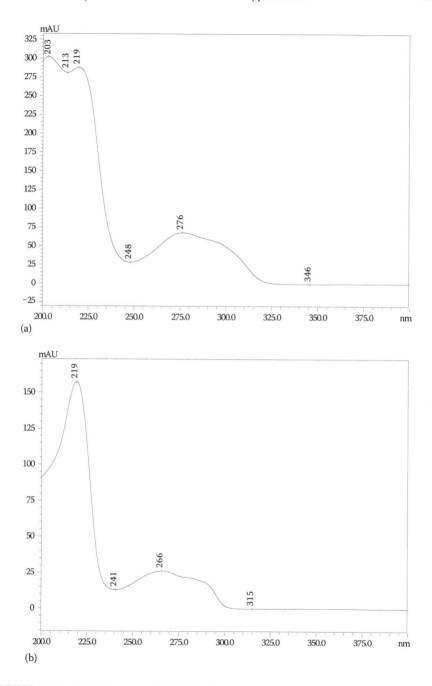

FIGURE 19.5 UV/VIS spectra of (a) 5-MeO-DMT and (b) 4-MeO-DMT.

possible to influence the degree of fragmentation (by changing the fragmentor voltage and the collision energy). As the preliminarily formed ions can be subjected to the further MS/MS analysis, the composition of not only pseudomolecular ions, but also other ions, can be investigated.

Low energy used to charge the molecule results in the ability to observe pseudomolecular ions in mass spectra. In such experiments, the low fragmentor voltage, e.g., 100 V, and no collision energy are used. The high accuracy of mass determination in this assay means the exact molecular formula of the compound can be determined. The problem concerns only isomers, which by definition have the same molecular mass, yet their distinction can be obtained from the MS/MS experiments, in which the initially formed ions are subjected to further fragmentation.

In contrast to the ionization of tryptamines under electron impact in GC-MS, the dissociation of the bond between carbon and nitrogen atom in the side chain is the characteristic cleavage for tryptamines under electrospray ionization, used in LC-QTOFMS. What is more, only the fragment containing nitrogen and the alkyl group substituted to this atom is detached from the molecule, and the remaining part is recorded. In the MS/MS experiments, it is possible to record the fragments originating from the indole moiety and its substituents. Difference in the cleavage of the side chain between GC-MS and LC-QTOFMS makes these techniques supplementary in the investigation of the structure of tryptamines. Combining data from both methods allows determination of the substituent in the alpha position and the number and kind of alkyl groups substituted to the nitrogen atom.

The fragmentation route of tryptamines is presented in Figure 19.6 and the characteristic ions for the selected representatives are given in Table 19.3. The QTOFMS spectrum of 4-MeO-DMT with the marked ions is presented in Figure 19.7. It is very important to notice that knowledge of the fragmentation route can be used for the prediction of the mass spectra of new tryptamines.

19.4 OTHER ANALYTICAL METHODS USED IN IDENTIFICATION OF TRYPTAMINES

19.4.1 CAPILLARY ELECTROPHORESIS

Capillary electrophoresis (CE) is a technique widely used in biosciences. The main application of CE in forensic science is DNA fingerprinting. This assay is also commonly used in pharmaceutical sciences for the analysis of drugs and related compounds, because it is an attractive technique for chiral separations. The presence of stereoisomers is also an issue when illicit drugs are considered. CE characterizes by high selectivity, which means that it provides a good resolution to the separations. Nonetheless, utilization of the CE methods for the analysis of drugs of abuse, including tryptamines, is comparatively underrepresented in comparison with chromatographic techniques, with the exception of several papers.

Tryptamine and its eight derivatives of forensic interest, including 5-methyltryptamine, DET, 5-MeO-DMT, DMT, 6-MeO-tryptamine, serotonin, bufotenine, and 5-MeO-DIPT, were separated by CE combined with UV-laser-induced fluorescence (LIF) detection (λex = 266 nm) [10]. The composition of the separation electrolyte

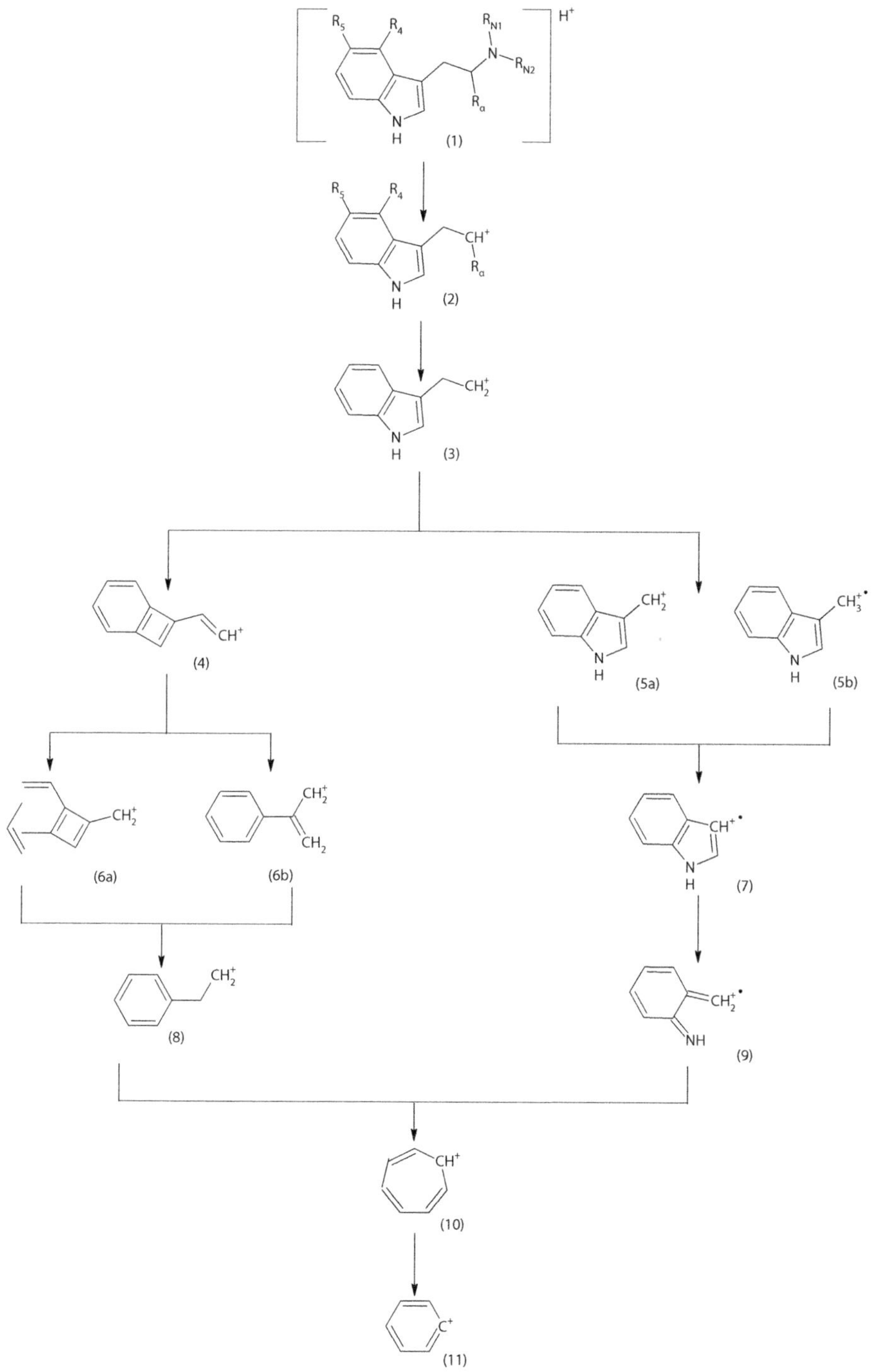

FIGURE 19.6 Main fragmentation route for tryptamines under ESI-QTOFMS.

TABLE 19.3

Structures and Masses of Characteristic Ions

Ion No	Substituent	Ion Structure	Formula and Mass
(2)	$R_4 \leftrightarrow R_5 = -OH$	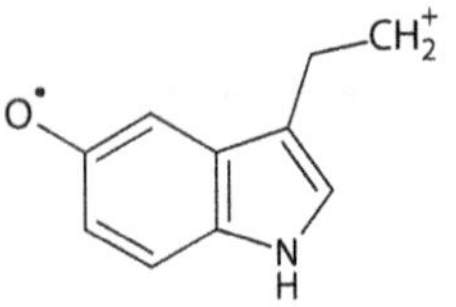	$C_{10}H_9NO$ 159.0678 Da
	$R_4 \leftrightarrow R_5 = -OCH_3$	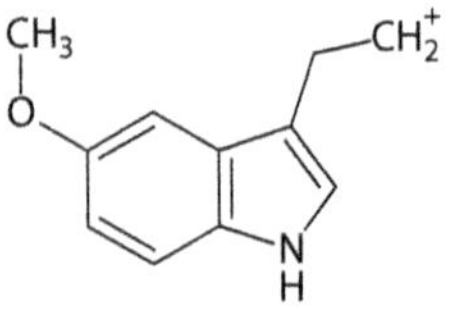 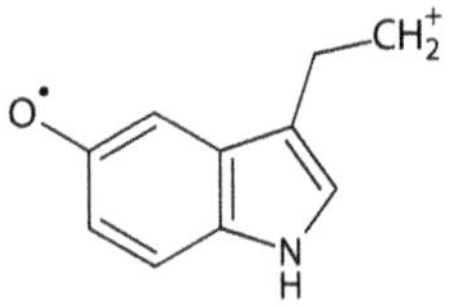	$C_{11}H_{12}NO$ 174.091 Da $C_{10}H_9NO$ 159.0678 Da
	$R_4 = -(=O)-C-CH_3$	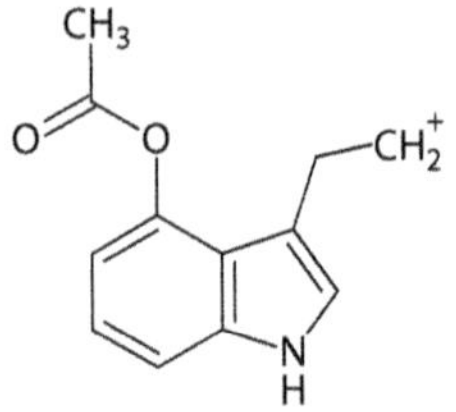 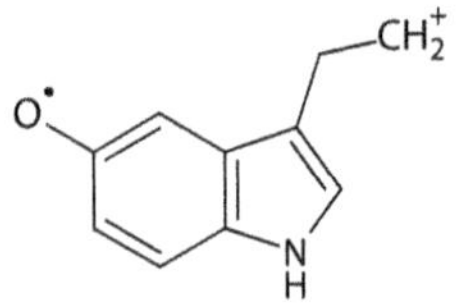	$C_{12}H_{12}NO_2$ 202.0862 Da $C_{10}H_9NO$ 159.0679 Da
	$R_\alpha = -CH_3$	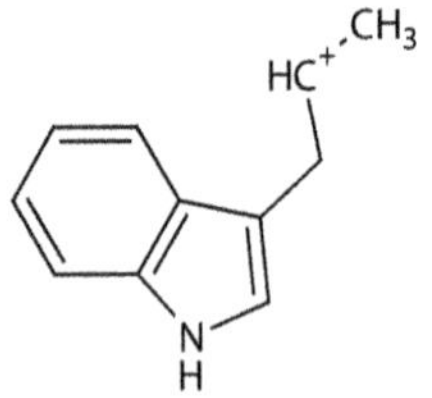	$C_{11}H_{12}N$ 158.0964 Da

(*Continued*)

TABLE 19.3 (CONTINUED)
Structures and Masses of Characteristic Ions

Ion No	Substituent	Ion Structure	Formula and Mass
(3)			$C_{10}H_{10}N$ 144.0808 Da
(4)			$C_{10}H_7$ 127.0542 Da
(5a)			C_9H_9N 131.0729 Da
(5b)			C_9H_8N 130.0651 Da
(6a)			C_9H_7 115.0542 Da
(6b)			C_9H_8N 130.0651 Da
(7)			C_9H_9 117.0699 Da
(8)			C_8H_9 105.0699 Da

(Continued)

TABLE 19.3 (CONTINUED)
Structures and Masses of Characteristic Ions

Ion No	Substituent	Ion Structure	Formula and Mass
(9)			C_7H_7N 105.0573 Da
(10)			C_7H_7 91.0542 Da
(11)			C_6H_5 77.0386 Da

was optimized with respect to the resolution of the solutes of interest and to the sensitivity of fluorescence detection. Native α-cyclodextrin was employed as a complex-forming modifier of the electrophoretic separation and the fluorescence-enhancing agent. With the help of a stacking procedure, limits of detection of 0.1–6 µg L^{-1} for all analytes were obtained. The repeatability for the peak area was less than 2.3% RSD. The accuracy of the method was verified by cross-validation with an independent HPLC-DAD method. The limits of detection obtained by the CE-(UV)-LIF method are lower than those obtained by the alternative HPLC-DAD method. It was proved that the CE-(UV)-LIF method enabled a fast baseline separation of nine tryptamines, and their determination in samples of forensic interest, such as illicit synthetic drugs, but also pharmaceutical formulation.

An ability of capillary electrophoresis to separate and determine tryptamine derivatives was compared with gas chromatography and high-performance liquid chromatography by Wang et al. [11]. AMT, DMT, 5-MeO-AMT, DET, DPT, DBT, DIPT, 5-MeO-DMT, and 5-MeO-DIPT were selected as model compounds. The limit of detection (S/N = 3) obtained by GC/MS and LC/UV-absorption ranged from 0.5 to 15 µg mL^{-1} and from 0.3 to 1.0 µg mL^{-1}, respectively. In contrast to this, based on the CZE/UV-absorption method, the limit of detection was determined to equal 0.5–1.0 µg mL^{-1}. However, when the sweeping-MEKC mode was applied, it dramatically improved to 2–10 ng mL^{-1}. In the case of GC, HPLC, and CE, migration times of the nine standards ranged from 11 to 15 minutes and from 8 to 23 minutes by GC and HPLC, respectively, and from 20 to 26 minutes by sweeping-MEKC. The order of migration of DMT, DET, DPT, and DBT follows the molecular weight, whereas the order of migration of AMT and 5-MeO-AMT (primary amines), DIPT (an isomer of DPT) and 5-methoxy-tryptamines (5-MeO-AMT, 5-MeO-DMT, and 5-MeO-DIPT) can be altered by changing the separation conditions. The studies were continued in the following years in order to obtain the optimal separation conditions and an online sample concentration of the aforementioned tryptamines using the micellar EKC (MEKC) with UV-absorbance detection [12]. The LODs (S/N = 3) for MEKC

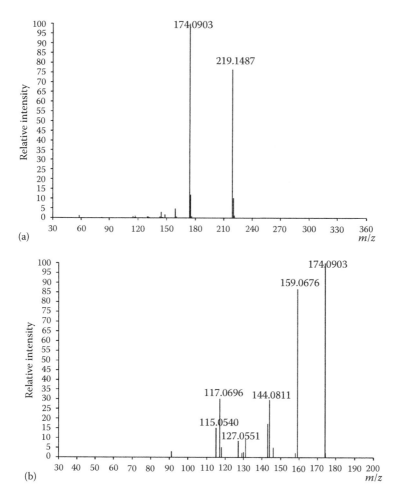

FIGURE 19.7 ESI-QTOF mass spectra of 4-MeO-DMT (a) recorded in MS mode and (b) fragmentation of the main ion (m/z = 174) in the MS/MS mode.

ranged from 1.0 to 1.8 μg/mL. Use of online sample concentration methods, including sweeping-MEKC and cation-selective exhaustive injection-sweep-MEKC (CSEI-sweep-MEKC), improved the LODs to 2.2–8.0 and 1.3–2.7 ng mL^{-1}, respectively.

19.4.2 MISCELLANEOUS METHODS

Colorimetric reactions, such as spot tests and thin-layer chromatography (TLC) with spray to produce a color, are used for screening or a preliminary identification of seized materials and residues extracted from biological materials, because it is easy to observe color changes without the use of special instruments. All tryptamines contain a nitrogen atom in their molecular structures, thus colorimetric reagents that react with nitrogen are usually used for detection tests. Spratley et al. [13] investigated five tryptamines, including 5-MeO-DIPT, 5-MeO-AMT, 5-MeO-MIPT, DPT, and 5-MeO-DMT, and verified two common mixtures used to detect spots on the TLC plates, that is, the Marquis reagent and the Ehrlich's reagent. In the presence of the Marquis reagent, each tryptamine analogue produced the same color change from yellow to black, except for DPT, which gave a yellow color only. Testing each of the tryptamine analogs with the Ehrlich's reagent produced the same change in color from purple to blue, except for DPT, which produced a violet color change, and 5-MeO-MIPT, which changed from purple to a faint blue. The problem of common TLC reagents is that they have poor selectivity or sensitivity. The detection limits for tryptamines were investigated by Nakamato et al. and it was shown that with Ehrlich's and Marquis reagents the limits ranged from 10 to 50 µg mL^{-1} depending on the substance, while they were at a level of 0.5–2.0 µg mL^{-1} with tetrabromo-phenolphthalein ethyl ether [14]. The other commonly used chemical, that is, the Simon's reagent, is selective to secondary amines, with which it produces a blue to purple color, but as psychoactive tryptamines are usually primary or tertiary amines, no color reaction is observed. Some others, e.g., p-dimethylaminocinnamaldehyde (DACA), paraformaldehyde reagent [15], the Van Urk–Salkowski reagent [16], and the acidified anisaldehyde reagent [17], have been reported to be used for detecting indole derivatives; however, these reagents are not convenient because using each one of them is time-consuming, they are toxic and very reactive chemicals, and their sensitivity is still limited. Recently, the rapid and sensitive TLC/fluorescence detection of psychoactive tryptamines and their metabolites in human urine samples was demonstrated [18]. The developed silica gel and RP-18 TLC plates were treated with sodium hypochlorite, hydrogen peroxide, or potassium hexacyanoferrate(III)–sodium hydroxide as an oxidizing reagent. Visualization of the fluorescent products was obtained at a wavelength of 365 nm and the LOD values were established in the range of 0.01–0.06 µg on the plate. Although the TLC-based applications have disappeared in recent years, the value thereof must not be underestimated for routine analysis. TLC is less sensitive and quantitatively worse-performing compared with GC/MS or LC/MS, but it has several advantages, including low cost and versatility.

ACKNOWLEDGMENTS

The author would like to thank Zuzanna Sobol for her support in the investigation of tryptamines by various analytical methods and in the identification of ion structures.

REFERENCES

1. Jones, R.S. 1982. Tryptamine: A neuromodulator or neurotransmitter in mammalian brain? *Prog. Neurobiol.* 19(1–2): 117–139.
2. Potter, J.V. *Substances of Abuse.* AFS Publishing Co., Redding, California, 2008.
3. Shulgin, A., Shulgin, A. *TIHKAL: The Continuation.* Transform Press, Berkeley, California, 1997.
4. Seetohul, L.N., Maskell, P.D., De Paoli, G., Pounder, D.J. 2012. Deaths associated with new designer drug 5-IT. *BMJ* 345: e5625.
5. Cody, J.T., Schwarzhoff R. 1993. Fluorescence polarization immunoassay detection of amphetamine, methamphetamine, and illicit amphetamine analogues. *J. Anal. Toxicol.* 17(1): 23–33.

6. Albers, Ch., Lehr, M., Beike, J., Köhler, H., Brinkmann, B. 2002. Synthesis of psilocin hapten and protein–hapten conjugate. *J. Pharm. Pharmacol.* 54: 1265–1267.
7. Hsiao, Y., Liu, J.T., Lin, C.H. 2009. Simultaneous separation and detection of 18 phenethylamine/tryptamine derivatives by liquid chromatography-UV absorption and -electrospray ionization mass spectrometry. *Anal. Sci.* 25(6): 759–763.
8. Zuba, D. 2012. Identification of cathinones and other active components of legal highs by mass spectrometric methods. *TrAC—Trends Anal. Chem.* 32: 15–30.
9. Yukiko, N., Kenji, T., Kenji, K., Tatsuyuki, K., Yuko, I.T., Kazuna, M., Fumiyo, K., Hiroyuki I. 2014. Simultaneous determination of tryptamine analogues in designer drugs using gas chromatography–mass spectrometry and liquid chromatography–tandem mass spectrometry. *Forensic Toxicol.* 32: 154–161.
10. Huhn, C., Pütz, M., Martin, N., Dahlenburg, R., Pyell, U. 2005. Determination of tryptamine derivatives in illicit synthetic drugs by capillary electrophoresis and ultraviolet laser-induced fluorescence detection. *Electrophoresis* 26(12): 2391–401.
11. Wang, M.J., Liu, J.T., Chen, H.M., Lin, J.J., Lin, C.H. 2008. Comparison of the separation of nine tryptamine standards based on gas chromatography, high performance liquid chromatography and capillary electrophoresis methods. *J. Chromatogr. A* 1181(1–2): 131–136.
12. Wang, M.J., Tsai, C.H., Hsu, W.Y., Liu, J.T., Lin, C.H. 2009. Optimization of separation and online sample concentration of N,N-dimethyltryptamine and related compounds using MEKC. *J. Sep. Sci.* 32(3): 441–445.
13. Spratley, T.K., Hays, P.A., Geer, L.C., Cooper, S.D., McKibben, T.D. 2005. Analytical profiles for five "designer" tryptamines. *Microgr. J.* 3: 54–68.
14. Nakamoto, A., Namera, A., Yahata, M., Kuramoto, T., Nishida, M., Yashiki, M. 2007. A systematic toxicological analysis for hallucinogenic tryptamines in seized and biological materials (in Japanese with English abstract). *Hiroshima Igaku Zasshi* 55: 1–14.
15. Toneby, M.I. 1974. Thin-layer chromatographic fluorimetry of indole derivatives after condensation by a paraformaldehyde spray reagent. *J. Chromatogr.* 97(1): 47–55.
16. Tonelli, D., Gattavecchia, E., Gandolfi, M. 1982. Thin-layer chromatographic determination of indolic tryptophan metabolites in human urine using Sep-Pak C18 extraction. *J. Chromatogr.* 231(2): 283–289.
17. Zhou, L., Hopkins, A.A., Huhman, D.V., Sumner, L.W. 2006. Efficient and sensitive method for quantitative analysis of alkaloids in Harding grass (*Phalaris aquatica* L.). *J. Agric. Food Chem.* 54(25): 9287–9291.
18. Kato, N., Kojima, T., Yoshiyagawa, S., Ohta, H., Toriba, A., Nishimura, H., Hayakawa, K. 2007. Rapid and sensitive determination of tryptophan, serotonin and psychoactive tryptamines by thin-layer chromatography/fluorescence detection. *J. Chromatogr. A* 1145(1–2): 229–233.

20 Toxicological Analysis of New Opioids

Karina Sommerfeld-Klatta, Artur Teżyk,
and Bogna Geppert

CONTENTS

20.1 INTRODUCTION

Opioids (ATC code N02A) are a group of substances that employ the mechanism of the stimulation of opioid receptors. They comprise opiates (phenantrene alkaloids, such as codeine and morphine, and isoquinoline alkaloids, such as papaverine and noscapine); endogenous morphines (endorphins) like encephalines and dynorphines; semisynthetic opioids (i.e., chemically modified opiates) like heroin and oxycodone; and synthetic opioids—among others, pethidine, methadone, and fentanyl and its analogs. Opioids stimulate δ (OP1), κ (OP2), and μ (OP3) opioid receptors situated in the brain, spinal cord, and peripheral tissues. For most of them, the analgesic effect is produced via the stimulation of μ (subclasses μ_1 and μ_2) and δ receptors. The stimulation of the μ_2 receptor causes central respiratory depression, bradycardia, constipation, and physical dependence on opioids; of the κ_1 receptor—among others, miosis, while of the κ_2 receptor—psychomimetic effects and dependence. Opioids have various affinity for opioid receptors, and therefore they are divided into agonists (morphine, codeine, fentanyl, methadone), partial agonists (buprenorphine), and mixed agonist–antagonist substances (pentazocine). A classical opioid analgesic medication is morphine, acting as a pure agonist of μ opioid receptors [1,2]. Schemes 20.1 through 20.3 illustrate chemical structures of morphine, oxycodone, and fentanyl.

Opioids differ not only in their affinity to various types of opioid receptors, but also in terms of receptor kinetics, potency, and time of action, as well as routes of elimination (via the digestive tract and kidneys). They are used with special caution in patients with liver and kidney damages, since their active metabolites,

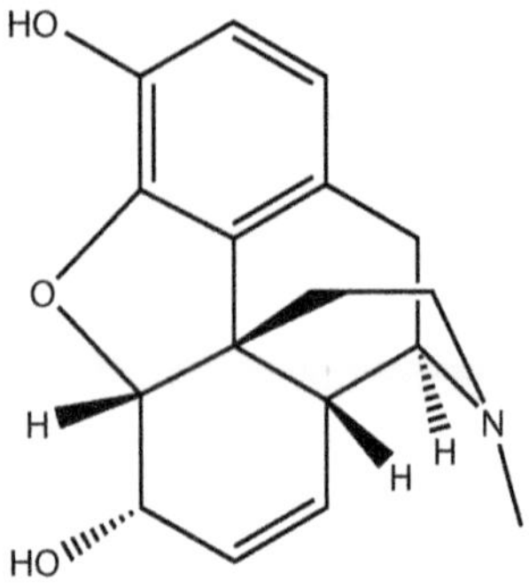

SCHEME 20.1   Morphine.

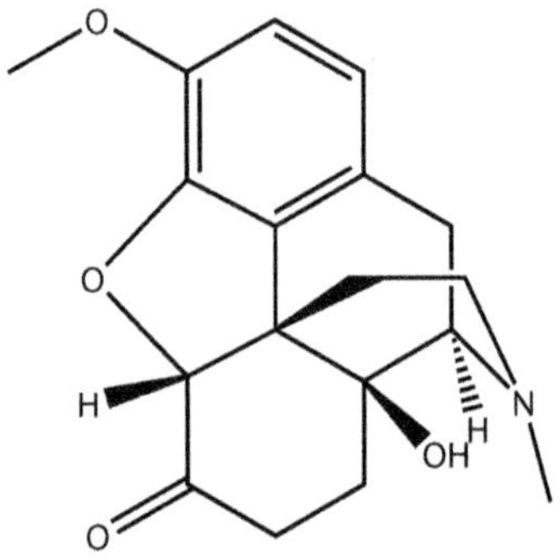

SCHEME 20.2   Oxycodone.

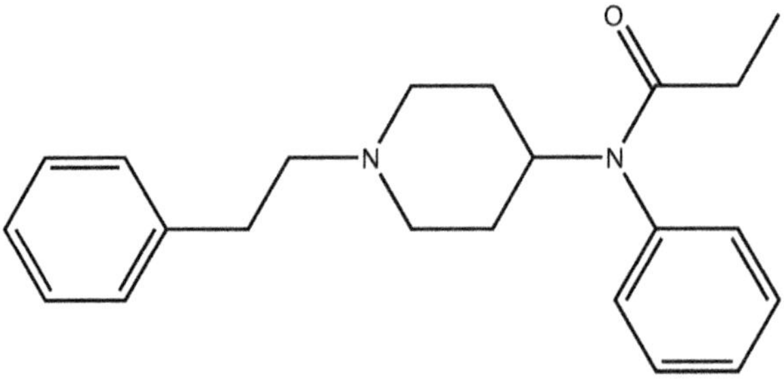

SCHEME 20.3   Fentanyl.

like morphine glucuronides or oxycodone metabolites, may accumulate in the kidneys, which may in turn produce the symptoms of poisoning [2].

In practice, opioids are used in the form of oral delayed-release preparations (controlled-release morphine in 20–30 mg doses every 12 hours) as well as in the form of transdermal therapeutic systems (TTS), such as fentanyl (12–25 µg/hour dose). It is also allowed to administer morphine intravenously in 1–2 mg doses every five–ten minutes in order to reach the proper level of analgesia or until the adverse reactions occur (drowsiness, emesis, nausea) [3].

For many years, opioids have remained under control, and the risk level connected with their use is defined, among others, by the American Drug Enforcement Administration (DEA), which divides them in relation to their addictive potential (heroin) and therapeutic usefulness (fentanyl) [4]. Table 20.1 shows medical use of opioids.

As it is clear from the official data of the European Monitoring Centre for Drugs and Drugs Addiction (EMCDDA), approximately 32 million people worldwide are addicted to opioids, a million of them being European [5]. Although heroin has dominated the opioid market for over 40 years, its users are becoming more and more interested in fentanyl, methadone, and buprenorphine. Nevertheless, heroin invariably remains the main cause of fatal poisonings in the world [5,6].

Most opioid poisonings have characteristic clinical symptoms (sometimes called the opioid toxic syndrome): pinpoint pupils, drowsiness passing into coma (central nervous system depression), and breathing disturbances passing into apnea (respiratory depression). Long-term results of opioid abuse are tolerance and physical dependence [7]. Therefore, it is essential to control the presence of opioids in the course of pain management as well as in poisonings, particularly due to a large number of fatalities.

The identification and quantitative analysis of opioids must employ specific methods, primarily including chromatographic methods combined with various types of detectors. The most frequently described methods are gas chromatography (GC) and liquid chromatography (LC) in connection with mass spectrometer (MS). Other known methods used for determination of opioids are gas chromatography with flame ionization detector (FID), nitrogen–phosphorus detector (NPD), or liquid chromatography with amperometric detection. There have been some reports on the use of capillary electrophoresis or thin-layer chromatography (TLC). It is important that all the discussed methods include the use of separation methods for the isolation of the targeted compounds from the biological and nonbiological samples. The most common ones are: liquid–liquid extraction (LLE), solid-phase column extraction (SPE), mixed-phase extraction (SPE) after prior hydrolysis, and solid-phase column microextraction (SPME) [8].

TABLE 20.1
Medical Uses of Opioids

Opioids	Medical Use
Morphine, methadone, fentanyl, pethidine, oxycodone	Analgesia
Alfentanil, sufentanil	Anesthetics
Diphenoxylate, loperamide	Antidiarrheals
Codeine, dextromethorphan	Antitussives

Source: Trescot, A.M. et al. 2008. *Pain Physician*, 11: 133–53.

20.2   NOVEL OPIOID DESIGNERS

Following the immense popularity of synthetic cannabinoids and canthinones, synthetic opioids have now appeared on the narcotic market, including not only fentanyl and its derivatives (acetyl fentanyl, butyryl fentanyl, furanyl fentanyl), but, most of all, new synthetic opioids, varying in their chemical structure, among which the following substances have been identified: AH-7921, MT-45, U-47700, U-50488, W-15, W-18, and, recently, U-51754.

Since 2008, the group of new psychoactive substances (NPSs) has grown, according to the data from EMCDDA, by 450 new compounds from the following classes: phenylalkylamines, tryptamines, synthetic cannabinoids, synthetic cathinones, benzodiazepines, and opioids [9]. The lack of legal solutions and routine tests detecting NPSs favors experiments with new compounds belonging to opioid receptors agonists. Until 2012, several of them had been known, mainly fentanyl derivatives.

The first one in this group was a preparation sold under the name of *Krypton*, which was a mixture of O-desmethyltramadol (the active metabolite of tramadol) and the extract of the plant from *Mitragyna species*, known as *Kratom* [10]. Schemes 20.4 and 20.5 illustrate chemical structures of tramadol and O-desmethyltramadol.

A new agonist of opioid receptors was first identified in Europe in May 2012– a compound labelled as AH-7921. Its presence was confirmed again in Japan in 2013 [11,12]. By the middle of 2015, AH-7921 had become a cause of a considerable number of poisonings and deaths in four countries [13–15]. AH-7921 belongs to the group of dimethylaminocyclohexane derivatives. As a selective agonist for κ receptors, it

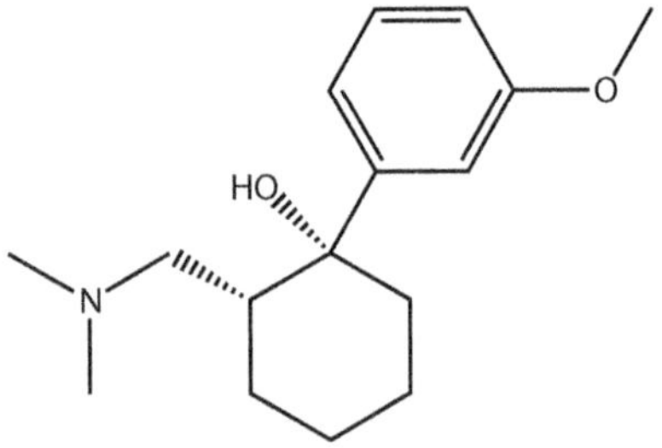

SCHEME 20.4   Tramadol.

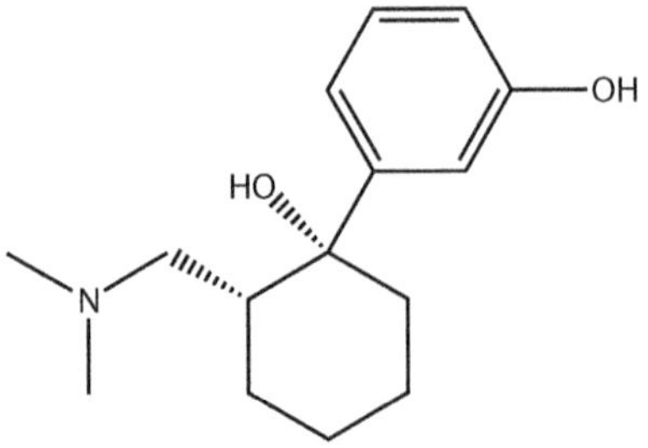

SCHEME 20.5   O-desmethyltramadol.

does not resemble morphine structurally, but bears a very slight similarity to fentanyl and phencyclidine. First synthetized in the 1970s, it has an analgesic effect (higher than codeine) [16]. Scheme 20.6 illustrates the chemical structure of AH-7921.

Initially, new opioids were sold as components of synthetic cannabinoids and/ or cathinones [12]. Nevertheless, they quickly became a new trend among the users of psychoactive substances and each month they grew in popularity among those who treated these new opioids as legal opioids. Table 20.2 shows the formal names, molecular formulas, and molecular weights of novel designer opioids.

Presently, in the cases of fatal poisonings, the most frequently detected compound belonging to the group of new synthetic opioids is U-47700, used as a substitute of heroin, and it is becoming more and more popular among the users of the designer drugs [6]. U-47700 belongs to 1,2-cyclo-ß-aminoamides and it is a structural isomer of AH-7921. It has an agonist action against μ-opioid receptors and is several times more potent than morphine. Known for more than 40 years, U-47700 is responsible for considerable tolerance (similarly as with the use of fentanyl) and euphoria (even greater than that evoked by heroin) [17]. Scheme 20.7 illustrates the chemical structure of U-47700.

The latest EMCDDA report for 2016 mentions 18 substances from the group of novel opioids monitored by the Agency between the years 2009 and 2015 [6]. Use of them by NPS consumers is connected with a significant risk of symptoms typical for these poisonings, i.e., respiratory depression, considerable tolerance, and major addictive potential.

Another substance on the list of opioid-like compounds is a piperazine derivative called MT-45, whose analgesic effects are primarily related to MT-45 (S)-isomer [(R)-isomer, influencing most probably σ receptors]. MT-45 first appeared on the narcotic market at the end of 2013, and in the next years it caused several dozen deaths, among others in Sweden [9]. Schemes 20.8 and 20.9 illustrate chemical structures of (R)-(−) MT-45 and (S)-(+) MT-45.

Selective agonists of κ and μ receptors also include U-50488, identified in 2015 as a cause of poisoning and known as an opioid analgesic since the 1970s. It was preliminarily tested on animals and has diuretic, antitussive, and anticonvulsive properties [18]. Scheme 20.10 illustrates the chemical structure of U-50488.

Moreover, attention is drawn to the compounds W-18 and W-15, originating from Canada (in the 1980s), which are less popular, although they have been monitored by EMCDDA since the middle of 2013 [5]. Schemes 20.11 and 20.12 illustrate chemical structures of W-15 and W-18.

The latest discovery is a compound described as U-54754, first identified in Slovenia at the beginning of 2017. Structurally, it resembles U-47700, and it also

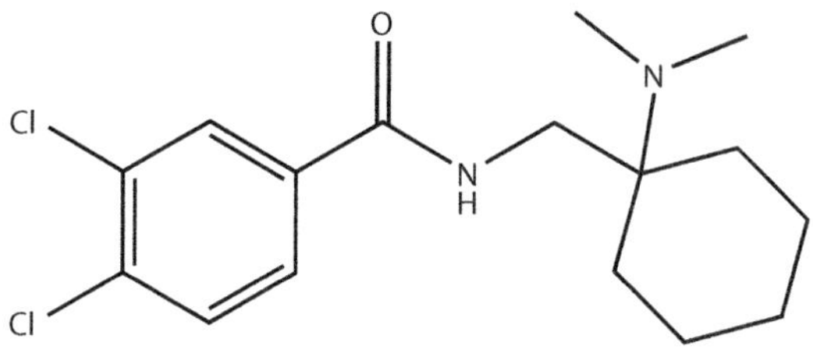

SCHEME 20.6 AH-7921.

TABLE 20.2

Formal Name (Abbreviation, Origin), Molecular Formula, and Formula Weight of Novel Designer Opioids

Compound/Origin/Abbreviation/ Other Names	Formal Name	Molecular Formula	Formula Weight [g/mol]	First Identification Poisonings/Death
AH-7921 (Allen and Hanbury Ltd.); AH, Doxylam, Doxylan)	3,4,-dichloro-N-[(1-dimethylamino) cyclohexylmethyl] benzamide	$C_{16}H_{22}Cl_2N_2O$	329.26	[13]
U-47700 (Upjohn Company; Fake morphine, U4)	trans-3,4-dichloro-N-(2-(dimethylamino) cyclohexyl)-N-methylbenzamide	$C_{16}H_{22}Cl_2N_2O$	329.26	[17]
MT-45 (Dainippon Company; IC-6, 1-C6, CDEP)	1-cyclohexyl-4-(1,2-diphenylethyl) piperazine	$C_{24}H_{32}N_2$	348.50	[32]
U-50488 (Upjohn Company; U50,488)	trans-3,4-dichloro-N-methyl-N-[2-(1-pyrrolidinyl) cyclohexyl]-benzeneacetamide	$C_{19}H_{26}Cl_2N_2O$	405.79	[26]
U-51754 (Upjohn Company)	2-(3,4-dichlorophenyl)-N-[2-(dimethylamino) cyclohexyl]-N-methylacetamide	$C_{17}H_{24}Cl_2N_2O$	379.75	[19]
W-18	4-chloro-N-[1-[2-(4-nitrophenyl) ethyl]-2-piperidinylidene]-benzenesulfonamide	$C_{19}H_{20}ClN_3O_4S$	421.9	[5]
W-15	4-chloro-N-[1-(2-phenylethyl)-2-piperidinylidene]-benzenesulfonamide	$C_{19}H_{21}ClN_2O_2S$	376.9	[5]

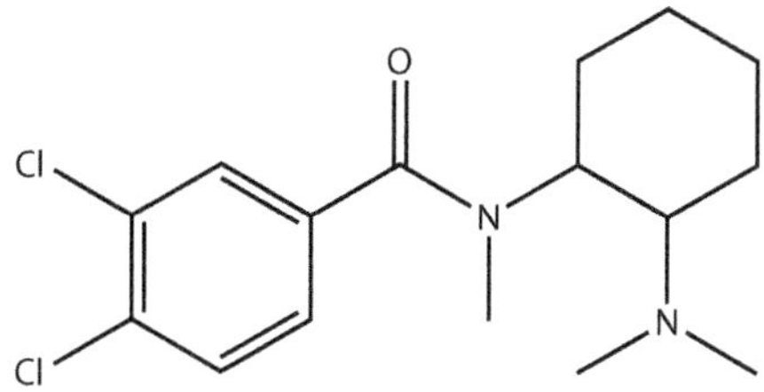

SCHEME 20.7 U-47700.

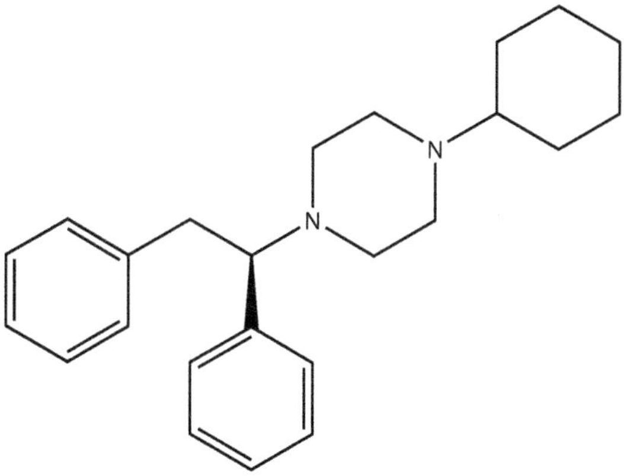

SCHEME 20.8 (R)-(−) MT-45.

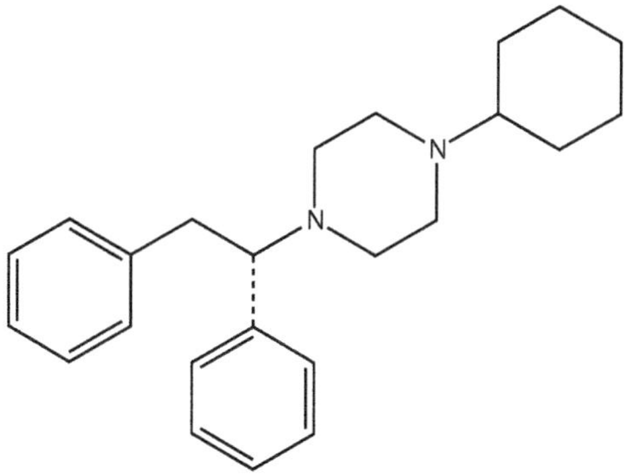

SCHEME 20.9 (S)-(+) MT-45.

belongs to the group of 1,2-cyclo-β-aminoamides [19]. Scheme 20.13 illustrates the chemical structure of U-54754.

Furthermore, users are put at risk of mixed poisonings that include not only exposure to synthetic opioids, but to the combination thereof with benzodiazepines, antidepressants, and other analgesics. An additional factor for the greater risk of deaths may be lower tolerance to opioid drugs, particularly in those who had abused them before and/or are already addicted [20].

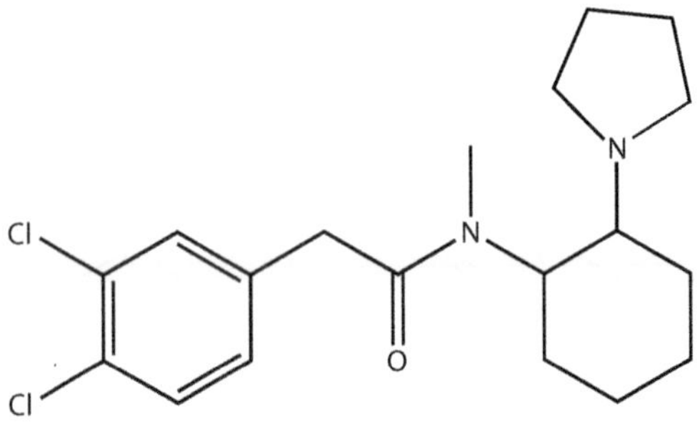

SCHEME 20.10 U-50488.

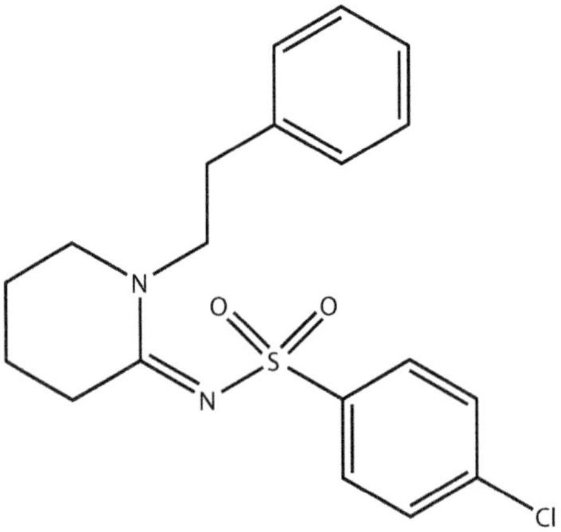

SCHEME 20.11 W-15.

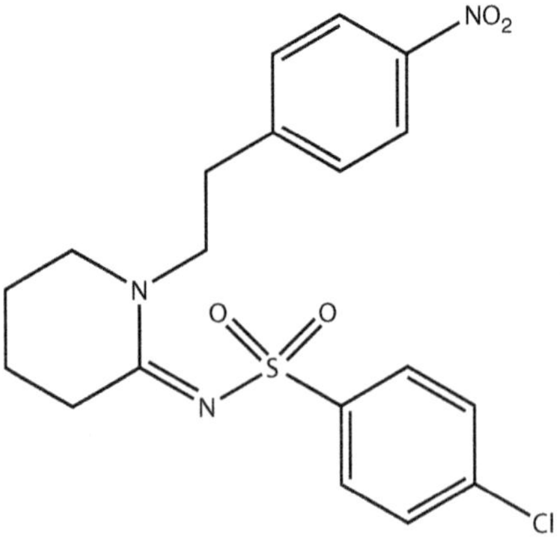

SCHEME 20.12 W-18.

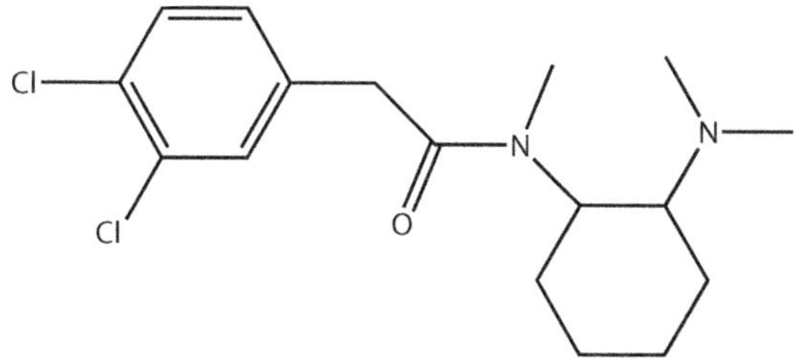

SCHEME 20.13 U-54754.

In all the cases of opioid-like NSPs discussed above, there are no known data obtained from clinical trials including patients, and only partial information is available (i.e., in the case of MT-45, W-15 and W-18, U-50488, and U-47700), pertaining to the action and possible effects that may be expected, all of which originate from animal testing or *in vitro* experiments [21,22].

20.3 METHODS OF FORENSIC ANALYSIS OF DESIGNER OPIOIDS—PRELIMINARY METHODS FOR OPIOID DETECTION

As far as the cases of deaths due to overdosing are concerned, the preliminary examination of the collected materials (samples postmortem, rarely collected from the patient during hospitalization) mostly includes the screening analysis for alcohol, benzodiazepines (or other hypnotics, i.e., barbiturates or Z-drugs), stimulants (amphetamine and its derivatives), opiates and opioids, tetrahydrocannabinol (THC), or finally antidepressants, anxiolytics, or anticonvulsants. Nevertheless, despite the significant role of preliminary examinations in forensic toxicology, they are insufficient to identify new opioids. No cross reactions to any novel opioid-like substances have been reported in the case of regularly used preliminary methods, such as enzyme-linked immunosorbent assay (ELISA), enzyme-multiplied immunoassay technique (EMIT), or immunochromatographic rapid tests for drugs of abuse, detecting morphine, its metabolites, and other opiates. Only Schneir et al. put forward a hypothesis of falsely positive urinalysis (Roche ONLINE DAT plus performed on a Cobas 6000 analyzer, Roche Diagnostics International) for the presence of benzodiazepines (cutoff reached 100 ng/mL), which revealed the cross reaction for U-47700 [23]. At the same time, with the use of liquid chromatography–tandem mass spectrometry (LC-MS/MS), with the limit of quantification reaching 20 ng/mL, and liquid chromatography time-of-flight (LC-TOF) high-resolution mass spectrometry (HR-MS) prepared by Chindarkar et al., the presence of the most common benzodiazepines and their metabolites (i.e., alprazolam, clonazepam, diazepam), as well of other opioids (i.e., fentanyl, tramadol, hydromorphon), was excluded [23,24].

Amidst classic chromatographic methods, the qualitative analysis of evidence (plant extracts) may employ such techniques as TLC. An example of such use is in a paper by Uchiyama et al., presenting an analysis of plant materials that contained novel synthetic cannabinoids, among which the previously unseen opioid AH-7921 was found [12].

Thus, there is a need for an extended screening analysis for novel substances, including opioids, with the use of the methods of ultra-performance liquid chromatography–tandem mass spectrometry (UPLC-MS/MS) and liquid chromatography quadrupole time-of-flight spectrometry (LC-QTOF-MS), since these coupled methods make it possible to detect unknown compounds that may be essential to establish the cause of death due to, for instance, acute respiratory failure. Moreover, vast collections of data (mass spectral libraries) may be insufficient (lacking novel substances) to establish the exact composition of the analyzed samples. In these cases, several factors come into importance, such as the knowledge of molecular masses, reliable medical history, information from the site, and collected evidences, since all of them may facilitate the verification and identification of new opioid substances.

20.4   ISOLATION AND ANALYSIS METHODS OF OPIOIDS FROM BIOLOGICAL SAMPLES

Blood (and, less often, urine) is the most frequently analyzed biological material in the reported cases of poisonings with novel opioids. There are few reports on the analysis of alternative materials, i.e., hair, saliva, or other tissues. There are single reports on the analysis of such materials as the liver, kidneys, gastric washings, lungs, or brain [15].

The first analytical data on the detection of novel opioids, as in the case of AH-7921, inform of the use of known and commonly used methods of biological sample extractions. The aforementioned fact results from the relatively simple chemical structure of the analyzed compound. The first quantitative analyses for AH-7921, described by Vorce et al., were based on the method of electron ionization (EI) selected ion monitoring (SIM) mode by gas chromatography mass spectral (GC-MS), preceded by the liquid–liquid alkaline extraction of blood using 1-chlorobutane [15]. The quantitative analysis in the discussed case concerned also the examination of urine, completed with solid-phase extraction columns, with the use of full-scan EI mass spectra. The results obtained by Vorce et al. seemed to suggest considerable differences in the final concentrations of AH-7921 in the peripheral blood and the blood collected from the heart (3.9 mg/L versus 9.1 mg/L), which interrupted their identification with the remaining samples [15]. It is a very rare phenomenon when the concentration of a drug in the peripheral blood is higher than in the heart (the most probable cause seems to be the dilution of the blood in the heart).

In order to check the stability of NPSs (among others, AH-7921), Soh et al. utilized the method of high-performance liquid chromatography with diode-array detection (HPLC-DAD) [25]. Next, they employed liquid-chromatography–tandem mass spectrometry and finally they used ultra-high-performance liquid chromatography with high mass accuracy quadrupole time-of-flight spectrometry (UHPLC-QTOF-MS). The scientists applied double liquid–liquid extraction with 1-chlorobutane in the alkaline medium, and then back-extraction with sulfuric acid. As it might have been expected, apart from confirming the presence of AH-7921, they detected two metabolites: *N*-desmethyl- and *N,N*-didesmethyl-AH-7921 in the blood and urine collected postmortem [25]. Table 20.3 shows the analysis methods of AH-7921 in blood and the other specimens.

TABLE 20.3

Identification of AH-7921 in Blood and Other Specimens, Instrumentation, Sample Preparation, Method Validation, and Results

Compound	Matrix	Sample Preparation	Column and Mobile Phase	Method	Method Validation	Application	Concentration [mg/L or mg/kg]	Coexist Drugs	Ref.
AH-7921	Heart blood	Alkaline LLE	J&W DB-5MS (20 m × 0.18 mm, 0.18 µm) with helium as the carrier gas	GC–MS EI SIM mode	Calibration range 0.05–2.0 mg/L	Analysis of fatal poisonings involving AH-7921	3.9	Dextromethorphan (only heart blood),	[15]
	Peripheral blood						9.1	4'-methyl-α-pyrrolidino hexiophenone (only urine)	
	Urine						6.0		
	Liver						26		
	Kidney						7.2		
	Spleen						8.0		
	Heart						5.1		
	Lung						21		
	Brain						7.7		
	Bile						17		
	Stomach content						120 mg (125 mL submitted)		
	Herbal-type products	Extracted with methanol under ultrasonication	Acquity UPLC HSS C18 (150 mm × 2.1, 1.8 µm) Gradient A—10 mM ammonium formate in water, pH 3.0) B—0.1% formic acid in Acetonitrile (ACN)	UPLC–ESI–MS GC–MS ESI+ mode		The study is the first reported case in which AH-7921 has been detected in an illegal product.			[12]

(Continued)

TABLE 20.3 (CONTINUED)
Identification of AH-7921 in Blood and Other Specimens, Instrumentation, Sample Preparation, Method Validation, and Results

Compound	Matrix	Sample Preparation	Column and Mobile Phase	Method	Method Validation	Application	Concentration [mg/L or mg/kg]	Coexist Drugs	Ref.
	Whole blood and plasma	Alkaline LLE with back extraction	Phenomenex Synergi Fusion (150 mm × 2 mm, 4 µm) Gradient A—70% ACN with triethylammonium bicarbonate (TEAP) buffer B—TEAP buffer	HPLC-DAD		This study is the first time to investigate the stability of AH-7921 in biological samples.			[25]
			Phenomenex Gemini (150 mm × 2 mm, 5 µm) Gradient A—70% ACN with 1% formic acid B—1 mM ammonium formate with 1% formic acid	LC-MS/MS					

Furthermore, the use of various types of extraction has been described for the identification of U-47700 and U-50488. The most current report is that by Mohr et al., published in 2016, which describes the detailed analysis performed by means of LC-MS/MS (positive electrospray, multiple-reaction monitoring mode) that made it possible to detect U-47700 at the level of 0.5 ng/mL in the blood in 11 fatal poisonings [26]. The blood underwent solid-phase extraction (130 mg Cleen Screen DAU extraction columns UCT). The average concentration of U-47700 in the blood reached 253 ng/mL (ranging from 17 to 490 ng/mL), and its identification was often accompanied by the presence of furanyl fentanyl [26]. Table 20.4 shows the analysis methods for U-47700 and U-50488 in blood. Moreover, there have been the reports of the methods of identification of U-47700 and its metabolites in urine. Fleming et al. proposed a quantitative analysis by LC coupled to tandem mass spectrometer (positive electrospray and multiple-reaction monitoring mode) [27]. For the purpose of qualitative analysis with the use of LC-QTOF, urine was diluted without previous hydrolysis, which made it possible to identify the second-phase metabolites (without reference standards for U-47700). For the purpose of quantitative analysis with the use of LC-MS/MS, the urine samples underwent solid-phase extraction with ß-glucuronidase and sodium acetate buffer, and then they were separated with PSCX cartridges. Fleming's paper confirmed the presence of N-demethylated and N,N-didesmethylated metabolites in urine [27]. Table 20.5 shows the analysis methods for U-47700 in urine.

Armenian et al. published a paper in which they described the case of an accidental poisoning with fentanyl and U-47700 contained in the forged preparation called *Norco* (supposedly it should contain fentanyl, acetaminophen, and hydrocodone) [28]. The confirmed blood concentration of U-47700 reached 7.6 ng/mL [28].

The analysis of alternative materials according to McIntyre included U-47700 in the blood, liver, vitreous humor, urine, and gastric system by means of the modified method used to determine fentanyl, that was GC coupled with an MS and SIM procedure, after the previous liquid–liquid extraction (alkaline with 1-chlorobutane) [29]. Additionally, the paper completed the information on the distribution of this opioid, determining the central blood/peripheral blood ratio as 1.8, and the liver/peripheral blood ratio as 8.9, which confirmed the modest potential of postmortem redistribution for U-47700 [29]. Table 20.6 shows the analysis methods for U-47700 in the alternative specimens.

In the paper discussing nine cases of fatal poisonings, in the course of which the presence of AH-7921 was confirmed (reported in Sweden during a six-month period), Kronstrand employed the method of isolating blood via precipitation with 0.075% formic acid in the mixture of acetonitrile and ethanol (90:10) [13]. Similar to the results obtained by Soh et al., in this study there were several cases of detecting AH-7921 metabolites with four more hydroxyl derivatives (unfortunately, the peak intensity was low, <1% of the matrix peak surface, without a possibility to obtain spectra) [13,25]. Liquid–liquid extraction with the use of ethyl acetate and heptane in the alkaline medium (borate buffer solution, pH 11) was also used in order to establish the cause of fatal poisonings with AH-7921 in combination with the other substances (4-fluoromethamphetamine, 3-methylmetcathinone and 2-(3-methoxyphenyl)-2-(ethylamino)cyclohexanone) [14]. Table 20.7 shows the analysis methods for AH-7921 contained in blood.

TABLE 20.4

Identification of U-47700 and U-50488 in Blood, Instrumentation, Sample Preparation, Method Validation, and Results

Compound	Matrix	Sample Preparation	Column and Mobile Phase	Method	Method Validation	Application	U-47700 Concentration [ng/mL]	Coexisting Drugs in Blood	Ref.
U-47700 U-50488	Femoral, peripheral and aorta blood	SPE (130 mg Clean Screen DAU extraction columns)	Agilent Zorbax Eclipse Plus C18 (4.6 × 100 mm, 3.5 μm) Gradient A—0.10% formic acid in water B—0.10% formic acid in methanol	LC-MS/MS ESI+ MRM mode	The limit of detection: 0.5 ng/mL, correlation coefficients of 0.9995 (U-47700) and 0.9999 (U-50488); calibration range: 1–100 ng/mL (U-47700) and 30–500 ng/mL (U-50488)	Analysis of 16 fatal poisonings involving U-47700, U-50488 not detected	382	Stimulants	[26]
							17	Alcohol, fentanyl derivative	
							217	Stimulants, benzodiazepine	
							334	—	
							252	Antidepressant	
							453	Stimulants, anticonvulsant	
							242	Stimulants	
							103	Antihistamine	
							299	Other opioid, benzodiazepine	
							311	Other opioid antidepressant	
							487	Antihistamine Benzodiazepine	
							59	N-phenethyl-4-piperidone (ANPP), quinine	
							135	Alcohol, N-phenethyl-4-piperidone (ANPP), quinine	
							167	Other opioid N-phenethyl-4-piperidone (ANPP), quinine	
							490	N-phenethyl-4-piperidone (ANPP), quinine	
							105	N-phenethyl-4-piperidone (ANPP), quinine	

TABLE 20.5

Identification of U-47700 in Urine, Instrumentation, Sample Preparation, Method Validation, and Results

Compound	Matrix	Sample Preparation	Column and Mobile Phase	Method	Method Validation	Application	Concentration in Urine [ng/mL]	Ref.
U-47700	Urine hydrolized with β-glucuronidase	SPE separated using cartridges PSCX	Agilent Poroshell 120 EC-C18 (2.1 mm × 100 mm, 2.7 µm) Gradient A—5 mM ammonium acetate and 0.01% formic acid B—0.01% formic acid in methanol	LC-MS/MS ESI+ MRM mode	Calibration range: 1–2500 ng/mL, linear response with an R^2 of 0.985 or greater	Two cases of U-47700 identification in urine	140 224	[27]

TABLE 20.6

Identification of U-47700 in Blood and Other Specimens, Instrumentation, Sample Preparation, Method Validation, and Results

Compound	Matrix	Sample Preparation	Column and Mobile Phase	Method	Method Validation	Application	Concentration [ng/mL]	Coexist Drugs in Blood	Ref.
U-47700	Peripheral blood	Alkaline LLE	Zebron ZB-5MS, 15 m, 0.25 (15 m × 0.25 mm, 0.25 μm) with helium as the carrier gas	GC-MS SIM mode	The limit of detection: 5 ng/mL; the limit of quantitation: 20 ng/mL; calibration range: 20–500 ng/mL	A fatality related to U-47700 in alprazolam abuser	190	Benzodiazepine, other opioid, psychostimulant	[29]
	Central blood						340		
	Liver						1700		
	Vitreous humor						170		
	Urine						360		
	Gastric contents						<1 mg		

TABLE 20.7

Identification of AH-7921 in Blood, Instrumentation, Sample Preparation, Method Validation, and Results

Compound	Matrix	Sample Preparation	Column and Mobile Phase	Method	Method Validation	Application	Concentration μg/g	Coexist Drugs in Blood	Ref.
AH-7921	Whole blood from the femoral vein	Alkaline LLE	Acquity UPLC HSS T3 (2.1 × 100, 1.8 μm) Gradient A—10 mM ammonium formate buffer pH 3.1; B—methanol	UPLC-MS/MS ESI+ MRM mode	Calibration range 0.033–0.66 mg/L $R^2 > 0.997$; QC samples results less than 14% deviation from nominal values	Analysis of two fatal accidental poisonings involving AH-7921	0.43 0.33	Other opioids, analgesic and stimulant Dissociative drug and benzodiazepines	[14]

(Continued)

TABLE 20.7 (CONTINUED)
Identification of AH-7921 in Blood, Instrumentation, Sample Preparation, Method Validation, and Results

Compound	Matrix	Sample Preparation	Column and Mobile Phase	Method	Method Validation	Application	Concentration	Coexist Drugs in Blood	Ref.
	Femoral blood	Precipitation with formic acid in acetonitrile: ethanol	Agilent Zorbax Eclipse Plus C18 (2.1 × 50 mm, 1.8 µm) Gradient A—0.05% formic acid in 10 mM ammonium formate; B—0.05% formic acid in methanol	LC-MS/MS ESI+ MRM mode	Calibration range 0.01–1.0 µg/g Matrix effects investigated, a mean within 10% of the target value, repeatability and accuracy was 1–5% and 94–115% respectively.	Analysis of nine fatal poisonings involving AH-7921	0.81 µg/g 0.99 µg/g 0.03 µg/g 0.20 µg/g 0.30 µg/g 0.08 µg/g 0.16 µg/g 0.35 µg/g 0.43 µg/g	Anticonvulsant Stimulant, antipsychotic Alcohol, antidepressant Antihistamine, other opioid and analgesic, benzodiazepines Sedating antihistamine, benzodiazepines, neuroleptic, analgesic, Z-drug Alcohol, dissociative drug Stimulant Stimulant Other opioid and analgesic, benzodiazepines, anti-epileptic, antidepressant	[13]

In the first reported case of a fatal poisoning with U-47700, Elliott used the basic back extraction with 1-chlorobutane to isolate blood and urine [17]. Initially, the employed HPLC-MS methods (the method used to identify NPSs) and UHPLC-QTOF-MS (a nonspecific method) suggested the presence of AH-7921. Therefore, it became a challenge to differentiate AH-7921 from U-47700, with the knowledge that the latter one is an isomer of the former one, and as a result, there are no differences in their molecular mass. Differences that were significant for the analysis were noticed in the obtained retention times and UV spectra, analyzing them with the use of HPLC-DAD (U-47700 eluted at 7.85 minutes with maximum peak at 201.7 nm and AH-7921 at 8.26 minutes with two maximum peaks at 205.4 and 241.3 nm). The analysis with the use of triple-quadrupole/linear ion trap LC-MS confirmed that both compounds had common product ions (*m/z* 145, 173, 284); however, for U-47700, the product ions of interest were *m/z* 204 and 81, and for AH-7921, 190 and 95, respectively [17].

Figure 20.1 presents the HPLC-DAD data obtained from U-47700 and AH-7921. Figure 20.2 shows the enhanced product ion scans of U-47700 and AH-7921. The reported concentration of U-47700 in the deceased's blood, described by Elliott, was considerably higher from the concentrations (1.46 mg/L) published so far in the cases of fatal poisonings with, among others, morphine or AH-7921 [17]. Table 20.8 shows the analysis methods for U-47700 contained in blood and urine.

Elliott confirmed the presence of metabolites *N*-demethyl-U-47700 and *N,N*-didesmethyl-U-47700 (not sufficient signal intensity) with the use of QTOF-MS, although it should be emphasized that without the high-accuracy MS/MS system, metabolic identification is just a conjecture [17].

In order to isolate MT-45 from the blood, Papsun et al. used the method of liquid–liquid extraction in the alkaline medium with the mixture of *N*-butyl chloride/

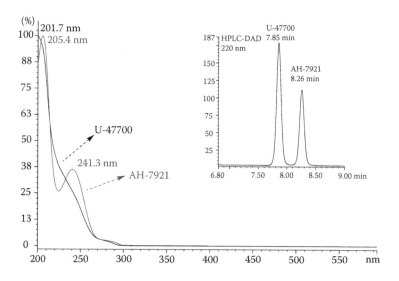

FIGURE 20.1 HPLC-DAD chromatogram and UV spectra of the U-47700 and AH-7921 standards. (From Elliott, S.P. et al. 2016. *Drug Test Anal.* 8: 875–9. With permission.)

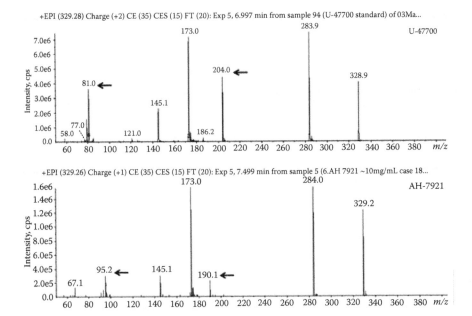

FIGURE 20.2 EPI scans of U-47700 and AH-7921. Both structural isomers were differentiated based on the distinct product ions as indicated by the arrows. (From Elliott, S.P. et al. 2016. *Drug Test Anal.* 8: 875–9. With permission.)

acetonitrile (4:1) [30]. The presence of MT-45 at the concentration of 520 ng/mL was confirmed in the blood of a woman who died as a result of taking the aforementioned opioid together with an illicit benzodiazepine, etizolam. Moreover, Papsun et al. tried to analyze MT-45 in alternative materials, such as urine, vitreous humor, and bile after a year from death, assessing the stability of the substance. The presence of MT-45 was confirmed; however, its blood concentration was 50% lower than in the first measurement. The authors employed the directed LC-MS/MS analysis in positive electrospray and MRM mode [30]. Table 20.9 shows the analysis method for MT-45 contained in blood.

Significant information on MT-45 is contained in a report prepared in 2014 by EMCDDA, which discusses not only the issues of detection, but also those of chemical and pharmacological properties, legislative aspects, and the influence on the health and life of potential users. The report enlists the methods enabling identification of MT-45, i.e., GC or LC coupled with spectrometry. The EMCDDA report describes also the use of Fourier-transformation or nuclear magnetic resonance (NMR) in order to analyze the compound. The ongoing presence of MT-45 on the market contributes to the fact that there are certified reference standards for both the racemic mixture and the enantiomers of MT-45. The report discusses more than 20 cases of poisonings with MT-45, among which the most numerous group are the papers by Helander et al. [31–33]. Uchiyama et al. also describes the HPLC-DAD method, revealing the ultraviolet and visible spectra of MT-45 [31].

TABLE 20.8

Identification of U-47700 in Blood and Urine, Instrumentation, Sample Preparation, Method Validation, and Results

Compound	Matrix	Sample Preparation	Column and Mobile Phase	Method	Method Validation	Application	Concentration in Blood [mg/L]	Coexist Drugs	Ref.
U-47700	Blood and urine	Alkaline LLE	Phenomenex Synergi Fusion (150 mm × 2 mm, 4 μm) 30% ACN with 25 mM TEAP buffer under isocratic elution	HPLC-DAD	Intra-day accuracy and precision values of <2% (at 0.5 and 2.5 mg/L), inter-day accuracy and precision values of <18% and <6%, respectively at 0.5 and 2.5 mg/L, LOD of 0.05 mg/L and LOQ of 0.3125 mg/L	The first reported fatality associated with U-47700	1.46 mg/L	Quetiapine, amphetamine, amitriptyline, mexedrone, ketamine in urine; quetiapine, amphetamine and naproxen in blood	[17]
			Phenomenex Gemini (150 mm × 2 mm, 5 μm) Gradient A—70% ACN with 1% formic acid B—1 mM ammonium formate with 1% formic acid	HPLC-MS					

TABLE 20.9

Identification of MT-45 in Blood, Instrumentation, Sample Preparation, Method Validation, and Results

Compound	Matrix	Sample Preparation	Column and Mobile Phase	Method	Method Validation	Application	Concentration in Blood [ng/mL]	Ref.
MT-45	Human whole blood	Alkaline LLE	Acquity UPLC BEH C18 (2.1 mm × 50 mm, 1.7 µm) Gradient A—0.1% formic acid in water B—0.1% formic acid in methanol	LC-MS/MS ESI+ MRM mode	Calibration range: 1–100 ng/mL, linear response with an R^2 of 0.999	Identification and quantitation of MT-45 in the decedent's blood after etizolam intoxication	520	[30]

20.5 ANALYSIS METHODS OF METABOLIC OPIOIDS

Currently, the greatest amount of information about biotransformation changes and their products—metabolites—occurs with regard to testing two substances—AH-7921 and U-47700. Wohlfarth et al. comprehensively describe the metabolism of AH-7921 in the course of the *in vitro* tests of cultured hepatocytes (human liver microsomes, HLM), as well as in *in vivo* and *in silico* tests [34]. The *in vitro* tests of HLM provide some precious information that reveals AH-7921 as a substance that the liver could easily process and metabolize (AH-7921 was ranked as a high-clearance and extracted drug). The incubation of the tested compound in human hepatocytes proved the presence of demethylated (the predominant) and didemethylated metabolites, in accordance with the results of the aforementioned papers by Soh et al., Vorce et al., and Kronstrand et al. [13,15,25,34]. Apart from the metabolites mentioned above, there was a hypothesis that *in silico* tests revealed the presence of hydroxyl derivatives, which partly confirmed the results obtained by Kronstrand et al. [13]. Moreover, there was a search for data concerning glucuronic metabolites, the presence of which in the tested samples was not confirmed in the tests performed by Kronstrand [13]. Thus, neither their structure nor their presence was established, and an attempt to do so was made in the urine sample, with and without hydrolysis. The results of urinalysis confirmed the presence of the majority of the expected metabolites (even the glucuronidated metabolite eluted after AH-7921); however, abandoning the issues of their intensity, the use of various types of ionization, and the influence of matrix effects. Table 20.10 shows the analysis methods for the AH-7921 metabolites *in vitro*, *in vivo* and *in silico*.

The results obtained by Soh et al., Elliott et al., and Fleming et al., have recently been confirmed by Jones et al., determining two U-47700 metabolites in the serum and urine, with the dominant dimethyl derivative, as well as some others: desmethyl, *N,N*-bisdesmethyl, desmethylhydroxyl, and *N,N*-bisdesmethylhydroxyl—four isomers in total [17,25,27,35]. Nevertheless, there was a lack of certainty with regard to the presence of second-phase metabolites—glucuronic acid derivatives [35]. Table 20.11 shows the analysis method for the U-47700 metabolites in blood and urine.

Metabolic data concerning the biotransformation of MT-45 may be derived from the performed tests of structurally similar compounds, originating as MT-45 from 1,2-dephenethylamines, comparing MT-45 to a simple structure of lefetamine [32]. So far, however, there have been no complex investigations of that matter.

20.6 CONCLUSIONS

1. The common features of novel opioid-like substances are, on the one hand, similar effects on opioid receptors, and, on the other hand, the lack of sufficient pharmacokinetic and pharmacodynamics data, unrecognized effects on the human system, as well as the lack of determined toxic range and lethal blood concentrations.

TABLE 20.10
Analysis of the AH-7921 Metabolites

Compound	Matrix	Column and Mobile Phase	Method	Ref.
Identification in vitro of 12 metabolites of AH-7921 (11 phase I metabolites generated by N-demethylation, hydroxylation or combination) and 1 phase II—the glucuronide conjugate.	Human hepatocytes	Kinetex C18 (100 mm × 2.1 mm, 2.6 µm) Gradient A—0.1% formic acid in water B—ACN	HPLC-Triple TOFMS ESI+ IDA mode	[34]
Identification in vivo of 11 of 12 metabolites of AH-7921 (11 phase I metabolites generated by N-demethylation, hydroxylation or combination) without glucuronide conjugate.	Authentic urine forensic			

Note: MetaSite Software (in silico) predicted 17 metabolites generated by N-demethylation, aliphatic or aromatic hydroxylation, N-oxidation, carbonylation, N-dealcylation, oxidative or reductive dechlorination.

TABLE 20.11
Analysis of U-47700 Metabolites

Compound	Matrix	Column and Mobile Phase	Method	Ref.
Identification in vivo of metabolites of U-47700 (phase I metabolites generated by N-demethylation)	Blood, urine	Agilent Zorbax Eclipse Plus C18 (100 mm × 2.1 mm, 1.8 μm) Gradient A—100% ACN with 1% formic acid B—aqueous solution of 1% formic acid	UHPLC-HR QTOF-MS Positive mode scanning with and without auto MS/MS fragmentation ESI	[17]

2. The key issues for the detection of novel opioids are properly chosen methods of isolation (the most frequently used extraction method is liquid–liquid extraction in the alkaline medium), and the directed analysis combining several analytical techniques (liquid chromatography–tandem mass spectrometry and liquid chromatography time-of-flight high-resolution mass spectrometry).

3. Current problems faced by forensics seem to be incomplete databases (mass spectral libraries), as well as problems with availability of certificated standards. What is available is the method of nuclear magnetic resonance that allows unambiguous identification of unknown substances in cases when the reference standard is missing.

4. Another problem seems to be a lack of performed investigations with regard to the knowledge of metabolic processes of novel opioids, which makes it impossible to make use of the already identified metabolites (as in the case of AH-7921 and U-47700) as the exposure markers or a sufficient evidence of their prior use.

5. Therefore, it is recommended to continuously develop analytical methods (also with the tests on alternative materials), to more strongly emphasize the greater number of tests estimating toxicity *in vitro* and *in vivo*, to use the *in silico* prediction methods for these compounds, and, finally, to actively report the greatest as possible number of acute and fatal poisonings.

REFERENCES

1. Karch, S.B. 2007. *Drug Abuse Handbook*. Boca Raton: CRC Press.
2. Trescot, A.M., Datta, S., Lee, M. et al. 2008. Opioid pharmacology. *Pain Physician*, Opioid Special Issue 11: 133–53.
3. World Health Organization. 1996. *Cancer Pain Relief*. (2nd ed.). Geneva, Switzerland. http://apps.who.int/iris/bitstream/10665/37896/1/9241544821.pdf (accessed 12 February 2017).

4. National All Schedules Prescription Electronic Reporting Act of 2005. A review of implementation of existing state controlled substance monitoring programs. 2007. Center for Substance Abuse Treatment. https://congress.gov/congressional-report/109th-congress /senate-report/117/1 (accessed 12 February 2017).

5. European Monitoring Centre for Drugs and Drug Addiction. 2013. EMCDDA. *Trends and Developments*. http://www.emcdda.europa.eu/system/files/publications/964/TDAT 13001PLN2_rev.pdf (accessed 2 February 2017).

6. European Monitoring Centre for Drugs and Drug Addiction. 2015. EMCDDA. New psychoactive substances in Europe: An update from the EU Early Warning System. http://www.emcdda.europa.eu/system/files/publications/65/TD0415135ENN.pdf (accessed 2 February 2017).

7. Hoffman, R.S., Howland, M.A., Lewin, N.A. et al. 2016. *Goldfrank's Toxicologic Emergencies*, eds. Nelson, L.S. and Olsen, D. McGraw-Hill Companies, 10th Edition, chapter 38.

8. Bogusz, M.J. 2008. *Forensic Science Handbook of Analytical Separations*. Copyright© 2017 Elsevier B.V. New York. Chapter 1: 1–77.

9. European Monitoring Centre for Drugs and Drug Addiction. 2015. European Drug Report Trends and Developments. http://www.emcdda.europa.eu/attachements.cfm /att_239505_PL_TDAT15001PLN.pdf (accessed 2 February 2017).

10. Philipp, A.A., Meyer, M.R., Wissenbach, D.K. et al. 2011. Monitoring of kratom of Krypton intake in urine using GC-MS in clinical and forensic toxicology. *Anal. Bioanal. Chem*. 400: 127–35.

11. European Monitoring Centre for Drugs and Drug Addiction. 2014. EMCDDA-Europol Joint Risk Assessment Report of a new psychoactive substances: 3,4-dichloro-N-[1-(dimetylamino)cyclohexyl]methyl)benzamide (AH-7921). http://www.emcdda.europa .eu/system/files/publications/774/TDAK14002ENN_480892.pdf (accessed 4 February 2017).

12. Uchiyama, N., Matsuda, S., Kawamura, M. et al. 2013. Two new-type cannabimimetic quinolinyl carboxylates, QUPIC and QUCHIC, two new cannabimimetic carboxamide derivatives, ADB-FUBINACA and ADBICA, and five synthetic cannabinoids detected with a thiophene derivative α-PVT and an opioid receptor agonist AH-7921 identified in illegal products. *Forensic Toxicol*. 31: 223–40.

13. Kronstrand, R., Thelander, G., Lindstedt, D. et al. 2014. Fatal intoxications associated with the designer opioid AH-7921. *J Anal. Toxicol*. 38: 599–604.

14. Karinen, R., Tuv, S.S., Rogde, S. et al. 2014. Lethal poisonings with AH-7921 in combination with other substances. *Forensic Sci. Int*. 244: 21–4.

15. Vorce, S.P., Knittel, J.L., Holler, J.M. et al. 2014. A fatality involving AH-7921. *J. Anal. Toxicol*. 38: 226–30.

16. Brittain, R.T., Kellet, D.N., Neat, M.L. et al. 1973. Proceedings: Anti-nociceptive effects in N-substituted cyclohexylmethylbenzamides. *Br. J. Pharmacol*. 49: 158–9.

17. Elliott, S.P., Brandt, S.D., and Smith, C. 2016. The first reported fatality associated with the synthetic opioid 3,4-dichloro-*N*-[2-(dimethylamino)cyclohexyl]-*N*-methylbenzamide (U-47700) and implications for forensic analysis. *Drug Test Anal*. 8: 875–9.

18. Cheney, B.V., Szmuszkovicz, J. Lahti, R.A. et al. 1985. Factors affecting binding of trans-N-[2-methylamino)cyclohexyl]benzamides at the primary morphine receptor. *J. Med. Chem*. 28: 1853–64.

19. European Monitoring Centre for Drugs and Drug Addiction. EMCDDA EU Early Warning System Formal Information: 2-(3,4-dichlorophenyl)-*N*-[2-(dimethylamino) cyclohexyl]-*N*-methyl-acetamide (U-51,754), email notification: 20 January 2017.

20. Druit, H., Strandberg, J.J., Alkass, K. et al. 2007. Evaluation of the role of abstinence in heroin overdose deaths using segmental hair analysis. *Forensic Sci. Int*. 168: 223–6.

21. Harper, N.J., Veitch, G.B., and Winnerley, D.G. 1974. 1-(3,4-dichlorobenzamidomethyl) cyclohexyldimethylamine and related compounds as potential analgesics. *J. Med. Chem.* 17: 1188–93.
22. Tortella, F.C., Robles, L., and Holaday, J.W. 1993. U50,488 a highly selective kappa opioid: Anticonvulsant profile in rats. *J. Pharmacol. Exp. Ther.* 264: 631–7.
23. Schneir, A., Metushi, I.G., Sloane, C. et al. 2016. Near death from a novel synthetic opioid labeled U-47700: Emergence of a new opioid class. *Clin. Toxicol.* 1–4. DOI: 10.1080/155563650.2016.1209764.
24. Chindarkar, N.S., Wakefield, M.R., Stone, J.A. et al. 2014. Liquid chromatography high resolution TOF analysis: Investigation of MSE for broad-spectrum drug screening. *Clin. Chem.* 60: 1115–25.
25. Soh, Y.N.A. and Elliott, S. 2014. An investigation of the stability of emerging new psychoactive substances. *Drug Test Anal.* 6: 696–704.
26. Mohr, A. Friscia, M., Papsun, D. et al. 2016. Analysis of novel synthetic opioids U-47700, U-50488 and furanyl fentanyl by LC-MS/MS in postmortem casework. *J. Anal. Toxicol.* 1–9. DOI: 10.1093/jat/bkw086.
27. Fleming, S.W., Cooley, J.C., Johnson, L. et al. 2016. Analysis of U-47700, a novel synthetic opioid, in human urine by LC-MS-MS and LC-QToF. *J. Anal. Toxicol.* 1–8. DOI: 10.1093/jat/bkw131.
28. Armenian, P. Olson, A., Anaya, A. et al. 2017. Fentanyl and a novel synthetic opioid U-47700 masquerading as street "Norco" in central California: A case report. *Ann. Emerg. Med.* 2017 Jan, 69(1):87–90. DOI: 10.1016/j.annemergmed.201.06.014.
29. McIntyre, I.M., Gary, R.D., Joseph, S. et al. 2017. A fatality related to the synthetic opioid U-47700: Postmortem concentration distribution. *J. Anal. Toxicol.* 41(2): 158–60.
30. Papsun, D., Krywanczyk, A., Vose, J.C. et al. 2016. Analysis of MT-45, a novel synthetic opioid, in human whole blood by LC-MS-MS and its identification in a drug-related death. *J. Anal. Toxicol.* 40: 313–17.
31. Uchiyama, N., Matsuda, S., Kawamura, M. et al. 2014. Identification of two new-type designer drugs, piperazine derivative MT-45 (I-C6) and synthetic peptide Noopept (GVS-111) with synthetic cannabinoid A-834735, cathinone derivative 4-methoxy-alpha-PVP and phenethylamine derivative 4-methylbuphedrine from illegal products. *Forensic Toxicol.* 32: 9–18.
32. European Monitoring Centre for Drugs and Drug Addiction. 2014. EMCDDA Risk Assessment Report of a new psychoactive substance: 1-cyclohexyl-4-(1,2-diphenylethyl) piperazine (MT-45). http://www.emcdda.europa.eu/system/files/publications/70/MT-45 _Risk_Assessment_Report_485096.pdf (accessed 4 February 2017).
33. Helander, A., Backberg, M., and Beck, O. 2014. MT-45, a new psychoactive substance associated with hearing loss and unconsciousness. *Clin. Toxicol.* 52: 901–4.
34. Wohlfarth, A., Scheidweiler, K.B., Pang, S. et al. 2015. Metabolic characterization of AH-7921, a synthetic opioid designer drug: *In vitro* metabolic stability assessment and metabolite identification, evaluation of *in silico* prediction, and *in vivo* confirmation. *Drug Test Anal.* DOI: 10.1002/dta.1856.
35. World Health Organization, U-47700 Critical Review Report Agenda Item 4.1 Expert Committee on Drug Dependence, 38th Meeting Geneva, 2016. http://www.who .int/medicines/access/controlled-substances/4.1_U-47700_CritReview.pdf?ua=1 (accessed 2 February 2017).

21 Toxicological Analysis of Designer Benzodiazepines

Artur Teżyk, Karina Sommerfeld-Klatta,
and Bogna Geppert

CONTENTS

21.1 INTRODUCTION

Benzodiazepines are a numerous group of anxiolytic, muscle relaxant, hypnotic, anticonvulsant, and sedative compounds. An abundance of derivatives and the multidirectional pharmacological effect render them one of the most frequently prescribed drugs in developed countries. It has been estimated that approximately 10–20% of adults in Western countries have had some contact with benzodiazepines [1]. According to data provided by the International Narcotic Board, worldwide consumption of anxiolytic benzodiazepine derivatives reached 19.8 billion S-DDD (expressed as defined daily doses, S-DDD, for statistical purposes), while the respective number for sedative and hypnotic benzodiazepines was 7.7 billion S-DDD [2]. Pharmacologically used benzodiazepines are under international legal control pursuant to the Convention on Psychotropic Substances. Some derivatives used therapeutically in just several countries, i.e., phenazepam, flutazolam, and etizolam, are not subject to international legal control. The first benzodiazepine derivative introduced in medical treatment was chlordiazepoxide and it was launched onto the market under the name of Librium. Chronologically, the next one was diazepam (Valium) and its active metabolite, i.e., oxazepam. Since that time, this group of drugs has developed rapidly. Approximately 3,000 derivatives have so far been synthesized and pharmacologically tested. The most abundant group of benzodiazepines are derivatives of 1,4-benzodiazepine, that is, 5-phenyl-1H-benzo[e][1,4] diazepine-2(3H)-one, presented in Figure 22.1. Substituents at the R^1, R^3, R^7, and $R^{2'}$

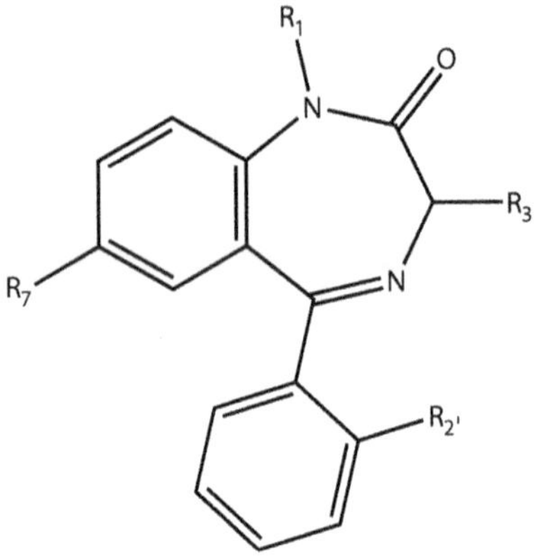

FIGURE 21.1   Basic structure of 1,4-benzodiazepine.

positions determine the various pharmacodynamic and pharmacokinetic properties of benzodiazepines [3].

Derivatives of 1,4-benzodiazepine may be categorized into several basic groups, including 2-keto derivatives (e.g., diazepam, chlorodiazepoxide, and prazepam), 3-hydroxy derivatives (e.g., lorazepam, lormetazepam, and oxazepam), and 7-nitro derivatives (clonazepam, flunitrazepam, and nitrazepam). Moreover, there are also triazolobenzodiazepines (alprazolam, estazolam, and triazolam) and imidazoloben-zodiazepines (midazolam and loprazolam). Some compounds also exist that are not included in the above-mentioned groups, e.g., 1,5-benzodiazepine derivatives (cloba-zam), or thienodiazepine derivatives (brotizolam) [4].

The widespread medical use of this group of drugs as well as their properties are the reasons why those medications are frequently abused. The nonmedical use of benzodiazepines comprises, among others, long-term self-medication as well as use for other purposes, such as inducing intoxication or relieving the effects of other substances. Benzodiazepines are rarely abused as single substances, since they are mostly combined with other compounds, e.g., opioids. As medications with a wide therapeutic index, they are relatively safe, and fatal intoxications with benzodiaz-epines are rare. Used in combination with alcohol or other central nervous system depressants, they act synergistically, thus being a frequent cause of poisonings. Consumed benzodiazepines undergo intensive metabolic reactions, mainly hydrox-ylation, demethylation, and conjugation with glucuronic acid. Many metabolites reveal a high pharmacological activity, often comparable with the parent compound. Active metabolites and parent compounds with a long half-life period may be accu-mulated in the system, which may evoke toxic effects with a repeated use. Apart from health- and life-threatening intoxications, a serious problem connected with benzodiazepine abuse is impairment of the ability to drive, which may increase the risk of accidents (driving under the influence of drugs, DUID). Moreover, drugs from this group are often recorded in drug-facilitated crimes (e.g., sexual assault) [5].

Therefore, the analysis of this group of medications and their metabolites in various biological materials (blood, urine, tissues and alternative materials, such as hair, sweat, or saliva) is especially compelling to clinical and forensic toxicologists. Over the years, many methods of analysis of those compounds have been developed, and chromatographic methods seem to be dominant among them.

A quick and cost-effective screening method is thin-layer chromatography (TLC) [6]. Expansions of this simple method are ultra-thin-layer chromatography (UTLC) and atmospheric pressure matrix-assisted laser desorption/ionization mass spectrometry (AP-MALDI-MS) [7].

A series of analytic methods have been formulated that may be used for analyzing benzodiazepines in biological materials and gas chromatography with different types of detectors, like nitrogen–phosphorus detector (NPD), electron capture detector (ECD), electron ionization (EI), or chemical ionization (CI) mass detector, which all count among the techniques employed for this particular purpose [8–11].

Due to a varied polarity of benzodiazepines and their generally low volatility, their analysis is based mainly on liquid chromatography combined with various different detection techniques. Earlier reports mostly employed ultraviolet detection (UV) or diode array detection (DAD) [12–15]. Presently, liquid chromatography combined with various types of mass detectors, such as single-quadrupole or triple-quadrupole mass spectrometry (MS/MS), time-of-flight mass spectrometry (TOF-MS), or high-resolution mass spectrometry (HRMS), are the standard techniques in this respect [16–19].

21.2 BENZODIAZEPINES AS NEW PSYCHOACTIVE SUBSTANCES (NPSs)

In recent years, new psychoactive substances have appeared on an unprecedented scale on the global narcotic market. According to the data provided by the European Monitoring Centre for Drugs and Drug Addiction (EMCDDA), in 2014 there were 101 new substances, while in 2015 there were 98 [20]. Despite the fact that they act similarly to the controlled analogs, they remain uncontrolled and are introduced on the market via various distribution channels (e.g., by sale on the Internet, traditional shops, etc.). These constantly reappearing new substances pose a serious challenge for clinicians, forensic toxicologists, and legislators. As well as numerous synthetic cannabinoids and cathinones, benzodiazepine derivatives also appear on the market, a fact which complicates the phenomenon of therapeutic drug abuse. Many of them are drugs candidates never approved for medical use.

The first to appear on the market were derivatives that had been locally used in treatment and therefore were not included in the Convention on Psychotropic Substances, and in 2007 this group consisted of phenazepam, and then etizolam and flutazolam. Next, in chronological order, appeared pyrazolam, flubromazepam, diclazepam, meclonazepam, deschloroetizolam, flubromazolam, nifoxipam, clonazolam, adinazolam, metizolam, nitrazolam, 3-hydroxyphenazepam, fonazepam, 4-chlorodiazepam, bromazolam, and norfludiazepam (their structures are given in Figure 21.2), which have never been used in medical treatment. Since that time, these groups of drugs have developed rapidly. In 2013, the reports of EMCDDA included benzodiazepines as a separate NPS category [20–21]. Until now, the occurrence of 20 new benzodiazepine derivatives has been analytically confirmed. However, one

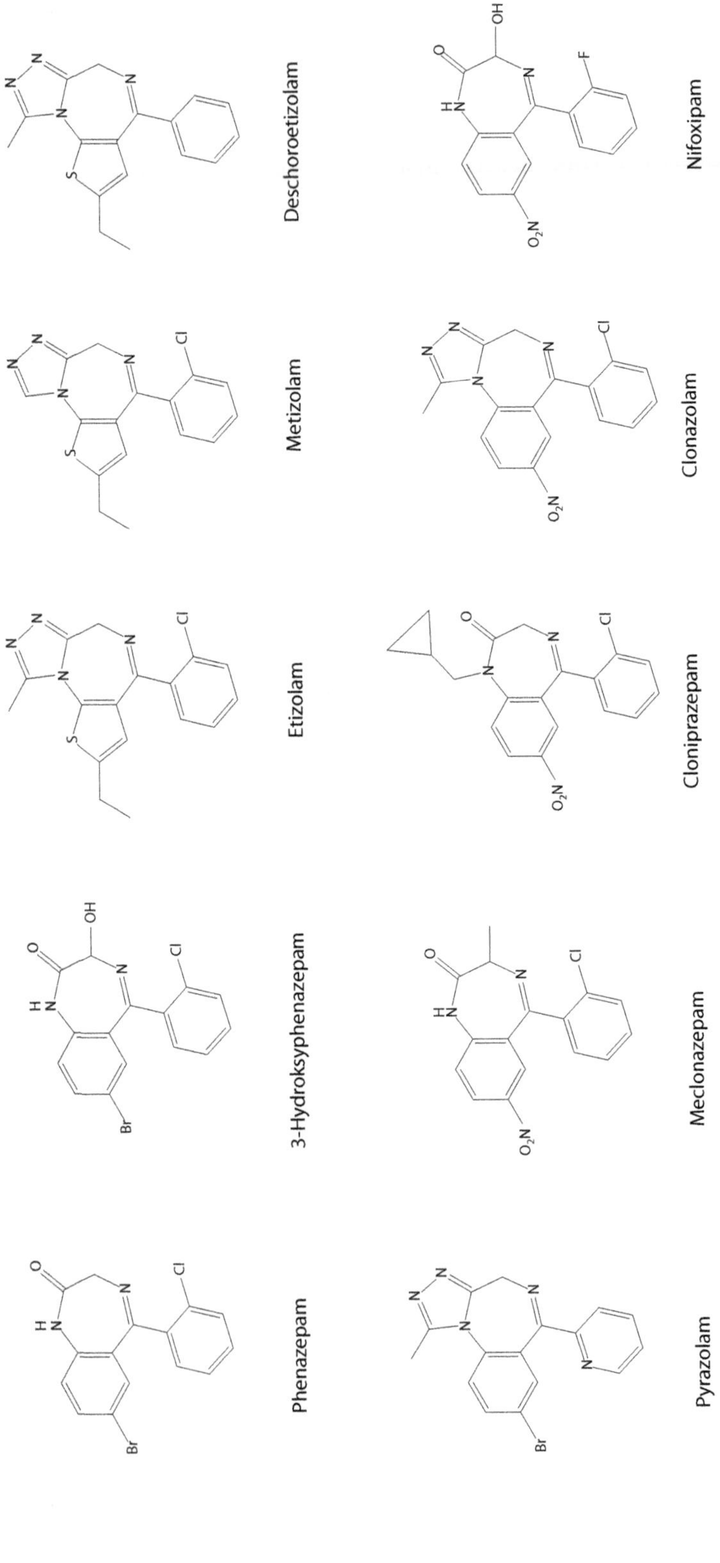

Deschoroetizolam

Nifoxipam

Metizolam

Clonazolam

Etizolam

Cloniprazepam

3-Hydroksyphenazepam

Meclonazepam

Phenazepam

Pyrazolam

(Continued)

FIGURE 21.2   Structures of designer benzodiazepines (DBZDs).

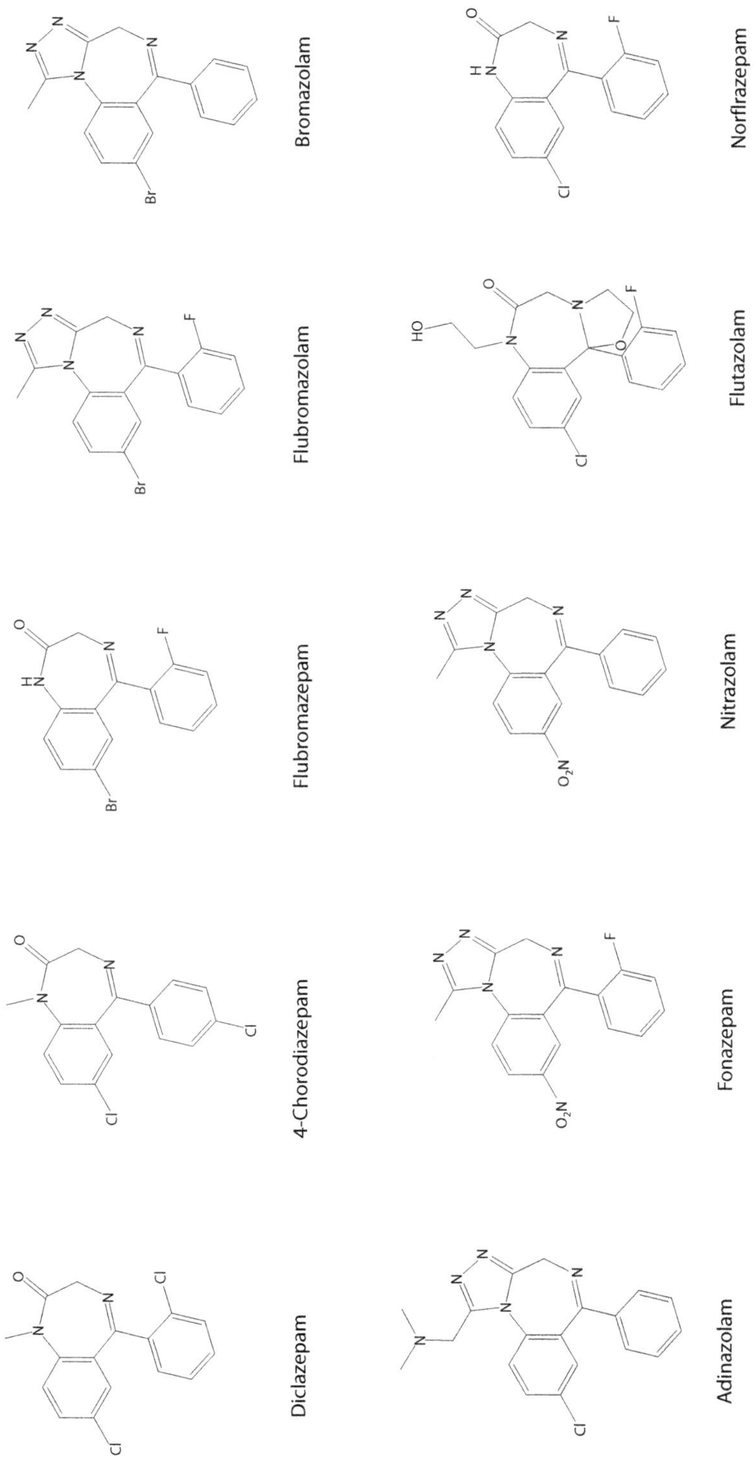

FIGURE 21.2 (CONTINUED) Structures of designer benzodiazepines (DBZDs).

Bromazolam

Norflrazepam

Flubromazolam

Flutazolam

Flubromazepam

Nitrazolam

4-Chorodiazepam

Fonazepam

Diclazepam

Adinazolam

should not overlook the fact that, on the NPS market, plenty of new compounds have appeared that have not yet been analytically confirmed, such as mexazolam, flutoprazepam, or flunitrazolam.

21.3   ANALYSIS OF SEIZED MATERIAL

Definite identification of NPSs is essential to differentiate the compounds that are scheduled from those that may be legally possessed. A serious problem faced by laboratories in the course of research on NPS (including benzodiazepines) is that reference materials are not readily available. In the identification tests of new psychoactive substances in seized materials, a fundamental role is played by gas chromatography–electron impact mass spectrometry (GCEI-MS) and liquid chromatography quadrupole time-of-flight mass spectrometry (LC-QTOF-MS). If it is necessary to differentiate positional isomers, nuclear magnetic resonance (NMR) is indispensable. Additional methods used in the identification process are infrared spectroscopy (IR or FTIR) and Raman spectroscopy.

The benzodiazepine NPSs are offered in various dosage forms, i.e., as pellets, capsules, tablets, blotters or powders, which, apart from an active substance, may also contain impurities or adjuvants.

Detailed analysis of tablets seized from an illegal production was described by Lim et al. [22]. The preparation known as Erimin and originally containing the controlled substance nimetazepam was subjected to comprehensive testing. Identification of an active substance in the methanolic extracts of tablets was performed by means of GC-MS EI, and on the basis of the obtained mass spectrum, the presence of undeclared phenazepam was confirmed. The quantitative test was performed by means of the validated method of ultra-performance liquid chromatography (UPLC) with a diode array detector (DAD). Using the TLC method, the dyes used in the forged products were identified. The IR method was used to identify the adjuvants.

At an initial stage of research on designer benzodiazepines and due to the lack of reference materials, standard diclazepam, pyrazolam, or flubromazepam were identified in tablets purchased in online shops, using the thin-layer chromatographic method. After homogenization, tablets were dissolved in a boron buffer (pH 9) and extracted with 1-chlorobutane. The organic extract was loaded on the TLC plate (silica gel 60, 10 × 20 cm, Merck F256) and developed with a mobile phase of acetic acid (99%), deionized water, methanol, ethyl acetate (2:15:20:80, v/v/v/v). After separation, the band of interest was extracted with ethanol and then identified by means of GC-MS, LC-MS/MS, LC-Q-TOF-MS, and by means of NMR in order to confirm purity [23–25]. No methods for the detection and determination of the other DBZDs in the seized materials (like powders or tablets) have been described.

21.4   METHODS OF FORENSIC ANALYSIS OF DESIGNER BENZODIAZEPINES

The published methods of analysis of DBZDs in biological samples mostly include the analysis of single compounds in the context of testing their pharmacokinetics and biotransformation [23–26], or the analysis of intoxication cases [10,27,28]. Few papers discuss simultaneous analysis of a greater number of compounds [29,33]. The analyses were performed from the varied biological material, that is, whole blood, serum, urine, internal organs, and tissues, and from alternative materials, such as hair, saliva, sweat, and exhaled air. The analyses of metabolism were performed *in vitro* on human microsomes and hepatocytes, or *in vivo* on mice, and in authentic cases of human urine samples, or in human urine after self-administration. A summary of the methods used for quantitative analysis of the designer benzodiazepines is presented in Table 21.1, while Table 21.2 presents the methodology used to identify the metabolites.

In the course of toxicological analysis of the so-called classic benzodiazepines, a wide range of immunoenzymatic screening methods are used. Due to the similar structure of the benzodiazepine derivatives, a high cross-reactivity is observed. The analysis of 13 DBZDs performed by Pettersson Bergstrand et al. also showed high cross-reactivity of the derivatives as well as their metabolites [40]. An exception was low reactivity of flutazolam, which may result from the fact that its structure is significantly different from the remaining derivatives.

Preparation of a sample is a key stage of analysis, during which endogenic substances that may interfere in the determination are removed. In the course of DBZD analysis, isolation from biological matrices is performed mainly via liquid–liquid extraction (LLE) after sample alkalization. For instance, clonazolam, diclazepam, flubromazepam, pyrazolam, and etizolam were extracted from urine alkalinized with borate buffer (pH 11) with a mixture of ethyl acetate and heptane (4:1, v/v) [29]. Phenazepam and 3-hydroxyphenazepam were extracted from the tissue homogenates with a mixture of hexane and ethyl acetate after alkalinization with the 0.2 M Na_2CO_3. As far as urinalysis is concerned, isolation is usually preceded by enzymatic hydrolysis with the use of ß-glucuronidase, in order to release parent compounds or their metabolites from the conjugation with glucuronic acid [33]. Apart from LLE, another technique used to analyze benzodiazepines is solid-phase extraction (SPE). For example, phenazepam was isolated from blood and urine by means of Bond Elut Plexa columns, containing a nonpolar polymeric sorbent [28]. Etizolam and its metabolites (α-hydroxyetizolam and 8-hydroxyetizolam) were isolated with the use of the Oasis HLB columns [34].

The use of UPLC methods with high-resolution chromatographic columns combined with quick and selective mass spectrometry detectors allowed a considerable reduction in the complexity of the process of preparing samples by precipitation of proteins (e.g., with use of acetonitrile) in the case of blood or tissue homogenates [35], or by direct injection after centrifugation in the case of urine [33,39].

Isolation from alternative materials (such as hair, saliva, sweat, and an exhaled air) was described by Kintz et al. Metizolam from the hair and saliva was extracted via the LLE method with the mixture of dichlorometane/n-heptane/isopropanol (25:65:10, v/v) after being alkalinized with a saturated ammonium chloride buffer (pH 9.5) [43]. Sweat was collected with the PharmCheck™ sweat patch, while the exhaled air was treated with the ExaBreath® DrugTrap device, and, next, metizolam was extracted with methanol.

There are certain reports available on analysis by means of gas chromatography–mass spectrometry (GC-MS), and they comprise mainly the compounds that have been known for years and are locally used as drugs (i.e., etizolam, phenazepam

TABLE 21.1
Quantitative Analysis of DBZDs in Biological Material

Compound	Matrix	Sample Preparation	Column and Mobile Phase	Method	Method Validation	Application	Ref.
Phenazepam, 3-hydroksyphenazepam	Post-mortem blood, urine, vitreous humor, brain, liver, muscle	LLE	Gemini (150 × 2 mm, 5 μm) Gradient A—0.1% formic acid B—0.1% formic acid in ACN	LC-MS/MS ESI+ MRM mode	Linearity 0.7–200 and 16–1000 ng/mL, LOD 0.3 and 7 ng/mL, LOQ 0.7–16 ng/mL, recovery 38–72 and 45–78% (for phenazepam and 3-OH phenazepam, respectively), accuracy, precision and ME within ± 15% for all matrices. selectivity	Analysis of 29 autopsy cases in which phenazepam was detected	[27]
Phenazepam	Serum, urine	SPE	Luna Phenyl-Hexyl (50 × 3 mm, 3 μm) Gradient A—ACN B—10 mM ammonium acetate in 0,1% formic acid	LC-MS/MS ESI+ MRM mode	Linearity 3–100 and 50–1000 ng/mL, LOD 1.4 and 5 ng/mL, LOQ 3.1 and 16 ng/mL for serum and urine respectively, expanded measurement uncertainty 31%	Analysis of 141 DUID cases	[28]
Phenazepam	Post-mortem blood	LLE	DB-5 HT (30 m × 0.32 mm, 0.1-μm film thickness)	GC-NICI-MS SIM mode	Linearity 10–400 ng/mL, R²-0.999, LOQ 10 ng/mL, accuracy 1.3–5.65%, precision 6.56–18.2%, extraction efficiency 80.8±7.3%, selectivity (no interferences)	Analysis of phenazepam in 17 medico-legal autopsy cases	[10]

(Continued)

TABLE 21.1 (CONTINUED)
Quantitative Analysis of DBZDs in Biological Material

Compound	Matrix	Sample Preparation	Column and Mobile Phase	Method	Method Validation	Application	Ref.
Clonazolam, diclazepam, flubromazepam, flubromazolam, pyrazolam, etizolam	Blood	LLE	Acquity UPLC BEH phenyl A—5 mM ammonium bicarbonate buffer pH 8 B—MeOH	UPLC-MS/MS ESI+ MRM mode	Calibration curves linear with $R^2 > 0.999$ Administrative cut-offs 14–37 ng/mL QC samples results within 15% nominal values	Analysis of 77 DUID cases	[29]
Pyrazolam, flubromazepam, diclazepam	Serum, urine	LLE	Synergi 4u Polar (150 × 2 mm, 4 μm) Gradient A—0.1% formic acid B—0.1% formic acid in MeOH	LC-MS/MS ESI+ MRM mode	Calibration range (0.1–1.0)– 100 ng/mL LOD, long-term stability 90–110% (for diclazepam only)	Pharmacokinetic study after self-administration	[23–25]
Etizolam and metabolites (α-hydroxyetizolam and 8-hydroxyetizolam)	Post-mortem blood	SPE	DB 5MS (30 m × 0.32 mm, 0.1-μm film thickness)	Ion-trap GC-MS/MS SIM mode	Linearity 5–50 ng/mL, R^2- 0.982–0.995 accuracy 5.9–9.3% recovery 103.7–120.6%	Analysis of two post-mortem blood specimens	[34]
Pyrazolam, diclazepam, flubromazepam, meclonazepam, etizolam, phenazepam, nifoxipam, deschloroetizolam, clonazolam, flubromazolam, flutazolam	Urine	Dilution, enzymatic hydrolysis and centrifugation	Acquity UPLC BEH Phenyl (50 × 1 mm, 1.7-μm) Gradient A—0.1% formic acid B—ACN	LC-MS/MS ESI+ MRM mode	Linearity, accuracy and precision LOD 1–10 ng/mL, selectivity, ME, dilution integrity, carry-over, stability Method-fulfilled European Medicines Agency validation criteria	Analysis of 390 clinical specimens positive in immunoassay for benzodiazepines	[33]

(Continued)

TABLE 21.1 (CONTINUED)

Quantitative Analysis of DBZDs in Biological Material

Compound	Matrix	Sample Preparation	Column and Mobile Phase	Method	Method Validation	Application	Ref.
Flubromazolam	Post-mortem blood, urine, muscle, liver, brain	Protein precipitation	Acquity UPLC BEH C18 (100 × 2.1 mm, 1.7-μm) Gradient A—0.1% formic acid B—ACN	UPLC-MS/MS ESI+ MRM mode	Imprecision and accuracy, < 20% (0.05 ng/mL), < 15%, (100 ng/mL) Extraction efficiency 69–74%, linearity 0.05–50 ng/mL ME 11–4%	Analysis of two post-mortem blood, urine, and tissue specimens	[35]
Metizolam	Urine	LLE	Acquity UPLC BEH C18 (100 × 2.1 mm 1.7 μm) Gradient A—0.1% formic acid B—ACN	UPLC-MS/MS ESI+ MRM mode	Linearity 0.05–50 ng/mL, $R^2 = 0.999$ LOD 25 pg/mL, within batch precision < 20%, ME< 20%, Selectivity (no interferences)	Analysis of elimination parameters after self-administration study	[36]
	Saliva, hair, sweat, exhaled breath	LLE	Acquity UPLC BEH C18 (100 × 2.1 mm, 1.7-μm) Gradient A—0.1% formic acid B—ACN	UPLC-MS/MS ESI+ MRM mode	Linearity and LOQ 50-1000, 2 pg/mL, 0.1–10, 0.1 pg/ mg, 5–500, 5 pg/patch, 5–50, 5 pg/filter for saliva, hair, sweat, and exhaled breath, respectively $R^2 > 0.99$, precision < 25%, ME < 20%		[43]

(Continued)

TABLE 21.1 (CONTINUED)
Quantitative Analysis of DBZDs in Biological Material

Compound	Matrix	Sample Preparation	Column and Mobile Phase	Method	Method Validation	Application	Ref.
Fonazepam, norfludiazepam, and 16 other benzodiazepines	Urine	Centrifugation	Zorbax SB C 18 (100 × 2.1 mm 3.5 µm) Gradient A—0.2% acetic acid and 2 mM ammonium trifluoroacetate B—2 mM ammonium trifluoroacetate in ACN	LC-MS/MS ESI+ MRM mode	LOD 3 ng/mL, LLOQ 10 ng/mL, Linearity 10–100 (400) ng/mL, R0z^2 > 0.99 precision <11.8%, accuracy <10%, No significant matrix effect	Analysis of 21 forensic urine samples	[39]
Fonazepam and 26 other benzodiazepines	Blood and urine	SPE	Acquity C18 (50 × 2.1 mm, 1.7-µm) Gradient A—10 mM ammonium bicarbonate pH 9 B—MeOH	LC-MS/MS ESI+ MRM mode	LOD 0.05 ng/mL, LOD 2 ng/mL, Linearity 2–500 ng/mL, ME 76.5–86.2%, recovery 70.4–88.1%, precision 2.5–15.6%, accuracy −15.4–15.2%	Analysis of 10 blood and urine forensic cases	[46]

TABLE 21.2

Analysis of DBZD Metabolites

Compound	Matrix	Sample Preparation	Column and Mobile Phase	Method	Application	Ref.
Clonazolam, deschloroetizolam, flubromazolam, meclonazepam, adinazolam, cloniprazepam, fonazepam, 3-hydroksyphenazepam, metizolam	pHLM	—	Kinetex 5F (100 × 2.1 mm, 2.6-μm) Gradient A—water with 1% ACN 0.1% formic acid and 2 mM ammonium formate B—acetonitrile (ACN) with 0.1% formic acid and 2 mM ammonium formate	LC-QTOF-MS ESI+ Full-scan and bbCID mode LC-MS/MS ESI+ MRM mode EPI mode	Identification of *in vitro* metabolites	[26,30]
Clonazolam, meclonazepam, nifoxipam	Urine	Dilution and centrifugation	PepMap RSLC Acclaim C18 (150 mm × 50 μm, 2-μm) Gradient A—0.1% formic acid B—0.1% formic acid in ACN	nanoLC-HRMS/MS ESI+ Full-scan and TMS^2 mode	Identification of *in vivo* metabolites	[31]
		Dilution	YMC UltraHT Hydrosphere C18 (100 × 2 mm, 2 μm) Gradient A—10 mM ammonium formate and 0.005% formic acid B—water/MeOH 90/10 (v/v) with 10 mM ammonium formate and 0.005% formic acid	UHPLC-HRMS/MS ESI+ Full-scan and TMS^2 mode		

(Continued)

TABLE 21.2 (CONTINUED)
Analysis of DBZD Metabolites

Compound	Matrix	Sample Preparation	Column and Mobile Phase	Method	Application	Ref.
Flubromazepam, diclazepam, etizolam, deschloroetizolam, flubromazolam, clonazolam, nifoxipam, meclonazepam	pHLM	SPE	Atlantis T3 (150 × 2.1 mm, 3-μm) Gradient A—2 mM ammonium formate buffer (pH 3) B—2 mM ammonium formate buffer (pH 3) ACN (10/90 v/v)	LC-MS/MS ESI+ MRM mode EPI mode	Identification of *in vitro* metabolites	[32]
Pyrazolam, flubromazepam, diclazepam	Serum, urine	LLE, protein precipitation	Synergi 4u Polar (150 × 2 mm, 4 μm) Gradient A—0.1% formic acid B—0.1% formic acid in MeOHl Acclaim RSLC 120 C18 (100 × 2.1 mm, 4 μm) Gradient A—water/MeOH 90/10 (v/v) with 5 mM ammonium formate and 0.01% formic acid B—MeOH with 5 mM ammonium formate and 0.01% formic acid	LC-MS/MS ESI+ EPI mode LC-QTOF-MS ESI+ Full-scan and bbCID mode	*In vivo* metabolism study after self-administration	[23–25]

(Continued)

TABLE 21.2 (CONTINUED)
Analysis of DBZD Metabolites

Compound	Matrix	Sample Preparation	Column and Mobile Phase	Method	Application	Ref.
Flubromazolam	pHLM	Protein precipitation	Acquity HSS C 18 (150 × 2.1 mm, 1.8 μm) Gradient A—0.1% formic acid B—0.1% formic acid in ACN	UHPLC-HRMS/ MS ESI+ Full-scan and PRM mode	Metabolism *in vitro* study	[35]
Metizolam	pHLM human, hepatocytes, mouse and human urine	Centrifugation and dilution	Acquity HSS T 3 (150 × 2.1 mm, 1.8 μm) Gradient A—0.1% formic acid B—0.1% formic acid in ACN	UPLC-QTOF-MS ESI+ Full-scan mode	*In vivo* and *in vitro* metabolism study	[36]
	pHLM, urine	SPE	Acquity HSS C 18(150 × 2.1 mm, 1.8 μm) Gradient A—0.05 formic acid in 10 mM ammonium formate B—0.05% formic acid in ACN	UPLC-QTOF-MS ESI+ Full-scan mode	Metabolism study *in vitro* and *in vivo* after self-administration	[37]
Meclonazepam	pHLM, human hepatocytes, mouse and human urine	Centrifugation and dilution	Acquity HSS T 3 (150 × 2.1 mm, 1.8 μm) Gradient A—0.1% formic acid B—0.1% formic acid in ACN	UPLC-QTOF-MS ESI+ Full-scan mode	*In vivo* and *in vitro* metabolism study	[38]

[10,34]), or those that are metabolites of controlled benzodiazepines (i.e., fonazepam). Amidst the published testing methods of designer benzodiazepines, the most frequently used are liquid chromatography methods, from high-performance liquid chromatography [23–25,31,39], through ultra-performance liquid chromatography [29,33,35,36,43], to nano-liquid chromatography (nanoHPLC) [31]. The use of UPLC columns with sub-2-μm particle size allows obtaining a very good separation with a very short time of analysis. Separation of 11 designer benzodiazepines, obtained by means of the UPLC-MS/MS method on the Acquity UPLC BEH phenyl column (50 × 1 mm, 1.7 μm), is presented in Figure 21.3.

In order to quantify DBZDs, tandem mass spectrometers equipped with the electrospray sources (ESI) working in the positive mode are mostly employed. Tandem mass spectrometers usually work in the multiple monitoring reaction mode (MRM), whose benefits are high selectivity and sensitivity of the analysis.

As far as identification of the metabolites of the designer benzodiazepines is concerned, analogical chromatographic techniques are used, often combined with high-resolution mass spectrometers. Meyer et al. studied the metabolites of clonazolam, meclonazepam, and nifoxipam *in vivo* using the nanoHPLC apparatus coupled with a high-resolution mass spectrometer (nanoHPLC-HRMS) [31]. The mass spectrometer worked in the full scan mode and in TMS^2 (targeted MS^2) scanning, in order to search for the precursor masses of the anticipated marked I and II phase metabolites. El Balkhi et al. have analyzed the metabolites of the eight designer benzodiazepines, which after incubation with human liver microsomes (HLM) were then isolated by means of the SPE method [32]. In order to identify the metabolites, the authors

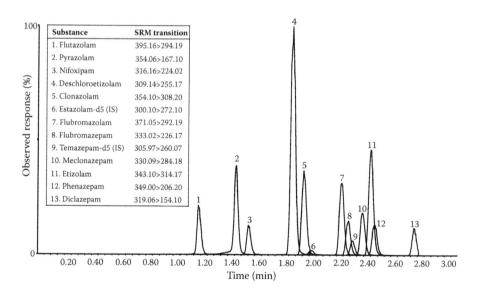

FIGURE 21.3 Chromatogram showing 11 designer benzodiazepine analytes and 2 internal standards (IS) covered by the LC-MS/MS method. (From Pettersson Bergstrand, M. et al. 2016. *J. Chromatogr. B.* 1035: 104–10. With permission.)

used UHPLC coupled with the triple-quadrupole linear ion trap mass spectrometer. Masses of hypothetical metabolites were scanned in the above-mentioned experiments by means of enhanced product scanning (EPI).

In particular cases connected with drug-facilitated crimes, it is very important to be able to confirm the exposure to a given compound after a long time, and, in such cases, it is essential to know the metabolic transformation pathways of the analyzed compounds and the detection times in biological material. Knowledge of the metabolites is helpful when confirming exposure to a given compound in the absence of the parent compound.

Some DBZDs are active metabolites of benzodiazepines used in therapy. For instance, fonazepam (also known as desmethylflunitrazepam, or norflunitrazepam) and nifoxipam (3-desmethylflunitrazepam) are active metabolites of flunitrazepam [41], 3-hydroxyphenazepam is an active metabolite of phenazepam, and norfludiazepam (also known as norflurazepam) is an active metabolite of flurazepam. Cloniprazepam is a prodrug out of which, via demethylation, clonazepam is formed. The hydroxylation reaction of diclazepam results in the formation of lormetazepam, while its demethylation results in the formation of delorazepam, both of which, via demethylation and hydroxylation respectively, give rise to lorazepam [25]. Next, adinazolam undergoes biotransformation via subsequent demethylation reactions to estazolam and α-hydroxyalprazolam [42].

Tandem mass spectrometers (quadrupole—linear ion trap, ion trap, quadrupole time-of-flight, or linear ion trap-orbitrap) become increasingly more common in toxicological analysis, as they can work in the data-dependent acquisition (DDA) mode. In the case of detection of a given precursor ion, mass spectrometry working in this mode performs its fragmentation and collects the product ion spectrum (PIS), which, compared with the spectrum library, can be used together with the other data to identify the compound. The DDA mode allows collection of the MS/MS spectra for many compounds during one analysis, which makes it possible to perform the so-called systematic toxicological analysis (STA). An identification possibility is limited to the earlier established list of compounds. Mollerup et al. presented the method of targeted and nontargeted analysis for blood analysis [44]. Blood samples spiked with 11 designer benzodiazepines in concentrations between 0.1 and 100 ng/mL, after protein precipitation with use of acetylonitrile and centrifugation, were analyzed by the UPLC-QTOF method. The proposed nontarget screening analysis of the data collected for unidentified chromatographic peaks allowed identifying 3 out of 11 designer benzodiazepines in blood at the level of 5 ng/mL, and 9 out of 11 at the level of 100 ng/mL.

A problem that has not yet been investigated is the stability of the designer benzodiazepines in biological material, especially in postmortem samples. Considering the fact that 6 out of 20 designer benzodiazepines are 7-nitro derivatives, one should be aware of instability of these compounds in the material. According to tests performed by Robertson et al., in postmortem blood contaminated with bacteria, nitrobenzodiazepines (clonazepam, nitrazepam, and flunitrazepam) undergo almost complete decomposition at a temperature of 22°C within 8 hours [45].

21.5   CONCLUSIONS

A large number of benzodiazepines and their metabolites being of varied polarity as well as low concentrations in the biological materials contributes to the fact that the analysis of designer benzodiazepines requires proper separating techniques, the best of which seems to be high-performance liquid chromatography combined with highly selective mass detectors.

In the case of DBZDs, one should take into consideration their close connection to the scheduled derivatives, used therapeutically. Due to the possibility of an *in vivo* synthesis of the controlled compounds, particular care is required while interpreting the analysis of benzodiazepines. Methods used for the identification and quantification of DBZDs should cover as many parent compounds and metabolites as possible. Due to the lack of metabolites standards, mass spectrometry data should be applied for identification.

REFERENCES

1. Karch, S.B. 2007. *Drug Abuse Handbook*, CRC Press, Boca Raton.
2. International Control Narcotic Board (ICNB). 2015. https://www.incb.org/documents /Psychotropics/technical-publications/2015/Technical_PSY_2015_ENG.pdf (accessed 21 January 2017).
3. Brandenberger, H., Maes, R.A.A. (Eds.) 1997. *Analytical Toxicology for Clinical, Forensic and Pharmaceutical Chemists,* Walter de Gruyter, Berlin.
4. Szatkowska, P., Koba, M., Kośliński, P., Wandas, J., Bączek, T. 2014. Analytical methods for determination of benzodiazepines. A short review. *Cent. Eur. J. Chem.* 12: 994–1007.
5. Persona, K., Madej, K., Knihnicki, P., Piekoszewski, W. 2015 Analytical methodologies for the determination of benzodiazepines in biological samples. *J. Pharm. Biomed. Anal.* 10:113: 239–264.
6. Jain, R. 2000. Utility of thin layer chromatography for detection of opioids and benzodiazepines in a clinical setting. *Addict Behav.* 25: 451–454.
7. Salo, P.K., Vilmunen, S., Salomies, H., Ketola, R.A., Kostiainen, R. 2007. Two-dimensional ultra-thin-layer chromatography and atmospheric pressure matrix-assisted laser desorption/ionization mass spectrometry in bioanalysis. *Anal. Chem.* 79: 2101–2108.
8. Bravo, F., Lobos, C., Venegas, K., Benites, J. 2010. Development and validation of a GC-NPD/micro-ECD method using dual column for the determination of benzodiazepine in human whole blood and plasma. *J. Chin. Chem. Soc.* 55: 454–457.
9. Pujadas, M., Pichini, S., Civit, E., Santamariña, E., Perez, K., de la Torre, R. 2007. A simple and reliable procedure for the determination of psychoactive drugs in oral fluid by gas chromatography-mass spectrometry. *J. Pharm. Biomed. Anal.* 44: 594–601.
10. Gunnar, T., Ariniemi, K., Lillsunde, P. 2006. Fast gas chromatography–negative-ion chemical ionization mass spectrometry with microscale volume sample preparation for the determination of benzodiazepines and alpha-hydroxy metabolites, zaleplon and zopiclone in whole blood. *J. Mass Spectrom.* 41: 741–754.
11. Papoutsis, I.I., Athanaselis, S.A., Nikolaou, P.D., Pistos, C.M., Spiliopoulou, C.A., Maravelias, C.P. 2010. Development and validation of an EI-GC-MS method for the determination of benzodiazepine drugs and their metabolites in blood: Applications in clinical and forensic toxicology. *J. Pharm. Biomed. Anal.* 52: 609–614.

12. Borges, K.B., Freire, E.F., Martins, I., de Siqueira, M.E.P.B. 2009. Simultaneous determination of multibenzodiazepines by HPLC/UV: Investigation of liquid–liquid and solid-phase extractions in human plasma. *Talanta* 78: 233–241.
13. Rouini, M.R., Ardakani, Y.H., Moghaddam, K.A., Solatani, F. 2008. An improved HPLC method for rapid quantitation of diazepam and its major metabolites in human plasma. *Talanta* 75: 671–676.
14. Uddin, M.N., Samanidou, V.F., Papadoyannis, I.N. 2008. Development and validation of an HPLC method for the determination of six 1,4-benzodiazepines in pharmaceuticals and human biological fluids. *J. Liquid Chromatogr. Relat. Technol.* 31: 1258–1282.
15. Mercolini, L., Mandrioli, R., Amore, M., Raggi, M.A. 2008. Separation and HPLC analysis of 15 benzodiazepines in human plasma. *J. Sep. Sci.* 31: 2619–2626.
16. Laloup, M., Ramirez Fernandez, M.D.M., De Boeck, G., Wood, M., Maes, V., Samyn, N. 2005. Validation of a liquid chromatography–tandem mass spectrometry method for the simultaneous determination of 26 benzodiazepines and metabolites, zolpidem and zopiclone in blood, urine, and hair. *J. Anal. Toxicol.* 29: 616–626.
17. Hayashida, M., Takino, M., Terada, M., Kurisaki, E. Kudo, K., Ohno, Y. 2009. Time-of-flight mass spectrometry (TOF-MS) exact mass database for benzodiazepine screening. *Leg. Med.* 11: 423–425.
18. Nakamura, M. 2010. Analyses of benzodiazepines and their metabolites in various biological matrices by LC-MS(/MS). *Biomed. Chromatograph.* 25: 1283–1230.
19. Vogliardi, S., Favretto, D., Tucci, M., Stocchero, G., Ferrara, S.D. 2011. Simultaneous LC-HRMS determination of 28 benzodiazepines and metabolites in hair. *Anal. Bioanal. Chem.* 400: 51–67.
20. European monitoring center for drugs and drug addiction (EMCDDA) (2016) EU drug market report. http://www.emcdda.europa.eu/system/files/publications/814/TDAN14001ENN_475519.pdf (accessed 21 January 2017); http://www.emcdda.europa.eu/system/files/publications/2880/TDAS16001ENN.pdf (accessed 21 January 2017); http://www.emcdda.europa.eu/system/files/publications/1018/TDAN15001ENN.pdf (accessed 21 January 2017).
21. Moosmann, B., King, L.A., Auwärter, V. 2015. Designer benzodiazepines: A new challenge. *World Psychiatry* 14: 248.
22. Lim, W.J., Yap, A.T., Mangudi, M., Koh, H.B., Tang, A.S., Chan, K.B. 2017. Detection of phenazepam in illicitly manufactured Erimin 5 tablets. *Drug Test Anal.* 9: 293–305.
23. Moosmann, B., Hutter, M., Huppertz, L.M., Ferlaino, S., Redlingshöfer, L., Auwärter, V. 2013. Characterization of the designer benzodiazepine pyrazolam and its detectability in human serum and urine. *Forensic Toxicol.* 31: 263–271.
24. Moosmann, B., Huppertz, L.M., Hutter, M., Buchwald, A., Ferlaino, S., Auwärter, V. 2013. Detection and identification of the designer benzodiazepine flubromazepam and preliminary data on its metabolism and pharmacokinetics. *J. Mass. Spectrom.* 48: 1150–1159.
25. Moosmann, B., Bisel, P., Auwärter, V. 2014. Characterization of the designer benzodiazepine diclazepam and preliminary data on its metabolism and pharmacokinetics. *Drug Test Anal.* 6: 757–763.
26. Huppertz, L.M., Bisel, P., Westphal, F., Franz F., Auwärter, F., Moosmann B. 2015. Characterization of the four designer benzodiazepines clonazolam, deschloroetizolam, flubromazolam, and meclonazepam, and identification of their in vitro metabolites. *Forensic Toxicol.* 33: 388–395.
27. Crichton, M.L., Shenton, C.F., Drummond, G., Beer, L.J., Seetohul, L.N., Maskell, P.D. 2015. Analysis of phenazepam and 3-hydroxyphenazepam in post-mortem fluids and tissues. *Drug Test Anal.* 7: 926–936.

28. Kriikku, P., Wilhelm, L., Rintatalo, J., Hurme, J., Kramer J., Ojanpera, I. 2012. Phenazepam abuse in Finland: Findings from apprehended drivers, post-mortem cases and police confiscations. *For. Sci. Int.* 220: 111–117.

29. Høiseth, G., Skogstad Tuv, S., Karinen, R. 2016. Blood concentrations of new designer benzodiazepines in forensic cases. *For. Sci. Int.* 268: 35–38.

30. Moosmann, B., Bisel, P., Franz, F., Huppertz, L.M., Auwärter, V. 2016. Characterization and in vitro phase I microsomal metabolism of designer benzodiazepines— An update comprising adinazolam, cloniprazepam, fonazepam, 3-hydroksyphenazepam, metizolam and nitrazolam. *J. Mass Spectrom.* 51: 1080–1089.

31. Meyer, M.R., Bergstrand, M.P., Helander, A., Beck, O. 2016. Identification of main human urinary metabolites of the designer nitrobenzodiazepines clonazolam, meclonazepam, and nifoxipam by nano-liquid chromatography-high-resolution mass spectrometry for drug testing purposes. *Anal. Bioanal. Chem.* 408: 3571–3591.

32. El Balkhi, S., Chaslot, M., Picard, N., Dulaurent, S., Delage, M., Mathieu, O., Saint-Marcoux, F. 2017. Characterization and identification of eight designer benzodiazepine metabolites by incubation with human liver microsomes and analysis by a triple quadrupole mass spectrometer. *Int. J. Legal Med.* DOI: 10.1007/s00414-017-1541-6.

33. Pettersson Bergstrand, M., Helander, A., Beck, O. 2016. Development and application of a multi-component LC–MS/MS method for determination of designer benzodiazepines in urine. *J. Chromatogr. B.* 1035: 104–110.

34. Nakamae, T., Shinozuka, T., Sasaki, C. et al. 2008. Case report: Etizolam and its major metabolites in two unnatural death cases. *For. Sci. Int.* 182: e1–e6.

35. Noble, C., Mardal, M., Bjerre Holm, N., Stybe Johansen, S., Linnet, K. 2016. In vitro studies on flubromazolam metabolism and detection of its metabolites in authentic forensic samples. *Drug Test Anal.* DOI: 10.1002/dta.2146.

36. Wohlfarth, A., Vikingsson, S., Roman, M., Andersson, M., Kugelberg, F.C., Green, H., Kronstrand, R. 2016. Looking at flubromazolam metabolism from four different angles: Metabolite profiling in human liver microsomes, human hepatocytes, mice and authentic human urine samples with liquid chromatography high-resolution mass spectrometry. *For. Sci. Int.* DOI:10.1016/j.forsciint.2016.10.021.

37. Kintz, P., Richeval, C., Jamey, C., Ameline, A., Allorge, D., Gaulier, J.M., Raul, J.S. 2016. Detection of the designer benzodiazepine metizolam in urine and preliminary data on its metabolism. *Drug Test Anal.* DOI: 10.1002/dta.2099.

38. Vikingsson, S., Wohlfarth, A., Andersson, M., Gréen, H., Roman, M., Josefsson, M., Kugelberg, F.C., Kronstrand, R. 2017. Identifying metabolites of meclonazepam by high-resolution mass spectrometry using human liver microsomes, hepatocytes, a mouse model, and authentic urine samples. *AAPS J.* DOI: 10.1208/s12248-016-0040-x.

39. Jeong, Y.D., Kim, M.K., Suh, S.I., In, M.K., Kim, J.Y., Paeng, K.J. 2015. Rapid determination of benzodiazepines, zolpidem and their metabolites in urine using direct injection liquid chromatography–tandem mass spectrometry. *For. Sci. Int.* 257: 84–92.

40. Pettersson Bergstrand, M., Helander, A., Beck, O. 2016. Detectability of designer benzodiazepines in CEDIA, EMIT II Plus, HEIA, and KIMS II immunochemical screening assays. *Drug Test Anal.* DOI: 10.1002/dta.2003.

41. Katselou, M., Papoutsis, I., Nikolaou, P., Spiliopoulou, C., Athanaselis, S. 2017. Metabolites replace the parent drug in the drug arena. The cases of fonazepam and nifoxipam. *Forensic Toxicol.* 35: 1–10.

42. Fraser, A.D., Isner, A.F., and Bryan, W. 1993. Urinary Screening for adinazolam and its major metabolites by the Emit® d.a.u.™ and FPIA benzodiazepine assays with confirmation by HPLC. *J. Anal. Toxicol.* 17(7): 427–431.

43. Kintz, P., Jamey, C., Ameline, A., Richeval, C., Raul, J.S. 2016. Characterization of metizolam, a designer benzodiazepine, in alternative biological specimens. *Toxicol. Anal. Clin.* DOI: 10.1016/j.toxac.2016.09.004.
44. Mollerup, C.B., Dalsgaard, P.W., Mardal, M., Linnet, K. 2016. Targeted and non-targeted drug screening in whole blood by UHPLC-TOF-MS with data-independent acquisition. *Drug Test Anal.* DOI: 10.1002/dta.2120.
45. Robertson, M.D., Drummer, O.H. 1998. Stability of nitrobenzodiazepines in postmortem blood. *J. Forensic Sci.* 43: 5–8.
46. Verplaetse, R., Cuypers, E., Tytgat, J. 2012. The evaluation of the applicability of a high pH mobile phase in ultrahigh performance liquid chromatography tandem mass spectrometry analysis of benzodiazepines and benzodiazepine-like hypnotics in urine and blood. *J. Chromatogr. A.* 1249: 147–154.

22 Thin-Layer Chromatography in the Analysis of Designer Drugs

Joseph Sherma, Teresa Kowalska,
and Mieczysław Sajewicz

CONTENTS

22.1 INTRODUCTION

Thin-layer chromatography (TLC) and high-performance TLC (HPTLC) are major analytical methods used in the analysis of bulk drugs and pharmaceutical products. The 1,039-page book with 51 chapters titled *Thin Layer Chromatography in Drug Analysis* in the Chromatographic Science Series [1] attests to that importance. Chapters in Part I describe materials and methods of TLC for drug analysis, including sorbents and layers; mobile phase preparation and optimization; sample application and layer development; detection and identification of analytes; and quantification by densitometry and videoscanning. Part II includes 37 chapters on applications to various classes of commercial over-the-counter and prescription drugs.

Bernard-Savary and Poole [2] explained that in modern HPTLC unit operations are carried out with separate devices affording high flexibility in offline and online operation. They discussed instrument platforms for sample application, layer development, layer evaluation with optional derivatization, photodocumentation, densitometric evaluation, and hyphenation with spectrometric detectors with an emphasis on the variety and performance of commercially available systems and their user flexibility for manual, semiautomated, or fully automated operation.

Another source of descriptions of state-of-the-art methods and instrumentation is the chapter on TLC in the fourth edition of Ewing's *Analytical Instrumentation Handbook* [3]. Covered in this chapter are sample preparation, stationary phases, mobile phases, application of samples, capillary flow and forced flow chromatogram development, zone detection, documentation of chromatograms, zone identification, quantitative analysis, combination with different modes of spectrometry, preparative-layer chromatography (PLC; also called preparative TLC), and thin-layer radiochromatography (TLRC).

This chapter will not include a general discussion of the principles, methods, materials, and instruments of TLC that are already recently well presented [1–3]. It will describe selected applications of classical and modern TLC to the forensic analysis of designer drugs, including the materials, techniques, and instruments used as described in each of the publications. Future prospects for the TLC analysis of designer drugs will also be suggested.

The terminology "new psychoactive substances (NPSs)" is being used more frequently in the literature in place of "designer drugs" to include drug-laced products available on the market that contain both man-made designer drugs as well as plant-based drugs. All of the papers cited in this chapter on the analysis of both of these types of products were identified by searching the literature (Web of Science and Chemical Abstracts Scifinder) with the keywords "designer drug" and "thin layer chromatography"; no hits were obtained by searching up to May 25, 2016 in either of these databases using the keywords "new psychoactive substance" and "thin layer chromatography." The terminology "designer drug" was used throughout this chapter.

22.2 APPLICATIONS

22.2.1 Analysis of Biological Samples for Designer Drugs

Lillsunde and Korte [4] developed a simple and selective comprehensive identification system using TLC for screening, and gas chromatography–mass spectrometry (GC-MS) for confirmation. It was developed in the National Public Health Institute, Department of Biochemistry laboratory, in Helsinki, Finland, for screening yearly about 2,000 urine samples in misuse, impaired driving, poisoning, and other forensic cases. About 300 drugs and metabolites, including barbiturates, benzodiazepines, amphetamines, phenothiazines, antidepressants, opiates, carbamates, and most commonly used analgesics, can be simultaneously detected with the method, including designer drugs. A common extraction method with ChemElut CE1020 EXTUBES containing diatomaceous earth and eluted with 2 × 15 mL of dichloromethane–isopropanol (90:10) was used to recover drugs from urine adjusted to pH 8–9 using solid disodium hydrogen phosphate, with special extraction methods for glucuronides (benzodiazepines, 11-nor-delta9-tetrahydrocannabinol-9-carboxylic acid (THC-COOH), and morphine), cocaine, and benzoylecgonine. Nine different TLC systems were designated and used: (A) barbiturates and glutetimide; (B) meprobamate, carbamates, and some analgesics; (C) opiates, methaqualone, and some analgesics; (D) phenothiazines and antidepressants; (E) amphetamines and related compounds; (Be) benzodiazepines; (THC) cannabinoids; (OP) opiates; and (Co) cocaine. Merck aluminum-backed TLC plates, 20 × 20 cm, 60 Å particles and 0.2-mm layer thickness, with fluorescent indicator (F plates) were used for systems A, B, C, THC, and Co, and without indicator for systems D, E, Be, and OP. Appropriate standards and extracts were applied to each plate, and development was carried out with the following mobile phases: chloroform–acetone (45:5) for system A; ethyl acetate–methanol–ammonia–water (43:5:0.5:1.5) for B, C, D, and Co ("ammonia" and "28% ammonia" designate concentrated ammonium hydroxide in mobile phases); methanol–ammonia (50:0.5) for E; toluene for Be; n-hexane-1,4-dioxane-methanol (35:10:5) for THC; and toluene–acetone–ethanol–ammonia (22:22:4:2) for OP. After mobile phase development, all plates were examined under 254 and 366-nm ultraviolet (UV) and then sprayed with the following detection reagents (multiple reagents were sprayed in succession and results interpreted after each step): mercuric sulfate for system A; furfuraldehyde and sulfuric acid for B; Dragendorff and iodoplatinate for C; ferric chloride, UV light, concentrated sulfuric acid, and concentrated nitric acid for D; Fast Black K salt and 1 M sodium hydroxide for E; sulfuric acid, sodium nitrite, ammonium sulfamate, and N-(1-naphthyl)-eyhylenediamine dihydrochloride for Be; Fast Blue B for THC; AMP (2-amino-2-methyl-1,3-propanediol) buffer (pH 9.3), potassium ferricyanide, 254-nm UV light, Dragendorff, and iodoplatinate for OP; and Dragendorff and iodoplatinate for Co. Portions of each extract were reserved for GC-MS analysis.

The Lillsunde and Korte method just described was slightly modified [5] to identify the amphetamine-related psychostimulant designer drug 4-methylaminorex (4-MAX), which has appeared on the clandestine market with the street name "U4Euh," in urine. The EXTUBE extract of urine was screened on two TLC plates using a combination of the original C and D procedures (above; designated C/D here) and procedure E (above). Aluminum-backed Merck silica gel 60 F plates were developed with the mobile phases ethyl acetate–methanol–ammonia–water (43:5:0.5:1.5) and methanol–ammonia (50:0.5) for C/D and E, respectively. The developed plates were first examined under 254 and 366-nm UV light and then sprayed with the following reagents: ferric chloride, UV examination, and iodoplatinate for C/D, and Fast Black K salt and NaOH for E. A GC-MS method for quantitative determination of *cis*- and *trans*-4-MAX as *tert*-butyldimethylsilyl derivatives in plasma, urine, and tissue was also given.

As reported in a review of amphetamine, methamphetamime, and amphetamine-derived designer drugs in blood and urine [6], the Toxi-Lab TLC system was applied in the determination of MDMA (3,4-methylenedioxy-methamphetamine; ecstasy) and other psychotropic, ring-substituted amphetamine derivative designer drugs in patient urine [7]. Use of Toxi-Tubes A allowed simultaneous specimen preparation for TLC screening and GC-MS confirmation. Toxi-Lab system A, as described by Jarvie and Simpson [8], was designed for identification of basic and neutral drugs and comprises Toxi-Tubes A containing sodium sulfate and bicarbonate to give a pH of 9; stationary phases made of glass microfiber and silica gel with holes for inserting discs, onto which standard compounds and extracts with unknown drugs are applied; the mobile phase methanol–water–ethyl acetate (2:1:58); and a five-stage detection procedure: preactivation with formaldehyde vapor, dip into vanadium salt-sulfuric acid reagent, dip into distilled water, visualization under 366-nm UV light, and dip into solution of potassium iodide, iodine, bismuth subnitrate, and acetic acid. There is also a system B for acidic and neutral drugs. The Toxi-Lab TLC system is commercially available currently in kit form from DRG Diagnostics, Marburg, Germany/Springfield, New Jersey.

Methcathinone (ephedrine, "CAT"), the N-methyl analog of the natural product cathinone ("khat"), and designer analogs were synthesized, and their stereochemical and analytical properties were studied [9]. These compounds can be prepared easily from readily available starting materials such as ephedrines and pseudoephedrines. TLC properties of methcathinone and cathinone were compared with those of phenethylamine drugs of abuse using solutions of each compound prepared in drug-free urine and the commercial Toxi-Lab system for drug screening in urine. Toxi-Lab A extraction tubes and mobile phase were used, and the R_F values were found to be in the 0.5 range, much higher than ephedrines, amphetamine, and methamphetamine. This fact as well as colors at four detection stages, i.e., concentrated sulfuric acid, water rinse, UV light examination, and Dragendorff reagent, showed that methcathinone and cathinone are not likely to interfere with TLC procedures for screening amphetamine-type stimulants.

An unknown compound was detected by TLC in the urine of an illicit drug abuser and then identified by GC-MS in the electron ionization (EI) and positive-ion chemical ionization (PICI) modes as chlorinated MDMA [10]. Urine sample was prepared by fractional diethyl ether liquid–liquid extraction and analyzed by TLC on Merck 10×20 cm silica gel 60 G (gypsum binder) glass plates with ethyl acetate–ethanol–ammonia (36:2:2) and benzene–methanol–isopropanol–ammonia (35:5:1:1) mobile phases. Zones were detected with Dragendorff, Marquis, and Fast Black K color-forming reagents.

The rapid and sensitive detection of six methylene dioxylated phenethylamines including MDMA in urine samples by TLC with fluorescence detection was proposed [11]. Urine was extracted by vortexing with ammonia and diethyl ether, and extracts were applied to Merck silica gel 60 TLC plates. After development with isopropanol–28% ammonia solution (95:5), acetone–toluene–28% ammonia solution (20:10:1), or *tert*-butanol–4 M ammonia solution (9:1), these compounds formed fluorophores on the developed plate following spraying with a reagent consisting of sodium hypochlorite, potassium ferricyanide, and sodium hydroxide and heating for three minutes at 100°C. Blue fluorescent zones were seen under UV light in the 250–400 nm range with a detection limit of 50 ng.

Methamphetamine (MA), MDMA, and 3,4-methylenedioxy-N-ethylamphetamine (MDEA) were separated and detected in spiked plasma samples by normal-phase (NP) and reversed-phase (RP) column high-performance liquid chromatography (HPLC)

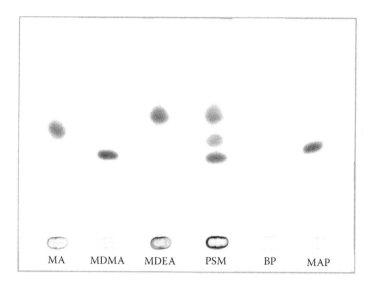

FIGURE 22.1 TLC separation of amphetamine analog chromogens on a silica gel plate. BP is blank plasma and PSM is plasma spiked with MA (R_{std} = 1.22), MDMA (1.00), and MDEA (1.57). (Reprinted from Oztung, A., Onal, A., and Toker, S.E. 2010, *J. AOAC Int.* 93: 556–561. With permission from Jennifer Diatz, AOAC INTERNATIONAL, Rockville, Maryland.)

and TLC based on purple chromogens formed by displacement reaction of these secondary aliphatic amine-containing drugs with 7,7,8,8-tetracyanoquinodimethane (TCNQ) at 80°C for 25 minutes [12]. For TLC, hexane-chloroform (1:9) and benzene–diethyl ether–petroleum ether–acetonitrile–methyl ethyl ketone (MEK) (20:35:35:5:5) were used as mobile phases for Alltech 0.25-mm silica gel 60 plates and Merck 0.2-mm cyano (CN) bonded plates, respectively. Chambers were presaturated with the mobile phase vapors for 30 minutes before development for a distance of 8 cm. Sample preparation comprised alkalization of plasma with sodium hydroxide, vortex extraction with chloroform–hexane (4:1), centrifugation, transfer of the organic phase into another tube, derivatization with TCNQ, evaporation to dryness, and dissolving of the residue in 20 μL chloroform that was applied to the TLC plate with a Desaga micropipette along with maprotiline (MAP) standard against which relative R_F values were calculated as R_{std}. Figure 22.1 shows a photograph of the separations of the purple TCNQ derivatives on silica gel, which gave better resolution than the CN plate. Limit of detection (LOD) values were 0.8, 0.6, and 1.3 μg mL⁻¹ in plasma for MA, MDMA, and MDEA, respectively, and double these values on CN plates.

22.2.2 Detection and Identification of Designer Drugs

Twelve derivatives (designer drugs) of frequently abused meperidine, including 1-methyl-4-phenyl-propionoxypiperidine (MPPP), 1-methyl-4-phenyl-1,2,3,6-terahydropyridine (MPTP), and 1-phenethyl-4-phenyl-4-propionoxypiperidine (PEPAOP), were synthesized, and any impurities were detected and structures were confirmed by TLC and GC-EI and CI-MS with isobutene reagent gas [13]. TLC was performed using Merck silica gel 60 F 0.25-mm glass plates and the mobile phases methanol–28% aqueous ammonia solution (100:1.5), benzene–dioxane–28% aqueous ammonia solution (50:40:5.5), chloroform–acetone (2:1), or chloroform–benzene–methanol (10:2:1). Zone detection was made by the following four methods: platinum chloride–potassium iodide reagent, Dragendorff reagent, 1% iodine–methanol solution, and UV absorption at 254 nm.

Data were collected to serve for forensic screening, identification, and quantification in seized tablets, crystals, and powders of methamphetamine, amphetamine, and six phenethylamine-type designer drugs, i.e., 2,5-dimethoxyphenethylamine, 2,5-dimethoxyamphetamine, 4-bromo-2,5-dimethoxyphethylamine (DOB), 4-bromo-2, 5-dimethoxyamphetamine, 4-iodo-2,5-dimethoxyphenethylamine, and 4-iodo-2,5-dimethoxyamphetamine [14]. The data included nuclear magnetic resonance (NMR) spectra, infrared (IR) spectra, retention times in Waters Acquity ultra-performance column liquid chromatography (UPLC) with a BEH-C18 (octyldecylsilyl) column, electrospray ionization (ESI) mass spectra without derivatization and with trifluoroacetyl (TFA) derivatization, and TLC. The TLC was carried out on Merck silica gel 60 F plates using seven mobile phases and detection of zones under 254-nm UV light, by spraying with fluorescamine reagent and viewing under 366-nm UV light, and with Simon's reagent for the secondary amine methamphetamine. The R_F values in the following mobile phases were listed so required separations could be designed: acetone–toluene–28% ammonia solution (20:10:1), 2-propanol–28% ammonia solution (95:5), MEK–dimethylformamide (DMF)–28% ammonia solution (13:1.9:0.1), chloroform–methanol (95:5)–28% ammonia solution (lower layer), chloroform–dioxane–ethyl acetate–28% ammonia solution (25:60:10:5), acetone–chloroform–methanol (3:1:1), and methanol–28% ammonia solution (100:1.5).

Thirty-one gelatinous capsules containing a white powder inside were seized by police in Sao Paulo, Brazil, in May 2005, and various analytical methods were used to identify the seized material, including a colorimetric Marquis test, HPTLC, MS, capillary zone electrophoresis (CZE), and IR spectrometry. HPTLC on Merck silica gel G plates with methanol–ammonia (100:1.5) mobile phase and detection of zones after spraying with ninhydrin reagent indicated that the capsules contained an amphetamine derivative. In combination with the other analyses, it was concluded that the capsules contained DOB [15].

22.2.3   DETERMINATION OF DRUGS THAT ARE NATURALLY PRESENT OR ARE ADDED TO HERBAL SAMPLES

A dietary supplement that claimed to be effective for the treatment of male erectile dysfunction was analyzed to identify illegal adulterants [16]. Adulterants in the encapsulated brown powder supplement were identified by TLC, UV spectrometry, HPLC-MS/MS, NMR spectrometry, and circular dichroism as tadalafil, sold under the brand name Cialis, and a designer analog first detected in this study and named aminotadalafil. Powder samples were extracted with 96% ethanol by ultrasonic shaking, and extract and tadalafil standard were simultaneously analyzed by a TLC method for which experimental details were not described. Two zones were separated, one of which was ascribed to tadalafil because of identical R_F values. The two zones were scraped and layer material extracted with 95% ethanol, and extracts were analyzed by the other methods for definite identification of the adulterants.

TLC and HPLC-photodiode array (PDA) detector–MS methods were developed for qualitative and quantitative analysis, respectively, of 36 commercial herbal dietary supplements suspected of containing eight phosphodiesterase type-5 inhibitors (PDE5-Is): sildenafil, hongdehafil, homosildenafil, hydroxyhomosildenafil, vardenafil, pseudovardenafil, tadalafil, and aminotadalafil [17]. Samples of products in the form of solids (e.g., tablets), soft gel capsules, and liquids were extracted with methanol, and extracts were filtered. Standard and sample solutions were applied with a CAMAG ATS4 autosampler to 20 × 10 cm Merck silica gel 60 F plates that were developed after drying in a desiccator containing phosphorus pentoxide with the mobile phase chloroform–ethyl acetate–methanol–water (40:40:15:11) in a saturated CAMAG twin trough chamber. Zones were detected under 254-nm UV light in a CAMAG viewing cabinet, and R_F values were compared between samples and standards for identification. Quantification by C18 column HPLC-PDA had $r^2 = 0.999$ linearity, limit of quantification (LOQ) of 0.04–0.09 µg mL^{-1}, 97.0–101% accuracy, and <1.32% relative standard deviation (RSD) for all PDE5-Is, and their MS fragmentation patterns were characterized for confirmation. Ten of the 36 supplements were found to contain one or more PDE5-Is.

A study was carried out to isolate and identify any synthetic cannabinoid disguised in scent bags known as "Funky Green Stuff" that are commercially available in Kuwait [18]. Extraction of the plant material was by percolation with methanol followed by filtration. The extract was chromatographed on 5 × 10 cm Miles Scientific (formerly Analtech) silica gel F plates with 0.25-mm layer thickness using the mobile phase toluene–methanol–diethylamine (8:1.5:0.5). The main compound was separated and detected as an orange zone after spraying with Dragendorff reagent. Flash column chromatographic separation was used to isolate this compound, and UV, IR, and NMR spectrometry identified it as AB-FUBINACA.

A 2015 review paper on the analysis of synthetic cannabinoids [19] cited a TLC paper by Logan et al. [20], who used TLC for the separation and identification of synthetic cannabinoids in herbal incense blends available in the U.S. (called "K2"). Methanolic extracts and combined acid/base extracts of the dried and crushed plant materials (flowers, stems, and leaves) were chromatographed on Whatman Partisil silica gel 60 LK6DF laned preadsorbent plates with 0.25-mm-layer thickness using toluene–diethylamine (9:1) and ethyl acetate–dichloromethane–methanol–ammonia (18.5:18:3:1) mobile phases. All 21 standard compounds studied exhibited UV absorption (fluorescence quenching of the layer indicator) at 254 nm, and some fluoresced white or yellow fluorescence under 366-nm UV light; in addition, the detection reagents Fast Blue B, fluorescamine, ninhydrin, 10% sulfuric acid, iodoplatinate, 50% nitric acid, mercuric sulfate, and 4-dimethylaminobenzaldehyde were used to compare standards to zones in sample extracts. Other analytical methods applied to analysis of various commercially available products were GC-MS, HPLC, and HPLC-MS, and the synthetic cannabinoids JWH-018, JWH-019, JWH-073, JWH-081, JWH-200, JWH-210, JWH-250, CP47,497 (C = 8; cannabicyclohexanol), RCS-4, RCS-8,

AM-2201, and AM-694, along with other noncannabinoid drugs including mitragynine (Kraton) were identified. Some of the products were analyzed quantitatively, and typical concentrations of identified drugs were in the range 5–20 mg g^{-1} or 0.5–2%. Whatman plates are no longer available commercially, but replacements for discontinued plates are discussed in a recent review of TLC stationary phases [21]; plates equivalent to Whatman LK6DF are sold by Miles Scientific as Catalog No. 44911.

Salvia divinorum is a hallucinogenic psychoactive herb of the mint family (known as "magic mint," among other colloquial names) that is local to Oaxaca in Central Mexico; its major active constituent is the selective opioid antagonist salvinorin A, which is usually the target analyte in analytical reports in the literature [22]. TLC has been used in combination with desorption electrospray ionization (DESI)-MS [23] and GC [24] for the analysis of raw dried leaves of *S. divinorum*. Online TLC-DESI-MS [23] of acetone extracts of leaves using a vortex mixer was performed on Merck silica gel 60 F plates developed with methyl *tert*-butyl ether (MTBE)–hexane (3:1). Salvinorin A, C, and B and divinatorin B were separated with respective R_F values of 0.49, 0.64, 0.95, and 0.85. The chemical image of the plate was recorded by rastering the surface in two dimensions with an Omni Spray automated DESI source in the positive-ion mode coupled to a Thermo Scientific LTQ linear ion trap mass spectrometer. The spectrum from the region of the plate corresponding to salvinorin A characteristically showed peaks located at m/z 433, 450, 455, 882, and 887. Semiquantitative analysis of salvinorin A in acetone extract was made using the integrated peak areas associated with the m/z 433 ion for standards and samples, and 1.6 mg mL^{-1} (0.8% w/w) in *S. divinorum* the leaf sample was estimated. TLC combined offline with GC-MS [24] detected salvinorin A in *S. divinorum*, 13 other *Salvia* species, as well as *Cannabis salvia L.* (marijuana). Optimum extraction solvents were chloroform and acetone at ambient temperature for GC-MS and methanol–chloroform (1:1) for TLC. Extracts and standard salvinorin A were applied onto Whatman Partisil K6 60 F 0.25-µm silica gel plates (Miles Scientific equivalent: Catalog No. 43911) and separated by development with ethyl acetate–hexane (1:1). Salvinorins were then detected as pinkish-purple zones by spraying with vanillin reagent. GC-MS employed a Hewlett-Packard (HP) 5890 Series II Plus gas chromatograph in combination with an HP MSD (mass selective detector) Series 5972 mass spectrometer, an HP 6890 Series injector, and a J&W Scientific HP-5 (5% phenyl–methyl siloxane) 15 × 0.25 mm id, 0.25-µm film thickness capillary column.

22.2.4 CHIRAL DETERMINATION OF DESIGNER DRUGS

A review article [25] on methods for the chiral determination of amphetamine and amphetamine-derived designer drugs in different matrices, including blood, hair, urine, medicaments, or standard solutions, cited three papers on TLC. Suedee et al. [26] resolved enantiomers of adrenergic drugs, including pseudoephedrine, ephedrine, norephedrine, and epinephrine, on molecularly imprinted polymers (MIPs) of (-)-pseudoephedrine and (-)-norephedrine that were prepared as chiral stationary phases (CSPs). The chiral template or print molecule is bound in a cross-linked polymer, and this is the method of preparation of MIPs selective to chiral compounds. Methacrylic acid and itaconic acid were used as functional monomers. Methanol or acetonitrile with added acetic acid was the mobile phase.

Synthetic polymers imprinted with quinine and calcium sulfate binder as the CSP were also used by Suedee et al. [27] to resolve enantiomers of pseudoephedrine, ephedrine, norephedrine, and epinephrine. It was found that during the imprinting process some residual print molecules were retained by the polymer with the result that the absorption of the background and of test substances on the plate at 366 nm was different, which was advantageous for zone detection. TLC plates were prepared on 76 × 26-mm glass microscope slides, and mobile phases were methanol or acetonitrile containing 1–10% acetic acid.

NP TLC enantioseparation of (+/-)-ephedrine and (+/-)-atropine on silica gel plates impregnated with optically pure L-tartaric acid and L-histidine, respectively, as chiral selectors, was reported by Bhushan et al. [28]. Mobile phases enabling successful resolution were different combinations of acetonitrile, methanol, and water, and zones were detected at 2 and 6 µg, respectively, in terms of the racemate with iodine vapor. It was shown that chiral interactions are affected by concentration of the impregnating agent, pH, and temperature.

22.2.5 QUANTIFICATION OF MDMA IN SEIZED TABLETS

An HPTLC-densitometry method was developed and fully validated for MDMA tablets seized in Chile [29]. Ground tablets were extracted for 15 minutes in an ultrasonic bath with methanol-4 M sodium hydroxide at pH 10, and extracts were syringe-filtered (Millex pore size 0.22 µm). Merck 20 × 10-cm silica gel 60 F plates were activated at 80°C for 30 minutes, standards and samples were applied in 3-mm bands with a CAMAG ATS-4 Automatic TLC Sampler, plates were developed in a CAMAG Automatic Developing Chamber ADC-2 for a distance of 70 mm with methanol–ammonia (100:1.5), and chromatograms were quantified with a CAMAG TLC Scanner-4 at 210 nm with slit dimension 4.00 × 0.3 mm at a speed of 20 nm/second controlled by winCATS Planar Chromatography Manager version 14.7. *In situ* spectra of each peak were recorded in the 190–400 nm range.

International Conference on Harmonization (ICH) validation of the method was performed and statistical parameters evaluated using STATGRAPHICS Centurion XVI.I software. Linearity of the method was established by the calibration curve between 51.0 and 510.0 µg band^{-1} (R = 0.9977), LOD was 12.1 µg band^{-1}, LOQ was 36.8 µg band^{-1}, precision of the method was <5.0% RSD, accuracy evaluated by recovery studies was 99.13%, and relative uncertainty was 6.66%. Real seized samples were found to contain between 18.15 and 59.84% by the method.

22.2.6 COUPLING OF TLC AND FOURIER-TRANSFORM (FT) IR SPECTROMETRY FOR ANALYSIS OF AMPHETAMINE-TYPE DESIGNER DRUGS

TLC on Merck 10 × 10 cm HPTLC silica gel 60 WRFs extra-thin plates designed for automated multiple development with the CAMAG AMD instrument was directly combined with FTIR spectrometry for identification and quantification of 3,4-methylenedioxyamphetamine (MDA), MDMA, and N-ethyl-3,4-methylenedioxyamphetamine (MDE) [30]. Samples were applied with a CAMAG Linomat 3; plate development was by a 30-step gradient employing methanol, ammonia, and dichloromethane; and *in situ* spectrometric measurement was carried out using a Bruker IFS 48 FTIR unit with a narrow-band detector combined with a diffuse reflectance FTIR (DRIFT) module. MDMA was assayed in confiscated samples by evaluation of Gram–Schmidt traces, and an average of 20.1% was found in agreement with GC and UV spectrometric confirmatory results.

22.2.7   ONLINE TLC-MS ANALYSES

In addition to the online TLC-DESI-MS procedure described above [23], designer drugs were analyzed in two papers by TLC-easy ambient sonic spray ionization MS (EASI-MS). LSD (lysergic acid diethylamide, known as "acid") and 9,10-dihydro-LSD were characterized by positive-mode EASI-MS directly from the surface of street drug blotters seized by several regulatory agencies in Brazil, followed by confirmation by HPLC-UV [31]. The combination of TLC with EASI-MS proved to be a most valuable tool in forensic investigations, with TLC used for simple screening of specific target drugs and then, in the positive samples, EASI-MS performed for undisputable confirmation. Blotter samples were cut into pieces of *ca.* 10 mg, and each piece was extracted with 10 mL of methanol under stirring. A volume of 10-µL extract was applied to a Merck silica gel 60 GF plate that had been dried for 30 minutes at 80°C and stored in a desiccator. The plates were developed with chloroform–acetic acid (20:80) for 8 cm in a CAMAG HPTLC horizontal chamber and then dried, and zones were detected under 254 and 366-nm UV light and then directly analyzed by EASI-MS (homemade EASI source and Shimadzu single quadrupole or Thermo Scientific ion trap instrument) with 10-second spectrum collection on each zone without any sample preparation. LOQ of LSD by TLC was 0.1 µg blotter[-1] at 365 nm and 0.5 µg blotter[-1] at 254 nm.

TLC-EASI-MS was also applied for the forensic analysis of 15 cocaine and crack cocaine samples seized in Rio de Janeiro, Brazil, all of which revealed positive results for cocaine as well as other drugs and adulterants such as lidocaine, caffeine, benzocaine, lactose, benzoylecgonine, and ecgonidine [32]. False positives and negatives judged by TLC analysis alone were corrected via zone characterization by online EASI-MS. Samples were dissolved in methanol and applied to Merck silica gel 60 GF plates that were then developed in a CAMAG horizontal chamber for 8 cm with methanol–chloroform–acetic acid (20:75:5). Dried chromatograms were detected at a limit of 2 µg and marked under 254 nm UV light. EASI-MS was performed with a single quadrupole Shimadzu LCMS-2010EV mass spectrometer equipped with a homemade EASI source described in detail earlier [33]. Acidic methanol (0.1% in volume, 20 µL min[-1]) and compressed nitrogen at a pressure of 100 psi were used to form the sonic spray. The capillary-surface and surface-spectrometer entrance angles were 45°, and spectra were accumulated for ten seconds. All impurity zones were scraped off of layers and further confirmed by GC-MS.

22.2.8   ISOLATION OF DESIGNER DRUGS BY PLC FOR THEIR IDENTIFICATION

Four examples of the use of PLC by two research groups for isolation of larger amounts of designer drugs prior to their identification are given in this section. PLC usually involves separation of a larger sample load on a thicker layer compared to analytical TLC, scraping off and collection of the zone of interest after its detection, and elution of the compound from the scraped sorbent with an appropriate solvent for further analysis [34].

Kneisel et al. [35] described a study in which a research microcrystalline chemical advertised as the cannabimimetic compound [(N-methylpiperidin-2-yl)methyl]-3-(1-naphthoyl)indole (AM 1220), purchased via an Internet trading platform, assayed as pure AM 1220 by GC-MS. However, when it was tested for purity by TLC, two zones were obtained, and after isolation by PLC and high-resolution MS and NMR spectrometry analysis, the azepane isomer of AM-1220 was confirmed to be present. Later, both substances were detected in several herbal mixtures purchased from different German Internet shops. The crystalline drug substance was dissolved in ethanol, and the herbal mixtures were extracted by vortexing with ethanol. Purity testing of extracts was carried out on Merck 20 × 20 cm silica gel 60 F aluminum plates developed with cyclohexane–diethylamine (9:1), and isolation on Merck 10 × 20 silica gel 60 F glass plates by developing with the same mobile phase followed by scraping of bands and elution with ethanol. Zones were detected on side lanes by dyeing with iodoplatinate reagent.

The same PLC layer, mobile phase, and detection method were used in the structural characterization of the synthetic cannabinoid 3-(1-adamantoyl)-1-pentylindole found in several "herbal incense" products purchased from a Dutch Internet shop [36]. Samples were vortex extracted with ethanol, and the major band after PLC of the extract was scraped and extracted for analysis by NMR spectrometry, high-resolution MS, and GC-MS/MS. The paper cautioned that the compound was misidentified by GC-MS before applying techniques that provided more structural information.

The cannabimimetic aminoalkyl naphthoyl indole derivative JWH-018 was identified by Uchiyama et al. [37] as a new adulterant in an herbal product being sold in Japan for its expected narcotic effect. GC-MS, UPLC-ESI-MS, direct analysis in real time (DART)–time-of-flight (TOF)–MS, and NMR spectrometry were performed on the compound isolated by PLC on 20 × 20-cm Merck silica gel 60 layers with 2-mm thickness. A methanol extract of herbal product was loaded on the plate, which was developed with hexane–acetone (4:1) mobile phase. The detected zone was scraped and eluted to give fraction 1, which was fractionated by further PLC with hexane–chloroform (1:20) to give the purified compound for analysis.

Uchiyama et al. [38] also identified the phenylacetyl indole JWH-251 and the naphthoylindole JWH-081 as designer drugs in illegal herbal products with the use of PLC, UPLC-MS, GC-MS, DART-TOF-MS, and NMR spectrometry. A chloroform ultrasonic extract of herbal product was placed on a Merck 20 × 20-cm silica gel 60 plate with 2-mm layer thickness, which was then developed using hexane–ethyl acetate (4:1). The portion of silica gel containing the target compound was scraped and eluted with chloroform–methanol (3:1) to obtain fractions 1 and 2. Each fraction was further purified by recycling preparative HPLC on a 500 × 20-mm i.d. JAIGEL-GS310 column with chloroform–methanol (1:1) to give the two compounds for analysis.

22.3   FUTURE PROSPECTS

It is to be expected that TLC will continue to be used in all aspects of designer drug analysis described in subsections of Section 22.2 above, especially for the identification and characterization of new drugs (e.g., [39,40]) and drug products in the marketplace [41]. HPTLC-densitometry should be much more widely used for the screening and quantification of active ingredients in designer drug products given its well documented advantages and wide applications in drug formulation and clinical analyses [1]. One reason for this is the 2015 call for methods to be included in a new general chapter in the US Pharmacopeia on adulteration of dietary supplements with drugs and drug analogs (<2251>). An example of a method developed in response to the call was published in the CAMAG CBS 114 [42] on screening of three PDE5-Is and eight of their analogs in lifestyle products (powdered pill, capsule contents, chewing gum, chocolate), including several levels of confirmation. The method included the use of Merck HPTLC silica gel 60 F plates, a CAMAG TLC Sampler 4 for bandwise application of sample and standard solutions, 70-mm development with the mobile phase *tert*-butyl methyl ether–methanol–28% ammonia (20:2:1) in a CAMAG ADC 2 with vapor saturation via a filter paper and 47% relative humidity, documentation with a CAMAG TLC Visualizer under 254 and 366 nm UV light, and recording of spectra from 190 to 550 nm using a TLC Scanner 4 and winCATS software. In addition to comparison of these *in situ* spectra and R_F values to reference standards, the identity of sample zones was confirmed using a CAMAG TLC-MS Interface and Advion Expression compact mass spectrometer (CMS) with ESI.

Indeed, the area most likely to grow appreciably in applications to designer drug analysis is TLC-MS, an example of which was just described among several others in Section 22.2.7. A feature article published in the journal *Analytical Chemistry* in 2016 [43] discussed ambient mass spectrometry as a powerful, rather universal, simple, fast, nondestructive, and robust tool in forensic analysis, including drug analysis after TLC separation. Particular MS methods likely to be most used in the future are EASI, DART, and single-quadrupole MS. The combination of EASI-MS with TLC was described by Haddad et al. [44], and its application to designer drug analysis is covered above [31,32]. A disadvantage is that the EASI sources reported in the literature have been homemade.

The Virginia Department of Forensic Science validates TLC drug identifications, made using silica gel plates; chloroform–methanol (18:1) and methanol–ammonia (100:1.5) mobile phases; and potassium permanganate, iodoplatinate, and ceric sulfate spray reagents, with an online JEOL AccuTOF-DART MS ionization source [45]. No designer drugs were reported to be analyzed in this paper, but the availability of a commercial source and <10 minutes DART confirmation time of TLC zones confidently portends such use in the future.

The most widely reported coupled TLC-MS approach uses the CAMAG TLC-MS Interface mentioned above [42]. The CAMAG TLC-MS Interface-1 was described in detail in a recent book titled *Planar Chromatography–Mass Spectrometry* [46]. Starting in 2013, Advion, Inc. and CAMAG joined in an exclusive OEM and distribution agreement to commercialize an integrated system including the CAMAG TLC-MS Interface-1 and Advion's single-quadrupole CMS offering unit resolution (this system is pictured in [47]). In 2015, Advion introduced their own automated TLC-Compact Mass Spectrometer system that is described in a paper [46] and shown in Figure 22.2. The same year, CAMAG introduced their TLC-MS Interface-2 that is illustrated in a biennial review of planar chromatography [48]. Like the TLC-MS Interface-1, this new version allows for rapid direct elution of TLC zones to a mass spectrometer, but, according to CAMAG, plate positioning is significantly simplified, the elution head has been modified, and an easily accessible, exchangeable

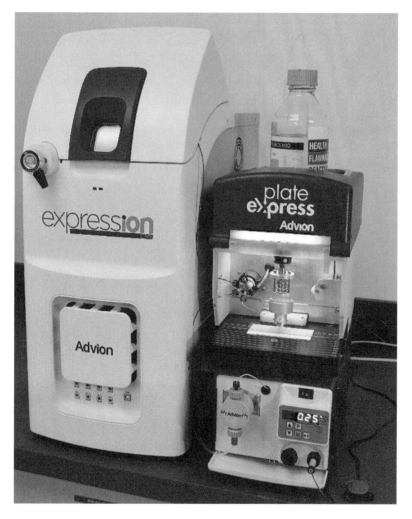

FIGURE 22.2 Advion Automated TLC-Compact Mass Spectrometry system. (From Jack Henion, Advion, Inc., Ithaca, New York. With permission.)

filter has been arranged in front of the valve to facilitate the cleaning of matrix particles from the elution path. It is inevitable that the CAMAG and/or Advion systems will be widely applied to designer drug analysis to confirm TLC zone identity, as demonstrated by Do et al. [49] in the screening of three PDE5-Is and eight of their analogs in finished products, including tablets, capsules, chocolate, instant coffee, syrup, and chewing gum. In this study, unknown TLC zones were identified by their *in situ* UV spectra compared to those of reference standards (Figure 22.3) and ESI mass spectra obtained using the CAMAG TLC-MS inter-face (Figure 22.4).

Note: Merck plates are sold in the United States and Canada by EMD Corp., 290 Concord Road, Billerica, Massachusetts, a division of Merck KGaA, Darmstadt, Germany.

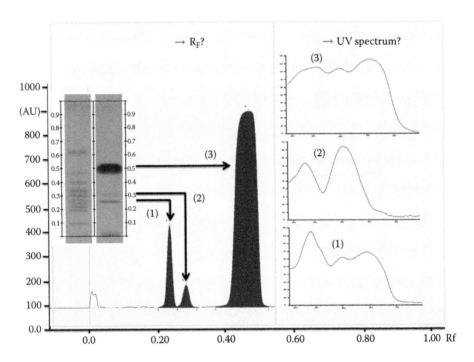

FIGURE 22.3   Strategy for identification of unknown PDEI-5s and their analogs on Merck silica gel 60 F Premium Purity HPTLC glass plates. Corresponding bands, densitometer scans, and *in situ* UV spectra are shown. Zones 1 and 3 displayed UV spectral features char-acteristic of thione analogs, while the spectrum of zone 2 was sildenafil-like. (Reprinted from Do, T.T.K. et al., 2015, *J. AOAC Int.* 98: 1226–1233. With permission from Jennifer Diatz, AOAC INTERNATIONAL, Rockville, MD.)

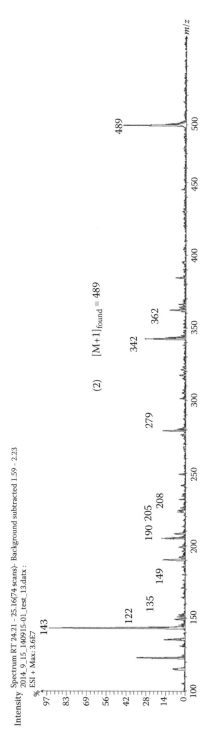

FIGURE 22.4 Mass spectrum of unknown zone (2) recorded using a CAMAG TLC-MS interface and Advion single-quadrupole mass spectrometer. The [M+H]$^+$ ion at m/z 489 suggested a formula of C23H32N6O4S, which could correspond to the dimethyl, propoxyphenyl, or isobutyl sildenafil analog. (Reprinted from Do, T.T.K. et al., 2015. *J. AOAC Int.* 98: 1226–1233. With permission from Jennifer Diatz, AOAC INTERNATIONAL, Rockville, Maryland.)

ACKNOWLEDGMENTS

Joseph Sherma thanks Karen F. Haduck, Interlibrary Resource Sharing Specialist, Skillman Library, Lafayette College, for her invaluable help in collecting the literature references cited in this chapter.

REFERENCES

1. Komsta, L., Waksmundzka-Hajnos, M., and Sherma, J., Eds. 2014. *Thin Layer Chromatography in Drug Analysis*, CRC Press/Taylor & Francis Group, Boca Raton, FL.
2. Bernard-Savary, P. and Poole, C.F. 2015, Instrument platforms for thin layer chromatography, *J. Chromatogr. A*. 1421: 184–202.
3. Ciesla, L., Waksmundzka-Hajnos, M., and Sherma, J. Thin layer chromatography, in *Ewing's Instrumentation Handbook, Fourth Edition*, Grinberg, N., Ed., CRC Press/Taylor & Francis Group, Boca Raton, FL, in press.
4. Lillsunde, P. and Korte, P. 1991. Comprehensive drug screening in urine using solid phase extraction and combined TLC and GC/MS identification, *J. Anal. Toxicol.* 15: 71–81.
5. Kankaanpaa, A., Meririnne, E., Ellermaa, S., Ariniemi, K., and Seppala, T. 2001. Detection and assay of *cis*- and *trans*-isomers of 4-methylaminorex in urine, plasma, and tissue samples, *Forensic Sci. Int.* 121: 57–64.
6. Kraemer, T. and Maurer, H.H. 1998. Determination of amphetamine, methamphetamine and amphetamine-derived designer drugs or medicaments in blood and urine, *J. Chromatogr. B*. 713: 163–187.
7. Gerhards, P. and Szigan, J. 1996. Determination of designer drugs and ecstasy, *GIT LaborMedzin*, 19: 212–214, 216.
8. Jarvie, D.R. and Simpson, D. 1986. Drug screening: Evaluation of the Toxi-Lab TLC system, *Ann. Clin. Biochem.* 23: 76–84.
9. DeRuiter, J., Hayes, L., Valaer, A., Clark, C.R., and Noggle, F.T. 1994. Methcathinone and designer analogues: Synthesis, stereochemical analysis, and analytical properties, *J. Chromatogr. Sci.* 32: 552–564.
10. Maresova, V., Hampl, J., Chundela, Z., Zrcek, F., Polasek, M., and Chadt, J. 2005. The identification of a chlorinated MDMA, *J. Anal. Toxicol.* 29: 353–358.
11. Kato, N., Fujita, S., Ohta, H., Fukuba, M., Toriba, A., and Hayakawa, K. 2008. Thin layer chromatography/fluorescence detection of 3,4-methylenedioxymethamphetamine and related compounds, *J. Forensic Sci.* 53: 1367–1371.
12. Oztung, A., Onal, A., and Toker, S.E. 2010. Detection of methamphetamine, methylenedioxymethamphetamine, and 3,4-methylenedioxy-*N*-ethylamphetamine in spiked plasma by HPLC and TLC, *J. AOAC Int.* 93: 556–561.
13. Shimizu, T., Lee, X.-P., Sato, K., Ohta, H., and Suzuki, S. 2001. Studies on designer drugs VI. Chromatographic discrimination of meperidine (pethidine) derivatives and their structure-activity relationship, *Jpn J. Forensic Toxicol.* 19: 228–235.
14. Kanai, K., Takekawa, K., Kumamoto, T., Ishikawa, T., and Ohmori, T. 2008. Simultaneous analysis of six phenylethylamine type designer drugs by TLC, LC-MS, and GC-MS, *Forensic Toxicol.* 26: 6–12.
15. da Costa, J.L., Wang, A.Y., Micke, G.A., Maldaner, A.O., Romano, R.L., Martins-Junior, H.A., Neto, O.N., and Tavares, M.F.M. 2007. Chemical identification of 2,5-dimethoxy-4-bromoamphetamine (DOB), *Forensic Sci. Int.* 173: 130–136.
16. Lin, M.-C., Liu, Y.-C., Lin, Y.-L., and Lin, J.-H. 2009. Identification of a tadalafil analogue adulterated in a dietary supplement, *J. Food Drug Anal.* 17: 451–458.

17. Cai, Y., Cai, T.-G., Shi, Y., Cheng, X.-L., Ma, L.-Y., Ma, S.C., Lin, R.-C., and Feng, W. 2010. Simultaneous determination of eight PDE5-IS potentially adulterated in herbal dietary supplements with TLC and HPLC-PDA-MS methods, *J. Liq. Chromatogr. Relat. Technol.* 33: 1287–1306.

18. Alsoud, A.R.A., Al-Tannak, N., Bojabarah, H., and Orabi, K.Y. 2015. AB-FUBINACA, a synthetic cannabinoid in 'Funky Green Stuff', *Int. J. Pharm. Pharmaceut. Sci.* 7: 111–115.

19. Znaleziona, J., Ginterova, P., Petr, J., Ondra, P., Valka, I., Sevcik, J., Chrastina, J., and Maier, V. 2015. Determination and identification of synthetic cannabinoids and their metabolites in different matrices by modern analytical techniques—A review, *Anal. Chim. Acta.* 874: 11–25.

20. Logan, B.K., Reinhold, L.E., Xu, A., and Diamond, F.X. 2012. Identification of synthetic cannabinoids in herbal incense blends in the United States, *J. Forensic Sci.* 57: 1168–1190.

21. Rabel, F. and Sherma, J. 2016. New TLC/HPTLC commercially prepared and laboratory prepared plates—A review, *J. Liq. Chromatogr. Relat. Technol.* 39(8): 385–393.

22. Smith, J.P., Sutcliffe, O.B., and Banks, C.E. 2015. An overview of recent developments in the analytical detection of new psychoactive substances (NPSs), *Analyst* 140: 4932–4948.

23, Kennedy, J.H. and Wiseman, J.M. 2010. Direct analysis of *Salvia divinorum* leaves for salvinorin A by thin layer chromatography and desorption electrospray ionization multi-stage tandem mass spectrometry, *Rapid Commun. Mass Spectrom.* 24: 1305–1311.

24. Jermain, J.D. and Evans, H.K. 2009. Analyzing *Salvia divinorum* for its active ingredient salvinorin A utilizing thin layer chromatography and gas chromatography/mass spectrometry, *J. Forensic Sci.* 54: 612–616.

25. Plotka, J.M., Morrison, C., and Biziuk, M. 2011. Common methods for the chiral determination of amphetamine and related compounds I. Gas, liquid and thin layer chromatography, *Trends Anal. Chem.* 30: 1139–1158.

26. Suedee, R., Songkram, C., Petmoreekul, A., Sengkunakup, S., Senkaswa, S., and Kongyarit, N. 1999. Direct enantioseparation of adrenergic drugs via thin layer chromatography using molecularly imprinted polymers, *J. Pharm. Biomed. Anal.* 19: 519–527.

27. Suedee, R., Songkram, C., Petmoreekul, A., Sangkunakup, S., Sankasa, S., and Kongyarit, N. 1998. Thin layer chromatography using synthetic polymers imprinted with quinine as chiral stationary phase, *J. Planar Chromatogr.-Mod. TLC* 11: 272–276.

28. Bhushan, R., Martens, J., and Arora, M. 2001. Direct resolution of (+/-)-ephedrine and atropine into their enantiomers by impregnated TLC, *Biomed. Chromatogr.* 15: 151–154.

29. Daffau, B.F., Rojas, S., Delgado, L.A., and Jofre, S. 2015. High performance thin layer chromatography method for analysis of 3,4-methylenedioxymethamphetamine in seized tablets, *J. Pharm. Pharmacog. Res. (JPPRes)* 3: 162–170.

30. Kovar, K.-A., Ensslin, H.K., Frey, O.R., Rienas, S., and Wolff, S.C. 1991. Applications of on-line coupling of thin layer chromatography and FTIR spectroscopy, *J. Planar Chromatogr.-Mod. TLC* 4: 246–250.

31. Romao, W., Sabino, B.D., Bueno, M.I.M.S., Vaz, B.G., Junior, A.C., Maldaner, A.O., de Castro, E.V.R., Lordeiro, R.A., Nascentes, C.C., Eberlin, M.N., and Augusti, R. 2012. LSD and 9,10-dihydro-LSD analyses in street drug blotters via easy ambient sonic-spray ionization mass spectrometry (EASI-MS), *J. Forensic Sci.* 57: 1307–1312.

32. Sabino, B.D., Romao, W., Sodre, M.L., Correa, D.N., Rocha Pinto, D.B., Alonso, F.O.M., and Eberlin, M.N. 2011. Analysis of cocaine and crack cocaine via thin layer chromatography coupled to easy ambient sonic-spray ionization mass spectrometry, *Am. J. Anal. Chem.* 2: 658–664.

33. Hirabayashi, A., Sakairi, M., and Koizumi, H. 1995. Sonic spray mass spectrometry, *Anal. Chem.* 67: 2878–2882.

34. Kowalska, T. and Sherma, J., Eds. 2006. *Preparative Layer Chromatography*, CRC Press/Taylor & Francis Group, Boca Raton, FL.

35. Kneisel, S., Bisel, P., Brecht, V., Broecker, S., Mueller, M., and Auwärter, V. 2012. Identification of the cannabimimetic AM-1220 and its azepane isomer (*N*-methylazepan-3-yl)-3-(1-napthoyl)indole in a research chemical and several herbal mixtures, *Forensic Toxicol.* 30: 126–134.

36. Kneisel, S., Westphal, F., Bisel, P., Brecht, V., Broecker, S., and Auwärter, V. 2012. Identification and structural characterization of synthetic cannabinoid 3-(1-adamantoyl)-1-pentylindole as an additive in "herbal incense," *J. Mass Spectrom.* 47: 195–200.

37. Uchiyama, N., Kikura-Hanajiri, R., Kawahara, N., and Goda, Y. 2009. Identification of a cannabimimetic indole as a designer drug in a herbal product, *Forensic Toxicol.* 27: 61–66.

38. Uchiyama, N., Kawamura, M., Kikura-Hanajiri, R., and Goda, Y. 2011. Identification and quantitation of two cannabimimetic phenylacetylindoles JWH-251 and JWH-250, and four cannabimimetic naphthoylindoles JWH-081, JWH-015, JWH-200, and JWH-073 as designer drugs in illegal products, *Forensic Toxicol.* 29: 25–37.

39. Gambaro, V., Casagni, E., Dell'Acqua, L., Roda, G., Tamborini, L., Visconti, G.L., and Demartin, F. 2016. Identification and characterization of a new designer drug thiothinone in seized products, *Forensic Toxicol.* 34: 174–178.

40. dos Santos, P.F., Souza, L.M., Merio, B.B., Costa, H.B., Tose, L.V., Santos, H., Vanini, G., Machado, L.F., Ortiz, R.S., and Limberger, R.P. 2016. 2-(4-Iodine-2,5-dimethoxyphenyl)-N-[(2-methoxyphenyl)methyl]etamine or 25I-NBOMe: Chemical characterization of a designer drug, *Quim. Nova* 39: 229–237.

41. Gine, C.V., Vilamala, M.V., Espinosa, I.F., Liadanosa, C.G., Alvarez, N.C., Fruitos, A.F., Rodriguez, J.R., Salvany, A.D., and Fornell, R.D. 2016. Crystals and tablets in the Spanish ecstasy market 2000–2014: Are they the same or different in terms of purity and adulteration?, *Forensic Sci. Int.* 263: 164–168.

42. Camag Laboratory. 2015. Screening of three PDE5-inhibitors and eight of their analogs in lifestyle products, *Camag Bibliography Service (CBS)*, 114: 9–10.

43. Correa, D.N., Santos, J.M., Eberlin, L.S., and Teunissen, S.F. 2016. Forensic chemistry and ambient mass spectrometry: A perfect couple destined to a happy marriage?, *Anal. Chem.* 88: 2515–2526.

44. Haddad, R., Milagre, H.M.S., Catharino, R.R., and Eberlin, M.N. 2008. Easy ambient sonic-spray ionization mass spectrometry combined with thin layer chromatography, *Anal. Chem.* 80: 2744–2750.

45. Howlett, S.E. and Steiner, R.R. 2011. Validation of thin layer chromatography with AccuTOF-DART detection for forensic drug analysis, *J. Forensic Sci.* 56: 1261–1267.

46. Sherma, J. 2016. The Camag TLC-MS interface, in *Planar Chromatography-Mass Spectrometry*, Kowalska, T., Sajewicz, M., and Sherma, J., Eds., CRC Press/Taylor & Francis Group, Boca Raton, FL, Chapter 3.

47. Hao, C.T., Sousou, N., Eikel, D., and Henion, J. 2015. Thin layer chromatography/mass spectrometry analysis of sample mixtures using a compact mass spectrometer, *Am. Lab.* 47(4): 24–27.

48. Sherma, J. 2016. Biennial review of planar chromatography: 2013–2015, *J. AOAC Int.* 99: 321–331.

49. Do, T.T.K., Theocharis, G., and Reich, E. 2015. Simultaneous detection of three phosphodiesterase type 5 inhibitors and eight of their analogs in lifestyle products and screening for adulterants by high performance thin layer chromatography, *J. AOAC Int.* 98: 1226–1233.

Index

Page numbers followed by f and t indicate figures and tables, respectively.

Printed and bound by CPI Group (UK) Ltd, Croydon, CR0 4YY

24/10/2024

01778304-0020